T0276183

Color Ontology and Color Science

Life and Mind: Philosophical Issues in Biology and Psychology
Kim Sterelny and Robert A. Wilson, editors

Cycles of Contingency: Developmental Systems and Evolution
Susan Oyama, Paul E. Griffiths, and Russell D. Gray, editors, 2000

Coherence in Thought and Action
Paul Thagard, 2000

The New Phrenology: The Limits of Localizing Cognitive Processes in the Brain
William R. Uttal, 2001

Evolution and Learning: The Baldwin Effect Reconsidered
Bruce H. Weber and David J. Depew, editors, 2003

Seeing and Visualizing: It's Not What You Think
Zenon W. Pylyshyn, 2003

Organisms and Artifacts: Design in Nature and Elsewhere
Tim Lewens, 2004

The Mind Incarnate
Lawrence A. Shapiro, 2004

Molecular Models of Life: Philosophical Papers on Molecular Biology
Sahotra Sarkar, 2004

Evolution in Four Dimensions: Genetic, Epigenetic, Behavioral, and Symbolic Variation in the History of Life
Eva Jablonka and Marion J. Lamb, 2005

The Evolution of Morality
Richard Joyce, 2006

Evolutionary Psychology as Maladapted Psychology
Robert C. Richardson, 2007

Describing Inner Experience? Proponent Meets Skeptic
Russell T. Hurlburt and Eric Schwitzgebel, 2007

The Native Mind and the Cultural Construction of Nature
Scott Atran and Douglas Medin, 2008

Color Ontology and Color Science
Jonathan Cohen and Mohan Matthen, editors, 2010

Color Ontology and Color Science

Jonathan Cohen and Mohan Matthen, editors

A Bradford Book
The MIT Press
Cambridge, Massachusetts
London, England

This book was set in Stone Serif by Toppan Best-set Premedia Limited.

Library of Congress Cataloging-in-Publication Data
Color ontology and color science / edited by Jonathan Cohen and Mohan Matthen.
p. cm.—(Life and mind : philosophical issues in biology and psychology)
"A Bradford book."
Includes bibliographical references and index.
ISBN 978-0-262-51375-3 (pbk. : alk. paper)—ISBN 978-0-262-01385-7 (hardcover : alk. paper)
1. Color vision. 2. Color—Psychological aspects. 3. Color (Philosophy) I. Cohen, Jonathan, 1971–
II. Matthen, Mohan.

QP483C643 2010
612.8′4—dc22

2009032012

To C. L. Hardin, who got this going

Contents

Introduction

I.1 Background: The Structure(s) of Color Space

Philosophical and psychological theorizing about color has traditionally made much of the idea that colors (or perhaps color appearances) are essentially, that is, by their own nature, ordered in ways that are phenomenally evident to a perceiver in virtue merely of her experience of them. In 1921, for example, the Cambridge logician W. E. Johnson intriguingly suggested that our grasp of "adjectives" such as number, color, and shape is governed by a characteristic structure. Johnson started by noting that there is no one character common to the members of such classes: there is, for instance, no "adjectival character" shared by red and blue, or by red things and blue things—none in virtue of which red and blue are *colors*, and red things and blue things are *colored*. The colors fall into a single class, then, not because they share something, but because of a "special kind of difference which distinguishes one color from another; whereas no such difference exists between a color and a shape." This "special kind of difference" is an ordering, Johnson thought: within such classes, we are able to generate certain subclasses simply by our grasp of the relevant ordering relations.

Number is Johnson's favored example:

1. The numbers do not share any characteristic, but the class of numbers is ordered by the relation "more than" (and its converse, "less than").
2. This ordering relation is constitutive both of number itself and of our grasp of it: we could not properly understand number without grasping "more than" as it applies to the numbers.

Finally,

3. Our grasp of number ordering enables us to construct subclasses such as {*x*: *x* is a number, and *x* is less than 4}: because we possess an adequate understanding of the number concept, we know immediately that the extension of this class is {1, 2, 3}. Johnson implies that we do not learn about the numbers less than 4 separately from learning about number.

In contrast, consider what Johnson calls "substantives": grasping a superordinate concept, such as MAMMAL, involves knowing what is in common to all mammals, but (by contrast with 3 above) gives us no specific knowledge about contained subclasses (subordinate concepts), such as the class of dogs (the concept DOG)—or even knowledge that such a subclass (subordinate concept) exists. Johnson's claim, then, is that ordering relations are a constitutive part of certain concepts, and that they afford us the means by which to construct certain subclasses of these concepts. He calls these superordinate classes "determinables," and the subclasses so generated "determinates."

Johnson thought that color is ordered in this manner; like number, it is a determinable. Here, he was drawing on an ancient tradition. The idea that colors are structured by ordering relations has always struck philosophers as intuitive. However, there was no consensus, either in philosophy or in psychology, of what the ordering relations might be. (See Kuehni, chapter 1, on various alternatives.) Darker-lighter is a partial ordering relation that caught Aristotle's attention, and it has stayed in the running through the centuries; saturated-desaturated entered the discourse at some point—but even taken together, these are not sufficient to generate all the determinate colors.

Philosophers in the mid- twentieth century likely thought colors were additionally ordered roughly as in the rainbow—or as in the Newtonian color wheel, perhaps—with brightness and saturation as additional dimensions, but the literature contains little that is explicit about these matters. (However, Wittgenstein had relevant observations among his musings about the phenomenology of color in the posthumous *Remarks on Colour* [1977].) More recently, psychologists and philosophers have been particularly interested in the ordering relations "is perceptually similar to" and "is perceptually distinguishable from." These can be defined (and psychophysically operationalized) over the domain of colors. Moreover, one can use these relations, together with mathematical tools, such as multidimensional scaling, to define a space of colors with a "distance" between any two colors. That is, plotting out the colors with the appropriate distances between them yields a multidimensional arrangement. The claim is that our perceptual grasp of colors is essentially ordered by this metrically organized space. This gives us a structure that conforms to Johnson's analysis. The distance measure corresponds to Johnson's ordering relation. Moreover, suppose that you are shown a sample of red—crimson, say. From this you can mentally generate brighter shades of red. You generate this subclass this simply by your grasp of color concepts.

As we mentioned earlier, philosophers were tempted by the idea that the ordering of colors was given by the Newtonian color wheel. By the 1950s, however, Leo Hurvich and Dorothea Jameson were urging the adoption of an alternative to the Newtonian color wheel. Using ideas foreshadowed in the work of Thomas Young and Ewald Hering, Hurvich and Jameson urged that hue consists of two dimensions—one running from blue to yellow, and the other from red to green. Adding darkness-lightness to hue (darkness-brightness in the case of luminous sources), we then get three dimensions of color. This proposal is stronger than the "similarity space of color" idea just mooted, because in addition to the idea that colors are ordered in three dimensions, the Hurvich-Jameson model identifies three privileged and simple dimensions. In this model, the primary colors are opposite poles of a continuous measure—for instance, blue is the maximum value in one direction of its axis, and yellow is the opposed extreme value. Every hue (including all the saturated hues) is a combination of some value along each of these privileged dimensions: for instance, orange is a combination of more or less equal amounts of red and yellow; various shades of orange have more or less of red and of yellow, and they are also arrayed along the bright-dark axis (a peachy orange being lighter than the color of Seville marmalade). Moreover, since these axes are simple, red is not a combination of purple and orange (even though it is in between them on the hue circle). Note also that in Hurvich-Jameson hue space, no color can be both reddish and greenish, or both bluish and yellowish, since these are opposite ends of a single dimension; this follows from the fact that each color occupies only one position in color space. (See Byrne and Hilbert, chapter 11, for more discussion.)

The Hurvich-Jameson model is closer to Johnson's notion of a determinable because it posits ordering relations that are separately known—not just an ordering in a three-dimensional space, but a space constructed from these separate dimensions. The Hurvich-Jameson hue circle is quite different from the ordering of colors in the rainbow (or the Newtonian color wheel), since red would be one right angle away from blue and yellow and two right angles away from green—whereas in the rainbow, green and yellow are relatively close together. (Incidentally, this talk of mixture should not be confused with that about the mixing of pigments, which has rules of its own. We are talking about color as it is experienced. The claim is that color is experienced as a combining of intensive values from scales just mentioned.)

The work of Hurvich and Jameson was relatively little known in philosophical circles until C. L. Hardin's influential book *Color for Philosophers: Unweaving the Rainbow* was published in 1988. Following this book, discussing color ordering in this way rapidly became standard. Through these authors' model of color space, Johnson's ordering relations came to possess a standard content, to the point where the "traditional color space" (TCS) is now routinely understood to include the Hurvich-Jameson hue circle. TCS is simultaneously thought of as generating the colors and as a graphical representation of their similarity relations—the closer two colors are in TCS, the more similar they are. Many philosophers of perception now adopt this view.

One point of disagreement obscured by the foregoing is whether the elements of TCS that are ordered by the various ordering relations under consideration are color appearances—construed as psychological states elicited by perceptual interaction with objects in the extramental world—or on the other hand colors themselves construed as properties of extramental surfaces, lights, films, and the like. (Reinhard Niederée takes them in the first way, for instance, in the bulk of chapter 4, though he clearly recognizes the distinction; Churchland, in chapter 2, takes them in the second way.) It is certainly possible to interpret subjects' reactions to stimuli, and therefore the relations based on these reactions, as revealing facts about those subjects' psychological states—color appearances—rather than about the distal stimuli. But since the relevant psychological states are typically produced by distal stimuli, or are representations of them, the very same reactions can be treated as inducing ordering (and other) relations on the distal stimuli—the colors. Of course, depending on what explanatory work one wants TCS to perform, extending this organizational scheme to distal properties may or may not meet one's needs, and therefore not everyone is interested in thinking about TCS as organizing the space of colors. Treating TCS as a space of color appearances is a weaker, more conservative option, since it is compatible with, but does not require, extending the organizational scheme to colors. On the other hand, this way of looking at things implies that perceivers possess awareness of distal similarity relations on the basis of their own perceptual experiences—and many philosophers want to avoid attributing this kind of awareness to ordinary perceivers. They prefer to say that in color vision, we experience colors themselves as ordered in certain similarity relations.

TCS is a source of several further scientific and philosophical questions. First, what is the full and correct description of TCS? Does the above characterization exhaust the dimensionality of color experience? What is the relationship of color dimensions to one another? Are there privileged axes in TCS such that all the hues are in some nonconventional sense composed of values along these axes? Scientifically, we want to understand the neural and psychological processes that underlie TCS in the visual system. Physical objects such as lights and colored surfaces send light to our eyes. The brain processes the light that arrives at our eyes, and experienced color somehow emerges as a result. Somewhere in this process, the light is stamped with the qualities that we experience—red-green, blue-yellow, and so forth. What are the relevant details of this process? What features of the world does it reliably track or preserve? Second, what is the nature of our knowledge of TCS and what it represents? Is our grasp of TCS innate or learned? Under what conditions ought one to say that it reveals reality? What aspect of reality does it reveal? What is the significance of the dimensions that order color experience?

Hardin, in particular, has been vocal in arguing that since TCS does not reflect a physical ordering of colors (in the way that even Newton's circle did, since it preserved the rainbow ordering, albeit with the two ends joined together), color vision and color experience do not correspond to any objective feature of the world. According to Hardin, TCS is an ordering imposed by sensory processing and simply does not have any objective significance. (Note, however, that, as noted above, saying that color similarity is experiential does not imply that the substrate on which it is imposed is also experiential.)

The chapters in this volume are all, in one way or another, about TCS and its significance. Interestingly, it emerges from these essays that the traditional picture is itself under some stress. The chapters fall under three broad (if somewhat overlapping) categories, which we outline below.

I.2 Color and Structure: Current Views

The first category is about the ordering itself. Is it a reasonable framework for describing experienced color? If it is, what is the significance of the ordering with respect to the physical counterparts of color out in the world? Some of the authors are content to operate within the traditional notion of a color ordering. Though they allow that it might be mistaken in some aspects, these authors are nonetheless confident enough of the broad outlines that they are willing to use it as a scaffold for constructing their ideas. To take one example, Paul Churchland argues against the idea that all naturally experienced colors fall into TCS—he shows how to generate certain experiences outside its limits. But he has no doubt that blue-yellow and red-green are the fundamental axes of TCS.

Others are skeptical: they argue that in one way or another, the traditional conception of the color ordering is misleading in fundamental ways. Here are two examples. Don MacLeod is worried by our lack of understanding of the neural basis of color phenomenology. At one time, TCS was viewed as a fairly straightforward outcome of "opponent processing," a hypothetical neural process that transformed the outputs of the three kinds of retinal cone cells, which underlie color vision, to maximize the capture of the information the cones provide. MacLeod, however, does not think that things are that simple. Kimberly Jameson is another skeptic. Not so much in her contribution to this volume, but in her work at large, Jameson argues for a much greater role for culture and learning than traditional opponent-processing theory is willing to concede. Her emphasis on the acquisition and construction of color is considerably at odds with TCS, which could be taken as committed to the claim that color is the product of innate processes, and that every experienced color has the same component structure, regardless of cultural influences.

In his contribution, Rolf Kuehni gives a magisterial summary of color-ordering conceptions, though he notes emphatically that these are all, in one way or another, mathematical abstractions, and that perceptual order is at best (but, significantly, often) on the ordinal level. In other words, don't take TCS models too literally, but do be guided by them in a qualitative way. The central problem concerning the sensory representation of color can be seen in this way. The amount of information contained in light incident on the eye is enormously greater than that which any organic color-vision system can analyze and represent. These systems reconfigure the information borne by light in a more compact form. One way of looking at this reduction is in terms of dimensionality. Any given light signal can be thought of as a spectral power distribution: an intensity of light in each of the wavelength intervals in the visible spectrum. If, for example, we sample a signal at 10 nanometer intervals, we get thirty-one values between 400 and 700 nanometers, each independent of the others—a thirty-one-dimensional representation. Color-vision systems need to reduce this dimensionality: they possess neither a receptoral array sufficiently sensitive to small differences in such a representation, nor the processing capacity to deal with such quantities of information, nor indeed the need for such exquisitely fine sensory response.

If we put the matter in this way, we can isolate two problem areas for a psychophysicist trying to describe sensory color ordering. First, how do you measure the dimensionality of color sensation? How, in other words, do you map out a color ordering given empirical observations of color-discrimination performance? Second, what is the nature of the dimension reduction between signal and percept? The limited number of kinds of retinal cone cell is one obvious source of dimensional reduction. Humans possess only three kinds, and each samples intensity over a broad interval (with varying sensitivity at different wavelengths). This effects a huge reduction of information. Is such a reduction strategic or circumstantial? That is, is it a well-designed sampling that records a large amount of relevant information, throwing away what is not needed? Or is it just leakage—information simply being lost as if by a careless housekeeper sweeping the mess off his employer's desk?

Kuehni implicitly suggests that, to a great extent, the reduction is simply a loss: "Experiments have shown that the ability to 'measure' absolute luminance is very poor, and relative luminance is, at best, estimated." (18) However, he seems to concede that color vision might be quite effective in normal ecological situations: "The full process used by the brain to assign a specific color experience to a given stimulus in a given visual field is as yet unknown, but it seems to be the result of relatively flexible interpretation of the [spectral power distributions] arriving at the eyes." (18)

We noted before that the principal hue axes of TCS induce an ordering quite different from that of the physical rainbow. In addition to this, the dimension reduction just discussed means that signals that are distinguishable in a large number of dimensions might be indistinguishable in an informationally impoverished sensation. This is the problem of metamerism (metamers being physically distinct signals that are perceptually indiscriminable). Paul Churchland says that "the apparently unprincipled diversity of metamers poses a genuine problem for a reductive account of objective colors." (39) Metamerism is a direct challenge to Churchland's desire that such an account provide a one-to-one correlation between our internal (phenomenal) representation of color and the set of physically distinct color signals. He wants to argue, nonetheless, that "there is a way to construe the initially opaque space of possible reflectance profiles so that its structural homomorphism with human phenomenological space becomes immediately apparent . . . thus the argument against color realism evaporates." (42)

Churchland's strategy consists in finding a mathematical transformation of the thirty-one- (or whatever) dimensional signal vector, the "CA ellipse." He claims that color sensation "tracks CA ellipses." In short, his claim is that there is a geometric construct in many-dimensional input space such that color sensation corresponds to its smoothed-out approximation in a few-dimensional TCS. Metamers do not constitute an "unprincipled diversity" according to Churchland; they are united by the fact that they share a CA ellipse. Color, then, is indeed a physically based ordering for Churchland: experienced color is the human color-vision system's best-shot approximation to a very rich reality, using CA ellipses as an approximation tool.

How does TCS function in cognition? Mohan Matthen takes TCS to be a semantic, or representational, vehicle by which visual sense orders real individual objects by color. His argument turns on the kinds of knowledge we have of the colors as opposed to the kinds of knowledge we have concerning what color belongs to what object. It is only recently that philosophers have distinguished the questions about the nature of objects—"What color is that object?"—from questions about the nature of colors—"What is the nature of orange? Is it, for instance, mixed or pure?" Matthen argues that answers to the second kind of question—questions about the colors themselves—are known with "Cartesian" rather than merely "empirical" certainty, and that this can be explained only by supposing that we possess innate knowledge of the representational system that our perceptual systems use to denote the color properties of environmental objects. He takes it to be characteristic of innate concepts that they are acquired developmentally, not by learning.

Churchland and Matthen are optimists about TCS; they are not committed to the details, but they do lean on the analyzability of color experience into independent dimensions that are not spontaneously chosen by the perceiver. Kuehni, too, grants the concept some validity and importance, though he demonstrates the multiplicity of theoretical approaches to the phenomenal structure of color. Others, such as Reinhard Niederée, are less sanguine.

To understand the nature of Niederée's challenge, we need first to distinguish two ways in which the dimensionality of color can be taken. The description of TCS given above assumes that TCS provides an answer to the following question: how shall we describe our phenomenal experience of the color of a thing? To answer this question, we can ask people to match and discriminate things by color and record how they respond; such experiments lead to a graphical representation of similarity relations something like TCS. But we can also ask another question: how does color experience vary with changes in objects in our environment? As it turns out, color experience depends not only on properties of the object to which color is attached, but also on the surroundings. For example, the color you attribute (on the basis of perception) to a fruit you are looking at depends not just on the fruit, but also on its background and surroundings. The fruit might, for instance, look luminously orange if it is placed in a dark surround, and somewhat flatter and yellower when it is placed in a light-colored surround. Niederée refers to this contextuality of color when he says, "there is a plethora of well-known spatial and temporal context effects (labeled as color contrast, assimilation, adaptation, and the like), which often even lead to 'new' colors not observed in isolated patches, such as (most shades of) brown or gray." (93) When you try to account for variations of color in context, Niederée shows, TCS is too weak an instrument.

Here is a simple example of what he has in mind. The color brown marks a region in TCS. Yet, it is a "contrast color": a color that is visible only when a contrast with other colors is available. (White and black are contrast colors, available only when there are brightness differences in a scene. Since brown is a blackish orange, yellow, or red, it, too, is dependent on brightness contrast.) Niederée analyzes the change of color appearance of a sample light as it is placed in surrounds differing in color and intensity. He shows that a three-dimensional ordering cannot account for the change of color appearance in such a situation, even conceding trichromacy. As he says, "this dimensionality result does not imply that color vision is based on four basic classes of retinal receptor types rather than three. Instead, higher dimensionality of the kind considered here obviously is the consequence of context effects, . . . [which] turn out to be more complex than commonly assumed." (95)

Rainer Mausfeld approaches TCS from a radically skeptical point of view. He starts by asking how a series of physical processes starting from an object in the physical world and ending up with a conscious percept could, in Russell's words, "suddenly [jump] back to the starting point, like a stretched rope when it snaps." How in other words, can we coherently suppose that the culminating percept represents the original object when everything in between is a transformation? Mausfeld's approach is to propose that the "perceptual system" has a coding and conceptual system that is used to tie the organism to its environment. (Here, he is in broad agreement with the "sensory classification thesis" of Matthen 2005, chapter 1 and passim.) And he maintains that there is no such thing as "color per se." One of the things Mausfeld means by this insistence is the denial of a certain strand of philosophical realism that takes the perceived world to be presorted into types that are somehow recreated by the perceptual system—the kind of realism that we find in Churchland, for example. He is also out of sympathy with the opinion, dominant among contemporary philosophers of perception, that sensation has only "nonconceptual content," an idea powerfully advanced by Gareth Evans (1982). Mausfeld posits a radical plurality of color-type parameters. Aperture colors do not signify the same kind of feature or situation as surface colors; lustrous colors "cannot be reduced to or understood from physical considerations of physical material properties" (139); and so on. Additionally, in claiming that there is no such thing as color per se, Mausfeld means to assert that the very construct of color is defined implicitly by the range of transformations carried out and uses within and outside the system to which it is put, and that it is not out in the world in this sense (just as is plausibly the case of, say, the syntactic property of being a control verb).

I.3 Color Spaces and Explanatory Spaces

A second broad class of questions concerns the relationships between colors and color appearances, as putatively conceptualized by TCS (or some inheritor conception) and various other aspects of color perception. These relationships matter because the significance of TCS lies largely in the explanatory projects relating to color perception in which it plays a role. The chapters in this section of the anthology are devoted to asking whether and how TCS can, in fact, serve various (psychological, neural, metaphysical) explanatory projects for which it has been enlisted.

For example, Kimberly A. Jameson is interested in the ways in which colors are partitioned into categories by perception and cognition. Although it is hard to imagine that this sort of color categorization is completely independent of the organization imposed on colors by TCS, there is plenty of room for debate about just how large a role this factor plays in color categorization. The flames of this debate were fanned by the celebrated results of Berlin and Kay (1969), which purported to demonstrate color categorization to be cross-culturally universal. Many have thought that the best explanation of these findings (and subsequent studies in the same tradition, such as the 2009 World Color Survey) lay in taking TCS to underlie color categorization. For example, it is at least somewhat suggestive that the color categories corresponding to all four poles of the chromatic axes distinguished by TCS (viz., red, green, blue, and yellow) were found to be fundamental rather than derived. And if, as many have held, TCS is itself more or less universal in adult human beings—perhaps even fixed innately somehow in the human biological-cognitive-perceptual endowment—then this would explain the cross-cultural convergence in color categorization reported by Berlin and Kay and others. In her contribution to the present volume, however, Jameson disputes the extent to which TCS can explain data about color categorization. While she does not doubt that there is an ordering of color appearances in human beings, she does doubt its universality, its independence from cultural influence, and its ability to explain color categorization by itself.

Austen Clark is concerned with another role that TCS has traditionally been assigned in connection with phenomenal consciousness. A view that finds broad acceptance among perceptual psychologists and philosophers of perception is that one cannot undergo the appearances that are organized by TCS without being in a conscious mental state. That is, on the standard view, phenomenal appearances cannot occur without perceptual consciousness. However, Clark argues that rejecting the standard view has explanatory benefits and instead holds that states found in early visual processing can have qualitative characters without being conscious. This raises further questions about what ingredients above and beyond qualitative character are necessary for a state to count as conscious. Clark's answer is that the state must be a locus of selective attention.

Just as with Jameson, Clark's contribution does not challenge the existence of TCS *per se*, though he implicitly challenges its connection to consciously experienced similarity. Instead, Clark is questioning some natural and widespread assumptions about what it means for a state of color (or other) appearance to be phenomenal, and what else this implies about the perceptual psychology of the subject in whom the state occurs. Thereby, he is challenging at least one of the explanatory roles traditionally assigned to TCS.

TCS finds another, quite different home in its use by some theorists to understand the metaphysics of color properties. This is the topic Jonathan Cohen confronts. Of course, there has been (increasingly in the last fifteen years) a large range of different proposals about how to understand the metaphysics of colors. One important class of such proposals that Cohen is concerned to defend, and that has its roots in great modern thinkers such as Locke and Boyle, has it that colors are fundamentally constituted in terms of relations to subjects. More specifically, proponents of such relationalist views typically hold that colors are what they are in virtue of the color appearances that they induce in subjects and that are organized by TCS. The traditional (so-called physicalist) rival of such relationalist views takes colors to be constituted in terms of their physical makeups, independently of the appearances they induce in subjects. The clash between relationalist and physicalist theories of color is not over whether TCS exists. Rather, it concerns whether TCS should be used as a basis for the ontology of color properties: relationalists think the relations colors bear to states organized in TCS are constitutive of colors, while physicalists think of these relations as interesting, and even important, but ultimately contingent features of colors—that is, features that are not part of the essences of those properties. (Recall that, as noted above, TCS can be construed as ordering just proximal color appearances or both proximal color appearances and distal color properties. Because relationalists construe the distal properties in terms of their relations to proximal appearances, they get the extension of TCS to distal properties for free, as it were.)

While Cohen spends some time arguing for relationalism over physicalism, he devotes most of the chapter to the project of defending relationalism from important criticisms, put forward by Hardin and others, that are independent of physicalism. These criticisms contend that by grounding colors in TCS, relationalism results in an insufficiently objective and insufficiently unified conception of color. Cohen responds to the worry about objectivity by arguing that we lack reasons to demand the objectivity of colors in any of the forms that color relationalism is unable to secure. And he argues that, although there is a good sense in which relationalism does not provide for the unity of colors by itself, this is a reflection of relationalism's schematic character: any more specific form of relationalism that would be counted an adequate theory of color will provide the ingredients needed to answer the unity demand. (He does not address, however, the epistemic objections to color relationalism put forward by Matthen.) Yet again, what hangs on Cohen's defense is not the existence or viability of TCS, but the viability of a certain use of TCS—specifically, the use of TCS to understand the metaphysics of color properties.

In contrast, Don MacLeod raises a rather more fundamental challenge to TCS itself and its neural underpinnings. MacLeod urges that our understanding of why TCS has the features it appears to have is (at least) badly incomplete in several important ways. For example, he argues that the standard three-receptor type explanation for trichromacy is empirically inadequate. Moreover, he contends that there is no viable neurophysiological explanation for the psychological color primaries that form the poles of TCS (which is not to say that the textbooks fail to offer explanations!). He urges that we are without an account of what neural mechanisms underlie the most basic relations in terms of which TCS is constructed—color discrimination. MacLeod goes on to claim that the standard (receptor-based) understanding of the differences between color appearances in normal and color-blind subjects makes false predictions. And he contends that we are almost completely ignorant about the neural basis for color constancy.

MacLeod's pessimism about TCS could be taken in different ways. On the one hand, the rough structure of TCS could turn out to be as we understand it, even if its mechanistic underpinnings have so far eluded us; if so, then MacLeod's list of concerns might be read as a twenty-first-century color-science analogue of David Hilbert's 1900 list of then-unresolved mathematical problems (David Hilbert, the famous German mathematician, not David Hilbert the American philosopher who cowrote chapter 11 in the present anthology). Alternatively, current views about the organization of TCS may need serious revision. Either way, MacLeod worries that serious gaps remain in some of the most fundamental aspects of our understanding of TCS. And either way, MacLeod's unresolved questions serve as warnings against premature triumphalism about our understanding of TCS and against overconfident appeals to TCS and its features in other explanatory projects, such as those pursued in the remainder of part II in this anthology.

Jonathan Westphal takes up the problem of explaining how certain exclusionary relations that appear to structure TCS (and that are at the heart of Wittgenstein's concerns in *Remarks on Colour*) are true about afterimage colors. As an example of one such exclusionary relation, consider the (now controversial—see Crane and Piantanida 1983) idea that, necessarily, nothing can be both red and green all over and at the same time. Westphal proposes that the best explanation of why ordinary surfaces cannot be both red and green all over and at the same time lies in the idea that being red and being green require incompatible kinds of interactions with incident light. Because the interactions in question are incompatible, no object can have both of them all over and at the same time; consequently, if being red and being green indeed require of things that they have such interactions, it is necessary that no object can be both red and green all over and at the same time. Although Westphal is generally sympathetic to this explanation of the incompatibilities, he points out that it cannot be applied straightforwardly to afterimages, since there is no light incident on afterimages (which are not surfaces). And yet, Westphal wants to insist, afterimages are colored, and the exclusionary relations at issue are no less true of them: just as with surfaces, no afterimage can be both red and green all over and at the same time. Westphal's question, then, is how to account for the relationships that appear to structure TCS in afterimages.

Taken together, these articles underline the importance of TCS by arguing about how (and in some cases, whether) it is related to a range of substantially disparate topics in color perception. But they also show up tensions in and challenges to the traditional assumptions about TCS.

I.4 Color Blindness

A final pair of chapters in the present volume debate the nature and extent of the differences between the color spaces of statistically normal trichromatic color perceivers, on the one hand, and those of perceivers with certain types of deficiencies—specifically, red-green dichromats—on the other. One way in which these matters interact with general questions about TCS is by addressing how the general shape of TCS varies systematically with different kinds of statistically abnormal alterations. These issues also shed light on questions about whether and to what extent our knowledge of TCS is mediated by certain kinds of experience that might be differentially available to different classes of observers—and, therefore, on the putative innateness of this knowledge.

Alex Byrne and David Hilbert argue for a revised version of a more or less orthodox position that they call the "reduction view," according to which the color space of red-green dichromats is a proper subset of TCS. They contrast the reduction view with the "alien view," which has it that the color space of red-green dichromats and TCS have no nontrivial overlap—rather, the elements of the former are just entirely different, or alien, with respect to TCS. Byrne's and Hilbert's contention is that if dichromatic vision is a reduction of normal trichromatic vision, then the (revised) reduction view is correct.

Their main argument in support of the reduction view depends on the Hurvich-Jameson-Hering–inspired idea that TCS is structured around just two chromatic opponent channels (red-green and yellow-blue). They note that, on the standard understanding, the value of the red-green channel depends entirely on the difference between the output of retinal L cones and that of retinal M cones, whereas the value of the yellow-blue channel depends on the output of three retinal cone types (L, M, and S cones). But since red-green dichromats have no functioning M cones (if deuteranope) or no functioning L cones (if protanope), Byrne and Hilbert reason that they should have a functioning blue-yellow channel but no functioning red-green channel. If that is true, then red-green dichromats should represent the chromatic properties of the world entirely through the contribution of their working yellow-blue channels, whereas normal trichromats represent the chromatic properties of the world through the contribution of both their yellow-blue channels and their red-green channels. This leads directly to the reduction view: red-green dichromats should represent a subset of the trichromat's TCS. As noted, Byrne and Hilbert endorse (again, assuming dichromatic vision is a reduction of normal trichromatic vision) not the standard reduction view, but a revised version of it. The revision comes from their contention that the yellow-blue channel carries information about not the determinable colors yellow and blue but the super-determinable colors yellowishness and bluishness. Adding this position to the mix turns Byrne's and Hilbert's reduction view into an alien view, because "normal trichromats never see this hue [yellowishness] without seeing more determinate hues like orange and yellow. On the revised reduction view, a red-green dichromat sees yellowishness unaccompanied." (280)

Justin Broackes thinks that this whole approach is wrong. He argues that the traditional view gives a poor account of how dichromats experience the colors they tend to confuse with one another. For, while the traditionalists construct "confusion lines" in trichromatic color spaces—lines that connect the colors that dichromats supposedly cannot distinguish—these confusion lines tell us little about how the supposedly confused colors actually look to dichromats. Do all the colors that lie along a confusion line look to the dichromat the way that one of those colors looks to the trichromat? (This would be Byrne and Hilbert's reduction view.) Or do they look fundamentally different from any of these colors? (This would be their alien view.) To Broackes, neither view is compelling. Dichromats simply do not fail to distinguish colors along a confusion line, at least not when they are presented with large fields and with surface colors (as opposed to just small fields of light seen through the tube of a colorimeter). They are less reliable than trichromats at detecting color differences among objects, but this does not mean that they are never able to do so, or that they do not make distinctions among trichromat colors. The whole business of "confusion lines" is overplayed, Broackes argues.

Broackes is out of sympathy with the view that TCS is somehow constructed out of cone inputs, with the consequence that the three-dimensional TCS of color normals is replaced in cone-deficient perceivers by a two- or enhanced two-dimensional system. From his point of view, TCS is specified (in humans at least) independently of cone-cell inputs. As he sees the problem for color vision, then, it is not to construct a color space commensurate with the number of input channels that cone cells provide. Rather, the problem is to make the discriminations needed to fill a color space given, to some extent, independently of receptors. Thus, dichromats, or at least many of them, do not end up with a lower-dimensional color space: for instance, lacking the red-green axis or with a radically different way of perceiving blue-yellow. Instead, they start out with TCS (or some other three-dimensional color space), and fill it by making the requisite distinctions with less information, though perhaps with a coarser grain and less reliability.

How can dichromats fill up TCS if they lack information from a whole class of cone cells? Here Broackes invokes an interesting maxim—that the information available to a reduced receptoral array may be enhanced by integrating information gathered over time and with movement of the perceiver. While the information gathered by a trichromat at a moment is a lot more color-sensitive than that which can be gathered by a dichromat, this deficit is to some extent mitigated over time. Consider the following: in the reddish light of evening, the sides of red objects that are turned toward the light are brighter than those of green objects of the same overall brightness. Again, as objects turn toward or away from the light, differently colored objects will display different darkening patterns. In these and other ways, dichromats can gather almost all the information that trichromats gather, and they can use this information to array objects in a three-dimensional color space.

Twenty-five years ago, both color science and philosophy of color enjoyed considerable certainty concerning the fundamental issues of color orderings and their role in color experience. Philosophers believed that the then prevailing views concerning color orderings conformed well with the schemes advanced by Johnson and others. Perceptual psychologists, for their part, felt that their ordering schemes pointed the way to discoveries about the neural organization of color processing. These views have been extended in some unexpected directions; they have also been challenged in unexpected ways. The time may be opportune for new ways of thinking about color and color perception. We hope that this volume will provide some of the material for new advances in color theory.

References

Berlin, B., and P. Kay. 1969. *Basic Color Terms: Their Universality and Evolution*. Berkeley: University of California Press.

Crane, H. D., and T. P. Piantanida. 1983. On seeing reddish green and yellowish blue. *Science* 221: 1078–1080.

Kay, P., B. Berlin, L. Maffi, and W. R. Merrifield. 2009. *World Color Survey*. Chicago: CSLI.

Evans, G. 1982. *The Varieties of Reference* (published posthumously, edited by John McDowell). Oxford: Oxford University Press.

Matthen, M. 2005. *Seeing, Doing, and Knowing: A Philosophical Theory of Sense Perception*. Oxford: Oxford University Press.

Wittgenstein, L. 1977. *Remarks on Colour*. University of California Press.

I Color and Structure: Current Views

1 Color Spaces and Color Order Systems: A Primer

Rolf G. Kuehni

1.1 The Idea of Color Space

The idea that the large number of color percepts humans can experience must fit into some kind of ordering system is old. Aristotle listed seven (or eight) color categories in a lightness-based scale of chromatic colors between white and black. But he despaired of sorting all color experiences he knew into a system. In the eleventh century, Avicenna and, in the twelfth, Theophilus described scales other than the Aristotelian one that range from white to black, which were later termed tint-shade scales (from a strongly chromatic color toward white and toward black). Beginning in the early seventeenth century, such scales were expanded to two-dimensional systems by connecting them to common white and black points (S. A. Forsius, F. Glisson), and to three dimensions in the mid-eighteenth century (T. Mayer, J. H. Lambert). Lambert's three-dimensional system was based on mixtures of three primary pigments—yellow, red, and blue—and demonstrated their modifications through mixture and with additions of white, black, or both. That the corresponding solid is located in Euclidean space was a natural assumption. A three-dimensional space was found necessary because at least three kinds of scales were known by Mayer's time: (1) tint-shade scales, where intense chromatic colorations from a particular pigment were lightened and darkened by increasing additions of a white and a black pigment, respectively; (2) tonal scales, where the chromatic pigment was toned down by increasingly adding a gray mixture of black and white pigments or of one or two other highly chromatic pigments; and (3) hue scales, in which colorations that appear perceptually sequential in hue can be arranged into a continuous circle. Usually, there was an attempt to make the steps between the grades of such scales perceptually uniform. Proposed forms of such three-dimensional solids varied from Mayer's double tetrahedron and P. O. Runge's sphere, to cones, cubes, pyramids, and irregular solids of various kinds.[1]

Attempts to fill the interior of the solids systematically raised the issue of the relationship between color stimulus and response, a relationship that, it was quickly realized, is complex. For purposes of color order, this relationship was intensively studied, beginning for lights with Newton and for colorants with Lambert. The development of psychophysics in the early nineteenth century resulted in closer attention being paid to the observer. Experimental psychology found many kinds of variability in the relationship between stimulus and response as a result of observation conditions. A small group of psychologists saw essential truths in perceptual variability as a function of observer and conditions of viewing. The view that the normal human color-vision system has a standard implementation and that all perceptual data are appropriately treated with normal statistical distribution methodology prevailed and, in the twentieth century, became the standard paradigm. It appears that this view has begun to shift toward a more complex conception of color vision and a realization of the limitations of applying simple mathematical-geometric principles to sensory representation. This chapter presents an overview of the critical issues regarding ordering of color percepts and the relationship between stimuli and percepts, as well as of the kinds of color order systems that have been developed, with their strengths and limitations.

1.2 Color Stimuli

All conventional color stimuli are lights.[2] Such light stimuli are generated either directly by light sources (sun, lamps) or by reflection from or transmission through objects (moon, banana, color chip, glass of red wine, etc). Light is defined as electromagnetic radiation of specific energy content. It can be considered to exist in waveform, with its energy expressed in terms of wavelength, or as quanta or photons, with the energy expressed in electron volt. Light visible to fauna ranges approximately from 300 nanometers (nm; 1 nm is a billionth of a meter) to 800 nm. The range visible to humans (called "visible range" from here on) is approximately from 380 nm to 730 nm. Light of the larger range is produced in abundance by the sun, much of it arriving on the surface of the earth and penetrating water to some depth. Lamps, for reasons of efficiency, are more and more configured to have output in the visible range only. Light is either in monochromatic form, consisting of one wavelength or a very narrow band only, or polychromatic, having radiation of multiple single wavelengths or more or less broad ranges of wavelengths. Isaac Newton experimentally demonstrated in 1668 that sunlight consists of a broad range of differently refracting wavelengths, resulting in different color experiences.

The quantitative range of light energy is very large, as is evident when comparing the intensity of candlelight with that of light leaving the surface of the sun. For technical purposes, lights are often described by their spectral power distributions (SPDs). SPD is a measure of the absolute or relative power of lights at all wavelengths of the visible range. Absolute spectral power is expressed in watts/m^2/nm, a measure of the amount of power per unit area and nanometer. Relative SPD is usually normalized to a value of 1.0 at 555 nm (approximately in the center of the human sensitivity range) on the ordinate, with wavelength in nanometers on the abscissa, thus relativizing light intensity. A light of some theoretical interest (no lamp of this type exists) is the equal energy spectrum (EE) light, having a relative value of 1.0 at all wavelengths of the visible range and thereby having no effect in calculations.

Material objects have properties that change the SPD of lights interacting with them. There are several kinds of interaction, but the most important are transmittance, reflectance, and dispersion. Transparent objects (filters, liquids) may contain materials that selectively absorb light quanta at given wavelengths, thus eliminating the quanta from the beam, or reducing their intensity. Transmittance is the spectral measure of the amount of light passing through them. Similarly, opaque objects may have materials on or near the surface that selectively absorb light. Reflectance is the spectral measure of the portion of light reflected from objects. At the interface of dispersing or refracting objects and, for example, air, light beams change their direction as a result of different optical densities of the media. The angle of refraction is a function of light energy (wavelength) and the optical densities (refractive indices) of the two materials involved. As a result, the wavelength components of polychromatic daylight are spatially separated when passing through a glass prism, because glass has a higher refractive index than air.

Light that is absorbed does not disappear, but the quanta lose some of their energy. They are emitted again at a lower energy level, or higher wavelength, usually in the infrared region, not visible to us. Intensity and SPD of lights can be measured with good accuracy. This means that the propensity of materials to absorb light energy, spectral transmittance or spectral reflectance, can also be accurately measured. Therefore, color stimuli arriving at the eye can be numerically specified with a degree of accuracy that considerably exceeds the differential color-vision threshold of humans.

The number of light stimuli that can be specified numerically is infinite. Systematic sampling of lights at 30 nm intervals from 400 nm to 700 nm and 0.1 units of spectral power increments already results in about 10 billion different stimuli. Of these, under given conditions, the average observer can, by direct comparison, distinguish only about 2 million. Presumably, they represent an equal number of distinct percepts.[3] Information contained in an SPD is, in mathematical terms, multidimensional. The information at each wavelength, the spectral power, represents one dimension. The dimensionality of an SPD thus depends on the number of wavelengths at which the relative spectral power is measured or expressed. At 1 nm intervals, an SPD from 380 to 750 nm is 371-dimensional; at 10 nm intervals (typical technical measurement) from 400 to 700 nm, it is 31-dimensional. Humans are incapable of conceiving of a solid with either number of dimensions. The highest space dimensionality most of us can comprehend is three, and even then, many people have some difficulties. As the ubiquity of maps indicates, few people have problems with two-dimensional representations. The problem that arises is how to reduce 31-dimensional SPDs to two or three dimensions to gain objective but comprehensible understanding of the relationships of multidimensional SPDs.

1.3 Color Stimulus Space as the Result of Mathematical Dimension Reduction

A widely used method of mathematical dimension reduction is the so-called principal component analysis. It objectively analyzes spectral functions for components that are present more often than others, resulting in a series of spectral component functions that, together, explain more and more of the variability in a set of SPD, spectral reflectance, or transmittance functions. Figure 1.1 illustrates the first three eigenvector functions derived from principal component analysis of 1,269 reflectance functions of Munsell color chips. When such chips are illuminated with EE light, the reflectance function is identical to a relative SPD function arriving at the eyes. Eigenvector function 1 explains the largest percentage of variability in the chip reflectance functions, function 2 the second largest percentage, and so on. The first function is a rough indicator of the average relative energy level of an underlying SPD; the second and third functions express spectral changes in SPDs, giving highest positive or negative

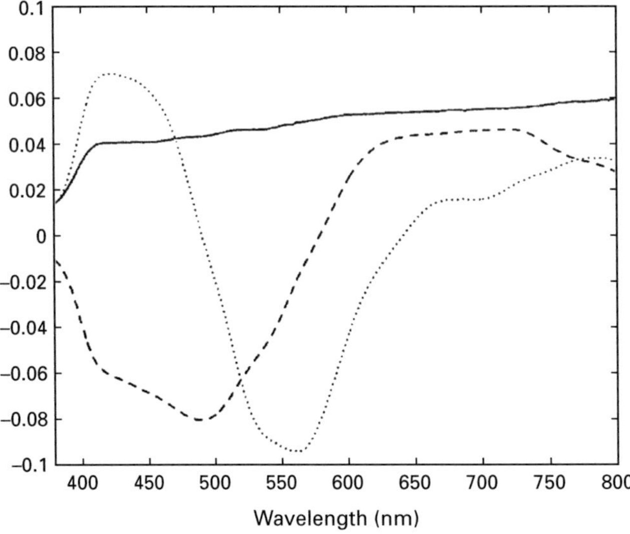

Figure 1.1
The first three eigenvectors from principal component analysis of reflectance functions of 1,269 Munsell color chips. Solid line, eigenvector 1; dashed line, eigenvector 2; dotted line, eigenvector 3 (Lenz et al. 1996).

weight to different wavebands. The latter two functions are expressive of the specific nature of the chip reflectance functions. The three functions are specific to the set of spectral functions from which they have been derived and explain only a significant portion but not all of the variability in the functions. Nearly all variability in the related set of reflectance functions is explained by the first five to seven eigenvectors; that is, with these functions and the resulting stimulus values, it is possible to reconstruct the original reflectance function (or SPD in the case of lights) at the 99+ percentage level. The first three functions can represent the axes of a three-dimensional space in which individual reflectance functions plot as points. The coordinates of the chips in this space are found by multiplying, at each wavelength, the reflectance value with the corresponding eigenvector value and summing the results for each of the three eigenvectors. When plotting the loci of a particular group of twenty Munsell reflectance functions, we find that they fall on an approximately elliptical continuous line (figures 1.2 and 1.3). The reflectance functions plotted in these figures represent every second color chip of a Munsell hue circle at constant value (lightness) and chroma (intensity of coloration). Of course, principal component analysis knows nothing of hue, value, and chroma. But, interestingly, we find that a (roughly) circular

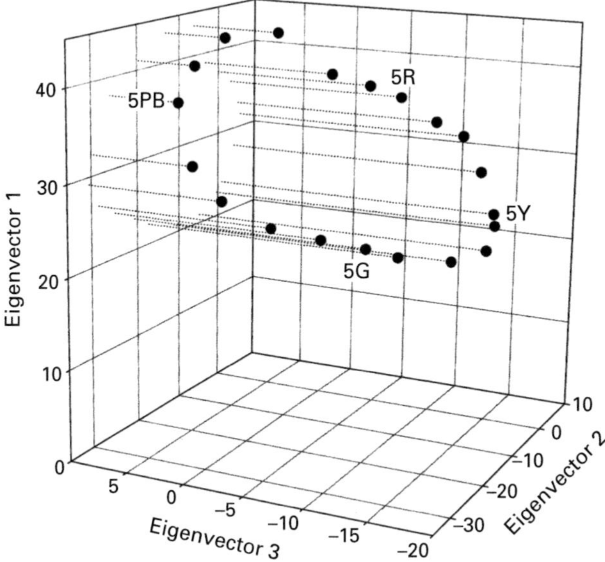

Figure 1.2
Plot of twenty Munsell color chips in the eigenvector space generated by the functions illustrated in figure 1.1. Every second hue of the Munsell hue circle is at value 6 and chroma 8. Four hue names are identified.

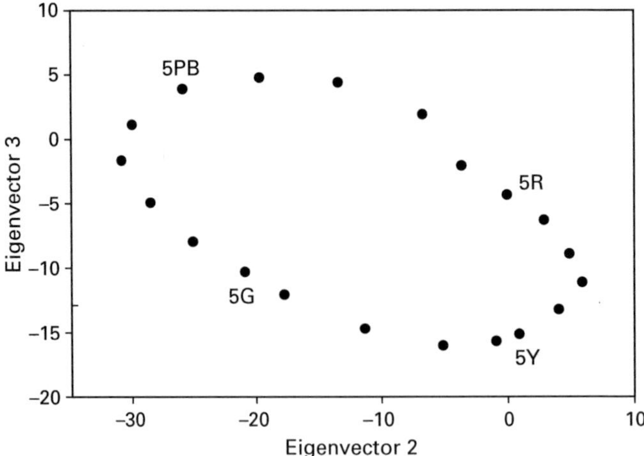

Figure 1.3
Projection of the figure 1.2 data on the plane formed by eigenvector functions 2 and 3.

arrangement of the data points is the automatic outcome of dimension reduction to three dimensions when using particular "filter" functions (in this case, the first three eigenvector functions), and that constancy of chroma and value are also roughly approximated by this procedure.[4] Therefore, these properties are fundamental indicators of information contained in the reflectance functions from which they are derived, and it is no surprise that they have found their expression in perceptual attributes. The complete set of the 1,269 reflectance functions, when plotted in their three-dimensional eigenvector space, have roughly the form of an irregular upright spindle.

An automatic result of mathematical dimension reduction of the type illustrated in figure 1.1 is what, in regard to color, has been named metamerism. This term refers to different SPDs or reflectance functions resulting in identical eigenvector values; that is, in the related three-dimensional space, they are indistinguishable. As alluded to above, an aspect of principal component analysis and other comparable mathematical operations is that the resulting component functions are of necessity functions of the SPD or reflectance function properties from which they have been calculated. In the case of color chips, the selection of chips and the reflectance functions of the pigments used in their preparation affect the form of the resulting component functions. What samples are found to be metameric depends on the mathematical dimension reduction procedure used and the spectral structure of the reflectance or SPD functions used in the calculation of the reduction functions. The three-dimensional eigenvector space is of particular interest because it meets all fundamental geometric requirements for a Euclidean three-dimensional space: it is automatically orthonormal (i.e., its three axes are properly perpendicular to each other), a situation that does not apply in many other color stimulus spaces, as will be discussed below. The three-dimensional eigenvector space is a mathematically closed ordering of any kind of higher-dimensional data, including color stimulus data; however, it involves some loss of information. Geometric distances between points in this space have defined mathematical meanings, but in the case of color stimuli, they are related at most on the ordinal (see section 1.10) level to perceptual distances.

The question of whether there are natural processes of dimension reduction arises, and the answer is positive. Some of our sensory processes are examples, the strongest of them being color vision, because three kinds of "filters," the cone types of the trichromat, have a dimension reduction effect on SPDs.

1.4 Dimension Reduction with Natural Light Sensors

The ability to sense the electromagnetic energy of light wavelengths is nearly universal among living organisms. The following discussion is limited to vertebrates in general and humans in particular. It is now known that evolution comparatively early developed five groups of visual pigments, some or all of which are expressed in all vertebrates (see, e.g., Zhang 2003). In genetics, the five related opsins (protein molecules to which the light sensitive molecules, such as retinal, are attached; they determine the wavelength sensitivity of the sensor in which they reside) are known as SWS1 (for short-wavelength sensitive) and SWS2, RH1 (for rhodopsin) and RH2, and LWS/MWS (for long- and medium-wavelength sensitive). Depending on their evolutionary descent and environmental needs, animals ended up with one or more of these, up to all five. In each of the groups, several possible expressions of the genes result in varying sensitivity ranges of the corresponding sensors. Peaks of SWS1 sensitivity are mostly in the UV range; those of SWS2 range from roughly 400 to 450 nm. The two rhodopsin sensor types have peak sensitivity around 500 nm, and the LWS/MWS sensor subtypes, from about 500 to about 570 nm. The rhodopsin sensors are highly sensitive and usually limited to night vision. The exact placement of the sensitivity maxima is genetically fine tuned. In some animals, the peak sensitivity ranges of the sensors are further tuned by the placement of colored oil drops in front of them (e.g., in chickens). Humans are comparatively impoverished by having only three types of sensors: SWS1, RH1, LWS/MWS. Our RH1 sensor is responsible for night vision capability and is generally taken not to contribute to color vision. The three sensors involved in human color vision are the SWS1 sensor, conventionally termed S, and the LWS/MWS sensors L and M. Average spectral sensitivity functions of the three daylight sensor types, known as cones, normalized for equal area under the curves, are shown in figure 1.4.

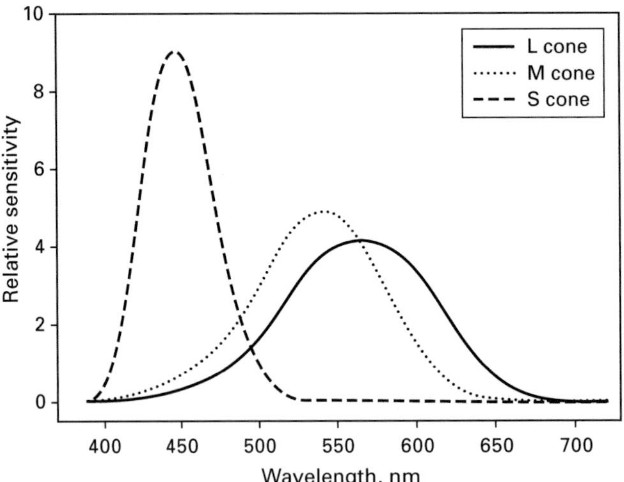

Figure 1.4
Spectral functions of the relative sensitivity of average human cones *L*, *M*, and *S*. The functions have been normalized for equal areas under the curves.

Addition of the M-cone variant is believed to be comparatively recent (an estimated 35 million years ago). M is expressed only in primates and Old World monkeys. Individual expression of the L and M cones varies to some degree, with peaks of the L cone ranging from 550 to 560 nm and those of the M cone from 530 to 536 nm (Neitz and Neitz 2004).

The three human cone types, L, M, and S, can be viewed as filters that reduce SPDs from, say, thirty-one dimensions to three, the three being the result of multiplication at each of the thirty-one wavelengths of cone response with the spectral power of the stimulus and summing the results in three values: the tristimulus values *L*, *M* and *S*. When taking the scales of the three tristimulus values as space axes, the result is a three-dimensional color stimulus space, with dimensions that are different from those of the eigenvector space discussed earlier because the filtering functions are different. An important difference is that cone space is not orthonormal; that is, it is not properly represented as a conventional Euclidean space, a situation generally overlooked in practice. Cone space can be assumed to have some relationship to perceived color because the cones (and rods) represent the only interface between visible electromagnetic energy and the brain.

Cone functions constrain the locations of points representing the SPDs in their three-dimensional space. The locus of spectral colors (lights of wavelengths from 380 to 730 nm) in this space is described by a curve beginning and ending at the origin, with a form somewhat resembling two slightly angled butterfly wings. Of particular

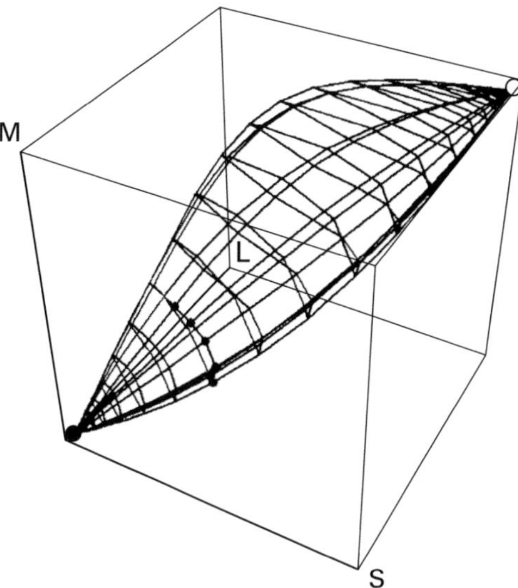

Figure 1.5
Optimal object color stimulus solid in the color stimulus space formed by the relative cone sensitivity functions of figure 1.4 (von Campenhausen et al. 2001).

interest is the solid in this space formed by all possible object color stimuli. If we make the simplifying assumption of using an equal energy light of intensity 1.0 at each wavelength to illuminate the objects, then in spectral representation and with exception of certain fluorescent objects, all possible object color stimuli fit into a rectangular diagram with wavelength on the abscissa and relative SPD from 0 to 1.0 on the ordinate. By sampling the possible SPDs appropriately and systematically, the surface of the implicit object color solid can be calculated (figure 1.5). It is a tilted spindle of irregular cross section. All possible object color stimuli (as viewed in EE light by the standard observer to whom the cone functions apply) fit on its surface or in its interior. SPDs can be metameric in this space; that is, certain groups of spectrally different SPDs are not distinguishable. It is apparent that, due to the shape of the cone sensitivity functions, cone stimulus space is filled with object color stimuli only to a comparatively small degree. Geometric distances in this color solid are also at most only in ordinal order of perceived distances.

1.5 Linear Transformation

In the middle of the nineteenth century, the German mathematician H. G. Grassmann proposed a mathematical framework for a limited aspect of the relationship between color stimuli and resulting perceptions, an aspect known as color matching. He postulated that the actions of cone sensors are univariant; that is, the qualitative response of the sensor to photons of light is independent of wavelength in the spectral region of the sensor's sensitivity. If this is valid, cone sensitivity functions can be subjected to linear transformation without loss of intrinsic information content. Consequently, the form of functions can, within the limits of such transformation, be changed to suit different purposes. Technically the most important transformation is the one promulgated by the Commission Internationale de l'Éclairage (CIE, International Commission on Illumination) for use in its colorimetric system. The CIE was interested in a transformation of cone functions into functions with greater relevance to color perception. One of the dimensions was expressive of brightness (for lights) or lightness (for objects). Linearity demanded that the luminance scale be additive: when adding two lights, the sum is also perceptually additive. This was found to be the case for only a specific experimental form of perceptual brightness-lightness determination: the so-called flicker method. Setting one of the three dimensions as luminance or luminous reflectance leaves the remaining two to express the chromatic content of colors. The normalized color-matching functions resulting from this transformation of the cone sensitivity functions of figure 1.4 (with identical intrinsic information content) are illustrated in figure 1.6.[5] SPDs or

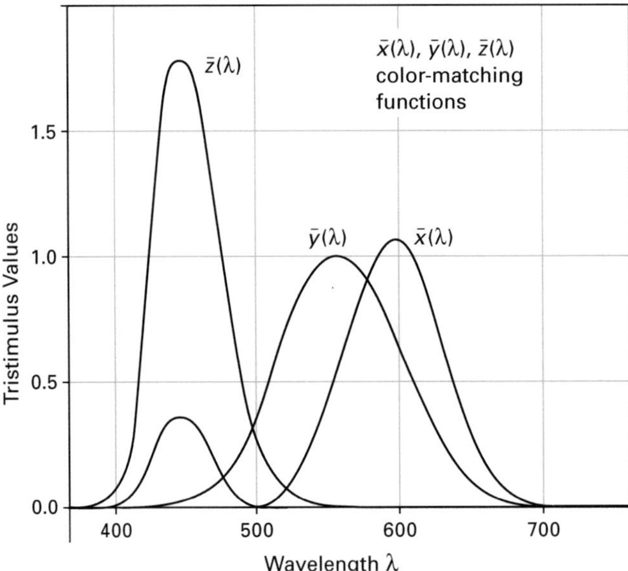

Figure 1.6
CIE 2° standard observer color-matching functions (Wyszecki and Stiles 1982).

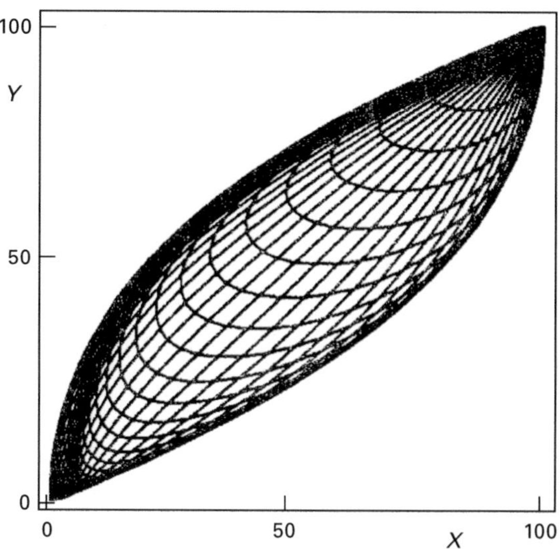

Figure 1.7
Projection of the optimal object color stimulus solid of the *X*, *Y*, *Z* tristimulus space onto the *X*, *Y* plane (modified from Koenderink and Kappers 1996).

reflectance curves can be dimensionally reduced into the corresponding tristimulus space, termed *X*, *Y*, *Z* space, in a manner comparable to what was mentioned for cone space. The resulting optimal object color space is illustrated in figure 1.7. It again forms a tilted spindle, with the vertical axis now indicative of luminance, or luminous reflectance.

D. L. MacAdam (1935) replaced EE light with standardized real lights when he calculated optimal object color solids for different CIE illuminants (light sources). He erected these solids over what is known as the CIE chromaticity diagram, itself a linear transformation of the CIE tristimulus space (figure 1.8).[6] The specific shape of such solids depends strongly on the SPD of the light in which the objects are viewed, a fact not in agreement with perceptual results.

Although one of the tristimulus values was selected to agree with the psychophysical dimension of luminance, neither in the *X*, *Y*, *Z* space nor in the *x*, *y*, *Y* space are geometric distances in more than ordinal order compared to perception. An important difference between the eigenvector functions derived from the collection of Munsell color chips and the cone sensitivity functions is that in the former case, two functions have positive and negative values (see figure 1.1). The question arises if there is a comparable physiology-based version of filters.

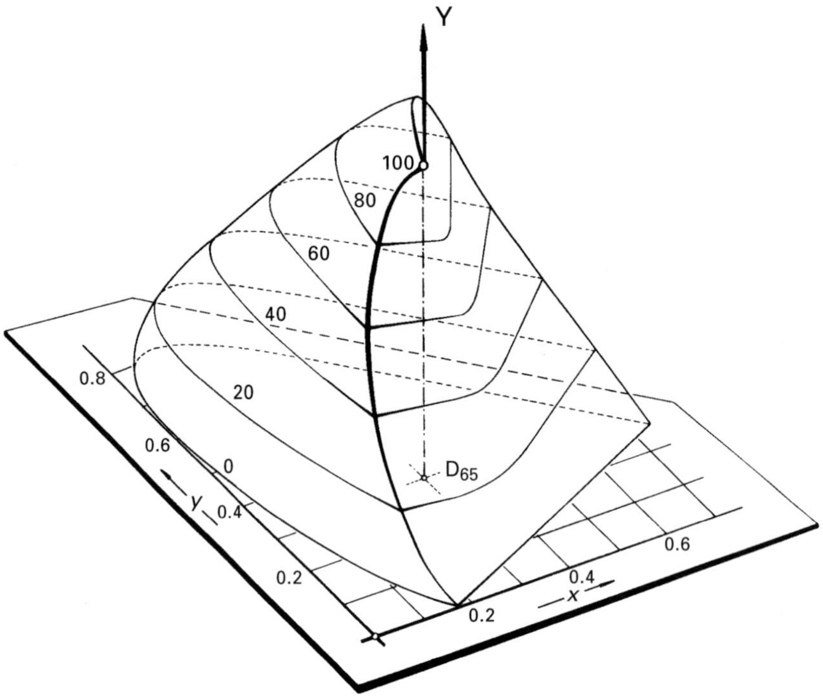

Figure 1.8
Projective view of MacAdam's optimal object color stimulus solid for the CIE 2° standard observer and illuminant D65 in the CIE *x*, *y*, *Y* stimulus space (Wyszecki and Stiles 1982).

1.6 Opponent-Color Space

In the later nineteenth century, the German physiologist E. Hering proposed a color perception theory radically different from the trichromatic theory, the latter a theory based on cone sensor functions and derived in part from the work of Newton, Young, and Maxwell, and formalized by Helmholtz. Hering's theory proposed three pairs of fundamental opponent-color perceptions—white-black, yellow-blue, and red-green—from components of which all human color perceptions are generated. The two chromatic pairs consist of the four chromatic Urfarben (fundamental colors), characterized by their hues, known as unique hues.[7] He, as well as Helmholtz, suggested early interpretations of this theory in terms of cone sensitivity functions. It was the German photoscientist R. Luther who in 1927 calculated a related object color solid (figure 1.9). The additional linear transformation used for this kind of solid has the advantage of normalizing the object color solid by righting it (vertical lightness axis) and by forming two pairs of opposing functions in balanced CIE form: $a = 2.3\ (X\text{-}Y)$, $b = Y\text{-}Z$,

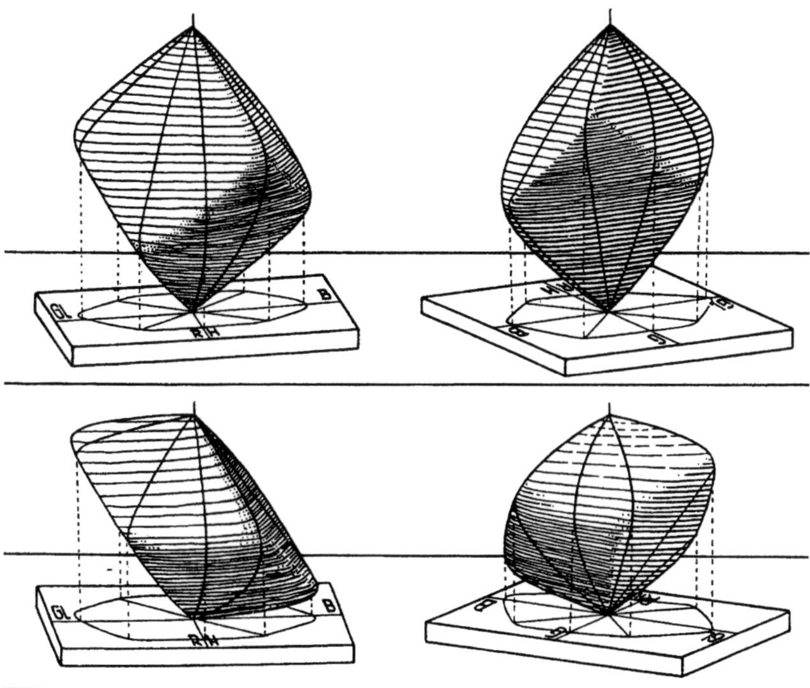

Figure 1.9
Two views each of slightly different versions of Luther's optimal object color solid based on opponent-color functions derived from linear transformation, similar to those of figure 1.10 (Luther 1927).

where *X*, *Y*, and *Z* are the CIE tristimulus values (figure 1.10).[8] These two functions clearly have some values positive and others negative across the spectrum. When combined with the luminance function Y of the CIE system, the three functions roughly resemble the three eigenvector functions of figure 1.1.[9] The *a*, *b* diagram, by design, is a chromaticity diagram of Cartesian form, with the achromatic stimulus at its origin. The luminance scale is erected perpendicularly to this plane. In nonlinear form, as the CIELAB color space (see below), it has, despite serious shortcomings, wide technical application. Such spaces are not orthonormal.

Cells of various kinds with opponent-color behavior were discovered in the retina and in other areas of the visual path in the second half of the twentieth century. A physiology-based linear opponent-color space of somewhat different form is the Derrington-Krauskopf-Lennie (DKL) space (1984). It represents the output space of parvocellular neurons in the lateral geniculate nucleus (LGN) of macaque (*M. mulatta*), which have, it is believed, a color-vision system quite similar to that of humans. The

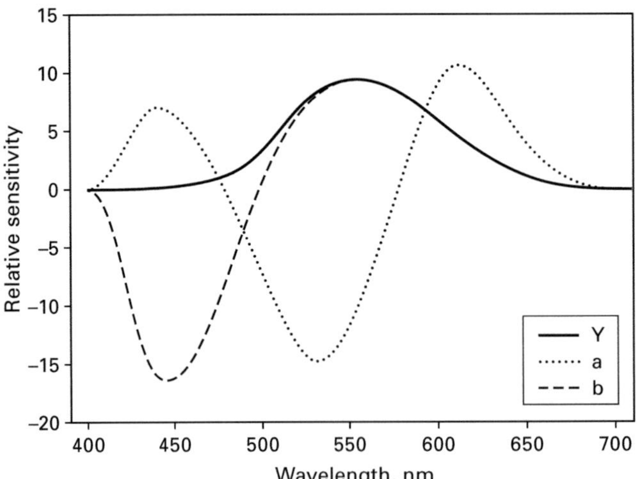

Figure 1.10
Luminosity function *Y* and balanced linear opponent-color functions *a* and *b*, calculated from the formulas given in the text.

DKL space's vertical axis also represents luminance, but its chromatic axes are rotated compared with those of the *a, b, Y* space. Somewhat grandly, its axes were described as "cardinal axes of color space." The LGN is an important waystation for visual data traveling from the eyes to the main visual center of the cortex at the back of the head. Since DKL space was first developed, it has become abundantly clear that considerable additional computation takes place in the cortex; the output of LGN parvocellular neurons is but a snapshot along a complex path from the eye to the as-yet-unknown locus of final assignment of a perceived color to a particular stimulus. As noted, the described spaces are not orthonormal, even though they are always represented as Euclidean spaces.

In 1982 J. B. Cohen and W. E. Kappauf offered an orthonormalized version of the CIE tristimulus space they termed fundamental color space (FCS). It also has two functions with positive and negative values, as illustrated in figure 1.11. Comparison with the three eigenvector functions of figure 1.1, derived on basis of principal component analysis, and Luther's opponent functions (figure 1.10) shows a degree of resemblance, particularly for the two "chromatic" functions (note arbitrary changes in polarity). One can conclude from this that the human cone functions were, to some degree, selected by evolution to extract an approximation of principal components from SPDs arriving at the eye. The object color solid for the functions of figure 1.11 was recently calculated by Koenderink and van Doorn (2004), and a view of it is shown in figure 1.12.

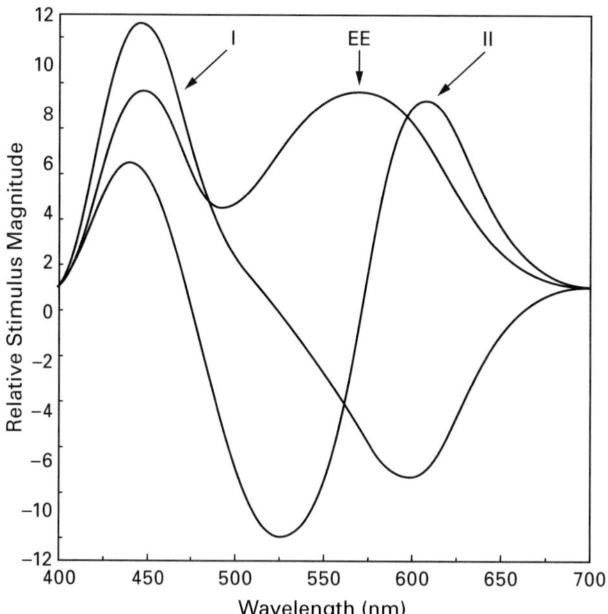

Figure 1.11
The three orthonormal canonical functions derived for the CIE 2° standard observer and the equal energy (EE) illuminant (Cohen 2001).

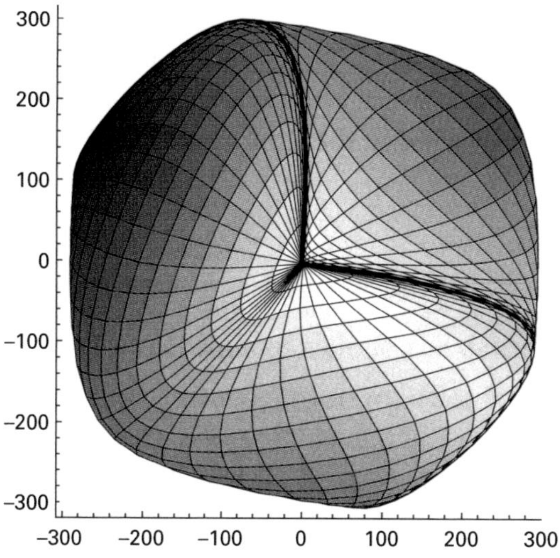

Figure 1.12
View against the white top of the optimal object color stimulus solid in the space formed by the functions of figure 1.11 (Koenderink and van Doorn 2004).

1.7 Human Color Perception

The construction of psychophysical color spaces of the kind discussed in the previous section implies that the human color-vision system has tools to determine SPDs of lights, absolutely or relatively, with a significant degree of accuracy, and to derive reflectance functions. But this is not the case. Experiments have shown that the ability to "measure" absolute luminance is very poor, and relative luminance is, at best, estimated. Elaboration of lightness perception from the information arriving at the eyes is already hugely complex and not yet known in all its details (Gilchrist 2006). Added to it are the complexities of chromatic color perception. SPDs in the visual field change as a function of alterations in the surround with every change in the direction of gaze. Depending on the angle of illumination and resulting shadows, an object has many different stimulus functions, a situation that raises the question of how the visual system would determine the "true" reflectance function of an object, if that were an objective. The full process used by the brain to assign a specific color experience to a given stimulus in a given visual field is as yet unknown, but it seems to be the result of relatively flexible and perhaps statistically-based interpretations of the SPDs arriving at the eyes.

The visual experiences we have as a result of exposure to color stimuli are private and mysterious. The best, if vague, general explanation is that they are the result of neural processing of the information contained in stimuli. The detailed organization of the visual-processing machinery in individuals considered to have normal color vision does vary to some degree (even more so for people with color-vision deficiencies). There are known genetic differences in the exact expression of the cone types. The relative numbers of the three types of cones in the retina vary widely in individuals without evident impact on color-vision ability (Neitz and Neitz 2004). In addition, perhaps somewhat analogous to language dialects we learn in childhood, early exposure to different color stimuli may also imprint our neural machinery in specific ways, with the result that we process color stimuli individually to some extent. A likely practical outcome, in persons with normal color vision, is the known wide variability of stimuli they select as representing the perceptions of unique hues (Kuehni 2004). Experiments have shown that these selections have high intraobserver reliability.

Beyond unique hues and the perceptions of black and white (also here, stimuli of "best" black and white objects differ individually), comparison of the qualitative aspects of color experiences from specific stimuli becomes more and more uncertain. What is possible are judgments of quantitative differences. In a three-dimensional space, an interior point can be different from another interior point in many ways. The question arises whether directions represent certain perceptual kinds. Such commonalities are known as attributes, like those of the Munsell system.

1.8 Color Attributes

That color perceptions can be described by three attributes was first suggested by Newton, who named them *color* (for hue), *luminosity* (for brightness), and *intenseness* (for saturation). These concepts and similar terms were later used by Maxwell and Helmholtz and became standard. The modern corresponding terms for objects are hue, lightness, and chroma. It is difficult to argue in a general way against them: hue and brightness/lightness are obvious and inescapable concepts. A large number of different color stimuli can be seen as having something qualitative in common, despite quantitative differences: their hue. Few people would argue this point when handed a constant-hue page of, say, the Munsell system atlas. Hue is, therefore, an attribute that allows classification of stimuli. Nothing has thereby been said about the subjective nature of the hue experience from a given stimulus. Similarly, there are few disagreements about the lightness sequence of object color percepts if they differ only along that dimension. Problems increase on a more global level: people vary considerably when asked to assess the lightness of objects varying widely in hue and chroma. Chromatic colors have a known perceived lightness component beyond lightness as measured by luminous reflectance. The corresponding effect, called the Helmholtz-Kohlrausch effect, complicates the relationship between stimulus and perceived lightness. Such kinds of effects also complicate the relationship between stimulus and hue.

Psychological testing has shown that, of the three attributes, chroma is the most difficult to judge: many people have trouble sorting the samples of a Munsell constant-hue page into variations in lightness and chroma (Melgosa et al. 2000). However, if hue and lightness are accepted as fundamental, their definition in a three-dimensional Euclidean space leaves room for only one additional attribute: the radial distance from the achromatic point. On a categorical level, as natural color language categories indicate, chroma clearly plays the least important role among the three. Object color order systems on the basis of Newton's three attributes were proposed for the first time in the early nineteenth century by G. Grégoire in France and M. Klotz in Germany. These attributes were rediscovered for the purposes of a color order system at the turn of the twentieth century by painter and art educator A. H. Munsell, who termed them hue, value, and chroma.

Hering recognized the obviousness of the hue attribute, but reportedly on a strictly perceptual basis, he chose for object colors the additional attributes whiteness and blackness to complete the attribute triple he found to be required. This was in part the necessary outcome of placing all prototypical hue examples (highest saturation) on a common plane, regardless of their lightness. The lightness scaling of the central vertical gray scale, as a result, is not indicative of the vertical dimension of the corresponding Hering space: the dimension is geometrically not well defined.

1.9 Stevens's Hierarchy of Measurement

Psychophysics, the science of the relationship between physical stimulus and perceptual response, was initiated in the first half of the nineteenth century by E. H. Weber and G. T. Fechner. They proposed that the relationship is logarithmic: the impact of a linear incremental change in stimulus results in a logarithmically graded change in response. In the mid-twentieth century, the psychophysicist S. S. Stevens (1975) developed data indicating that the relationship is best expressed by power functions. In 1946 he proposed a hierarchical scheme of psychophysical measurement and the resulting scales.[10] Scales at their most basic level are termed *nominal*. In the case of colors, the nominal scale refers to the names we assign to them, a categorical judgment. At the next level, scales are termed *ordinal*: here the scale value assigned to an item reflects the order in which it is placed in the scale. An ordinal lightness scale places samples in order of lightness. However, the scale numbers contain no information about the size of difference between successive items, or their magnitude. Ordinal scales place the samples in rank order. The physical and psychophysical spaces discussed earlier place the related psychological experiences in ordinal order: numeric or geometric distances are not indicative of perceptual distances. At the next, *interval*, level, psychophysical scales predict not only ordinality, but also the size of the perceptual differences between samples. Interval order does not express anything about the absolute magnitude of the percepts involved. This information is expressed at the *ratio* level, indicating not only order and interval, but also absolute magnitude of perception. Ratio scales are interval scales with a natural origin. While the Celsius scale of temperature is an interval scale, the Kelvin scale is a ratio scale, owing to its origin at the zero energy point. Ratio scales for colors are controversial. It is possible to consider a lightness scale a ratio scale because it originates in black; however, we are not used to considering one red to be twice as red as another. A hue scale, having no natural origin, cannot be considered a ratio scale.

1.10 Perceptual Color Scales

Several scaling methods are used to determine perceptual color scales. One of these is addition of thresholds. The four classical methods of threshold measurement have been described by Fechner (1860). In the method of limits, or of minimal change, to take one example, the experimenter presents stimulus magnitude changes in preset small increments, in ascending or descending order, starting with an imperceptible increment in the former and a clearly perceptible one in the latter. Stimulus changes continue until the observer indicates perceiving a difference in the former, or no longer a difference in the latter. The just noticeable difference is the average difference of ascending and descending trials.

Suprathreshold differences involve judgments. Several methods have been developed for this purpose, such as partition scaling, category scaling, paired comparison, minimally distinct border scaling, multidimensional scaling, and others. Different methods tend to produce more or less different results. As Laming (1997) pointed out, humans can make reliable ordinal rankings, but whether we can make highly reliable interval judgments with low interobserver variability is not evident. No convincing experimental data in any psychological field, including color perception, shows that we can do so. Large-scale perceptual experiments dealing with color scales have produced considerably different results for reasons that are not fully understood and may have to do with the lack of a specific tool in the brain supporting such judgments as well as with observer variability and differences in methodology (Kuehni 2003). Known variables are the size of the difference between the samples, and the relationship between the colors of the sample pair and the lightness and chromaticity of the surround. There are clear indications, however, that individual divergence in judging stimulus differences is the main contributor to interobserver variability. As a result, there are no generally agreed-upon quantitative scales of color attributes or color differences.

Uniform differences along Newtonian or Heringian attributes do not lead to an isotropic (uniform in all directions) color space and solid. For any point in the interior of the isotropic color solid, colors uniformly different from it are located on a sphere. One issue this raises is that a Euclidean solid cannot be filled completely with spheres. Strict isotropy is not possible in Euclidean geometry. On an isotropic sphere surface, only a small number of points express pure attribute differences. Most of the differences are of mixed attributes. The question of how different kinds of attribute differences add up perceptually, as simple sums or as Euclidean sums,is not fully answered yet and there are indications that there is a perceptual continuum between the two modes (Stalmeier and de Weert 1991). In one large experiment (Optical Society of America uniform color scales; see below), attribute differences have been replaced with general perceptual differences. Multiattribute differences are typical for color distinctions in the production of colored materials.

1.11 Bottom-Up or Top-Down?

Conceptually, a perceptual color order system can be built from accumulation of just noticeable differences (JND) in all three spatial dimensions or by dividing global color differences into smaller and smaller units. In a perfect Euclidean world, the two approaches mesh perfectly. However, the human color-vision system has developed various adaptations to interpret visual data in a life-sustaining manner. These adaptations include the hue superimportance effect (see next section) and the "crispening" effects, maximizing perceived differences of objects with small differences in reflectance compared with the surround. Such effects make the relationship between stimulus and perceived color highly nonlinear and dependent on the surround and the size of the unit perceived difference of the system.

Magnitude assessment of differences seems to become more difficult as the distance in color increases. Thus, in a top-down approach, it is difficult to objectively assign reference points in the system, such as the locations of black and white relative to those of the highest chroma yellow, red, blue, and green with unique hues. Hering simply made them equidistant from each other, a choice that is arbitrary.

On the other hand, the addition of threshold differences in a bottom-up approach is only applicable to a few multiples, perhaps six, until the results begin to diverge more and more from reality. Color order systems either are limited to ordinal scaling or, at the interval level, must have variable parameters.

1.12 Empirical Perceptual Color Order Systems

Under normal circumstances, all our color percepts are the result of visual stimuli. Since the foundation of psychophysics in the first half of the nineteenth century, most efforts in color order have been to elucidate the relationship between stimuli and color percepts. We now know that this is a very complex matter. However, as Hering proposed and as Munsell was interested in, we can consider color percepts alone, without being concerned with the stimuli that give rise to them. The relationship between percepts and stimuli is loose. We can conclude this from the fact that by physical measurement, we have no problems distinguishing many billions of different color stimuli. But the number of distinguishable color percepts is limited to perhaps 2 million. On that basis alone, it is evident that many different SPDs can give rise to identical color perceptions. Other important factors that limit the number of percepts versus the number of SPDs are metamerism, contrast, adaptation, and color constancy.

Grégoire and Klotz proposed the first systems of empirical perceptual color order based on color attributes. Hering developed the detailed structure of his system conceptually, on the basis of minimal empirical data. Only in the later twentieth century was a somewhat firmer empirical basis for this kind of system developed in the form of the Swedish Natural Color System (see below).

At the turn of the twentieth century, Munsell began work on an empirical perceptual system, developed for educational purposes and for determining rules of color harmony. But when a system is to be extensively colored according to certain principles, the question of the relationship between stimulus and percepts is inescapable. What was, in Munsell's system, originally a sphere developed into an irregular solid when Munsell realized that pigments have varying coloring power, just as spectral lights have varying saturation. Munsell liked the decimal system and decided to separate the hue circle as well as the lightness scale into one hundred perceptually equidistant steps (only forty hue grades and eleven lightness grades are illustrated in his atlas). The chroma scale, of necessity, was open ended. These decisions, arbitrary in numbers of steps, resulted in varied unit perceptual distances in the three scales. The system is cylindrical in nature, with hue circles centered on the achromatic points, color percepts varying in chroma located on radial lines from that point, and the value scale erected perpendicularly to the chromatic planes (figure 1.13). One of the geometric consequences is that the hue distance between percepts on neighboring constant hue planes is presumed to vary as a function of chroma up to a factor of ten or more. A much-reduced factor is found when comparing the actual sample pairs for hue differences.

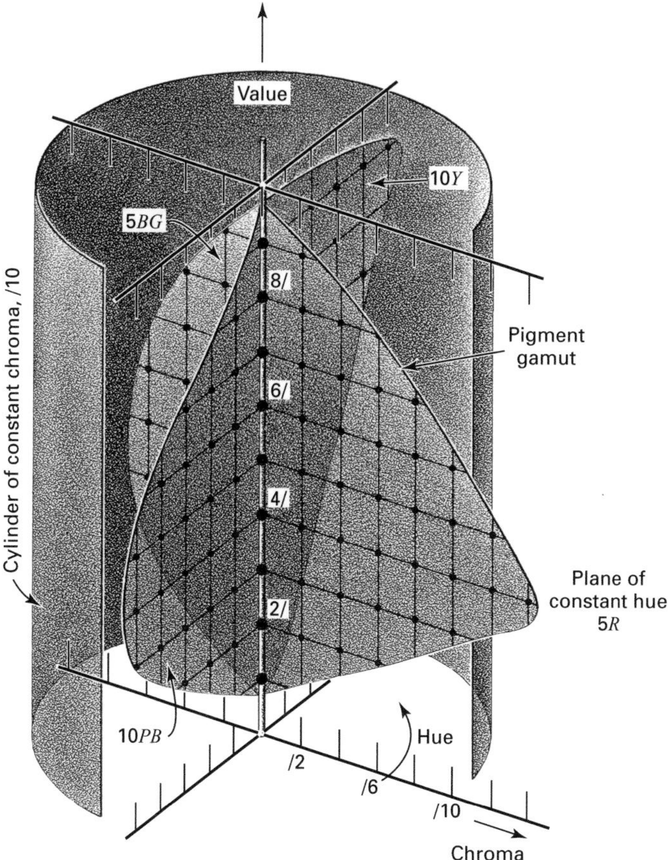

Figure 1.13
Schematic structural view of the Munsell system. Each dot represents a color chip. Only four of the forty constant hue planes of Munsell's atlas are shown (Wyszecki and Stiles 1982).

When D. Nickerson (1936) compared the perceptual magnitude of the three scales in an attempt to develop a color difference formula, she found that one unit of value difference is equal to two units of chroma difference and, at chroma 5, to three 100-step units of hue difference. In 1942, B. R. Bellamy and S. M. Newhall investigated the same distances in terms of JND and found one unit of value to be equal to eight JND of chroma and, at chroma 6, to twenty-two JND of hue, thus demonstrating that the relationship strongly depends on the magnitude of the unit difference.

D. B. Judd calculated the geometric properties of the implied color solid based on Nickerson's data and found that the total hue angle for all hue steps is slightly more than twice the geometric total hue angle of 360°. Judd termed this the hue superimportance effect (HSE; Judd 1969). In simplest terms, the HSE indicates that a smaller change in cone activation between two stimuli is required for a hue difference than for a chroma or a lightness difference of the same perceptual magnitude, resulting in a system particularly sensitive to hue differences. As a result, quantitative representation of the Munsell perceptual system (or any other more or less isotropic system) in a Euclidean space is not possible. The HSE is active at differences of any magnitude, from subthreshold to large, and is a fundamental fact of color perception.

Another shortcoming of the Munsell system is that it does not specify the surround in which the samples are to represent perceptual attribute equidistance. In the first edition of the atlas (1907), Munsell demonstrated the widely different appearance of color chips when mounted on white and on black backgrounds. The modern value scale of the system does not accurately apply to any kind of background.

Exemplifying a system in the form of an atlas inevitably raises the matter of dependence of percepts from given stimuli on the surround, as well as the degree of color inconstancy of percepts from given stimuli as a function of the light in which the atlas is viewed. The former issue has been sidestepped in all atlases, while efforts have been made to develop pigment formulations that are reasonably color constant in terms of the resulting perceptions when viewed in common light sources. Nothing is expressed, thereby, about the viewers' personal experiences of a given chip.

The Munsell system and atlas were developed over a fifty-year period. An unknown number of observers and several different experimental procedures were involved. The most significant changes in the system were made by the Optical Society of America committee, which developed the Munsell Renotations, a modification of the original system and the basis of the modern system. Aside from the fundamental problems, there are also doubts about the perceptual uniformity of steps around the hue circle, as well as of chroma steps, particularly along the yellow-blue axis. As discussed, it is not an isotropic system but merely approximately uniform within attributes. As such, it originally represented a significant advance in color order systems.

An isotropic system would have the highest information content and has remained a goal in the twentieth century. A large-scale attempt was made by the Committee on Uniform Color Space of the Optical Society of America over a twenty-seven-year period in the mid-twentieth century. Members of the committee realized at the beginning that the cylindrical form of the Munsell system is not suitable for an isotropic system, for the reasons mentioned above. Hue and chroma attributes were abandoned, and instead the magnitude of the total color difference between colored tile pairs was compared (triangular pair comparison) in an experiment involving more than seventy observers. Psychometric scales were calculated from accumulated individual results. Analysis by Judd indicated that the HSE is also active under these experimental conditions. The committee was aware that a true isotropic space is not possible because of the HSE effect and the fact that a Euclidean space cannot be filled solidly with spheres of unit distance around points in the space. Geometry dictates that solid space packing with the highest number of independent scales is possible only if the unit solid is a cubo-octahedron (figure 1.14). In this arrangement, twelve colors are geometrically equidistant from central color M, forming seven different cleavage planes in the solid (figure 1.15, plate 1). Rather than truly isotropic, such a color solid (were it possible) would be equidistant along six axes. Upon confirming the HSE in their data, the committee decided to change the goal to producing the Euclidean color space that comes closest to a uniform space, by mathematically modifying the data so that they fit into such a space. The result, curiously, was named Optical Society of America Uniform

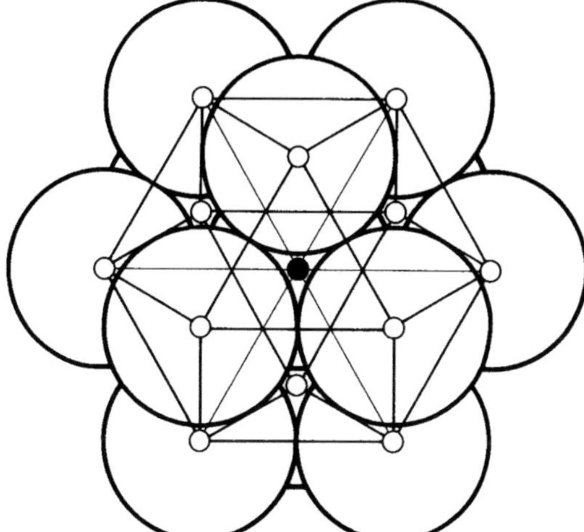

Figure 1.14

Schematic view of the cubo-octahedron forming the structural basis of the Optical Society of America Uniform Color Scales system, with the central color locus in black, and nine of the twelve possible equally distant color loci in white. The figure illustrates the incomplete space-packing of spheres (Gerstner 1986).

Figure 1.15
View of a Uniform Color Scales model showing different kinds of cleavage planes, which form scales of approximately equally distant colors. Horizontal planes represent colors of equal lightness, adjusted for the Helmholtz-Kohlrausch effect (Image courtesy D. L. MacAdam). See plate 1 for a color version of this figure.

Color Scales (OSA-UCS). The system consists of an atlas as well as a psychophysical formula modeling it. It is interesting to note that when Judd and Nickerson translated the findings of the OSA-UCS basis experiment into the Munsellian cylindrical format, they found large discrepancies between the Munsell system and the OSA-UCS, perhaps as a result of the different observer panels or scaling techniques (Kuehni 2003).

Hering's general format was used in the early twentieth century by W. Ostwald in a method that combined physics and psychophysics. He employed physical complementarity and the Weber-Fechner law of logarithmic relationship between stimulus and response to scale the solid. The result is far from isotropic but represented the first large-scale (in terms of number of samples) systematic internal scaling of the object color solid. Toward the middle of the twentieth century, Hering's approach was pursued in Sweden by Johansson and Hesselgren, who unsuccessfully tried to find a compromise between the Hering and Munsell systems. In response to this outcome, the Scandinavian Colour Centre decided in the 1960s to develop a color system and atlas based on a particular interpretation of Hering's conceptual system, employing limited perceptual scaling and inter- and extrapolations. The idea (as yet unsupported by extensive testing and likely affected by the large individual variability in selecting stimuli that represent unique hues) is that the average observer can scale with good reliability any color percept in terms of "percentages" of one or two unique hues, black, and white content. The results are placed in a double cone system, formed by assembling constant-hue equilateral triangles of the different hues along their common achromatic axis (figure 1.16, plate 2). The full color (maximal saturation) positions are often not occupied since they are considered unattainable with colorants. The four unique hues are located on the Cartesian axes, all on the same plane, regardless of lightness. Between planes of constant unique hue, the hue circle is scaled into ten equal "percentage" steps of the two adjoining unique hues. These are different from average perceptually equal steps, and they vary in absolute size between different unique hue pairs. The central gray scale is placed perpendicular to the chromatic plane. The resulting Natural Color System (NCS) makes no claims for being isotropic.

The number of object color percepts we can experience is finite, despite an infinite number of stimuli. We can consider the percepts representing a cloud of points, with that cloud fitting into any kind of three-dimensional space as long as the meaning of internal distances is immaterial. Just as orderly arrangements of object color stimuli result in several different three-dimensional spaces, depending on how the dimensionality of the stimuli is reduced, solids of object color percepts can have different forms depending on the principles used to arrange them. The finite number of object color percepts indicates that the same percept is achieved with many different stimuli as a function of illumination and surround. Only aspects of a posited universal object

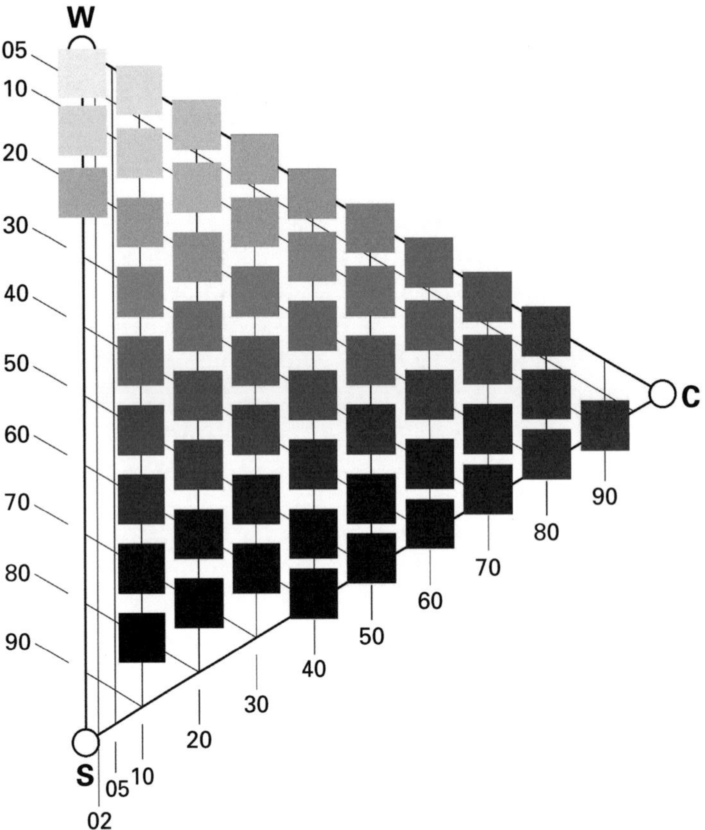

Figure 1.16
Image of the constant hue plane Y90R of the Swedish Natural Color System in the form of an equilateral triangle with white (W), black (S), and the full color (C) in the corners. Lines of constant blackness *s* and of constant chromaticness *c* are indicated (image courtesy NCS Color Center/USA). See plate 2 for a color version of this figure.

color percept space can be revealed in a given implementation, dependent on atlas sample selection, light source, surround, and observer.

1.13 Fitting Psychophysical Models to Perceptual Color Space

As demonstrated, the mapping from stimulus to percept is very complex and varies by observer, surround, and in the case of objects, illumination. For many observation situations, physical and psychophysical SPD ordering systems result in correct ordinal placement of percept attributes. The limits of ordinal placement accuracy have not yet been explored in detail.

Fitting an accurate psychophysical model to valid perceptual data requires accurate knowledge of the physiology of the color-vision process. Many details are still lacking. In the past, there were implicit assumptions that knowledge of the transduction process of electromagnetic energy into neurochemical energy in the retinal cones is sufficient for a description of the process. Although this process represents an important first step, it is insufficient given the apparent complexity of postreceptoral and cortical computations.

Accurate prediction of average perceptual differences between manufactured items that are supposed to have identical perceived color has been and is of considerable interest in manufacturing technology. Because all manufacturing processes are subject to variability, and because different materials or economic competition may require chemically different colorants, perfect accuracy is not possible. In addition, the question arises: for which observers and what viewing conditions should the results be valid? Observation conditions and the size range of judged differences are now well known to have a significant effect on perceived results. The uncomfortable conclusion is likely that a psychophysical model with good accuracy (say, in the 80–90 percent range) is only possible for particular groups of observers, limited size of difference, and certain surround, light source, and viewing conditions. However, experimental data to support firm conclusions are still largely missing.

The first model relating stimulus changes to discrimination (JND), based on cone sensitivity, was offered by Helmholtz, a so-called line element. Several additional line elements (including the empirical one by MacAdam) have been developed. In the end, quite clearly because of additional processing in the retina and the cortex, they were found to be inadequate. More complex global color-vision models, such as those by Müller-Judd, S. L. Guth, or R. L. De Valois and K. K. De Valois, have also been found inadequate for the purposes of color difference calculation.

Since the middle of the twentieth century, color technology has made considerable efforts toward developing accurate color difference models, resulting in many proposals. By the mid-1970s, more than a dozen formulas were in use in industries producing and using colorant, resulting in Babylonian confusion. To improve this situation,

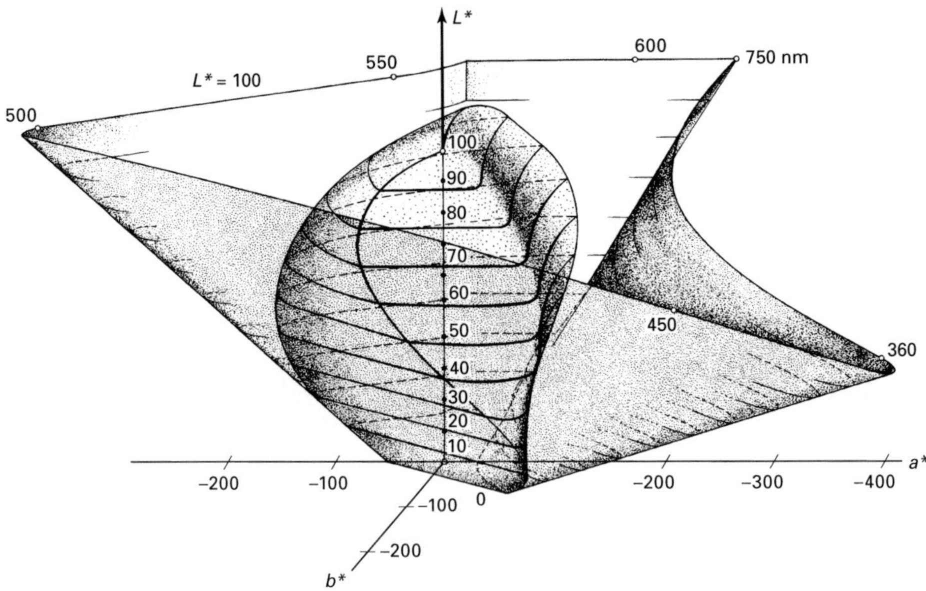

Figure 1.17
Projective view of the "cone" of spectral lights and the optimal object color solid as represented in the CIELAB perceptually approximate uniform color space; axes L^*, a^*, b^* are based on a cube-root relationship between stimulus and response (Wyszecki and Stiles 1982).

the CIE offered in 1976 two roughly uniform formulas, abbreviated CIELUV and CIELAB, derived from formulas dating back to the 1940s and already known then to be unsatisfactory. The former is a linear transformation of CIE tristimulus values, preferred by lighting engineers because of the additive nature of light mixture. The latter is a nonlinear transformation, generally used for object color differences. The spectral light cone mantle and the object color solid of CIELAB is illustrated in figure 1.17.

The issue of stimulus compression requires further brief exposition. Already in the eighteenth century it was known that the relationship between light intensity and constant perceived brightness intervals of light was not linear. As mentioned, in the mid-nineteenth century, Weber and Fechner developed the basis of experimental psychology and concluded that the general relationship between stimulus and percept was logarithmic. In the mid-twentieth century, Stevens produced various kinds of psychophysical data showing that the stimulus compression can be modeled with a power relationship, with the power differing as a function of psychological measure type (Stevens 1975). This fact has also been found to be (loosely) applicable to color scales (see figure 1.18). A frequently used power relationship for the compression of luminous reflectance (CIE tristimulus value Y) to lightness is cube root. This relationship was found to be a well-fitting model for the lightness scale of the Munsell renotations, averaged from perceptual data against three different surrounds (against gray surrounds, the relationship is more complex, not well fitted with a simple power compression). In CIELAB space, the cube-root relationship is assumed to apply to all three tristimulus values. The CIELAB color space and difference formulas do not take into account hue superimportance and various other perceptual effects. Since the CIELAB space's inception, at least three color atlases have been introduced containing colorations based on its intervals.

After 1976, the CIE and other organizations have promulgated various color difference formulas that take CIELAB space as a basis and make multiple adjustments. The latest of these is CIEDE2000 (Luo et al. 2001). In CIELAB the predictive accuracy (for the average observer) of the size of small suprathreshold color differences is approximately 50 percent; in CIEDE2000, it has risen to approximately 65 percent. This is not a satisfactory situation, but color difference calculation using such formulas is widely practiced in industry. CIELAB space can, for many reasons, not be considered isotropic, but it is closer to being so than X, Y, Z space.

1.14 Colorant and Light Stimulus Color Solids and Atlases

That a variety of hues (if differing in chroma) can be generated with three colorants has been known for 2,000 years or more. The first colored hue circles to appear in print (Anonymous 1708) illustrate the results of mixing three pigments in various ratios.[11] In the mid-nineteenth century, Maxwell demonstrated that the optimal pigments to achieve wide ranges of hue and lightness are not yellow, red, and blue but yellow, what came to be known as magenta, and cyan. In the late nineteenth century, corresponding colorants began to be used in halftone printing and later in color photography.

The first three-dimensional color order system based on primary colorants yellow, red, and blue was introduced in the eighteenth century by J. H. Lambert (triangular pyramid based on T. Mayer's earlier proposal of a double pyramid). It was followed by several other systems, including more recent ones based on halftone printing, using the primary colorants yellow, magenta, cyan, and black. Well known among these are the Hickethier color atlas and Kornerup and Wanscher's *Methuen Handbook of Colour* (1961). The interior structure often differs among such systems. The scaling ranges from attempted perceptual uniformity along attributes to percentage changes in colorant concentration, or halftone screen density.

Computer video displays and the related technologies of mixing three primary lights have resulted in corresponding color order systems. In software packages such as Adobe's Photoshop, several color system displays are available. In the RGB (red, green, blue) display, at the pixel level, the intensity of the display unit's light output of the

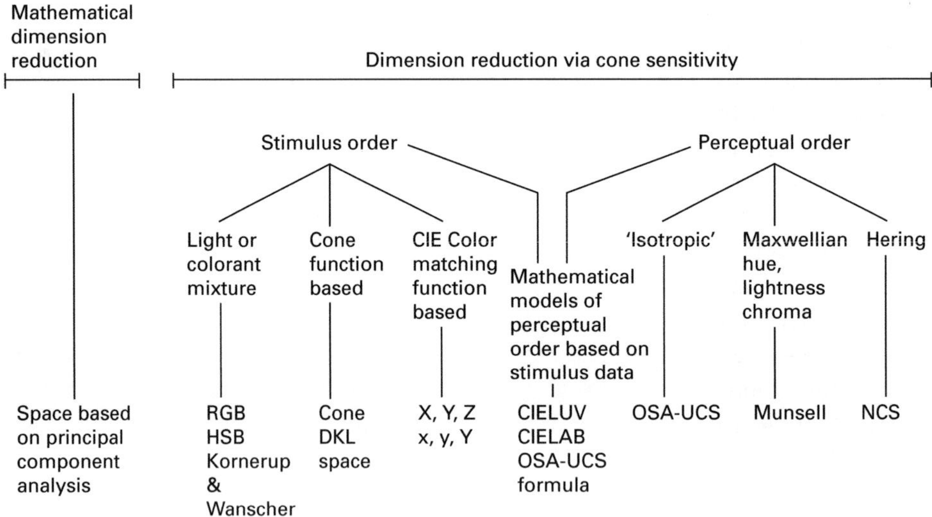

Figure 1.18
Classification scheme for color order systems, with typical examples.

three primaries is individually controlled in a relative range from 0 to 255. There is also an HSB (hue, saturation, brightness) display, where the arrangement is in the form of perceptually more salient attributes. Further, information on the displayed stimulus can be in implied units of the printing primary inks CMYK (cyan, magenta, yellow, black), or in CIELAB-like *L, a, b* scales. None of these displays, except the last one, makes an attempt at perceptual uniformity.

1.15 Classification of Color Order Systems

Color order systems can be classified into eight groups (figure 1.18). This classification does not offer space to all historical systems (numbering in the hundreds), some of them having little connection to reality. Many ordering systems belonging loosely to one of the two major groups are conceptual. They are systems representing a general idea of their author, often in the form of a simple geometric model, with little or no consideration of the internal coloring scheme, which requires recognizing that each stimulus or color perception must simultaneously be a part of three or more independent scales.

1.16 Conclusions

Color stimulus ordering systems provide an objective order. But to assume that this order is equivalent to perceptual order would be to commit Titchener's stimulus error.[12] The space and associated object color stimulus solid, based on mathematical dimension reduction of color stimuli data, have no connection to humans. Other ordering systems use average human cone sensitivity functions as basis functions for the space, either in direct or in linearly transformed shapes (transforms sometimes intended to represent neurological mechanisms). A further kind of stimulus ordering system is based on order of colorant concentration or light intensity increments. The relationship between order in any of these systems and perceptual order is at best (but, significantly, often) on the ordinal level. It appears that the three cone types and the physiological opponent-color system are natural methods to efficiently extract common features from SPDs. But the information contained in these features is variously modified by extensive, complex, and to a degree individualistic manipulation of the information in the cortex.

Perceptual order systems are of three kinds: (1) based on the Helmholtz attributes hue, brightness, and saturation; (2) based on the Hering attributes hue, blackness, and whiteness; or (3) quasi-isotropic systems, such as the OSA-UCS attempt. Helmholtzian and Heringian systems use different definitions of uniformity, the former based on perceptual distance along attributes, the latter based on uniform increments of the percepts of its six fundamental colors. If it is accepted that hue and lightness (for object colors) are meaningful attributes in Euclidean representation, logic requires a third attribute, such as chroma. In this form, the space cannot be isotropic because of the HSE and the representation of hue differences as a function of chroma that is far from the perceptual one. Why should the human visual system have striven to fit all requirements for color order into a Euclidean or an isotropic space? This appears to be a cultural idea, not met in reality. Species survival has favored hue difference detection as the most important kind, resulting in the HSE. Species survival has not pushed for an adaptation with the goal of accurate evaluation of perceptual distances between color stimuli at the interval level. This has been found of use only in the advanced industrial age. Our visual system cannot determine spectral reflectance or accurately measure SPDs. The location and form of objects in the field of vision are determined from the detection of contours, and then filled in with color percepts. Recognition of objects has been and remains of great importance. An endless variety of conditions of lighting and surround in the natural visual field have resulted in adaptations that normalize, incompletely, the effects of several such variables for conditions that our forebears encountered during life in jungle and savanna, thus making detection and identification of objects easier. The purpose was quite clearly to provide owners with useful, if not necessarily veridical, information about the world around them. All perceptual effects, including widely different individual choices of stimuli as representing the four unique hues, and individual variations in the visual system based on genetics or past experiences, act against a simple relationship between stimulus and percept, and in this manner against any kind of universal, natural color order on more than the ordinal level. As a result, color atlases are color stimulus collections, influenced to a degree by the workings of our color-vision system and to another by cultural or technical ideas, as well as by the assumption that Gaussian statistics apply.

With the metaphor of the color cloud in mind, color stimuli can be fitted into any kind of color solid and related space, but the perceptual meaning of the geometric distances changes from one to the next and is often minimal. The highest level of information would be carried by an isotropic space, found to be impossible in its full sense. It is an ideal aspired to for centuries, but seemingly not realizable.

Notes

1. For a more detailed history of the development of color order, see Kuehni (2003) and Kuehni and Schwarz (2008).

2. I am not concerned here about color perception in dreams, as a result of drug consumption, or from other unusual causes.

3. This is not certain because contrast effects may assign percepts to stimuli that under different conditions are assigned to other stimuli more distant from each other.

4. The elliptical shape could easily be converted to a more circular shape by multiplying the values of eigenvector 2 or 3 with a suitable constant.

5. The implied progression from cone functions to color-matching functions was actually in the opposite direction. Cone sensitivity functions are difficult to measure directly, and results of such measurements have been adjusted to agree with average color-matching function data.

6. A comparable solid, based on early implicit cone functions, was calculated by the German physicist S. Rösch (1928).

7. Unique red, for example, is a red hue that appears as having neither a yellowish nor a bluish cast, but is purely red. The same principle applies to the other three unique hues.

8. These formulas are somewhat different from Luther's but linearly related. "Balanced" in this context means that the sums under the positive and negative loops of the *a* and *b* functions are identical.

9. To assume that Luther's functions, or functions *a* and *b*, quantitatively represent Hering's red-green and blue-yellow, respectively, perceptual opponent pairs requires a large, and as it turns out unjustified, leap of faith. One reason is that stimuli reliably selected by color normal individuals as representing (for them) unique hues (and thereby perhaps their perceptual color fundamentals) vary to a considerable extent (Kuehni 2004).

10. Stevens's hierarchy has now been shown to represent a simplification of the real world, not always strictly applicable to the analysis of statistical data. However, it is useful as a categorical system for different kinds of color scales.

11. In a 1708 Dutch edition of a popular French work on miniature painting (in print since 1673), an addition on the subject of pastel painting appeared, written by an anonymous author and containing hand-colored seven- and twelve-color circles (Anonymous 1708). The reason for seven colors was that the author did not have a neutral red pigment and posited two red primaries, a yellowish fire red and a bluish carmesin red. In the twelve-color circle, they are mixed to form the primary red hue. Here, the two red pigments can be understood as the result of mixing the true red with the yellow and separately with the blue primary pigment.

12. English psychologist E. B. Titchener warned in 1905 against confusion of sensations with the related stimuli, for which he used the term *stimulus-error*.

References

Anonymous (CB). 1708. *Traité de la peinture en mignature*. den Haag: van Dole.

Bellamy, B. R., and S. M. Newhall. 1942. Attributive limens in selected regions of the Munsell color solid. *J. Opt. Soc. Am.* 32: 456–473.

Campenhausen, C. von, C. Pfaff, and J. Schramme. 2001. Three-dimensional interpretation of the color system of Aguilonius/Rubens 1613. *Color Research and Application* 26, supplement: S17–S19.

Cohen, J. B. 2001. *Visual Color and Color Mixture*. Urbana: University of Illinois Press.

Cohen, J. B., and W. E. Kappauf. 1982. Metameric color stimuli, fundamental metamers, and Wyszecki's metameric blacks. *American Journal of Psychology* 95: 537–564.

Fechner, G. T. 1860. *Elemente der Psychophysik*. 2 vols. Leipzig: Breitkopf und Härtel.

Gerstner, K. 1986. *The Forms of Color*. Cambridge, MA: MIT Press.

Gilchrist, A. 2006. *Seeing Black and White*. New York: Oxford University Press.

Judd, D. B. 1969. Ideal color space. *Palette* 29: 25–31; 30: 21–28; 31: 23–29.

Koenderink, J., and A. Kappers. 1996. *Color Space*. Report 16/96 of the research group Perception and the Role of Evolutionary Internalized Regularities in the Physical World, Bielefeld, Germany: Bielefeld University.

Koenderink, J., and A. J. van Doorn. 2003. Perspectives on colour space. In *Color Perception: Mind and the Physical World*, R. Mausfeld and D. Heyer, eds. Oxford: Oxford University Press, pp. 1–66.

Kornerup, A., and J. H. Wanscher. 1961. *Methuen Handbook of Colour*. London: Methuen. Reprinted in 1962 as *Reinhold Color Atlas*, New York: Reinhold.

Kuehni, R. G. 2003. *Color Space and Its Divisions*. Hoboken, NJ: Wiley.

Kuehni, R. G. 2004. Variability in unique hue selection. *Color Research and Application* 29: 158–162.

Kuehni, R. G., and A. Schwarz. 2008. *Color Ordered.* New York: Oxford University Press.

Laming, D. R. J. 1997. *The Measurement of Sensation.* New York: Oxford University Press.

Lenz, R., M. Österberg, J. Hiltunen, T. Jaaskelainen, and J. Parkkinen. 1996. Unsupervised filtering of color spectra. *J. Opt. Soc. Am. A* 13: 1315–1324.

Luo, M. R., G. Cui, and B. Rigg. 2001. The development of the CIE 2000 colour-difference fomula: CIEDE2000. *Color Research and Application* 26: 340–350.

Luther, R. 1927. Aus dem Gebiete der Farbreizmetrik. *Zeitschrift für technische Physik* 8: 540–558.

MacAdam, D. L. 1935. Maximum visual efficiency of colored materials. *J. Opt. Soc. Am.* 25: 361–367.

Melgosa, M., M. J. Rivas, E. Hita, and F. Viénot. 2000. Are we able to distinguish color attributes? *Color Research and Application* 25: 356–367.

Munsell, A. H. 1907. *Atlas of the Color Solid.* Malden, MA: Wadsworth-Holland.

Nickerson, D. 1936. The specification of color tolerances. *Textile Research* 6: 509–514.

Neitz, M., and J. Neitz 2004. Molecular genetics of human color vision and color vision defects. In *The Visual Neurosciences*, vol. 2, L. M. Chalupa and J. S. Werner, eds. Cambridge, MA: MIT Press, pp. 974–988.

Rösch, S. 1928. Die Kennzeichnung der Farben. *Physikalische Zeitschrift* 29: 83–91.

Stalmeier, P. F. M., and C. M. M. de Weert. 1991. Large color differences and selective attention. *J. Opt. Soc. Am. A* 9: 237–247.

Stevens, S. S. 1946. On the theory of scales of measurement. *Science* 103: 677–680.

Stevens, S. S. 1975. *Psychophysics.* New York: Wiley.

Titchener, E. B. 1905. *Experimental Psychology,* Vol. II. New York: Macmillan, pp. xxvi ff and lxiii ff.

Wyszecki, G., and W. S. Stiles. 1982. *Color Science.* 2nd ed. New York: Wiley.

Zhang, J. 2003. Paleomolecular biology unravels the evolutionary mystery of vertebrate UV vision. *Proceedings of the National Academy of Sciences* 100: 8045–8047.

2 On the Reality (and Diversity) of Objective Colors: How Color-Qualia Space Is a Map of Reflectance-Profile Space

Paul M. Churchland

2.1 Introduction to the Problem

At least since Locke, color scientists and philosophers have been inclined to deny any objective reality to the familiar ontology of perceivable colors, on grounds that physical science has revealed to us that material objects have no qualitative features at their surfaces that genuinely *resemble* the qualitative features of our subjective color experiences.[1] Objective colors are therefore dismissed as being, at most, "a power in an object to produce *in us* an experience with a certain qualitative character." Accordingly, colors proper are often demoted from being "primary properties" (i.e., objective properties of external physical objects) to the lesser status of being merely "secondary properties" (i.e., properties of our subjective experiences only).

To be sure, we are not logically forced to this eliminative conclusion by the failure of the first-order resemblances cited. A possible alternative is simply to *identify* each of the familiar external, commonsense colors with whatever "power within external objects" it is that tends to produce the relevant internal sensation. More specifically, we might try to identify each external color with a specific *electromagnetic reflectance profile* had by any object that displays that color. The objective reality of colors would then emerge as no more problematic than is the objective reality of the *temperature* of an object (which is identical to the mean kinetic energy of its molecules), or of the *pitch* of a sound (which is identical to the dominant oscillatory frequency of an atmospheric compression wave), or of the *sourness* of a spoonful of lemon juice (which is identical to the relative concentration of hydrogen ions in that liquid). These parallel properties also fail the first-order resemblance test imposed by Locke and other early modern thinkers. Nonetheless, their successful reduction to objective properties of material objects is an accomplished fact, both of science and of settled history. Locke's criterion for objective reality—a first-order resemblance to the qualities of our sensations—was simply ill conceived.

On the more modern reductive approach displayed in these examples, color may turn out to be, by the standards of uninformed common sense, a somewhat surprising sort of feature, namely, a profile of reflectance efficiencies across the visible part of the electromagnetic (EM) spectrum. But this is no more surprising than any of the other identities just cited. And no more surprising, perhaps, than is the identification of light itself with EM waves. Such identities may surprise the scientifically uninformed, but they leave the objective reality of light, temperature, pitch, and sourness entirely intact.

Unfortunately, this happy (reductive) accommodation would seem to be denied us in the case of colors in particular. For, it is often argued, there *is no* unique EM reflectance profile that corresponds to, and might thus be a candidate for identification with, each (or indeed any) of the familiar colors. On the contrary, to each of the familiar colors there corresponds an apparently unprincipled variety of decidedly different reflectance profiles. The scattered class of such diverse profiles, for each "objective" color, is called the class of *metamers* for that color, and they are indeed diverse, as the four profiles in figure 2.1 (plate 3) illustrate.

Four distinct material objects, each boasting one of the four reflectance profiles here portrayed, will appear identically and indistinguishably yellow to a normal human observer under normal illumination (e.g., in broad daylight). And these four profiles are but a small sample of the wide range of quite distinct reflectance profiles that all have the same subjective effect on the human visual system. The fact is, our rather crude resources for processing chromatic information—namely, the three types of wavelength-sensitive cone cells and the three types of "color-opponency" cells to which they ultimately project—are simply inadequate to distinguish between these metamers. Any object boasting any one of them will look to be a qualitatively uniform yellow, at least under normal illumination.

These examples concern the color yellow, but a similar diversity of same-looking metamers attends every other color as well. If one had hopes for a smooth reduction of each of the commonsense colors to a uniquely corresponding reflectance profile, those hopes are here frustrated: first by a real diversity of reflectance profiles corresponding to each visually distinguishable color, and second by our apparent inability to characterize what unifies the relevant class of diverse reflectance profiles, appropriate to each visually distinguishable color, independently of appealing to the qualitative character of the visual sensations they happen to produce in the idiosyncratic visual system of the human brain. If that is the only way in which we can specify what unites the class of metamers specific to any color, then we must either resign ourselves to a deflationary *relational* reconstrual (Cohen 2004) of what common sense plainly takes to be *monadic* properties of material objects, or we must resign ourselves to the elimination of objective colors entirely, as Hardin (1993, 300n2), coherently enough, recommends.

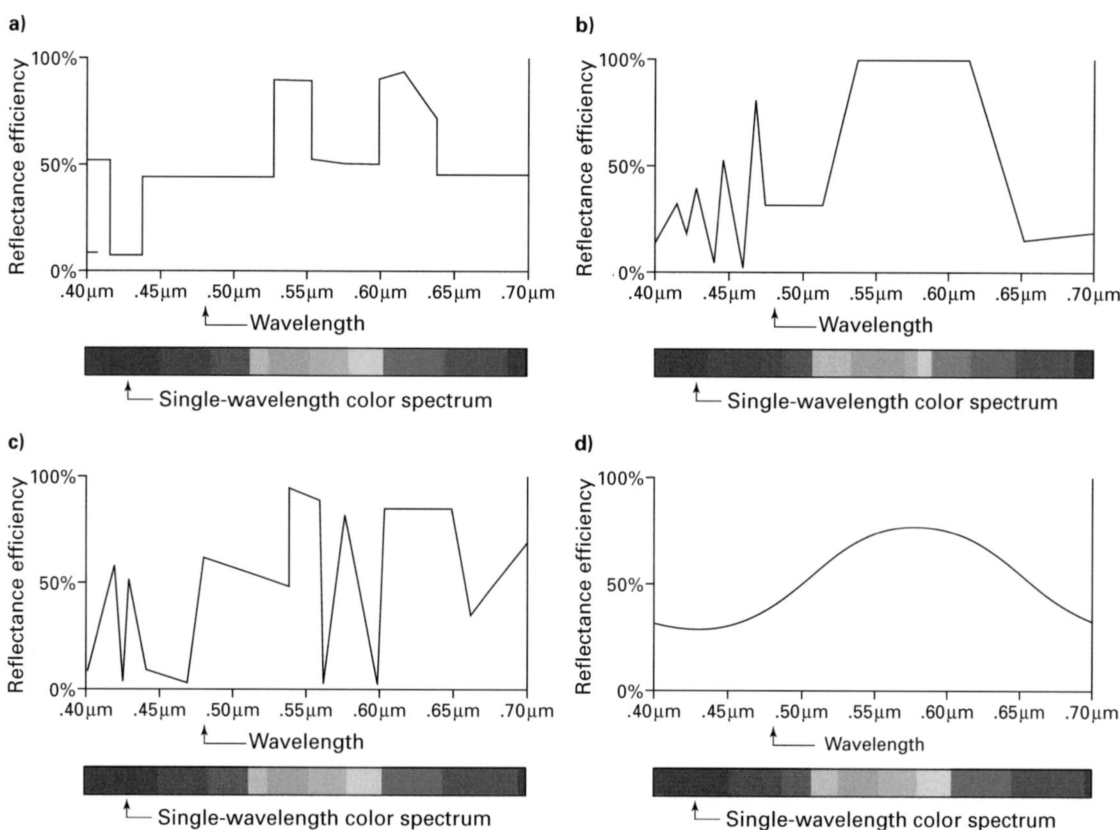

Figure 2.1
Reflectance profiles for four distinct material objects. See plate 3 for a color version of this figure.

2.2 Reformulating the Problem

That the apparently unprincipled diversity of metamers poses a genuine problem for a reductive account of objective colors can be seen from a second perspective, one of central importance for understanding how the brain portrays the external world. A promising general approach to understanding how the brain—or any of its various subsystems—represents the external world posits the brain's development, through learning, of a variety of (often high-dimensional) *maps* of the objective similarity structure of this, that, or the other objective feature domain. Through extended experience with the relevant objective feature domain, the relevant part of the brain can construct an internal map of that domain—of the range of possible faces, or the range of possible voices, or the range of possible reaching motions, or the range of possible colors, and so forth. Such internal maps represent the lasting, or fixed, structure of each external feature domain, and they constitute the brain's general knowledge of the world's objective structure, that is, of the entire range of possible features that the world might display at any given time and place.

Once these conceptual resources are established, the ongoing activity of the brain's various sensory systems will produce fleeting activations at specific locations within those acquired background maps, activations that code or index where, in the space of background possibilities comprehended by the map, the creature's current objective situation is located. For example, I am now looking at *my wife's* face; I am listening to *my wife's* voice; she is reaching for *a coffee mug*; and that coffee mug is *white*. In sum, I have a background conceptual framework—or rather, an interconnected system of such frameworks—and my sensory systems keep me updated on which of the great many possibilities comprehended by those frameworks are actualities here and now.

But the informational quality of such sensory indexings is profoundly dependent on the antecedent representational virtues of the background framework in which they fleetingly occur. The basic virtue of such background maps—as with any map—is a structural homomorphism between the map as a whole, on the one hand, and the entire feature domain that it attempts to portray, on the other. The family of *proximity relations* that configure the many map elements of the brain's internal map must have a relevant homomorphism with the family of *similarity relations* that configure the many landmark features within the domain to be portrayed.[2] Such homomorphisms, or *second-order* resemblances, on this view are the essence of the brain's representational achievements. One might call this account *domain-portrayal* semantics, to contrast it with such familiar doctrines as *indicator* semantics or *causal covariation* semantics.

I will not pause, in this chapter, to detail the many virtues of this unified approach to how the brain represents the world's general or background categorical structure, and how it represents the world's local configuration here and now.[3] I sketch it here because it provides the background for a powerful contemporary objection to the reality of external colors in particular. "How," it may be asked, "does the peculiar and well-defined three-dimensional structure of the human phenomenological color space (see the spindle-shaped solid in figure 2.2, plate 4) map onto the objective space of possible electromagnetic reflectance profiles displayed by material objects? What is the internal structure of that objective target feature domain in virtue of which the internal structure of our phenomenological color spindle constitutes an accurate map of that target domain?"

The objector's questions here are, of course, semirhetorical. Their point is to emphasize the presumed fact that *there is no* objective structure that nicely configures the range of possible reflectance profiles displayed by material objects. Collectively, they

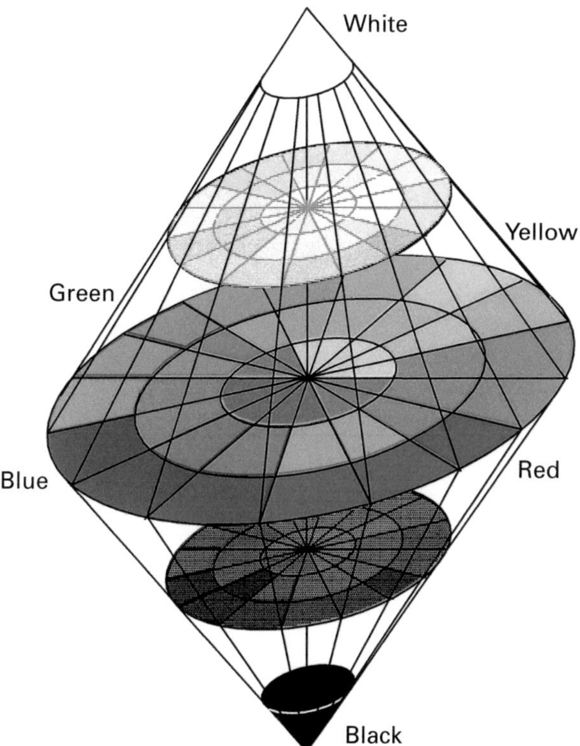

Figure 2.2
Three-dimensional spindle representing the human phenomenological color space. See plate 4 for a color version of this figure.

form a noisy and unprincipled scatter of possibilities. At the very least, if there is some structure within that range of possibilities, it fails to answer in any way to the very specific and demanding structure displayed in our phenomenological color spindle. Objective colors, one might therefore conclude, are a grand illusion. The objective reality, concerning the surfaces of physical objects, is distinctly and importantly different from the naïve assumptions of common sense, and from the crude and misleading deliverances of our native sensory equipment (Hardin 1993, 300n2; Thompson et al. 1992, 16).

This is, at least potentially, a very powerful argument *against* any commonsense view of colors as objective features of material objects. It appeals to the correct account of how objective feature domains get represented in and by a brain, and it points to an apparently massive failure of the required second-order or structural representation in the specific case at issue. The unreality of objective colors is the presumptive consequence.

Nonetheless, I shall presume to resist this argument, because I think it rests on a false premise. Despite a negative first impression, there is a way to construe the initially opaque space of possible reflectance profiles so that its structural homomorphism with human phenomenological space becomes immediately apparent. Accordingly, our color space does map an objective reality after all, I shall argue, and thus the argument against color realism evaporates.

We understand one-half of this "mapping conundrum"—namely, our phenomenological color space—quite well, both empirically and theoretically. The now-familiar Hurvich-Jameson opponent-process neural-network model of human color coding provides a compelling reconstruction of the empirical details of the spindle-shaped color solid of figure 2.2. Figure 2.3a (plate 5) portrays the connectivity of that network, and figure 2.3b portrays the wavelength-sensitivity profiles of the three types of input cones. If one calculates the full range of possible activation patterns across the three types of second-layer color-coding cells, given the details of the network's connectivity, that color-coding space turns out to have the shape portrayed in figure 2.3c. Evidently, it has the same dimensionality, shape, and representational organization of the empirical color spindle, wherein lies its claim to explain the organization of our phenomenological color space.[4] This half of our problem—namely, the nature and ground of our internal map—is stable and more or less settled. It is the nature of the external reality being mapped that needs to be importantly reconceived.

2.3 Reconfiguring the Space of Possible Reflectance Profiles

The conventional way of representing any given reflectance profile that is located within the narrow window of the visible spectrum (see figure 2.4a, plate 6) positively hides an important feature of the range of possibilities therein comprehended. Perhaps the first hint of an alternative mode of representing those possibilities arises from the fact that the phenomenological color that corresponds to any narrowly monochromatic stimulus varies continuously across the visible spectrum, but it tends toward the *same* color—as it happens, a sort of deep purple magenta—at each of the two opposite extremes: 0.40 μm at the extreme left, and 0.70 μm at the extreme right. It doesn't quite get there in either case, for no single wavelength of light will produce a sensation in the purple-magenta range. To get that (strictly nonspectral) range of colors, you need simultaneous retinal stimulations at two places in the visible spectrum, toward its left and right extremes. But purple magenta remains the missing color toward which each extreme tends. (Everyone since Newton has acquiesced in his constructing a continuous color wheel in which the nonspectral purples are interposed to fill in the "similarity gap" left open by the full range of single-wavelength stimuli.[5]) One's sense of rightful symmetry might therefore suggest that—as no more than an idle exercise, perhaps—one should pick up the planar figure in 2.4a and roll it into a cylinder so that its right-most vertical edge makes a snug contact with its left-most vertical edge, as in figure 2.4b. This converts the original planar space into a space that has no boundaries in the horizontal direction. It has boundaries only at the top and bottom of the space.

This trick turns the original reflectance profile itself (figure 2.5a, plate 7), whatever its idiosyncratic ups and downs, into a wraparound configuration that admits of an optimal approximation by a suitable planar cut through the now-cylindrical space. The locus of any such planar cut through the cylinder will always be an ellipse of some eccentricity or other (a circle in the limiting case of a planar cut that is orthogonal to the cylinder), as portrayed in figure 2.5b.

The peculiar ellipse produced by a specific cut will be said to be an optimal—or, as I shall say henceforth, a *canonical*—approximation of the original or target reflectance profile when it meets the following two defining conditions. The altitude of the ellipse must be such that the total area *A* above the canonical ellipse, but below the several upper reaches of the target reflectance profile, is equal to the total area *B* beneath the canonical ellipse, but above the several lower reaches of the target reflectance profile. (This condition guarantees that the total area under the target reflectance profile equals the total area under the approximating ellipse.) The angle by which the ellipse is tilted away from the horizontal plane, and the rotational or compass-heading positions of its upper extreme, must be such as to minimize the magnitude of the two areas *A* and *B*. (This condition guarantees that the approximating ellipse follows the gross shape of the target reflectance profile, at least to the degree possible.)

A suitably situated, tilted, and rotated ellipse that meets these optimizing conditions, for a given reflectance profile, will be said to be the *canonical approximation* of that profile. Note that an indefinite variety of distinct reflectance profiles can share the very same ellipse as their canonical approximation. That clustering population, I propose, constitutes the class of metamers for whatever "seen color" is produced by an object with a reflectance profile that displays their shared canonical approximation.

Equally important, for each and every individual reflectance profile, however jagged, there is a unique canonical approximation. (This is a consequence of sheer geometry, and of the definition provided above.) Note also that the canonical approximation for a given profile is an objective fact about that profile, and about the material object that possesses that profile. Its specification makes no reference to the human visual system, nor to the nature of its phenomenological responses to anything. The canonical approximation for the reflectance profile of a given material thing is an objective, mind-independent feature of that material thing. We can safely be realists about whether a given reflectance profile has a specified ellipse as its canonical approximation (for short, its CA ellipse), just as we can safely be realists about the reflectance profile thereby approximated.

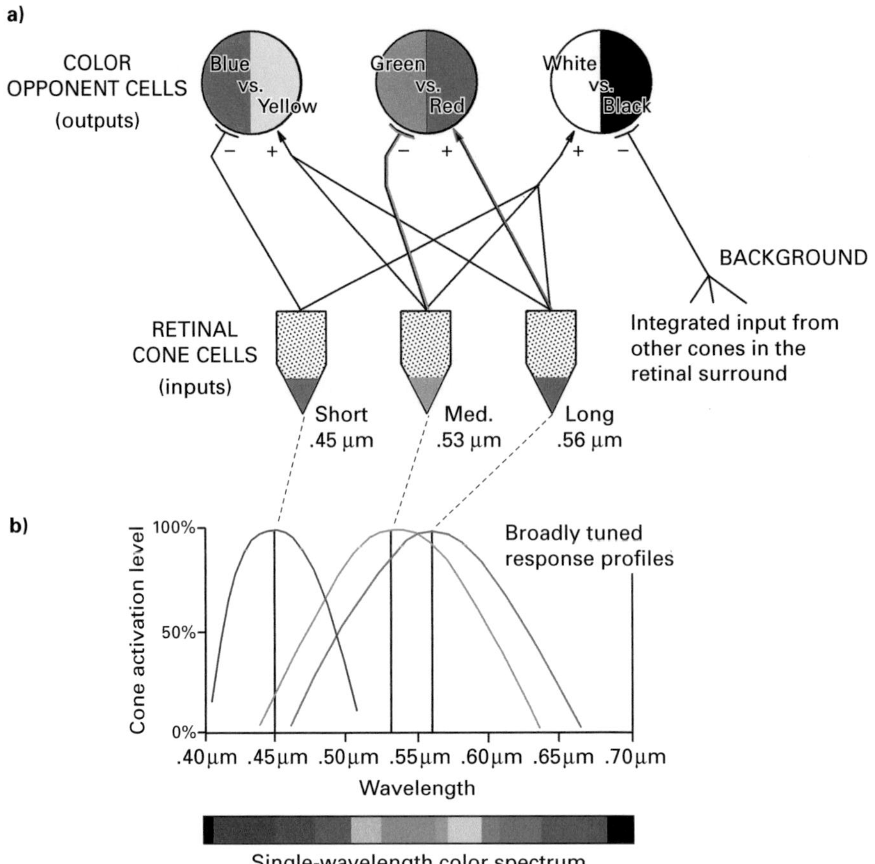

Figure 2.3
Empirical details of the Hurvich-Jameson model of human color coding: (a) connectivity of the neural network; (b) wavelength-sensitivity of input cones; and (c) color-coding space of possible activation patterns. See plate 5 for a color version of this figure.

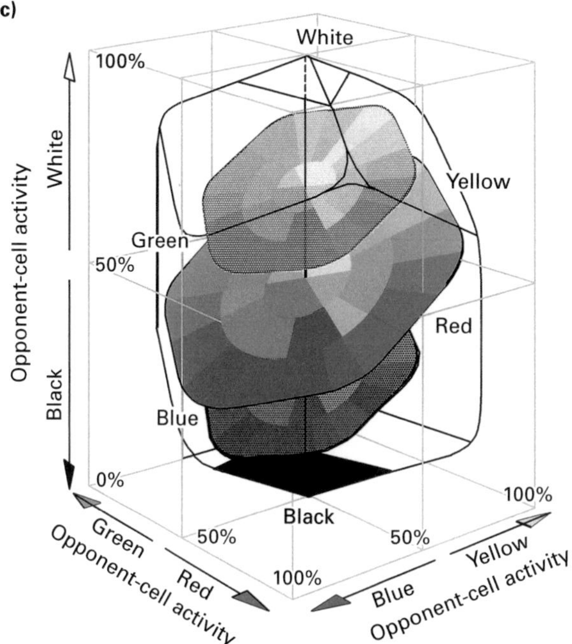

Figure 2.3
(continued)

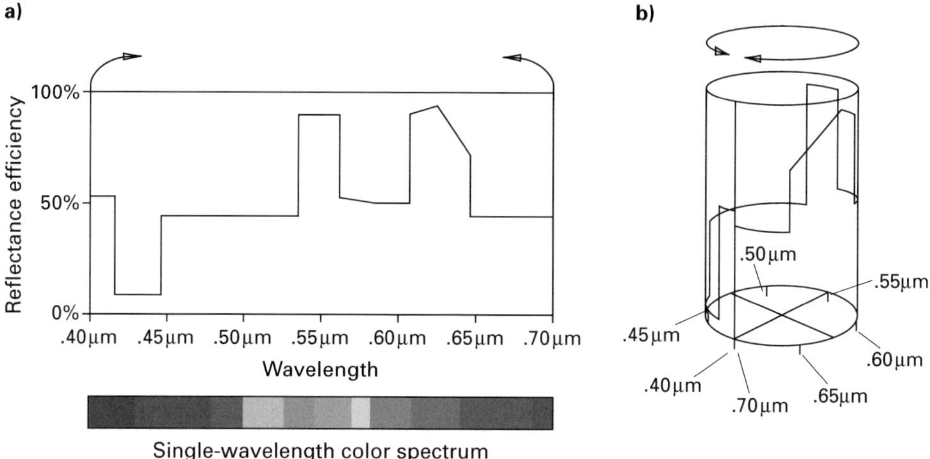

Figure 2.4
Reflectance profile within the visible spectrum: (a) planar and (b) cylindrical representations. See plate 6 for a color version of this figure.

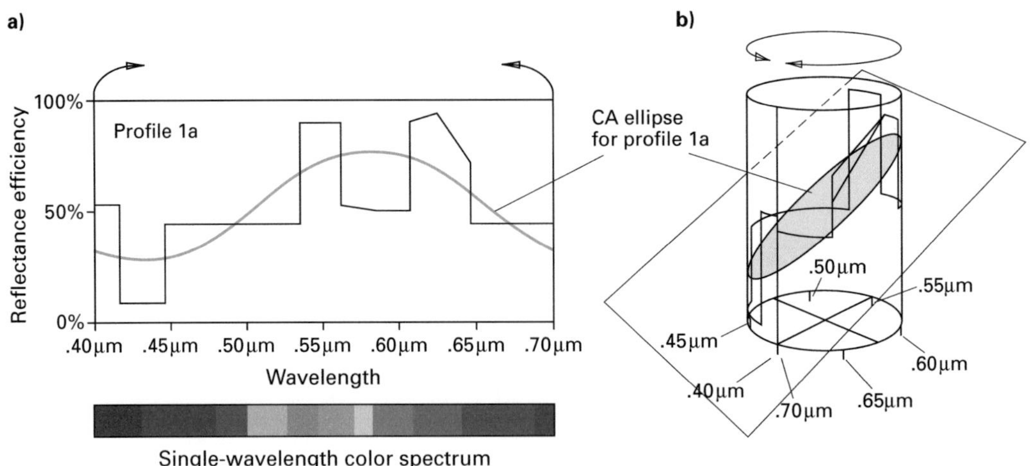

Figure 2.5
The visible spectrum reflectance profile (a), and the elliptical locus of a planar cut (b) through that profile's cylindrical representation. See plate 7 for a color version of this figure.

2.4 How the Human Visual System Tracks CA Ellipses

Having identified such an objective, mind-independent feature of material objects, we might be tempted, straightaway, to identify any objective color with the canonical approximation of the relevant material object's reflectance profile. But this is emphatically not my purpose. As will emerge, my aim is the more narrowly focused aim of identifying colors proper with the original, fine-grained reflectance profiles themselves, and not with their canonical approximations. But more of that in a moment. For the present, I wish to point out that the changing activities of the human visual system—as explored experimentally by generations of psychologists since Munsell, and as portrayed in the familiar Hurvich-Jameson network's (Hurvich 1981) theoretical reconstruction of our phenomenological color space (once again, see figures 2.2 and 2.3c, plates 4 and 5)—track the *canonical approximations* of the sundry reflectance profiles of various material objects very effectively indeed. Let me illustrate, and let us begin, by simply examining the global structure of the entire space of *possible* CA ellipses.

The first thing to appreciate is that the space of possible CA ellipses has three dimensions of variation: (1) the vertical position or altitude of the given ellipse's center point within the reflectance-profile cylinder of figure 2.6a (plate 8); (2) the degree to which that ellipse is tilted away from being perfectly horizontal; and 3) the rotational position, around the cylinder, of that ellipse's highest point. This three-space is clearly finite, and it boasts the global shape portrayed in figure 2.6b.

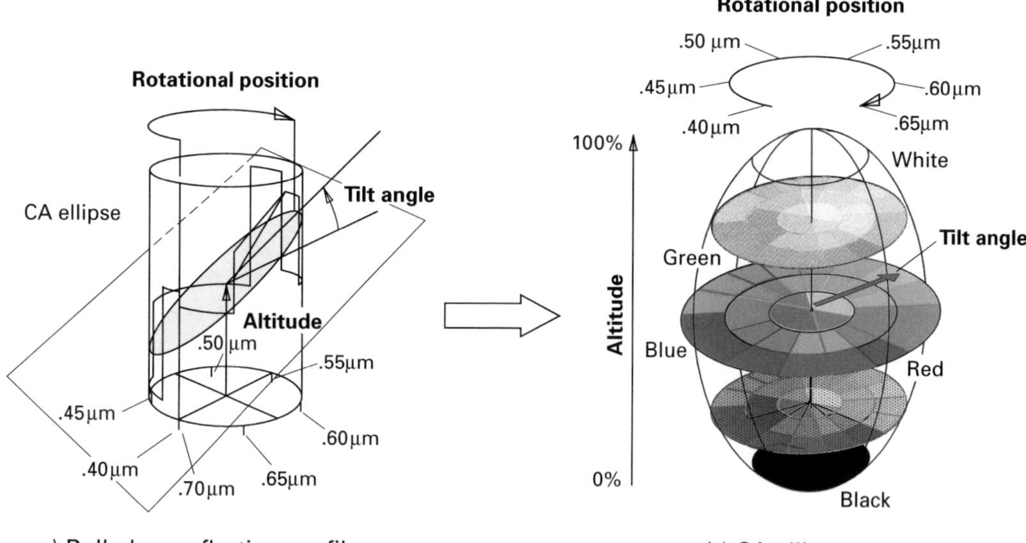

Figure 2.6
Three dimensions of variation in a space of possible CA ellipses. See plate 8 for a color version of this figure.

Note well its spindlelike, or football-like, configuration. The horizontal dimension (orthogonal distance away from the vertical central axis) shrinks sharply to zero as the extreme top and bottom of the space are approached. This reflects the fact that any CA ellipse in figure 6a will be progressively "forced" into an increasingly horizontal position as its altitude approaches the upper or lower extremes of the rolled-up reflectance-profile space. Its tilt must fall to zero as its altitude is forced ever closer to the ceiling or the floor of that cylinder. Accordingly, the horizontal dimension of the CA ellipse space, which represents that tilt, must also tend to zero at both the top and bottom of that space's vertical axis.

That tilt, recall, ultimately represents the degree to which an object's reflectance profile strongly favors some particular region of wavelengths over all of the other wavelengths in the spectral window 0.40 µm to 0.70 µm. And that dimension of variation corresponds very closely indeed to the dimension of color *saturation* displayed in the original *phenomenological* color solid of figure 2.2 (plate 4). That dimension, of course, also shrinks to zero at the top and bottom extremes of that original space, wherein reside the hueless maximally bright white and the hueless maximally dark black, respectively.

To continue, the vertical dimension of the CA ellipse space represents the altitude of a given ellipse's center point along the central axis of the reflectance-profile cylinder, which altitude ultimately represents the total area under the CA ellipse. That is, it represents the total energy of both the original reflectance profile itself and its CA ellipse (these, recall, are always the same). And that dimension of variation corresponds very closely indeed to the dimension of color *brightness* and *darkness* displayed in the original phenomenological color solid of figure 2.2 (plate 4). That dimension bottoms out at maximal black and proceeds through its central axis to progressively lighter shades of gray, until it tops out at maximal white. In between those extremes, and away from the central axis toward the phenomenological space's outer surface, the various hues proceed from dark and weakly saturated versions of each (i.e., muddy versions), through maximally vivid or saturated versions at the equator, through progressively lighter and more weakly saturated versions of each (i.e., pastel versions) as we move up the color spindle. Here again, we confront another salient dimension of variation within our phenomenological space that corresponds very closely, this time, to the vertical dimension of the CA ellipse space of figure 2.6b (plate 8). Brightness, evidently, is the objective feature therein represented.

Finally, there remains the dimension of angular position around the central axis of the CA ellipse space. This dimension of variation reflects the angular position of the objective *high point* of the given CA ellipse in the rolled-up reflectance-profile space, which corresponds, in turn, to the *seen hue* within the phenomenological color solid of figure 2.2 (plate 4). Evidently, a physical object's hue is the objective feature therein represented.

The CA ellipse space (figure 2.6b), let us remind ourselves, contains only points. (It is the rolled-up reflectance-profile space that contains the jagged profiles themselves, and the wobbling ellipses that variously approximate them.) But that CA ellipse space displays, immediately, exactly three dimensions, each one of which corresponds to a salient dimension of our antecedently appreciated subjective phenomenological color space, which also has three dimensions. Moreover, each of these two spaces displays the same global shape: something close to a spindle or a football. Additionally, both spaces code the brightest objects at the upper tip of their spindles and the darkest objects at the very bottom. Finally, both spaces code for the very same hues in their corresponding equatorial positions, in the same sequence as we proceed around that equator. Altogether, the internal structure of our subjective phenomenological color space provides an unexpectedly accurate map of the internal structure of the entirely objective CA ellipse space.

Exactly how accurate is it? Topographically speaking, it is the answer to a color realist's prayer: three dimensions, exactly two of which present themselves in polar coordinates; the same global shape; and apparently all of the same betweenness relations. But how accurate is it metrically? It is very good, but not perfect. First, our phenomenological map is metrically deformed, somewhat, in the green-yellow-orange-red region, where the human L-cone sensitivity curve and the M-cone sensitivity curve substantially overlap each other (figure 2.3b, plate 5). This idiosyncratic feature of the human visual system for detecting color samenesses and differences makes us slightly hyperacute in that region. Because of this overlap, our color-processing system is here more sensitive to small changes in the dominant incident wavelength than it is to wavelength changes elsewhere in the optical window: in the short-wavelength or blue region, for example. The result is that the system counts smallish wavelength changes in the green-to-red region as equal in magnitude to somewhat larger increments of wavelength change elsewhere. You can see this metrical deformation directly by looking at the familiar rainbowlike color bars underneath figure 2.1a through 2.1d (plate 3). Those bars mark off equal increments of wavelength, but the "seen colors" that correspond to them change only slowly in the blue region to the left, and rather more quickly in the green-to-red region toward the right.

Metrical deformations of some kind are a familiar feature of real-world maps. Think of the early modern maritime maps made of the Americas. These were fairly accurate in the vertical direction, since the map-making ship's latitude was easily reckoned by the maximum nighttime altitude, above the horizon, of familiar stars. But they were notably inaccurate in their horizontal dimension, since the earliest expeditions had no accurate clocks and thus no surefire way of determining their east-west, or longitude, position as they made charts of their target coastlines. The west coast of North America, for example, was occasionally misportrayed as tilting almost 45 degrees to the left of its actual profile, all the way up to Vancouver Island. Their inaccuracies aside, those maps were still maps. A more exaggerated example of metrical deformation is that displayed in any Mercator projection of the Earth's surface, such as those that still grace the walls of every grade-school classroom in America. As one approaches the north and south extremes of such maps, their metrical (mis)representation of east-west distances grows to absurd proportions. These gross metrical failings notwithstanding, the Mercator projection of the Earth's surface remains a paradigm example of a map, and a very useful one at that. Overall, and metrically speaking, our color map is much more accurate than a Mercator map of the Earth.

Second, and as is to be expected, our internal phenomenological map shows a nontrivial metrical deformation—this time in the vertical, or brightness, dimension—in the areas toward the extreme left and the extreme right of the optical window portrayed in figure 2.1 (plate 3), for this is where the absolute sensitivity of our S cones and our L cones falls to zero.[6] As with measuring instruments generally, the accuracy of our color-processing system plunges swiftly as one tracks its performance at the extreme limits of its proprietary range of sensitivity. Specifically, reflectance profiles with a substantial but isolated spike hard against either end of the 0.40 μm to 0.70 μm window will get (mis)represented as being essentially hueless, and as being much darker than they objectively are. In these narrow regions, the visual system fails to track accurately the objective tilt and altitude of a profile's CA ellipse, at least if the relevant ellipse owes its proprietary configuration to a large reflectance spike confined to that insensitive region. Such residual representational failures are inevitable. They represent genuine, if minor, defects in the human visual system for representing objective color, but they do not represent any defect in the claim that the human visual system *does* represent objective CA ellipses. For it remains true that, these minor defects aside, the phenomenological space in which our visual system codes its measurements plainly does constitute a recognizable map of the space of CA ellipses for objective reflectance profiles.

Moreover, and as if to make amends for its representational failures at, or very close to, the 0.40 μm–0.70 μm boundary of the rolled-up reflectance-profile space of figure 2.4b (plate 6), the human visual system does indeed make effective discriminations of the actual configuration of CA ellipses whose high-point lies anywhere close to that problematic boundary if, but only if, the reflectance profiles thereby approximated possess the bulk of their energies at two distinct wavelength spikes at some distance on either side of that discriminational "dead point." In fact, it is precisely such two-headed profiles that get coded, by the human visual system, with the familiar (but appropriately nonspectral) purples!

This idiosyncratic feature of human color coding has been familiar to color scientists for many years (Hardin 1988, 115, fig. 3.1). The CA ellipse story of what it is that our visual system is coding for accounts for this wrinkle very nicely. The fact is, it takes a reflectance profile containing two substantial energy peaks *straddling* that dead point (and little or no energy elsewhere in the spectral window) to yield a CA ellipse with a high point at that problematic boundary. The story also explains why maximally saturated purples are always so dark relative to the saturated versions of all of the other colors. A maximally saturated purple requires a strongly tilted CA ellipse whose high point is located at the dead-point boundary. But that high point is doomed to fall short of the maximal tilt possible elsewhere around the cylinder, for the more we concentrate the incident reflectance profile's two energy peaks toward the dead point, the feebler is the visual system's response. On the coding story here proposed, therefore, a maximally saturated purple is thus doomed to seem somewhat darker than any of the other saturated colors, at least to humans. And so it is.

All told, the structure of phenomenological space corresponds quite nicely to the structure of an antecedent space of specifiable objective features after all, namely, the space of possible CA ellipses. So long as we portrayed reflectance profiles as so many lines meandering across a flat and everywhere-bounded two-dimensional space, the manner in which they cluster into objective similarity classes was almost certain to remain opaque. But once we roll that space into a horizontally unbounded tube, such matters become much easier to see. My central proposal, therefore, is that the objective physical feature that unites all of the reflectance-profile metamers for any seen "commonsense" color is the peculiar CA ellipse that they all share as their best approximation.[7] And our phenomenological color space maps the range of *possible* CA ellipses very faithfully indeed, dimension for dimension, and internal location for internal location.

To see this directly, simply compare the space of possible CA ellipses portrayed in figure 2.6b (plate 8) with the long-familiar space of possible color sensations portrayed in figure 2.2 (plate 4) and with the space of neuronal coding triplets portrayed in figure 2.3c (plate 5). Evidently, the differences are minor. First, the equator of the CA ellipse space is not tilted up toward yellow, as is the equator of color-sensation space. This reflects, once again, the fact that the sensitivity curves of our three kinds of cone receptors are nonuniformly distributed across the human spectral window: the L and M cone curves overlap substantially. A saturated yellow sensation (which requires a near-maximal external stimulation of *both* L and M cones) will therefore seem brighter than any other saturated color sensation. And second, the CA ellipse space is plainly "bulgy," or more egg shaped, than is the phenomenological spindle, as drawn in figure 2.2 (plate 4). Figure 2.2 reflects the textbook orthodoxy of representing phenomenological color space as a double-*coned* spindle. But that portrayal is only a graphical convenience.

Phenomenological color space, too, is more bulgy than is conventionally portrayed (figure 2.2, plate 4), as has been known since Munsell first sought to portray it over a century ago. A more accurate portrayal would have it bulging outward somewhat, toward its top and bottom, which would bring its global structure even closer to the space of CA ellipses portrayed in figure 2.5a (plate 7). Finally, a mathematical reconstruction of the shape of the human color solid, based on the Hurvich-Jameson model network mentioned earlier (see again figure 2.3c, plate 5), also yields a space that is like the double-coned spindle of figure 2.2 but is rather bulgier toward the top and bottom extremes.[8]

In all, our internal phenomenological color space is evidently a systematic homologue of the space of objective CA ellipses. It is a reliable map of the global structure of that external feature space. Moreover, our ephemeral sensory indexings within that background map (i.e., our fleeting color sensations themselves) are moderately accurate indications of which CA ellipse we might be confronting at any given moment. Finally, and most important, those CA ellipses evidently constitute the *resolution limit* with which the human visual system can access the objective and often jagged reflectance profiles of objects. That resolution limit is fairly coarse, to be sure, but there is something objective which is being reliably, if rather fuzzily, resolved: reflectance profiles across the entire spectral window. We call them colors.

(I should mention that the story just outlined is not the first attempt to find systematic similarities between the structure of our phenomenological color space and the structure of objective or physical color space. In a recent paper, L. D. Griffin (2001) finds some notable similarities between the several "symmetry axes" of the color spindle of figure 2.2 (plate 4), and the symmetry axes displayed in the less familiar CIE space for objective colors widely used in the lighting industry. I believe the parallels he finds are entirely genuine, if less comprehensive than the systematic structural isomorphism discovered on the present analysis. My only criticism is that he has chosen, as his representational target, the wrong space for objective color. The CIE space is a space for representing and analyzing *illuminants*, not reflectance profiles. It is a space for predicting the seen color that will result from mixing light at three utterly specific and canonical wavelengths, those corresponding to the focal λ sensitivities of the human S, M, and L cones. It is a perfectly good and useful space, but it does not address the reality of the objective colors of of the vast majority of objects in our terrestrial environment, which are almost exclusively *reflectance* colors, not self-luminous colors. Moreover, it fails to represent the all-important dimension of objective lightness and darkness captured by the space of possible CA ellipses, as portrayed in figure 2.6b (plate 8). The CIE space has no room for black, for example, or for any of the darkish colors in the neighborhood of black. (The range of colors it comprehends corresponds most closely to a single horizontal cut through the equator of CA ellipse space, a plane of constant brightness.) Nonetheless, Griffin's psychological and physical parallels are entirely welcome, for the colors of self-luminous bodies are as objectively real as are the more common reflectance colors. More on self-luminous colors below, in section 2.7.)

2.5 Some Specific Tests

That the space of color sensation tracks (fairly closely) the space of CA ellipses is quite evident. But it is still a hypothesis—if a plausible one—that what unites the (uniform-illumination) metamers for any given humanly perceivable color is the CA ellipse that they severally share. (It is initially plausible because the coarse-grained resources of the human visual system typically *cannot tell the difference* between a given profile and its canonical approximation.) But let us quickly test the hypothesis against two salient examples of real metameric pairs, one drawn from Hardin (1988, 47), and one drawn from Fraser et al. (2003, 30). The first example appears in figure 2.7.

These two reflectance profiles are metameric pairs, according to Hardin, despite their evident differences. How do they compare with regard to their respective CA ellipses? To answer this question, I traced each of these profiles onto a separate transparency and rolled each into a cylinder. I then probed each profile (separately) with another rotatable cylinder slid inside it, a cylinder graduated with ellipses of varying tilt angles, until a closest match was achieved, according to the criteria set out at the end of section 2.3. (The relevant areas were measured by integrating over a

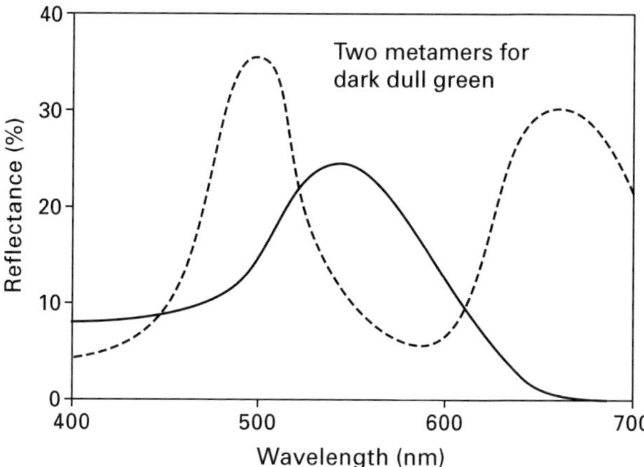

Figure 2.7
Reflectance profiles for metameric pairs per Hardin (1988, 47).

substantial number of narrow, vertically oriented rectangles.) This yielded a unique CA ellipse for each profile. The CA ellipse for the double-peaked profile has a peak at a rotational position R = 0.52 µm, a height H = 14 percent, and a tilt angle T = 17 percent of maximum. The CA ellipse for the single-peaked profile has a peak at a rotational angle R = 0.535 µm, a height H = 13 percent, and a tilt angle T = 16 percent of maximum.

The difference between these two CA ellipses is ΔR = 5 percent, ΔH = 1 percent, and ΔT = 1 percent. The difference is marginal, and both CA profiles (with peaks very close to 0.53 µm) will present as a dull and quite dark green—barely distinguishable, if they are distinguishable at all.

The next pair of metameric profiles also present to us as green, though a somewhat brighter and more saturated green than in the preceding example. The taller of these two profiles (figure 2.8) was probed in the manner described above, and proves to have a CA ellipse of R = 0.53 µm, H = 33 percent, and T = 33 percent of maximum. The second profile has a CA ellipse of R = 0.53 µm, H = 29 percent, and T = 35 percent of maximum. The difference between them is ΔR = 0% percent, ΔH = 4 percent, and T = 2 percent. Once again, the differences are marginal—at or close to the limits of human discrimination.

Given the systematic match already noted between our phenomenological color space (figure 2.2, plate 4) and CA ellipse space (figure 2.6b, plate 8), these singular matches should come as no surprise. But it is salutary to check out the hypothesis (that the class of same-seeming metamers for humans corresponds very closely to the class of reflectance profiles that share the same CA ellipse) against independent data.

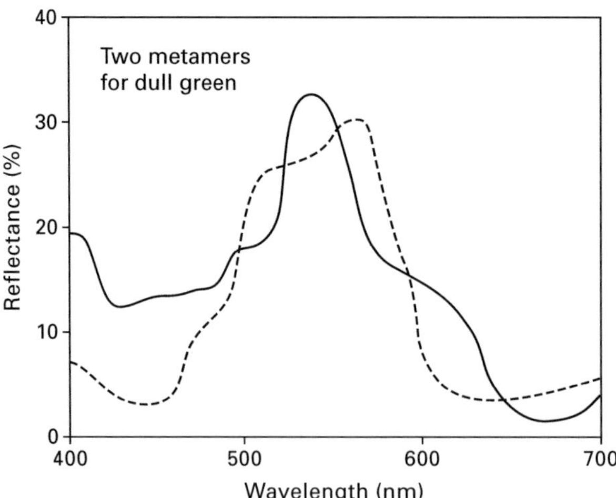

Figure 2.8
Reflectance profiles for metameric pairs per Fraser et al. (2003, 30).

2.6 An Important Objection

There remains a possible objection to my claim that our phenomenological color space is a (fairly high-resolution) map of CA ellipse space, and thus is a (rather low-resolution) map of the range of objective reflectance profiles. Hardin complains that our phenomenological color space displays an inescapable contrast between "unmixed" colors (such as red or blue) and "mixed" colors (such as orange or purple), a contrast that is completely absent in both the CA ellipse space *and* in the objective space of possible reflectance profiles. How, then, can we identify colors with the latter?

Let us agree, at least for the sake of argument, that both parts of Hardin's claim are correct. This situation does nothing to undermine the claim that our phenomenological color space accurately maps the space of possible CA ellipses, for the structure of the latter *is* plainly reflected, dimension for dimension, in the structure of the former. Hardin's antirealist argument here has the "onus of match" exactly backwards. It is not incumbent on *the domain portrayed* to have every feature displayed by its portraying map: maps can display all sorts of features that are incidental to their role as effective maps (a common street map crumples easily and weighs about an ounce, for example, in dramatic contrast to the urban domain that it portrays). The contrast between mixed and unmixed phenomenal colors is just such an incidental feature—an artifact, presumably, of the opponent-process architecture of our color system.[9]

What *is* required is that the relevant structure of the objective reality (namely, the three dimensions of variation for a CA ellipse, as portrayed in figure 2.6b, plate 8) finds itself reflected in some structural features of the map that purports to portray that objective reality (namely, our phenomenological color spindle, as portrayed in figure 2.2, plate 4). In the present case, that requirement is plainly met. That the map might have *other* features that happen *not* to correspond to external structures is irrelevant.

My critique of Hardin's eliminativist position can perhaps be clarified with the following parallel, drawn from another modality. The human nervous system responds to temperature with two anatomically distinct types of receptor neurons: one for registering temperatures *above* the skin's temperature, and another for registering temperatures *below* it. The first system produces a range of increasingly intense sensations, all of which are similar to one another, and so does the second. But the family of sensations for warmth, on the one hand, and the family for coldness, on the other, are qualitatively quite distinct *from each other*. (No surprise, given that they arise from anatomically and physiologically distinct systems.)

Now, are we going to deny that objective temperature is identical with mean molecular kinetic energy on grounds that the objective scale of molecular kinetic energies embodies no such objective *qualitative* distinction between the regions above and below human skin temperature? Of course not. Nor should we hesitate, for similar (bad) reasons, to identify objective colors with reflectance profiles, on the grounds that there is nothing in the domain of reflectance profiles that answers to the phenomenological distinction between pure and mixed colors.

2.7 What, After All, Are Colors?

Even so, it remains to be discussed exactly how our familiar *objective colors* should be fit into this emerging framework. Simply identifying the familiar range of colors with the evident range of CA ellipses is a very poor option, since only a negligible proportion of material objects have a reflectance profile that is actually identical with the Platonic perfection of a CA ellipse. Any CA ellipse, of course, projects back onto the original reflectance-profile space as a perfectly smooth, one-cycle *sine wave* of some altitude, amplitude, and left-right location within that space (see again figure 2.5, plate 7). But most objects will have a much noisier reflectance profile than a perfect one-cycle sine wave: the meandering metamers still dominate the reflectance profiles we actually encounter in the real world. Accordingly, identifying the various colors with the various reflectance profiles displayed by perfect CA ellipses would have the consequence that almost nothing in the world is colored.

A much better option is to identify the full range of objective colors with the full range of objective reflectance profiles—both the relatively rare perfect CA ellipses *and* the multitude of metameric profiles that severally cluster around them—and then acknowledge that we humans are able, with our native visual equipment, to perceive and discriminate those highly various reflectance profiles *only at a rather low level of resolution*. As we noted above, the CA ellipse constitutes the *limit of resolution* at which humans can discriminate sameness and differences between objective reflectance profiles. In particular, we are typically unable to discriminate between any of the many metameric reflection profiles that share *the same* ellipse as their canonical approximation. These mutually clustered metameric profiles will typically present themselves, to the casual human eye, as the same color, despite the residual but real differences between them.

This situation, however, is entirely unremarkable. The human auditory system, to take a related example, is skilled at recognizing and discriminating power-spectrum profiles within the acoustic spectrum. We are good at recognizing and discriminating the distinct voices of people familiar to us, the distinct voices of various musical instruments, birds, animals, and so forth. No one will deny that specific types of sounds are identical with specific power-spectrum profiles displayed in a propagating wave train, and no one will deny that our auditory skills reside in the cochlea's ability to respond to those various profiles in an appropriately discriminatory fashion.

And yet, our cochlea has a resolution limit as well. Clustered around the distinctive power-spectrum profile of a typical oboe's middle A lies a multitude of possible "acoustic metamers," all of them different from one another in ways that lie beneath the capacity of my cochlea to resolve. Despite their differences, they will all sound the same to me. And so also for any other familiar sort of sound. At a certain point, and inevitably, our discriminatory powers simply run out. Such acoustic metamers for familiar sounds are as real, and as inevitable, as are the electromagnetic metamers for familiar colors.

But these undoubted facts about acoustic reality provide no grounds for irrealism or eliminativism about our commonsense ontology of sounds. Nor do the parallel facts, concerning electromagnetic metamers, provide grounds for irrealism or eliminativism about colors. Indeed, the ontological advantage, if any, should lie with colors. Sounds are ephemeral: a bird, a musical instrument, or an animal emits a sound only occasionally, and the sound fades (as $1/r^2$) to nothing as it promptly flees its point of origin. By contrast, a material body's electromagnetic reflectance profile is a quasipermanent and stable property of that material body. It will change only if the molecular structure of the body's surface is modified in some way.

The stable solution then, to which we are thus attracted, is that the objective color of an object is identical with the electromagnetic reflectance profile of that object, within the window 0.40 μm to 0.70 μm. Our native ability to recognize and discriminate such profiles is limited to recognizing and discriminating the altitude, tilt, and rotation angle of the CA ellipse that approximates any given reflectance profile. But this native ability still gives us a highly reliable grip on an often-telling dimension of objective reality.

To be more specific, an object is a maximally saturated *red*, on this view, just in case its reflectance profile has a CA ellipse of altitude 50 percent, a maximum tilt, and a rotation position with the ellipse's highest elevation at 0.63 μm (see figure 2.9a, plate 9). An object is a somewhat dull *yellow*, on this view, just in case its reflectance profile has a CA ellipse of altitude 50 percent, a moderate tilt, and a rotation position with the ellipse's highest elevation at 0.58 μm (see figure 2.9b). And so forth for every other objective color, no matter what its lightness, degree of saturation, and peculiar hue (see figure 2.9c–f). We might think of this as the "wobbling penny" account of the space of possible reflectance profiles, for that is how the CA ellipse variously appears for diverse reflectance profiles.

This position has the consequence that two distinct objects, both of which are a maximally saturated yellow (or any other color) need not be exactly the *same* color, for they may sport distinct metamers included within the class *maximally saturated yellows*. They are both genuine instances of maximally saturated yellow, let us assume, but they may be *different* instances of what (as we now appreciate) is an interestingly diverse class. This description is, to be sure, a significant departure from our normal modes of speech, because common sense innocently assumes that there *are no* color differences underneath what our eyes can discriminate in broad daylight. But this naïve assumption must be let go. And the existence of diverse reflectance-profile metamers is precisely what demands its surrender. Even so, the existence of colors themselves, as objective features of objects, is not threatened. We simply have to acknowledge that there is slightly more to color than meets the human eye, even under the optimal conditions of broad daylight.

Interestingly, that hidden diversity and sameness of objective colors is not entirely inaccessible to the human visual system. A simple trick will make such matters visually available, even to one who is color-blind. Given two shirt buttons, of apparently the same yellow color to normal vision, one can determine whether they (1) are *exactly the same* color (that is, have identical reflectance profiles) or (2) merely have *distinct metameric variants* on a general yellow theme. We can do this by running both buttons, side by side, through the gauntlet of a rainbow projected on a wall (figure 2.10a, plate 10). If we project sunlight through a prism in an otherwise darkened room, as indicated, we will produce a fan of distinct monochromatic wavelengths to serve as diagnostic probes of each button's reflectance at any given wavelength of light. If the two buttons do share identical reflectance profiles, then their joint appearance to the human eye will vary, of course, as they are marched through the fan of distinct diagnostic illuminants. But at each position against the fan, they will always display the same appearance *as each other* (figure 2.10b). By contrast, if the buttons have distinct metameric variants on yellow, then at one or more points in their journey across the rainbow they will appear *different from one another* (figure 2.10c). They must. That they have distinct reflectance profiles entails that they will display differential reflectance

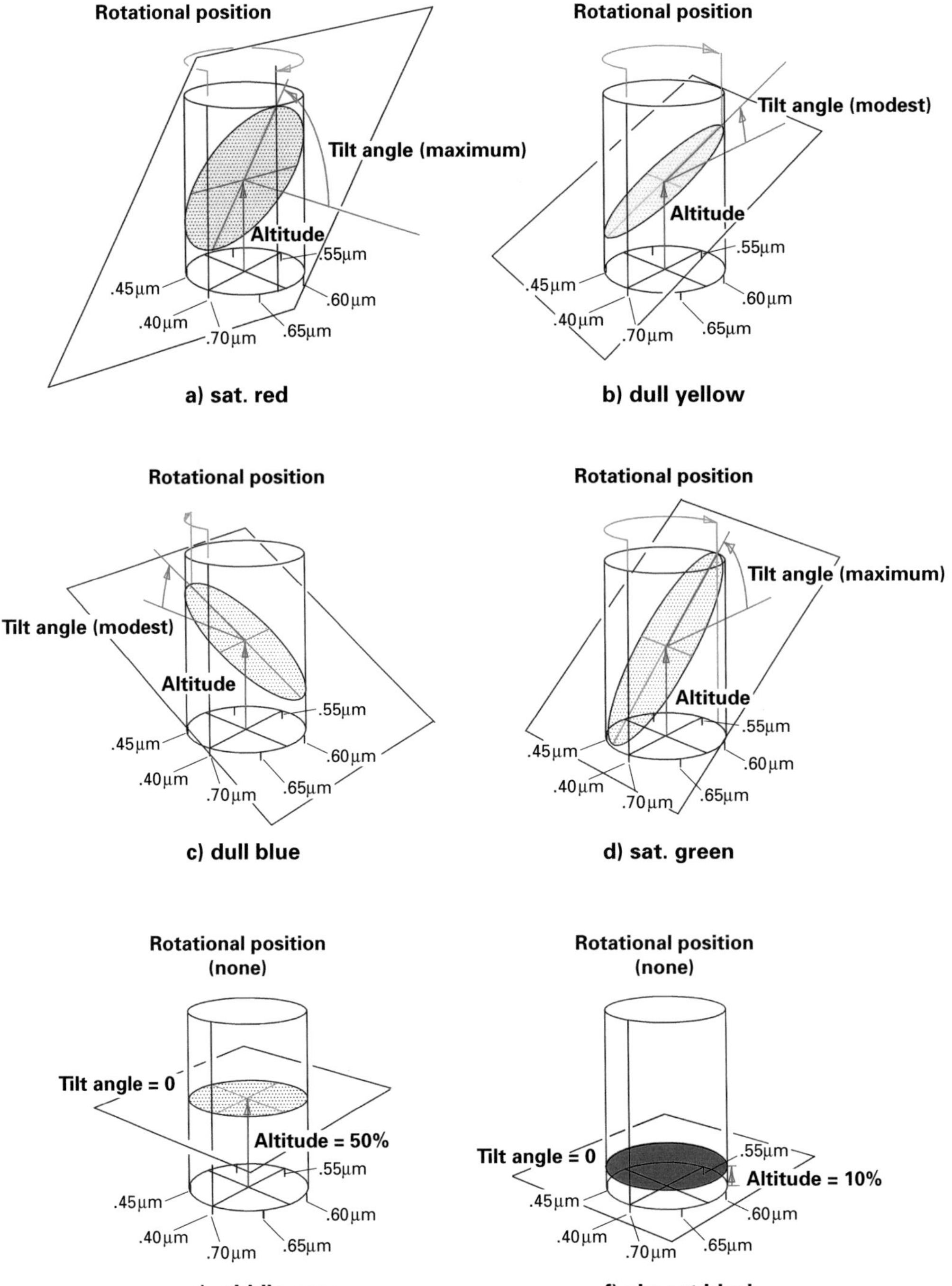

Figure 2.9
CA ellipse, tilt, and rotation position for sample color reflectance profiles. See plate 9 for a color version of this figure.

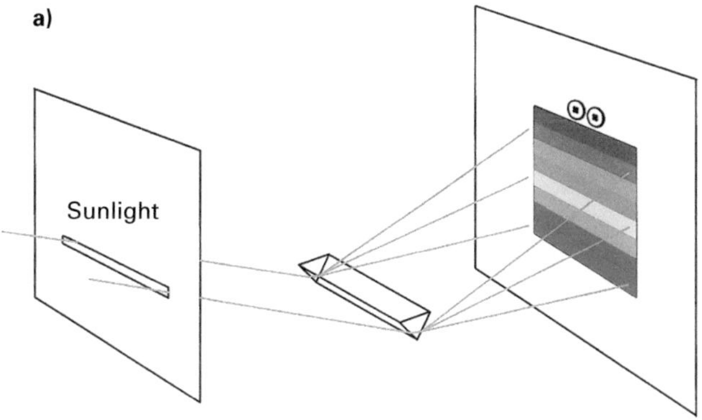

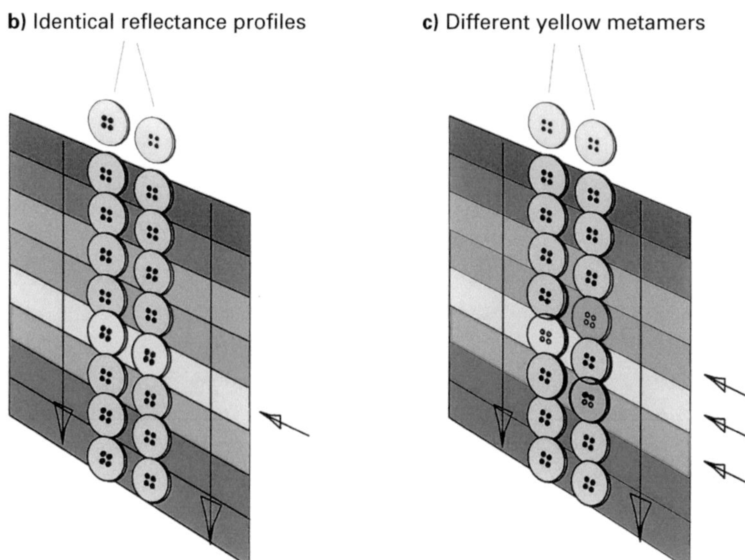

Figure 2.10

Using (a) a rainbow projected on a wall, we can compare two buttons of apparently the same color to determine if they are (b) identical or (c) different metamers. See plate 10 for a color version of this figure.

behavior at some one or more points within the visible spectrum. Even a color-blind person will detect such discrepancies in their objective reflectance behaviors, since they will still present themselves, to him, as visible differences in apparent *grayscale* brightness. In this way are the hidden color metamers made visible, even to people who are color-blind.

Collateral or background information can also be a reliable guide to judging whether two same-seeming objects really have identical reflectance profiles, or merely share the same ellipse as their canonical approximation. If one is viewing two visually identical dark red cherries, or two visually identical yellow bananas, for example, one can be confident that the two cherries have genuinely identical reflectance profiles, and so also for the two bananas. For one can be independently confident that the two ripe cherries have identical molecular constitutions at their surfaces, as do the two ripe bananas. Such identity in molecular constitution physically guarantees identity in their reflectance profiles. However, when background information suggests a quite different molecular constitution for two same-seeming objects—as with a purple plum and a patch of purple paint—the distinct-metamers hypothesis will have a better claim on the situation.

2.8 The Diversity of Objective Colors

It remains to highlight the contrast between the familiar *reflective* colors, as characterized in the preceding pages, and the less common *self-luminous* colors, as displayed in a fire, a star, an incandescent bulb, or a light-emitting diode (LED). The former is a matter of what profile of light an object *reflects*; the latter is a matter of what profile of light an object *emits*. Whatever common sense might think, these are entirely distinct properties. One and the same object can simultaneously possess incompatible "colors" of each kind, as when a stove-top heating element veridically presents its familiar occurrent color—a dark charcoal gray—when the kitchen lights are on; but when the lights are switched off and the room is plunged into darkness, the element reveals its self-luminous color of dull red (the relevant dial on the stove's control panel was set at "low" all along). Though we could not see its self-luminous color in the first condition—because the magnitude of the reflected light swamped the comparatively faint emitted light—the darkened condition allows that self-luminous color to become visible.

Self-luminous colors were an extremely rare occurrence in the evolutionary environment that gave birth to our current color vision. Only the sun, the stars, the occasional firefly, and the occasional forest fire *ever* displayed a self-luminous color. Accordingly, and apart from the character of solar radiation as a background illuminant, the self-luminous colors must have played a negligible role in the evolutionary selection of the enabling mechanisms for human color vision. In modern society, of course, the self-luminous colors have become commonplace. And in fact, where metamers are concerned, the self-luminous colors are somewhat better behaved than are the reflective colors. The colors of a *thermally incandescent* object almost always present a *smoothly varying* emittance profile, whose peak magnitude is tightly tied to the object's absolute temperature. And the self-luminous color of an object engaged in *spectral emission* (that is, in photon emission from electron-shell transitions) is almost always a matter of one or more *narrow spikes* of monochromatic light (as from an LED or a sodium streetlight), which color is a reliable guide to the object's peculiar atomic constitution. Metamers are entirely possible here, as elsewhere, but in fact they are much less common for self-luminous colors than for reflective colors.

The objective space for self-luminous colors (in the window 0.40 μm to 0.70 μm) is slightly but importantly different from the space for reflective colors. In particular, the vertical dimension of the relevant cylinder represents, not reflective efficiency (which tops out at 100 percent), but *emission intensity*, which has no upper limit. Nonetheless, our native representational space (specifically, the color-coding neuronal activation space of figure 2.3c, plate 5) does its best to represent this relatively new range of possible *emittance* profiles, using its existing resources of brightness, saturation, and hue. But it here encounters an anomalous situation, in that the *brightness* levels of typical self-luminous objects are much too high to be accounted for in terms of a 100 percent reflectance efficiency across the spectrum (i.e., as originating from a maximally reflective *white* object), and they often display a vivid *hue* (i.e., a nonwhite color) in any case. The human visual system responds by coding such anomalous (i.e., self-luminous) inputs at an appropriate place *on the ceiling* of our phenomenological color space, but *outside* the pointlike apex of the spindle-shaped volume that confines all of the representation points for the less dramatic reflective colors (figure 2.11, plate 11).

These are "impossible" positions as far as the reflective colors are concerned. (No *reflective* color can be as bright as the brightest possible white and yet be something other than white.) But by that very fact, those unusual ceiling positions serve as reliable diagnostic positions, in our preexisting neuronal activation space, to indicate the presence of a *self-luminous* object, and to indicate its peculiar hue and saturation. An information-processing system that was shaped by evolution to recognize and discriminate one kind of color turns out to be able to recognize and discriminate a second kind of color as well, and to do so in a manner that can sharply distinguish both.

Precisely because they are typically coded, by the visual system, *outside* the normal phenomenological color spindle, the self-luminous colors typically stand out like beacons against the darkness.[10] Their typical representational space is the *two-dimensional ceiling* of the opponent-cell activation space. Save for a single point at its center, this space is entirely distinct from the three-dimensional volume of the familiar spindle-shaped solid for representing the reflective colors. But it, too, maps moderately well

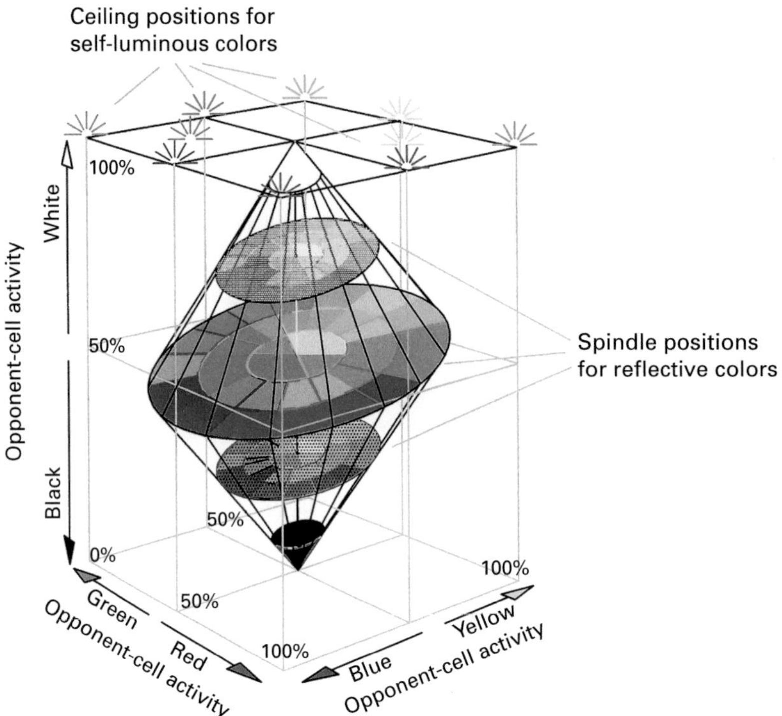

Figure 2.11
The different representation points of self-luminous colors and reflective colors within the human phenomenological color space. See plate 11 for a color version of this figure.

onto the objective range of its proprietary properties, namely, the various *emittance profiles* displayed by self-luminous bodies. (Exactly how the visual system discriminates between the different *brightness* levels of the self-luminous colors—note that the ceiling of the opponent-cell space has only two dimensions—is a matter that is still unclear on the present account. But I shall leave its pursuit for another occasion.)

I conclude that there are at least *two* quite distinct kinds of objectively real colors, the reflective colors and the self-luminous colors. The objective structure of each domain of properties becomes evident if one rolls the window of the visible spectrum into an abstract cylinder, and then examines the space of possible planar cuts through that cylinder, as providing the best approximation of the fine-grained details of the possibly noisy power-spectrum profile currently portrayed around its surface. For reflective colors, the space of possible planar cuts (i.e., the space of possible CA ellipses) is homomorphic with the spindle-shaped solid of our phenomenological space. And for the more ostentatious self-luminous colors, the space of their possible hues and saturations is homomorphic with the space of possible activations within the otherwise unused two-dimensional *ceiling* of the opponent-cell activation space. The colors, of both kinds, are thus entirely real, and are as objective as you please. We can see them both, for in most cases we can recognize and discriminate both reflectance profiles and emission profiles, and discriminate the one from the other. The only disappointment here is both negligible and inevitable: some of the fine-grained structure of those profiles lies beneath the resolution of our native visual system. But this is no argument for irrealism about those profiles (of course), nor is it an argument for irrealism about colors. For that is precisely what the two kinds of colors *are*: reflectance profiles and emission profiles, respectively.

2.9 Comparison with a Related View

The account of objective colors defended above shares many of the same motivations and some of the positive substance of the realist account of colors recently urged by Byrne and Hilbert (2003). But some important differences stand out, and I will close by bringing several of them to your attention. First, and perhaps least, those authors propose a new notion of a "determinate color" (for example, determinate red) whose extension is exactly the set of metameric reflectance profiles between which the human visual system is unable to distinguish. By contrast, I propose to identify distinct colors with distinct reflectance profiles, and then embrace the consequence that the human visual system cannot distinguish all color differences, since it cannot distinguish all reflectance profile differences. We see color similarities and differences only down to the resolution limit defined by the range of possible CA ellipses. There are color similarities and differences beneath what we can detect just by looking. Our existing color vocabulary, therefore, comprehends only a coarse partitioning of the objective reality. Nevertheless, that partitioning is still objective in character, and is highly useful as a guide to many of the causal properties of material objects.

Second, and much more important, Byrne and Hilbert acquiesce in the received wisdom that the family of metamers for any commonsense color displays no unifying intrinsic feature specifiable in purely physical terms. They should not have acquiesced to this claim, because we can indeed specify, in terms "of interest to a physicist," the feature that unites the family of metamers for a given commonsense color: they all share the identical reflectance-space ellipse as their canonical approximation. Moreover, this shared objective feature is precisely what gets mapped within the human subjective or phenomenological color space. Accordingly, we can see *how* the space of human color sensations counts as a structurally accurate map of an objective domain of properties, a real achievement by our visual system that remains either denied or unrecognized by their view.

Finally, Byrne and Hilbert attempt to salvage a *unitary* conception of color, by attempting to knit together the distinct features of reflectance, emittance, and transmittance into a single and deliberately more general notion of *productance*. By contrast, the view of this chapter tends in exactly the opposite direction. I claimed earlier that reflective colors are a family of properties genuinely distinct from the family of self-luminous colors. And I will say the same for a third family of color properties: the various profiles of *transmittance* displayed by transparent and translucent objects such as colored glass, colored liquids, and some gemstones. All three types of color are features to which the human visual system gives us some nontrivial perceptual access, and in each case, that perceptual access involves our capacity to distinguish the power-spectrum profiles of the electromagnetic radiation arriving to our eyes. But the three types of objective properties themselves (reflectance, emittance, and transmittance) are radically distinct from the point of view of informed physics. It would be folly to try to conflate them all into a single notion, especially when we already enjoy a perceptual system that allows us to spontaneously recognize the distinctions between them fairly reliably.

To add further force to this argument, the varieties of objective chromatic phenomena do not end with the three types just mentioned. We must reckon also with the range of *scatterance colors*, as instanced in the *blue* of the daytime sky. There are also *interference colors*, as displayed, for example, in various thin films, such as an oil-slick floating on water. There are also *refractance colors*, as get displayed when sunlight hits a prism, a many-faceted gemstone, or a spray of spherical water droplets. These three additional kinds of color *also* involve importantly *different* intrinsic properties, features, or mechanisms for interacting with light, each in their own characteristic ways.

If we want to respect the impulse toward objectivity implicit in the commonsense conviction that the colors are real features of the external world, we should respect the lessons of modern physics that color properties come in a substantial variety of objectively distinct families. We are merely stuck with a *single perceptual modality*—a trichromatic visual system—with which to ply access to all of them. But that is no grounds for conflating the distinct types of objective color themselves. They are importantly different from one another, even in the details of their visual appearance. Withal, and however various, those distinct types of color remain as real and as objective as you please, despite what commentators from Locke to Hardin have too hastily insisted. The colors—all six distinct families—deserve to be welcomed back into the fold of objectively real properties. We just need to understand them a little differently.

Acknowledgments

The central idea of this paper occurred to me while listening to a provocative talk on color given by Mohan Matthen during the Vancouver Conference on the Philosophy of Color, in October 2003. My thanks for his inspiration. The paper also reflects what I have learned over the years about color from Larry Hardin, Kathleen Akins, and Martin Hahn. My thanks to them also.

Notes

1. Locke (1689, book 2, ch. 8). For the analytic and exegetical case that Locke was indeed an eliminativist, rather than some sort of reductionist, about objective colors, see the thoughtful essay by Rickless (1997). To be sure, Locke's text admits of other interpretations.

2. Those landmark map elements will be prototypical *activation patterns* across the relevant neuronal population.

3. For a broad account, look for my *Plato's Camera: How the Physical Brain Grasps Abstract Universals* (forthcoming).

4. For details, see Churchland (2005).

5. See Hardin (1988), 115, figure 3.1, for a portrayal of exactly where that nonspectral gap lies.

6. My thanks to an anonymous referee for forcing my attention toward this particular imperfection in the human visual system's capacity to track similarities and differences among CA ellipses. Its misrepresentations here are fairly minor and highly localized, however, especially compared with those embodied in a Mercator projection, and they do nothing to undermine the claim that our phenomenal space is a moderately faithful *map* of CA ellipse space *as a whole*.

7. Well, *almost* all. Recall once more that the human visual system tracks CA ellipses increasingly poorly for reflectance profiles that display significant amounts of their energy in the narrow region of the dead point, where 0.40 μm abuts 0.70 μm This isolated failing can lead to (rare) profile pairs that share the same objective CA ellipse, yet look slightly different to us.

8. For the details of its derivation, see Churchland (2005).

9. See again the Hurvich-Jameson color-processing network of figure 3*a*.

10. Evidently, there is plenty of room, inside the human activation space for color-coding neurons, for coding vectors that lie *outside* the confines of the familiar color spindle (see again figure 2.2, plate 4). These rogue coding vectors—representing "impossible" colors—are explored at length in Churchland (2005).

References

Byrne, A., and D. R. Hilbert. 2003. Color realism and color science. *Behavioral and Brain Sciences* 26: 3–21.

Churchland, Paul M. 2005. Chimerical colors: Some phenomenological predictions from cognitive neuroscience. *Philosophical Psychology* 18 (October): 527–560.

Churchland, Paul M. Forthcoming. *Plato's Camera: How the Physical Brain Grasps Abstract Universals.*

Cohen, J. 2004. Color properties and color ascriptions: A relationalist manifesto. *Philosophical Review* 113 (4): 451–506.

Fraser, B., C. Murphy, and F. Bunting. 2003. *Real World Color Management: Industrial Strength Production Techniques*. Berkeley, CA: Peachpit Press.

Griffin, L. D. 2001. Similarity of psychological and physical color space shown by symmetry analysis. *Color: Research and Application* 26 (2): 151–157.

Hardin, L. 1988. *Color for Philosophers: Unweaving the Rainbow*. Indianapolis, IN: Hackett.

Hardin, L. 1993. *Color for Philosophers: Unweaving the Rainbow*. Expanded ed. Indianapolis, IN: Hackett.

Hurvich, L. M. 1981. *Color Vision*. Sunderland, MA: Sinauer.

Locke, John. 1689. *An Essay Concerning Human Understanding*. London: W. Tagg.

Rickless, Samuel C. 1997. Locke on primary and secondary qualities. *Pacific Philosophical Quarterly* 78: 297–319.

Thompson, E., A. Palacios, and F. Varela. 1992. Ways of coloring: Comparative color vision as a case study for cognitive science. *Behavioral and Brain Sciences* 15: 16–34.

3 Color Experience: A Semantic Theory

Mohan Matthen

In this chapter, I present a *semantic* theory of color *experience*, on which color experience *represents* or *denotes* color properties, and *attributes* these properties to visual objects. On such a theory, color experience informs us about the external world by means of a semantic relationship, *representation* or *denotation*, that it bears to external world properties, and it attributes these properties to external objects (that it also denotes, though this is not something that I will take up in this chapter).

The use of concepts such as *representation* is essential to any semantic theory. In the specific version of the theory that I offer,

1. the above semantic relationships operate through the vehicle of a systematic set of similarity relations.

I argue that

2. our grasp of these perceptual relations is (for the most part) innate; and
3. the similarity relations in question are *dynamic*, in the sense that they are constitutively linked to certain cognitive changes in the perceiver subsequent to color perception.

A central feature of the argument presented in this chapter is that it relies on distinctions concerning the degree of certainty that attaches to different kinds of propositions that we come to know through color perception.

3.1 The Evidential Status of Color Vision

Consider a statistically normal human color perceiver, Trich (so-called for her trichromacy), viewing a piece of fruit in reasonably good light. Suppose that to Trich this fruit looks orange. Her visual experience of the fruit, *O*, is *evidential* in a number of ways.

3.1.1 Object Knowledge

Trich can take *O* as strong support for the following proposition:

OK (3.1) That [fruit] is orange.

If Trich is in any doubt about (3.1), she could improve her viewing conditions: she might put on her glasses; she might take the fruit to the window, where the sunlight is stronger; she might try to eliminate any sources of nonstandard illumination; she might turn the fruit over in her hands or bring it closer to her eyes. At the end of this process, she would be in a state of *empirical* certainty about (3.1). By *empirical certainty*, I mean this: Trich has no *isolated* reason for doubting (3.1)—no reason that would not simultaneously lead her to doubt a whole raft of unconnected propositions.

Empirical certainty does *not* entitle her to dismiss skeptical doubt. She may be dreaming, deluded, or deceived by an evil demon; she may be a brain in a vat; her visual pathways may be subject to a strange form of transcranial interference. This kind of doubt is not, however, *isolated*. Skeptical doubt attacks the background conditions for empirical knowledge (e.g., the state of the observer's eyes or wakefulness); more generally, it attacks the epistemic status of a proposition without any grounds specific to that proposition. It undermines (3.1), but simultaneously many other propositions. For the possibilities just mentioned would lead Trich to doubt in turn whether the lemon next to the orange was really yellow, whether it was really her hand in which the fruit resided, whether it was really daylight, whether she was really awake, and so on. Skeptical doubt spreads by knocking away the foundations of empirical knowledge. Empirical certainty is subject to the security of these background conditions.

My point is that given careful visual inspection, Trich may still doubt (3.1), but only in an expansive way. After such inspection, she has no reason to doubt (3.1) that is not equally a reason to doubt other deliverances of her senses.

3.1.2 Color Knowledge

In light of her experience, *O*, Trich is also entitled to certainty regarding certain propositions about the *color* presented therein. To wit:

CK (3.2) The color presented in *O* is yellowish.
CK (3.3) The color presented in *O* is less reddish than that presented in another visual state *C* (which happens to be occasioned by a cherry).
CK (3.4) The color presented in *O* is less yellowish than that presented in another visual state *L* (which happens to be occasioned by a lemon).
CK (3.5) The color presented in *O* is more similar to that presented in *L* than it is to that presented in *C*.

These propositions concern *colors*, not the objects that possess them. (Note that I am speaking of the *propositions* expressed by 3.2–3.5—I am not claiming that Trish understands the meaning of the *sentences* written above, or of words such as "yellowish.")

Now, Trich's certainty with regard to CK propositions is actually higher than empirical certainty. For, it seems, that to know the above sort of fact about the colors, one needs only to experience them (cf. Johnston 1992; Hellie 2005). For instance, for

Trich to know that the color of the thing she seems to see is yellowish, as asserted in (3.2), she needs nothing more than her experience of that color. She does not need to know that the fruit actually is the color it seems to be; she does not even need to know that the fruit exists. Even if Trich's experience, *O*, is a dream or a hallucination, it is nonetheless an experience of a certain color, and this is enough for her to be sure that the color in question is yellowish in hue. Hallucinatory color experiences are no less probative regarding color knowledge of the sort contained in (3.2)–(3.5) than are normal veridical experiences. For this reason, Trich is entitled to dismiss even skeptical doubt with respect to the CK propositions. She is entitled to *Cartesian certainty* with regard to them. Cartesian certainty is certainty that is proof even against skeptical doubt.

How can Trich be entitled to so strong a level of certainty regarding the colors presented in her color experiences? You might think that one could possess such certainty only with respect to subjective facts—Cartesian certainty starts with that which can be introspected. (Famously, Descartes argued that it attaches to the proposition "I am," and to the content of ideas that are clearly and distinctly apprehended, such as that matter is spatially extended but consciousness is not.) However, color attribution is objective—whether or not the fruit Trich is looking at is orange does not depend just on her or her state of mind. Given that this is so, *orange* cannot be merely a subjective or private category. It isn't like "This fruit gives me pleasure," of which she can be certain because it is true in virtue of a subjective state that she can introspect. How can she be so certain of something that she doesn't determine in the same way?

The thesis advanced in this chapter is that this entitlement arises because color-knowledge propositions (3.2)–(3.5) are "quasi-analytic" consequences of Trich's innate representational scheme for colors. These propositions are about colors that exist independently of Trich, but her knowledge of the similarity relations among these colors and of their compositional structure arises out of this representational scheme.

3.1.3 Unexperienced Colors

Trich is able visually to imagine a "missing shade" of yellowy orange "in between" the color presented by this fruit and that presented by the lemon, even though she happens never to have actually encountered anything of such a color (cf. Hume's missing shade of blue, *Treatise* 1.1.1). She is not dependent on memory to know of such a color. Moreover, she can be induced to have experiences of "impossible" colors: colors that are both reddish and greenish (Crane and Piantanida 1983) and colors that lie outside the color solid (Churchland, this volume). In virtue of such experiences, she apprehends propositions analogous to the CK propositions above regarding colors that can never be instantiated in real objects. She is entitled to Cartesian certainty about these colors too, even though she never sees them "for real." Having imagined them, she is entitled to be certain of whether they are bluish or yellowish, bright or light, and so on.

It is perhaps worth emphasizing that unexperienced colors are not just idle artifacts of the philosophical museum. It is easy to imagine a rainy equatorial forest in which blue and saturated yellow are never visible, or some particularly sheltered environment in which blood is never shed, and men never see red. But even putting such fantasies aside, the dye-deprived cultures of antiquity did not have access to *all* the colors. Yet, even such chromatically deprived people were capable of "raising up" (as Hume put it) the ideas of the missing shades. And when they did, they would have been entitled to Cartesian certainty regarding facts analogous to those expressed by propositions (3.2)–(3.5).

Knowledge about colors that one has never encountered must be a priori. How can one have a priori certainty about things that exist independently of oneself?

3.1.4 Object Appearance Knowledge

Finally, in light of her experience *O*—and *O* alone—Trich possesses Cartesian certainty about one other proposition:

OAK (3.6) That [fruit] *looks* orange.

OAK propositions, like OK propositions, are about the fruit. They are evidence for object knowledge, but not strong enough evidence to override skeptical doubt regarding knowledge about objects.

Proposition (3.6) is trivial in my description of *O*, for I stipulated at the outset that *O* is the experience of the fruit looking orange. This triviality does not lessen the interest of the observation. Undergoing an experience of the *O* type—however that experience is to be described—gives Trich Cartesian certainty concerning (3.6). How can that be? This is related to the CK propositions (3.2)–(3.5). What does it say about the colors that we are entitled to certainty concerning certain facts about them? And with regard to (3.6): what is it about our knowledge of *orange* that makes us entitled to be certain that the fruit looks orange? Compare: "That fruit looks like a lemon." The latter demands that one knows not only what that fruit looks like, but also what a lemon looks like. Why is something like this not true of (3.6)? To be certain that something looks orange, one needs to be certain about relevant facts concerning the nature of the color *orange*. How does one come to be so? Since one cannot have Cartesian certainty of what *objects* look like even after viewing them, the source of one's certainty regarding the color *orange* must be different.

3.2 Secondary Quality Theories of Color

Some theories make it a part of the essence of color that it produce a certain kind of sensation—"secondary quality," or "response-dependence" theories, as they are called.

How would they account for the evidentiary status of color vision as outlined in the previous section?

3.2.1 Dispositionalism

According to dispositionalism, color is a disposition in things to evoke in "normal" perceivers a certain type of sensation in "normal" circumstances. According to this theory, something is orange because it possesses a disposition to create a certain type of color sensation, of which *O* is an example. Dispositionalists suppose that the Cartesian certainty of Trich's color knowledge arises from her incorrigible access to her own sensations, which have the properties in question—reddishness, yellowishness, and so forth. Facts concerning the compositionality of color as in (3.2)–(3.4) and similarity relations as in (3.5) ride on some quality of the sensation, according to the dispositionalist, and are thus knowable with certainty.

These theorists also have a partial account of Trich's merely empirical certainty with regard to object knowledge. When she first looks at the fruit, dispositionalists say, she is certain only of the sensation that the fruit evokes in herself in the particular circumstances of viewing. This is *some* evidence that the fruit possesses a disposition to evoke this sensation in normal circumstances, but not conclusive evidence. By examining the fruit closely in a variety of normal viewing conditions—by bringing it to the window, turning it over, and putting on her glasses—she expands the range of circumstances in which she has viewed the fruit. Thus, she increases her certainty about its dispositions with respect to her own sensations in normal circumstances of viewing. However, such certainty is subject to skeptical doubt—this kind of scrutiny does not rule out the possibility that she is systematically deluded about her circumstances.

At first sight, then, dispositionalists can account for the distinctive kinds of certainty possessed by the OK and CK propositions. There is, however, an obstacle on the dialectical path to object knowledge—Trich's empirical certainty regarding the color of something she sees. According to dispositionalists, the color of the fruit is not perceiver dependent; it is *in the fruit*. Its color is its disposition to create a certain kind of sensation not only in Trich, but in other normal perceivers as well. Looking at the fruit in all kinds of different circumstances certainly helps Trich know its propensities with regard to her own color experiences—but it does not help her know how it is disposed to other perceivers. Examining the fruit closely could not help her much with this; this procedure only gives her broader knowledge of the fruit's dispositions to evoke sensations in herself. (In fact, as we shall see in section 3.8, it is most likely *false* that things elicit exactly the same sensations in different perceivers.) To draw any conclusion about what sensations it creates in other perceivers requires other arguments—including, perhaps, an argument by analogy. No such argument is compelling enough to provide Trich with anything like empirical certainty.

So the dispositionalist cannot, in the end, account for the intuition I advanced in section 3.1.1 that Trich can arrive at empirical certainty about object knowledge.

A related problem arises for Cartesian certainty of knowledge concerning the apparent color of the fruit. Given her definition of *orange*, the dispositionalist glosses OAK (3.6) as follows:

(D3.6) That fruit looks as if it possesses a disposition to create an orange sensation in normal perceivers in normal circumstances.[1]

However, Trich's experience fails to give her empirical certainty about the experiences of others. Proposition (D3.6) falls short of empirical certainty for exactly this reason, and so it falls short of Cartesian certainty as well.

On the dispositionalist's account, object-appearance knowledge does not come out *more* certain than object knowledge—rather, it possesses exactly the same credibility. Moreover, both object knowledge and object-appearance knowledge regarding color fall short of empirical certainty. This is surely a problem.

3.2.2 Relationalism

The problem with dispositionalism is that it posits too great an evidential gap between sensation and color, as dispositionalists understand color. Relationalist accounts of color, such as those of Brian McLaughlin (2003) and Jonathan Cohen (2004, this volume) reduce this evidential gap to almost nothing. Consequently, they do better on these tests. But their position on the evidential relationship between object-appearance knowledge and object knowledge is quite implausible.

The same thing may look different colors in different conditions of viewing. When something looks white in white light and orange under a sodium bulb, is it really white or really orange? According to McLaughlin and Cohen, both options give undue precedence to one set of circumstances over the other. Relationalism is their way of unraveling this Gordian knot: on this position, *orange* is a many-place predicate that has as many parameters as the determinants of color experience. Thus, things are not orange as such, or white as such, but (simplifying somewhat) orange relative to an observer, given certain circumstances of viewing, and white relative to another observer, or relative to different circumstances of viewing. Thus, *orange* is a relation, not a simple unary predicate as it appears in ordinary language and thought. To say that a fruit is orange is syntactically incomplete in exactly the same way as it is syntactically incomplete to say that New York is in between. (In between what and what? Orange relative to whom and what circumstances?) When we say that something is orange *tout court*, we assume some way of filling in the extra parameters. For example, we might mean that it is orange *for* observer *O* (or some set of observers) *in* circumstances of viewing *C* (or some range of circumstances).

Relationalism runs into an immediate difficulty. For relationalists, the most specific way to take object knowledge (3.1) in the circumstances we have been considering is this:

(R3.1) That fruit is orange for Trich in the circumstances of viewing that obtained when she viewed it.

To determine the truth of *this* proposition, Trich needs nothing but her own experience. Her certainty should thus be Cartesian (too high by my reckoning). And of course, exactly the same is true of (3.6), which concerns the color *look* of the fruit. So the relationalist cannot account for the difference between (3.1) and (3.6).

McLaughlin and Cohen have another option, however, with respect to the fruit's looking orange. They *could* introduce normalcy in some such way as the following:

(R3.1′) That fruit is orange *to normal perceivers* in Trich's circumstances of viewing.

Or,

(R3.1*) That fruit is orange to normal perceivers in normal circumstances of viewing.

Or,

(R3.1!) That fruit is orange to Trich in normal circumstances.

And so on. By relativizing to normal perceivers, normal conditions of viewing, or objectively specified surrounds, rather than merely to Trich, relationalists can reduce the degree of certainty that attaches to object knowledge.

Now, (R3.1′) and (R3.1*) are not entitled to empirical certainty for reasons given earlier: examining the fruit more carefully does not give Trich certainty about how normal perceivers would experience it. (R3.1!), however, is more promising, because it corresponds quite closely to my account of object knowledge as far as empirical evidence goes, and Trich can adopt the same attitude to it as I earlier argued she ought to adopt with regard to OK (3.1). That is, when she examines it in a range of "good" circumstances of viewing, she acquires empirical certainty about it.

However, the corresponding reinterpretation of object-*appearance* knowledge (3.6)—the fruit looking orange—is this:

(R3.6!) That fruit *looks* orange to Trich in normal circumstances.

But this is simply equivalent to (R3.1!) on the relationalists' account, since "looks" is the same as "is" for them in this context. We've lowered certainty to the desired level for *being* orange, but at the cost of doing the same for *looking* orange. The close correspondence of being a color and looking a color makes it difficult for relationalists to differentiate the level of certainty that attaches to "looks orange" and "is orange."

Difficult, but not impossible. For Cohen (2007) suggests circumstances in which a relationalist wants to say that something looks but is not orange. Suppose that because of some strange quirk of Trich's brain, it so happens that the smell of an orange makes whatever happens to be in Trich's hand look orange. When drinking fresh-squeezed orange juice, an apple would look orange to her—and so, of course, would an orange. But a relationalist might want to hold that *this* orange appearance has the wrong pedigree to serve as a proper undergirding for being orange—even to her. In "deviant" circumstances such as these, neither the apple nor the orange *is* orange, even though both look that way.

Because she cannot rule out this kind of possibility, Trich cannot dismiss skeptical doubt regarding the fruit in her hand *being* orange—even when object knowledge is relativized to her alone, as in (R3.1). Thus, her certainty regarding the color of things falls short of Cartesian demands. Nevertheless, it might be that close examination leads her to empirical certainty regarding the fruit being orange to her. On the other hand, she never has reason to doubt that it *looks* orange—it looks that way even under deviant circumstances. By focusing on the difference between empirical doubt and skeptical doubt, the relationalist can thus arrive at the right balance of certainty between "looks" and "is."

This is a good result, but it is nonetheless doubtful that the relationalist can give a proper account of the relationship of looking orange to being orange. Let me put the matter in a way that brings out the difference between their motivations and mine. I have been urging that object knowledge and object-appearance knowledge are about two quite different things—"is orange" has to do with a property of an object, while "looks orange" has to do with a perceiver's visual state with regard to that same property and that same object. The second can normally be read off the perceiver's own state, but it is evidence for the first. This accounts for the difference in Trich's epistemic entitlements with respect to these claims.

Relationalists do not recognize such a distinction—for them, the difference between "is orange" and "looks orange" is a contextually defined difference about the same factual category. McLaughlin and Cohen follow the ancient Sophist Protagoras in matters of color. One could summarize them thus:

Color Protagoreanism In nondeviant circumstances, man is the measure of all colors: of what is orange that it is orange, of what is not orange that it is not orange.

Thus, they are unsympathetic to the very attempt to distinguish categorically object knowledge from object-appearance knowledge. Further, they do not think that something's looking orange is evidence, in the normal way, for its being orange. Bracketing the weird situations of which Cohen takes note, looking orange is for them constitutive of being orange, not evidence for it.

From my perspective, this is a weakness. Intuitively, attributing a color to an external object is to attribute to that object a characteristic it possesses independently of how it might look in a particular circumstance—a characteristic that influences how it looks but is nonetheless invariant with appearance. This is the intuition that relationalists simply throw away.

3.3 Primary Quality Theories of Color

According to primary quality theories, color experience is no part of the essence of color. An example of this kind of position is physicalism, according to which color is definable in the language of physics. I will not attempt a precise definition here, since it is somewhat tricky. Let us say, just so we know what kind of property we are talking about, that according to a primary quality account, color is a wavelength-related property. In the case of lights, it will be related to the relative strengths of wavelengths that objects emit; in the case of transparent or translucent filters, it will be related to the proportion of each wavelength that such filters transmit as a proportion of the light that falls on them; in the case of opaque surfaces, it will be related to the proportion of light of each wavelength reflected. I shall unite these under the term *wavelength productance*, or *productance* for short (cf. Byrne and Hilbert 2003). As we shall see, the term is a placeholder: it is not definable in advance.

Now, I am sympathetic to primary quality theories, and to physicalism in particular. In my view, the colors are indeed productances, or something of this sort. (A more complicated account is needed when we include properties that modify color, such as *luminescent*, *shiny*, *matte*, or *velveteen*—see Mausfeld, this volume—but here I will consider only the colors that are described in terms of hue, saturation, and lightness.)

Here is an argument in support of primary quality theories:

Color-vision is allied, in all organisms that have it, with cognitive mechanisms that lead animals to form expectations of finding other properties that have earlier been found associated with colors. For example, if they are fed from yellow dishes and given water from green ones, then they will come to anticipate that they will find food in yellow dishes and water in green ones.

Color vision serves these animals in such contexts only because such associations are indeed predictive.

But they could not be predictive if their essence lay merely in their relationship to experience, for experience varies with circumstances of viewing.

Therefore, color vision must engage experience-independent properties.

This argument is certainly not conclusive, but it is a broad motivation for a view of color that makes it a property of material objects independent of conditions of viewing.

However that might be, primary quality theories need elaboration if they are to accommodate adequately the evidentiary status of color experience. Consider in particular the color knowledge incorporated in CK (3.2)–(3.5) above. If no more is said about color than that it is physical wavelength productance, it is a mystery how we could have Cartesian certainty about it (cf. Johnston 1992).

Consider, for example, (3.5)—the proposition that the color of the orange is closer to that of a lemon than to that of a cherry. How can Trich be certain in virtue of her own experience that the productances are related by these relations of relative similarity? In fact, a statistically normal human trichromat's experiences are positively misleading with respect to physical similarity. For such a color perceiver, violet is more similar to red—since both are reddish—than either is to saturated green. Yet as far as wavelengths go, green is in between violet and red. A monochrome green light would for this reason be, from the physics point of view at any rate, more similar to monochrome red or violet lights of the same brightness than the latter are to each other. So, if colors were physical, apparent similarity would not be a reliable guide to the kind of similarity that is supposed to count for the primary quality theorist: it could be misleading that the color of an orange is more like that of a lemon than like that of a cherry, at least as far as its physical characteristics are concerned. This violates the Cartesian certainty attributed to color knowledge.

The problem is acute for missing shades. Suppose that *Y* is a missing shade of yellow that I visually imagine but have never seen. According to 3.1.3, I am entitled to Cartesian certainty about certain compositional and relational properties of *Y*. Yet, how can I possess any kind of certainty about primary qualities I have never encountered? And what about impossible shades? How can something be a primary quality if it *cannot* be instantiated in anything that exists in the real world?

There is another daunting (though, as we shall see, illusory) difficulty. In physics, there are wavelengths and various properties associated with wavelength. Psychologists often note, however, that aside from particular sorts of experience, it means nothing to say that a particular wavelength is orange, or blue, or green. These categories depend on how light is experienced and have no physical significance independently of experience. Thus, while one may agree that *light of 600 nanometers (nm)* is a primary quality, it is unclear how *orange* could be so. The essence of *orange* is essentially connected with the color experience of the orange type: if light of 570 nm elicited *this* kind of experience, then *it* would be orange. But it is certainly possible for light of 570 nm to elicit this experience in creatures only a little differently constructed than we.

Thus, it seems to follow:

(3.7) It is possible that *light of 600 nm* is not identical with *orange*.

However, by a well-established truth of modal metaphysics,

(3.8) (Possibly x is not identical with y) implies (x is not identical with y).

So,

(3.9) *Light of 600 nm* is not identical with *orange* (or any of its determinates).

This contradicts physicalism.

The questions for the primary quality theorist, then, are first, how knowledge concerning *light of 600 nm* can possess the Cartesian certainty of color-knowledge propositions (3.2)–(3.5), and second, how experientially identified properties such as *orange* could be primary qualities.

3.4 A Projective Semantics for Color Experience

3.4.1 Color-Experience and Denotation

In a semantic theory, color experiences *denote* colors. Just as the word 'cat' denotes the property that cats share, so Trich's *O* experience denotes, or represents, a property that (some) orange things share.[2] This gives color experience a role quite different from that envisaged by secondary quality theories. It is a *symbol* internal to the workings of the mind, a token by which the color-vision system passes to other epistemic faculties, and to the perceiver herself, the message that by means of the mind's own data processing, it has determined the color of this visual object to be orange.

The *O* experience is, I am suggesting, the semantic marker of this message—semantic not in language that the perceiver herself uses (such as English or Malayalam), but in the signaling system that is used by the perceiver's cognitive apparatus. The relationship between the *O* experience and *orange* is conventional in the following sense: the experience is a *marker*, or *label*, for that color; any other experience would have served as well for the same purpose. It is important to internal functioning that this type of experience consistently be used to mark this particular color, but it is a matter of historical and genetic happenstance that this particular type of experience, rather than any other kind, is used to pass messages about this particular color property. The system could have used some other kind of experience in place of the one it in fact uses: the pairing of the *O* experience type and *orange* is in this sense arbitrary. (See Matthen 2005, part 4, for a more detailed treatment of these points.)

3.4.2 Color-Experience and Sense

A semantic theory relies on a mind-world relationship fundamentally different from those employed by the other theories we have been considering. In some of these theories, perceivers have to know something outside the mind before they can come to know color: in dispositionalism, perceivers have to know the starting point of a causal relation that (in normal circumstances) vectors inward from world to color experience; in the case of primary quality theories, they have to know a physical quantity. Relationalism does not fall prey to this problem, but it doesn't countenance *any* mind-world relation in color experience: in this theory, for something to be colored is simply for it to be experienced this way.

Semantic theories, on the other hand, rely on an *outbound* relation between symbol and denotation. One grasps something of what a symbol signifies when one knows its sense, or meaning. This is what semantic theories require of color experience. But since it is possible to grasp the sense of a symbol without being able to identify its referent or denotation, the semantic theory does not require knowledge of anything outside the mind.

3.4.3 Projective Symbols

Some symbols refer (or denote) *projectively*. Consider an air-traffic controller working at a radar display. A particular aircraft on her screen is shown as a red dot. She refers to this aircraft as "the red plane." Of course, she does not mean or imply that the aircraft itself is red. She has no idea what color it is. She is using a property of the symbol to refer to the object that it denotes. I call this use of "red" *projective*.

Some projective symbols form a structured denotational system. Think of a map with intervals between equally spaced vertical grid lines marked A, B, C, and so forth, from left to right, and intervals between equally spaced horizontal lines marked 1, 2, 3, and so forth, from top to bottom. Let us stipulate that this is a projective system in which "Pemberton Street is in G3" means that the Pemberton Street (the external world object) is in the geographic region denoted by grid location G3 (and *not* that the map representation of the street is in that grid location on the map). Here, location is projectively denoted by the denotative system formed by the symbols on the map.

Certain facts about location are implied by the structure of this projective scheme. Consider:

LK (3.10) C3 is a greater distance away from A1 than from A3.

As stipulated in the preceding paragraph, LK (3.10) is about locations in the real space represented by the map in question—it is *location* knowledge, not *map* knowledge. Yet, we know that (3.10) is true, not from measuring the distance, but a priori as a result of knowing how the denotational system works (and Pythagoras' theorem). Proposition (3.10) is known a priori, but it is about the external world.

Notice that (3.10) is more certain than

OK (3.11) Pemberton Street is in G3.

OK (3.11) requires not only knowledge of the denotational system, but also empirical information about Pemberton Street. The difference between LK (3.10) and OK (3.11) is reminiscent of the difference between object-knowledge proposition OK (3.1) and color-knowledge propositions CK (3.2)–(3.5).

3.4.4 Projective Symbols and Color Experience

My thesis is this:

Color experiences constitute a structured projective denotational system.

Color experience is organized around three axes: bright-light, red-green, and blue-yellow. Every color experience is a combination of values of each axis. These experiences are arrayed around these axes as a similarity ordering: the more similar two experiences are, the closer together they are in this system. The qualities of these experiences are projected onto what they denote. To say that something is yellow is to say that it has the color denoted by the experience we recognize as of the yellow type.

In this view, some color knowledge is projected from our knowledge of what it is like to experience color, and from the projective topology of color experience. The *color knowledge* contained in (3.2)–(3.5) is accounted for (as I shall argue in section 3.8) by innate knowledge of the denotative space of color, as is our knowledge of missing shades. For instance, CK (3.2) assigns a somewhat determinate color—the color of the fruit—to an extensive region within Trich's color-similarity space—the yellowish region. Her knowledge is analogous to location knowledge of the following sort: "C3 is contained in the regions between C2 and C4," which is about location, but is known a priori as a consequence of knowing the denotational system of grid location. Trich's knowledge of CK (3.2) is implicit in her very perception of the fruit as orange. This perceptual state cannot be wrong about the relation stated by (3.2), though it might be mistaken with respect to the fruit. Propositions like (3.2) are *quasi-analytic*: all that is needed to grasp them is knowledge of the denotative scheme—and this, I propose, is innate. Despite their quasi-analyticity, such propositions are *substantive*—they are about relations among the colors (cf. Williamson 2007, especially chs. 3 and 4). (The view that I am advancing here is evidently kin to Kant's view that space and time are "forms" of sensible intuition, and that propositions regarding them are, as a result, known in a special way. My view is that analogously, color-knowledge propositions arise out of the form of color intuition. Of course, Kant disavows any taint of analyticity here.)

This accounts for *object color-appearance* knowledge too, as in OAK (3.6). The content of (3.6) is simply that Trich's color-visual state denotes a certain color. That it denotes this color is part of what the experience *means*.

Object knowledge as in (3.1) is explained by general facts about how we use our senses to probe the world. When we look at something in different and demanding ways, our senses converge on a determination that is immune from all but skeptical doubt—that is, from doubt that does not infect unrelated propositions about our own position or about the condition of our sensory apparatus. Our knowledge of object color depends on determining occurrent facts; this is why it is inherently subject to skeptical doubt.

Our knowledge of impossible shades lies a little further afield. In any compositional language, there is always the possibility of impossible compounds—"square circle" is an example. However, it was for a long time thought impossible that a sensory system could produce such compounds; it was thought to lack the conceptual freedom that discursive thought enjoys. For this reason, it was not only thought impossible that things in the world should be both reddish and greenish, but also that one could have such an experience. But it has been shown that under certain circumstances, the visual system can be induced to produce tokens of this sort.

3.5 Color Essences

I have been discussing truths that arise from the system of sensory denotation—call these *representation-relative truths*. Other propositions concerning the colors are necessary as a consequence of what these colors are—call these *color-essence truths*. Representation-relative truths are necessary because of how the colors are denoted. Color-essence truths are necessary not merely *under some description* but without qualification. For instance, it is necessary of the colors that they are related to the wavelength of electromagnetic radiation in certain ways. This is not necessity in virtue of how the colors are represented in sensation, but necessity in virtue of physical constitution—that is, because of what color is. In fact, experiencing the colors does not afford us access to these necessary truths.

This distinction is useful to primary quality theories of color. Let's once again take the wavelength-productance view as our paradigm of this approach. According to this theory, the color denoted by *O* is a productance related to *light of 600 nm* (*S*), say. Such a productance exists independently of us. *O* denotes *S*; *O* is our sensory way of identifying it—indeed *O* is our *only* way of identifying *S* prior to physics. But the physical character of *S* is not revealed by *O*.

Note that *O* need not have denoted *S*. *O* would not have stood in this relation to *S* if our visual systems had been constructed slightly differently. If they had been so constructed that light of 590 nm elicited *O*, then *O* would have denoted a productance related to *light of 590 nm* (*S′*). In that case, *O* would have been our way of identifying *S′*.

This shows what is right and what is wrong with the argument expressed at the end of section 3.3. On the one hand, we have knowledge of color that is dependent on the representational structure of color experience. This knowledge includes some quasi-analytic truths. We are entitled to Cartesian certainty regarding these. However, these truths are necessary only given the relationship between color experience and color. They are not necessary concerning colors taken in themselves. On the other hand, there are truths necessary of colors taken in themselves—but we are not entitled to certainty about these through our experiences. Indeed, we may not know them at all. That color is wavelength-productance is one such truth.

Consider, then, the following propositions about color experience and color properties. (In the hope that it will help the reader's intuitions, I'll place a linguistic parallel in parentheses.) Given that *O* denotes *S* (*domestic cat* denotes *felis sylvestris catus*) one may say,

(3.12) The color that *O* denotes is identical with *S* (*domestic cat* is identical with *felis sylvestris catus*). (Important note: Dropping the quotation marks around "domestic cat" and italicizing the words enables us to talk about the property denoted by the term. It is this property that is identical with the species, not—obviously—the term itself. Similarly, it is the color denoted by experience *O*, not *O* itself, that is being identified with the productance.)

From (3.12), it follows that

(3.13) The color that *O* denotes is *necessarily* identical with *S* (*domestic cat* is necessarily identical with *felis sylvestris catus*).

But (3.13) should not be confused with

(3.14) *O* necessarily denotes *S* ("domestic cat" necessarily denotes *felis sylvestris catus*).

Proposition (3.14) is false. If our color-vision system had been constructed differently, *O* would not have denoted *S* but *S*′. But this is not sufficient to contest (3.12) or (3.13). Though *O* might have denoted something else, it is nonetheless true of the color denoted by *O* that it is identical with *S*.

The proposition expressed by (3.13) describes the necessary properties of something independently of how we denote that thing. On the other hand, (3.14) describes the relationship between a meaningful symbol and something in the world. It implies nothing about this worldly thing beyond the relationship in which it stands to the symbol. Proposition (3.13) is about something beyond the relationship.

Primary quality theories of color are still in the running, given the semantic conception. The powerful-sounding argument at the end of section 3.3 is mistaken. That argument assumes that *orange* must be different from the productance associated with *light of 600 nm* because the former but not the latter is essentially connected with experience. This commits a kind of *de dicto/de re* confusion: that something is necessary *given* how the color-vision system denotes it is confused with necessity *tout court*.

3.6 Defining Color Similarity

What about the color categories defined by experience? These are founded on similarity relations among the color experiences. What guarantee do we have that these similarity relations correspond to a *real* similarity measure? We experience *red* as more similar to *violet* than to *green*, but what reason do we have for affirming any physical basis for such a similarity grading? I have introduced a role for color experience in grasping color properties. But shouldn't a semantic theorist eschew appeal to *experienced* similarity in favor of *real* similarity? Since experience is a symbol for color, it is a bit like saying that cats are similar to rats because the word "cat" is similar to the word "rat." The similarity of color experiences seems an unimportant and inconsequential basis for a real-world similarity relation that is meant to be cognitively important.

One position that one might take here is that nature has chosen to reveal the similarity of colors to us in color experience. Color experiences are experiences *of* color; color-similarity experience is experience *of* their similarity. Experiencing two colors as similar is not a matter of having similar experiences of them. In other words, color-similarity experience does not consist merely in transferring an incidental similarity relation from experiences over into the realm of colors. It is rather an experience that tracks similarity. This is a more meaty position than the one outlined in the preceding paragraph, namely that the similarity relation revealed by color experience is nothing more than a projection from a similarity relation on color experiences.

As it stands, however, this more meaty account is unsatisfying. The claim is that color experience tracks an independently valid similarity relation among physical colors. What similarity relation is that? We do not want to end up saying, *x* and *y* are experienced as similar whenever they stand in physical relation *P*; therefore, *P* is the relation we experience as color similarity. Such a theory does not advance beyond the account outlined at the start of this section. What is needed is some independently motivated explanation, something that could make us understand *what* real relation experiential similarity tracks.

Purely physical accounts do not suffice for this purpose. In section 3.3, we encountered one important reason for thinking that wavelength-based and other physical accounts are inadequate: color experience distorts the physical spectrum, making violets closer to red than to green. Another reason is the phenomenon known as *metamerism*: colors that are physically quite different from one another can look the same because they have the same effect on the color receptors in the retina.

I would like to propose an alternative account of the basis of sensory similarity relations: that these similarity relations are the substrate of conditioning. Let's suppose that feature *F* is associated with feature *G*. Given certain other conditions about the naturalness of *F* and *G*, an animal that has been exposed to this association will act in the presence of *G* as if *F* were presented. In other words, the subjective probability that the animal attaches to *F* rises when it senses *G*—its response is conditioned by the association of *G* with *F*. It is important to note that the conditioning function—the function that determines how incoming data modifies responses—is itself an unconditioned response. That the animal adopts a conditioned response to *F* when it finds *F* associated with G is itself an unconditioned, or innate, response to the association. The perception of F together with G results in a modification of the animal's *epistemic state*.

Now, what happens when G', which is slightly different from G, is presented? The answer is this: the rise in the subjective probability in F will depend on how similar G' is to G. The more similar it is, the greater the probability that the response conditioned on F will be triggered. In fact, this is how similarity space is measured in animals: G is similar to G' to the degree that conditioned responses to G pass over to G'. I shall take this as a definition of similarity—call it similarity for conditioning. Thus:

Similarity for Conditioning For any animal x, G is similar to G' to the degree that x's unconditioned responses to G would overlap, over a long series of trials, with its unconditioned responses to G'.

Here, I should emphasize again that forming a conditioned response—that is, the act of forming such a response—is itself an unconditioned response. The latter unconditioned response is *dynamic*: it is a change in epistemic state (i.e., a change in subjective probabilities) consequent to a perceptual state.

This conception of color-similarity is based on the unconditioned propensity to form conditioned associations. I am assuming that *experienced* similarity is innately set up to track similarity for conditioning (not physical similarity). This is not analytically true, or even a priori true. It *could* have happened that experienced similarity had contours quite different from those of similarity for conditioning. Nevertheless, it is empirically plausible that experienced similarity closely follows similarity for conditioning. In the first place, humans seem to form associations unconditionally based on conscious sensory experience. For example, if an experimenter delivers an air puff to your eye just after activating a flashing red LED, a subsequent activation of the LED will trigger an involuntary blink even if no air puff is forthcoming—here, your experience of the LED is connected by association with the air puff. Moreover, as I said earlier, this is what students of comparative vision assume when they measure the similarity relations of animals that cannot speak. The assumption could in principle be tested by figuring out whether the two measures of similarity diverge for humans.

The proposal that color similarity is similarity for conditioning may strike some as not very different from the weaker proposal mentioned at the beginning of this section, namely, that it is merely experienced similarity. Indeed, I have just reinforced this impression by suggesting that similarity for conditioning probably coincides with experienced similarity. Is experienced similarity merely subjective, then, in the way that experience itself is regarded as merely subjective?

I do not think that this is an objection. My proposal *explicates* experienced similarity. To say merely that two colors are similar to one another when they are so experienced gives us no idea what the content of the experience is. As such, it leaves it open that color similarity is an isolated phenomenon that has nothing at all to do with any extra-experiential realm. The proposal that experienced similarity is similarity for conditioning does better than this. It makes clear what the stakes are in experiential similarity. Two things are experienced as similar to the degree that certain learned expectations regarding the one spread to the other. So experiential similarity represents a certain color-associated clustering of properties in external things. To experience two things as similar is to experience them as sharing other properties as well. Similarity for conditioning is a dynamic interpretation of similarity: it links similarity of color to certain processes within the organism.

3.7 Pluralistic Realism

One important corollary of identifying color-similarity relations with similarity for conditioning is that it opens up the possibility that these relations could vary from species to species. Animals of different species live in different environments and undertake different activities in order to live in these environments. They also vary in how they receive color information: they have different numbers of color-sensitive receptors, which may be tuned differently; some have filters; in invertebrates, the eye itself is differently structured than in vertebrates. These dissimilarities make for significant differences in the domain and contour of color-similarity relations among animals (Matthen 1999). Not only do they receive different information from the environment (because their receptors are different), but they also process this information differently to best serve the data-processing needs of their styles of living. Further, their unconditioned responses to sensory data vary—in terms of both innate behavioral response and how conditioning works.

The variability of color across species has an important consequence. One cannot say a priori what color is—it is different things to different kinds of animal. Of course, it involves discrimination based on the wavelength of light. But beyond this are important differences. For example, information about the polarization of light is visible to birds and bees "in color"; it is more or less invisible to humans. How do secondary quality theories handle this? According to these theories, color is the property of normally caused specific sensations. Which sensations? There is no guarantee that polarization sensations in birds are the same as those (if any) in bees, nor that these sensations correspond to those by which humans perceive color. So this approach to defining color runs into difficulties with identifying bird color. A properly comparative approach to color vision should not face such difficulties.

Similarity for conditioning defines color similarity from the organism's point of view. Let's call it *primary similarity*. It is possible to investigate the physical correlates of primary similarity. For instance, psychophysicists are interested also in discovering the physical similarities that hold between objects that are united by primary similarity. In what physical respect do two things have to resemble each other when they elicit the same conditioning responses? In what physical respect must they differ (despite other similarities) when they elicit different conditioning responses? The answers to these questions will determine the physical *denotation* of the internal representation of color. I'll call this *secondary* color similarity (cf. Matthen 2005, 232–234). By the psychophysical investigation of secondary similarity, we arrive at a specification of productance.

The relationship between primary and secondary similarity is somewhat analogous to that between Frege's sense and denotation: primary similarity describes color similarity from the perspective of the organism; secondary content identifies the real-world realization of the organism's similarity relations. The properties and relations denoted by color vision exist independently of the organism—they are objective, and in this way, the account I have offered is a primary quality account. On the other hand, our grasp of these primary qualities is through similarity relations realized by our mental dispositions. In fact, this is our *primary* grasp of such qualities, even though we could in principle come to know of these qualities in a physics class, it is nonetheless true that physicists identify colors by means of human experiences. The problem raised at the end of section 3.3 is solved by noting that *orange* is simply the physical variable that a particular type of color sensation denotes. That sensations are in this way epistemically indispensible for identifying the colors accounts for why color has been thought to be response dependent, or a secondary quality.

My position is *realistic*. But on my position, primary color similarity varies across species, as does secondary similarity. In virtue of this variation, the position is pluralistic—it allows that color is realized in different ways in different organisms. The position can be entitled *pluralistic realism* (cf. Matthen 1999, 2005).

3.8 Similarity Spaces and Color Knowledge

A *similarity space* is a graphical representation of comparative similarity relations on some domain. In the case of color, the points in this space are colors—or productances. The distance between two points in this space is inversely proportional to the degree of similarity between them. Sometimes, a similarity space is represented as a geometric solid. The advantage of this representation is that certain lines or axes of such a solid are thought to isolate some fundamental parameter of variation in color. Color has often been represented as a double cone (cf. Churchland, Kuehni, this volume), with the polar axis representing lightness and darkness, radial distance from this axis standing for saturation, and points on the equator standing for saturated hues. Most color scientists take this to be at best an approximation (or perhaps worse: Mausfeld, MacLeod, Niederée, this volume). Nothing in my presentation depends on a strong interpretation of the color solid.

My thesis is first of all that color vision assigns each object to a place in this space. Each such place corresponds to an experience, which denotes a productance (or class of productances). Each such place also links to conditioning: when a place in the color-similarity space is found coinstantiated with some other sensory feature, a conditioning link is established; any occurrence of either will increase the subjective probability of the other.

The propositions that we know by color experience have to do with relations among colors in this similarity space. For example, consider again,

CK (3.2) The color presented in *O* is yellowish.

This tells us that the productance in question is contained in that part of color-similarity space that is denoted by experiences of the yellowish kind. The containment here is revealed in a special way:

Color *X contains Y* if *Y* is one pole of a hue dimension (*blue* or *yellow*; *red* or *green*) or *Y* is *light* or *dark*, and *X* is located in the part of color-similarity space denoted by *Y*.

This use of "contains" is no doubt encouraged and influenced by some quality of color experience—for example, the experience of *turquoise* has discernible elements of bluishness and greenishness. The important thing to understand here is that we are not compelled to treat containment as either a *primary* relationship that exists independently of color experience or a *secondary* one that simply transfers to colors from experience as the result of some cognitive illusion. Containment can be true of colors in virtue of a cognitively important relation between experience and reality.

Now to the evidentiary status of color experience. I suggested at the outset that color-vision maps distal objects into an *innate* similarity space. Famously, Noam Chomsky proposed that the deep syntactic structure of language is innate. A few years after human infants have been exposed to some language, they have a nearly comprehensive grasp of the grammatical principles of this language. They understand sentences they have never heard before; they understand how to transform sentences in interrogative form into sentences in indicative form; they know when a verb they have never heard before agrees or fails to agree with its subject, and so on.

The view that I am proposing is that, analogously, we possess innate knowledge of the similarity space of color. This is implied by the idea that similarity for conditioning is unconditioned, and hence innate. The innateness of color space accounts for our knowledge of the quasi-analytic propositions (3.4)–(3.7). Such knowledge emerges in development, and a certain amount of experience of the right kind is needed for that emergence. Nevertheless, the relations of relevance that exist between the kind of experience needed for the emergence of innate knowledge and that knowledge are not sufficient for calling the former *evidence* for the latter—the stimulus is insufficient to the kind of knowledge that emerges from it. The requisite color experience is merely a causal condition for the expression of something "in the genes."

This view accounts particularly well for our knowledge of missing and impossible shades. Consider Trich's knowledge of a missing or impossible shade, or of similarity relations such as that mentioned in (3.7). She knows these by knowing the similarity space of color and its component structure. Knowing how these components combine makes it possible for her to combine them in a way that produces images of colors she has never before experienced; decomposing them allows her to know the structure of impossible shades though they can never be instantiated in the world. Since a perceptual grasp of the similarity space of color is a precondition of her sensing the color of the fruit in the first place, it is not an additional burden for the account.

Finally, on this point, consider the variability of color perception from individual to individual within a single species. It commonly occurs that one person sees a particular color chip *B* as saturated blue with no admixture of red or green (i.e., as "uniquely blue"), while another sees it as a somewhat greenish blue. (This, by the way, is one reason it is wrong to suppose that things elicit from all other perceivers the same sensations as they do from her.) Such a difference is often accompanied by agreement with respect to color matching—the two persons may mostly agree when asked whether two chips are of the *same* color, and this may be so across the whole range of colors. Asked whether *B* is the same or different from other chips, they mostly agree—though they continue to disagree about whether *B* is uniquely blue. The question is, what is the content of this perceptual difference? What are two people disagreeing about when one sees *B* as uniquely blue, and the other sees it as possessing an admixture of green?

On the secondary quality view, the disagreement is about the external world. In one perceiver, *B* is disposed to create a sensation of unique blue, and hence has the color disposition associated with this sensation; relative to the other, it is differently disposed. Thus, it appears a different color to these different perceivers. Given that such differences are possible, neither can possess certainty about whether anything produces a particular sensation in perceivers other than themselves. Thus, neither perceiver can be certain of the color of the fruit. And this uncertainty does not go away with careful examination. Examining a fruit closely, turning it over, taking it over to the window, putting on one's glasses—none of these strategies will elevate one's level of confidence concerning the fruit's color, at least not as far as what sensation it evokes in other perceivers. Similarly, on a primary quality account, there is real disagreement: according to one perceiver, a thing of *B*'s productance is unique blue; according to the other, it is not (cf. Tye 2006; Byrne and Hilbert 2007). Yet, it is hard to say what exactly the disagreement is about. What is it for a productance to be unique blue in itself, that is, independently of anybody's perception of it?

With a semantic theory, the problem goes away. In one perceiver, the color of the fruit is denoted by the experience that occupies one position in color-similarity space; in the other, it is denoted by the experience that occupies another. This is not a difference in what *color*, what productance, is denoted. Rather, it is a difference in how it is denoted—by what kind of sensation. Though I have been talking about interspecies differences, this kind of variation can occur even among individuals of the same species. Certain transformations of internal representational schemes make no difference to the information that is contained in them, and thus they do not imply a difference of the information carried by experiences. For unknown neurological reasons, such insignificant transformations occur quite frequently. Thus, it sometimes happens that one representational scheme has a blue-yellow hue dimension that is slightly rotated relative to another. Consequently, an object that falls directly on this axis in one lies slightly off it in the other. In the on-axis case, the chip is seen as uniquely blue. When the chip falls off-axis, it is seen as not uniquely blue. Though the experiences are different, they denote the same productance.

3.9 Conclusion

The view that I have been propounding is that the relationship between color experience and color is semantic. Such a view is quite natural if one takes perceptual states to be either true or false. Since propositions are either true or false, the view that perceptual states have intentional content implies that they are directed in some way toward propositions. Now, propositions contain *properties* as constituents. Hence, it is natural to suppose that perceptual states are directed in some way toward properties. This is the consequence of the view that I have been developing: color perceptual states denote color properties. Since the view that perceptual states are either true or false is pretty widely held, it is somewhat surprising that the semantic view of color experience has not been more thoroughly explored.

Acknowledgments

Many thanks to Stephen Biggs, Jonathan Cohen, and Reinhard Niederée for detailed comments, which influenced the content of this paper.

Notes

1. Proposition (D3.6) is the dispositionalist's gloss on (3.6). I'll use this kind of notation in what follows.

2. The *O* experience is more determinate than *orange*—this is why not all orange things evoke this precise kind of experience.

References

Byrne, A., and D. R. Hilbert. 2003. Color realism and color science. *Behavioral and Brain Sciences* 26: 3–21.

Byrne, A., and D. R. Hilbert. 2007. Truest blue. *Analysis* 67 (293): 87–92.

Cohen, J. 2004. Color properties and color ascriptions: A relationalist manifesto. *Philosophical Review* 113 (4): 451–504.

Cohen, J. 2007. The relationalist's guide to errors about color perception. *Noûs* 41: 335–353.

Crane, H. D., and T. P. Piantanida. 1983. On seeing reddish green and yellowish blue. *Science*, New Series 221 (4615): 1078–1080.

Hellie, B. 2005. Noise and perceptual indiscriminability. *Mind: A Quarterly Review of Philosophy* 114 (455): 481–508.

Johnston, M. 1992. How to speak of the colors. *Philosophical Studies: An International Journal for Philosophy in the Analytic Tradition* 68 (3): 221–263.

Matthen, M. 1999. The disunity of color. *Philosophical Review* 108: 47–84.

Matthen, M. 2005. *Seeing, Doing, and Knowing: A Philosophical Theory of Sense-Perception*. Oxford: Clarendon Press.

McLaughlin, B. P. 2003. Colour, consciousness, and colour consciousness. In *Consciousness: New Philosophical Perspectives*, Q. Smith, A. Jokic, eds. Oxford: Oxford University Press.

Tye, M. 2006. The puzzle of true blue. *Analysis* 66 (291): 173–178.

Williamson, T. 2007. *The Philosophy of Philosophy*. Oxford: Blackwell.

4 More than Three Dimensions: What Continuity Considerations Can Tell Us about Perceived Color

Reinhard Niederée

4.1 Introduction

This chapter presents a novel approach to the issue of dimensionality in color vision that, in my view, is apt to fundamentally challenge certain still widely held mainstream assumptions concerning the three-dimensionality of color space in normal human color vision. In referring to a space of colors, we have to take into account right from the start what could be termed a *Janus-facedness* of the concept of color: It may be understood as referring to some property of real objects "out there" (external aspect), or to the color appearances evoked by such objects in a human perceiver, that is, to *perceived color* or, as I prefer to call it, *phenomenal color* (internal aspect). These two aspects may, in fact, be viewed as potentially complementing each other. Needless to say, the conceptual status of both is highly controversial. This chapter does not address issues pertaining to the external aspect at all, so I will not be concerned here, in particular, with corresponding philosophical controversies such as that between color realists and antirealists. Instead, this chapter deals with the psychophysics of color, that is, with phenomenal color and its relation to the proximal stimuli—physical light patterns impinging on the retina—by which they are typically evoked.[1] Since the concept of phenomenal color, too, has been under attack in part of the literature, it may be worth noting that the core dimensionality arguments presented below refer to rather basic relations among phenomenal colors such as identity and diversity and to corresponding perceptual judgments (match vs. mismatch), so they even allow a behaviorist reading. Nevertheless, I will freely refer to "phenomenal colors" here as "colors." Note that this is merely a terminological simplification for the purposes of the present analysis, a simplification that is unproblematic since I am concerned here only with the internal aspect of the concept of color.

From a psychophysical point of view, colors (i.e., phenomenal colors) are commonly envisaged as points in a three-dimensional color space, to be coordinatized, say, in terms of hue, saturation, and brightness. Accordingly, achromatic colors are assumed to vary in the latter dimension only; they are assumed to constitute a one-dimensional subspace, depicted geometrically as the *achromatic axis*. This is a pretty well-established concept for special cases such as the colors evoked by isolated patches of light (trichromacy). Often, however, it is assumed that this account naturally generalizes to colors emerging in more complex situations. Simplifying matters, this general position will for now be termed the *mainstream view*. Such a general three-dimensional conception of color—and a corresponding one-dimensional conception of achromatic color—has been subjected to criticism by a number of researchers throughout the past century, such as Katz (1911/1935), Bühler (1922), Gelb (1929), Evans (1974), and Heggelund (e.g., 1974), to name only some of them. A review of the controversial dimensionality discussion is beyond the scope of this discussion, however.

Instead, this chapter is centered on a rigorous and simple mathematical argument that, while being in tune with trichromacy, in my view clearly militates against the mainstream view. It makes use of a novel method: an analysis of *continuous paths* in a space of colors that draws on considerations of how evoked colors vary as the underlying stimulus parameters are varied gradually. It is based on rather weak assumptions that are supported by well-established empirical observations concerning color vision in center-surround displays (which in what follows means a light stimulus composed of a central spectrally homogeneous disc in an annular spectrally homogeneous surround, which typically is embedded in a dark outer surround). This approach implies that at least four "dimensions" are required to fully characterize the colors evoked by such stimuli. To state this result rigorously requires clearly specifying what is meant here by a "dimension," to avoid the conceptual confusions so often found in discussions of "the dimensionality of (the) color space." In fact, there is no such thing as *the* color space, let alone *the* dimensionality of it (cf. section 4.5.1 below), so a very careful use of these terms is required.

In the psychological color-vision literature, in general, and in debates about dimensionality of color, in particular, one may, very crudely, distinguish between a *sensation*-oriented and a *perception*-oriented perspective. In many discussions, however, both perspectives tend to be merged in one way or another. By a sensation-oriented theoretical perspective, which traditionally prevails in the classical psychophysics of color, I simply mean an approach that starts out with the concept of a (phenomenal) *field of colors*. This *phenomenal output* (percept) is considered to be evoked by suitable patterns of retinal stimulation (physical input). Needless to say, these two levels have to be carefully distinguished in what follows. The stimuli studied range from single patches of light in dark surround, via center-surround configurations, to more complex patterns. As mentioned before, for small homogeneous patches of light and the colors they evoke, there is a well-established concept of three-dimensionality in so-called normal observers, based on color-matching tasks, namely, *trichromacy*. Three-dimensionality would readily generalize to more complex stimulus configurations if in those situations there simply were a local input-to-output (wavelength-to-color) relation. However, there is a plethora of well-known spatial and temporal context effects (labeled as color contrast, assimilation, adaptation, and the like), which often even lead to "new" colors not observed in isolated patches, such as (most shades of) brown or gray. These effects are already found in center-surround stimuli, where for instance the color pertaining to the central disc at the output level is influenced by the physical composition of both the light pertaining to the central disc and that pertaining to the surround at the input level. Therefore, on a priori grounds, it is not clear whether the idea of a three-dimensional color space, as established by trichromacy, meaningfully generalizes to more complex configurations. Proponents of a mainstream account assume it does, but various students of color vision have opposed this view. Evans (1974), for instance, advanced the thesis that an appropriate description of the color appearance pertaining to the central disc of center-surround stimuli should require two times three (i.e., six), and in the case of double surrounds (perhaps) even nine, psychophysical quantities. Regarding the mainstream account, he noted that "only a persistent desire to keep the system three-dimensional (so it can be visualized?) can explain the circumlocutions that have been resorted to, to make it so appear" (Evans 1974, 136f).

Sometimes the issue is trivialized by simply *defining* color in terms of the three perceptual attributes of hue, saturation, and brightness (assuming, erroneously I believe, that these attributes are unequivocally defined, or at least consistently definable, also for the colors evoked by more complex stimulus configurations such as center-surround displays). In contrast, whenever color is addressed as part of a field of colors, I adopt here the less restrictive definition of color in Wyszecki and Stiles (1982). They introduce color as "that aspect of visual perception by which an observer may distinguish differences between two structure-free fields of view of the same size and shape, such as may be caused by differences in the radiant energy concerned in the observations" (487). Note that this concept of color—or color "sensation"—refers to local phenomenal colors in a field of colors, which may, however, causally depend on more than one location in the proximal *stimulus*. So it should not be confused with concepts of sensation like Helmholtz's notion of *Empfindung*, which he often used with regard to local stimulation.

A *perception*-oriented approach to color vision, in contrast, conceives of the percept as some kind of (phenomenal) 3-D spatial arrangement of (phenomenal) "colored objects," the relevant notion of color here being that of perceived object or surface color. Again, one is interested in how such a richly structured visual percept depends on the proximal stimulus. Typically, the latter is considered, in turn, as being causally determined by some physical "outer scene" composed of surfaces and objects under some illumination, key physical features being reflectance distributions of surfaces, energy distributions characterizing lights, and their interaction. In standard accounts of perception, the phenomenal "inner scene," then, is considered as a more or less successful *representation* of certain aspects of this outer scene, emerging as the output of some complex visual processing starting out from the proximal stimulus (input). Of special interest in this connection is the phenomenon of approximate *color constancy*, the often encountered (more or less perfect) stability of perceived object or surface colors despite changes of physical illumination and environment. In the perception-oriented context, an additional type of dimensionality conception comes into play, namely, three-dimensional "spaces" of surface color (color solids). This idea of a *trivariance* of perceived surface color is in fact compatible with my position in this chapter, though it needs to be carefully qualified (for instance, by additionally taking into account perceived illumination features).

The dimensionality arguments that follow are formulated in a sensation-oriented setting. From a modern perception-oriented perspective, this might at first sight look like somewhat old-fashioned psychophysics. However, for one thing, a sensation-oriented perspective seems well suited for the stimuli considered, namely, simple center-surround stimuli. For another, in view of the complexities of the phenomenology of color, a sensation-oriented approach and a perception-oriented approach in my opinion should be considered as complementary rather than mutually exclusive. Last but not least, I believe these results and methods bear on perception-oriented issues as well. Some of these issues will be discussed in the concluding discussion section.

So, as a start, let us adopt a sensation-oriented perspective and focus attention on center-surround stimuli: a central disc in an annular surround, made up of spatially homogeneous lights a and b, presented in a completely dark outer surround (compare figure 4.1).

Here I am interested in the evoked *color pertaining to the central disc* (for short, the *focal color*), which is known to depend on both *a and b*. Two such stimuli will be said to *match* if the focal colors are the same, that is, if the two central regions are perceptually *indistinguishable*. If this holds for two such stimuli with different annular surrounds, one speaks of *asymmetric color matching*. Note that the criterion of being perceptually indistinguishable implies here that only *proper* matches will be called a color match, as opposed to many empirical asymmetric color-matching studies where approximate matches (i.e., the subjectively best approximation subjects could achieve in a given setting) tend to be reported as "matches."

In what follows the focus will be on N-dimensional color codes $h(a, b) = (x_1, \ldots, x_N)$, where $x_1, \ldots, x_N$ denote real numbers. These are to code the focal color in such a way that if two stimuli (a, b) and (a', b') match, then the associated codes $h(a, b) = (x_1, \ldots, x_N)$ and $h(a', b') = (x'_1, \ldots, x'_N)$ are identical and vice versa. Note that this is a very flexible concept. The components of these vectors may themselves denote perceptual attributes (such as hue, saturation, brightness) or (univariant) neural codes at some level. But they may equally well denote some abstract psychophysical coordinatization of focal colors in center-surround stimuli. For them to be meaningful, however,

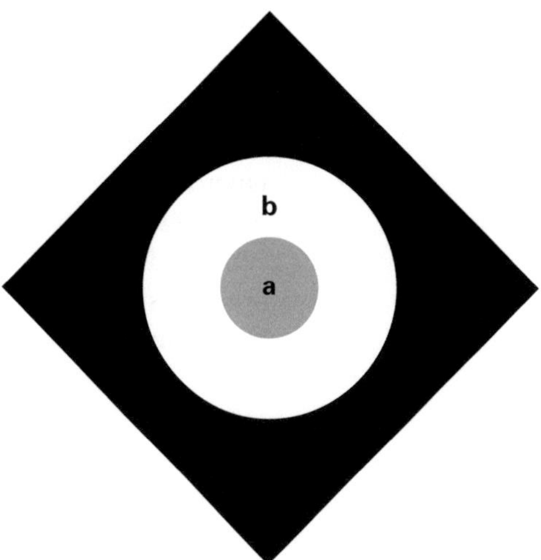

Figure 4.1
Basic layout of a center-surround stimulus in dark outer surround. The latter is supposed to extend over the entire outer visual field.

some additional restriction must be imposed on these codes. In my analysis, this is a *continuity requirement*: I assume that if the stimuli are varied gradually, then the associated codes vary gradually, too.[2] This seems a natural requirement from a psychophysical viewpoint, and virtually all models of asymmetric color matching found in the literature satisfy this condition explicitly or implicitly. The central question then is, what is a minimal number N for which such a continuous color code exists? I call this the *global dimensionality* of the color space considered (i.e., in the case at issue, the space of focal colors evoked by center-surround stimuli). Most models are inspired by the mainstream view and assume that the answer is $N = 3$. In contrast, I will discuss elementary empirical observations that prove to be mathematically inconsistent with this mainstream assumption, implying instead that $N \geq 4$. This dimensionality result does not imply that color vision is based on four basic classes of retinal receptor types rather than three. Instead, higher dimensionality of the kind considered here obviously is the consequence of context effects (i.e., of how the surrounding light b influences the evoked focal color), and these simply turn out to be more complex than commonly assumed.

The result just mentioned will be captured by two *dimensionality theorems*, one being based on stronger empirical assumptions—which might be doubted by most proponents of the mainstream view—and the other being based on much weaker empirical assumptions concerning pretty obvious mismatches. The argument underlying these theorems is based on the analysis of continuous paths in the respective color spaces considered. Similar arguments could in principle be applied to various situations (actually, *mutatis mutandis*, also outside the domain of color vision). One could, for instance, restrict attention to achromatic situations, where analogously $N \geq 2$ is implied for center-surround stimuli. Or, considering the general case again, one might wish to sustain the hypothesis that $N \geq 5$, and so on. So considerations of that kind might perhaps turn out eventually to lend support to Evans's aforementioned dimensionality hypotheses that $N = 6$ and $N = 9$ for double surround stimuli (and so forth?). But this is by no means automatically implied by the analysis in this chapter, for no such result (including our claim that $N \geq 4$) can be derived in an a priori fashion. Rather, we need to plug in sufficiently strong empirical observations.

For the sake of conceptual clarity, in this chapter I keep apart as clearly as possible (a) conceptual issues such as a formal statement of the relevant mathematical notion of dimensionality (section 4.2), (b) a clear description of the empirical observations needed for our arguments (section 4.3), and (c) the announced dimensionality theorems based on these observations, along with their proofs (section 4.4).[3] The mathematical proofs will be presented in a somewhat simplified manner that can be easily visualized; no advanced mathematical knowledge will be presupposed. These three sections form the core of the chapter, which then is supplemented by a discussion section (section 4.5), which allows us to view these results from a somewhat broader perspective by relating them to other concepts of dimensionality and to a perception-based approach to color vision. I am aware that the price to pay for the stepwise mode of presentation chosen here is that it may perhaps impede a more holistic synopsis of how all of this is interconnected; this might require a second reading of parts of the chapter.

4.2 Conceptual and Notational Preliminaries

In what follows, the variables a, b, c (possibly with sub- or superscripts) are used to denote spectrally homogeneous lights impinging on an area of the retina. A light a is characterized by its (radiant or irradiant) spectral energy distribution, understood (for simplicity) as a function a_λ of the wavelength λ, which may vary over the visual spectrum. The zero spectral energy distribution (no light) is denoted by 0. Two basic operations on lights are *additive mixture* $\oplus$ and *scalar multiplication* $*$. Here $a \oplus b$ denotes the physical superposition of a and b (its spectral energy distribution being $a_\lambda + b_\lambda$ across the visual spectrum), while for nonnegative real numbers t, the term $t * a$ denotes a constant physical intensity change of light a by factor t (i.e., its spectral energy distribution is $t \cdot a_\lambda$). Two lights a and b are called *metameric* ($a \sim b$) if they evoke identical color percepts under certain standardized viewing conditions. Typically, this means a foveal presentation in the form of comparatively small patches—say, 2° of visual angle—in an otherwise dark surround (no light), and suitable preadaptation (e.g., dark adaptation). I call such stimuli *isolated patches* (IPs), and the colors evoked by them *α-colors* (for reasons that will become obvious below). In corresponding color-matching experiments, both lights are usually simultaneously displayed in a single bipartite field, but for the present purposes, a discussion of experimental details is unnecessary. Note that while $\oplus$ and $*$ are purely physical concepts, metamerism is a psychophysical concept insofar as it refers to the evoked colors at the phenomenal level (or at least to perceptual judgments).

4.2.1 Trichromacy

A well-established concept of three-dimensionality in color vision is *trichromacy*. Trichromacy essentially means that any light a (in a dark surround) can be matched by a mixture of three (suitably chosen) fixed *primary lights* b_1, b_2, b_3, while fewer than three primary lights would not suffice. This is understood in the sense that, given the three primary lights, for each light a there are real numbers t_1, t_2, t_3, such that

$$a \sim t_1 * b_1 \oplus t_2 * b_2 \oplus t_3 * b_3.$$

As is well known, whatever set of primary lights is chosen, there will always be some lights a where one or two factors t_i on the righthand side will become negative, meaning that the corresponding components have to be transposed to the lefthand side to obtain a match (e.g., $a \sim 2 * b_1 \oplus (-3) * b_2 \oplus 4 * b_3$ stands for $a \oplus 3 * b_2 \sim 2 * b_1 \oplus 4 * b_3$).

In conjunction with some elementary laws—the so-called Grassmann laws[4]—the criterion of trichromacy implies the existence of some *linear code* ϕ, which assigns to each light a a triple $(\phi_1(a), \phi_2(a), \phi_3(a))$ of real numbers in such a way that for all lights a, b,

$$a \sim b \text{ if and only if (iff) } \phi(a) = \phi(b) \tag{4.1}$$

(i.e., iff $\phi_i(a) = \phi_i(b)$ for $i = 1, 2, 3$). Linearity here means that $\phi_i(a \oplus b) = \phi(a) + \phi(b)$ and $\phi(t * a) = t \cdot \phi(a)$. Condition (4.1) simply means that (ideally) each individual code $(\phi_1(a), \phi_2(a), \phi_3(a))$ is associated with exactly one α-color (for each observer and for the respective single-patch situation considered; note in passing that the term "code" is used here to denote both a vector-valued function, such as ϕ, and individual values $\phi(a)$). A linear code with this property will be called a *primary color code*, and given another such code, some linear mapping (i.e., technically speaking, a nonsingular 3 × 3 matrix) always carries one into the other. In this way, given a fixed primary color code ϕ, the set of lights is mapped onto a three-dimensional convex cone C in the vector space Re^3, which simply is the set of all vectors $\phi(a)$. This account goes back to Grassmann (1853) and was worked out and generalized in a precise measurement-theoretical setting by Krantz (1975; compare also Suppes et al. 1989). Note that this concept of dimensionality is related to the algebraic notion of dimensionality pertaining to vector spaces. All this is well understood. It is the basis of CIE's (Commission Internationale de l'Éclairage's) widely accepted colorimetric *tristimulus coordinates*, for example, the X, Y, Z and the R, G, B system (cf. Wyszecki and Stiles 1982).

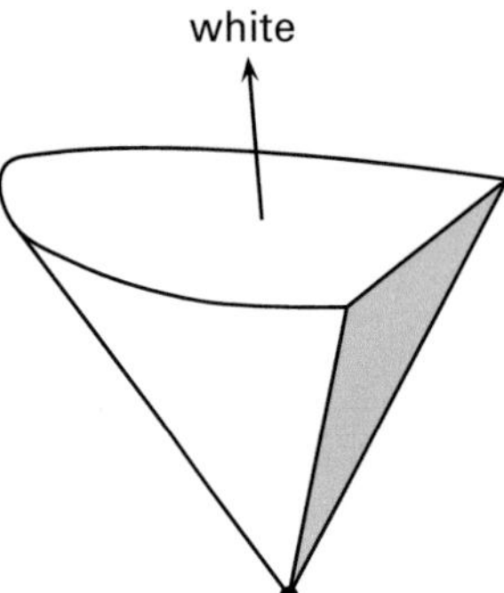

Figure 4.2
A schematic illustration of the (truncated) convex cone C of primary color codes.

At the physiological level, trichromacy is usually explained by the Young-Helmholtz hypothesis, which assumes that color vision relies on the output of three types of primary retinal receptors (plus some neural processing based on these receptor outputs). In this view, certain primary codes (e.g., the Smith-Pokorny fundamentals; see Wyszecki and Stiles 1982) have the distinctive property of describing the input-output behavior of these receptors (approximately, and up to a monotonic transformation; but see, e.g., MacLeod 1986 and this volume).

The concept of trichromacy not only remains unchallenged by the following arguments, but it provides a convenient basis for my analysis. So, in what follows, ϕ always denotes a fixed primary color code, and C the convex cone generated by ϕ (see figure 4.2). The arguments do not depend on the code ϕ chosen.

4.2.2 Extended Trichromacy

The focus of this discussion is on center-surround stimuli: a focal disc made up by some spectrally homogeneous patch of light a, but now surrounded by an annulus filled with light b in an otherwise dark surround (which includes the possibility of a and/or $b = 0$ as limiting cases). For simplicity, both the central disc and the annulus are assumed to cover only a few degrees of visual angle. Such a stimulus will be denoted by (a, b), and given another such stimulus (a', b'), the notation $(a, b) \sim (a', b')$ will stand for a match, that is, a *perfect focal match* in the sense that the focal colors pertaining to the central discs are the same (perceptually indistinguishable). The concept of trichromacy applies here to such stimuli in two ways, based on two assumptions, both generally assumed to hold empirically (within certain limits, that is, as is the case for the standard IP-based concept of trichromacy, too). In fact, they usually would not even be considered separate assumptions. First, it is assumed here that trichromacy applies to stimuli of the form (a, a) just as well, which simply are larger homogeneous patches obtained by taking the same light for the center and the surround. These stimuli will be termed *large isolated patches* (LIPs). More precisely, I assume that

$$(a, a) \sim (b, b) \text{ iff } a \sim b.$$

In terms of primary codes, this condition then simply reads

$$(a, a) \sim (b, b) \text{ iff } \phi(a) = \phi(b) \tag{4.2}$$

To simplify this discussion, I additionally adopt the common assumption that the colors evoked by LIPs are the same colors as those evoked by IPs, even though this is not strictly needed for the present arguments.[5] So I assume that the space of colors generated by LIPs is just the space of α-colors introduced above.

The second generalization of the original concept of trichromacy concerns stimuli that consist of multiple patches of light, giving rise to a percept composed of a field of colors. The concept of trichromacy as introduced above does not directly apply to such situations. This limitation can be overcome if one assumes that replacing each of the lights in the (proximal) stimulus by a metameric one leaves the percept invariant. For center-surround stimuli this means that if

$a \sim a'$ and $b \sim b'$

then

$$(a, b) \sim (a', b').$$

In terms of primary color codes, the if clause just says that the vectors $\phi(a)$ and $\phi(a')$, on the one hand, and the vectors $\phi(b)$ and $\phi(b')$, on the other, are the same, so for our purposes each stimulus (a, b) can be completely characterized by the corresponding pair of primary codes $(\phi(a), \phi(b))$. Trichromacy combined with the two standard extensions just mentioned will henceforth be termed *extended trichromacy*. Two points are worth mentioning in this connection. First, in accepting extended trichromacy as a basic principle, our account will *not* imply that a television engineer, a producer of color films, a photo-realist painter, a book printer, and the like would have to employ more kinds of "basic color stuff" (types of pigments, etc.) for a perceptual simulation of a given scene "out there" than is commonly supposed. For, at least ideally, this simulation essentially amounts to replacing the retinal pattern of lights generated by the scene with a locally metameric pattern of lights generated by the simulating object (picture, etc.). Second, extended trichromacy does not imply, of course, a local wavelength-to-color relation. For it allows for the possibility of arbitrarily complex context effects. All it says is that these context effects are invariant regarding the replacement of retinal light patches with metameric ones. So it does not predict which color will actually be evoked, or whether the central patches of two center-surround stimuli (a, b) and (a', b') will match in color. Fortunately, for the practical simulation projects just mentioned, this need not be known. For the purposes of vision science, however, this gap should be filled by some model.

4.2.3 Continuous Focal Color Codes

Typically, such a model takes the form of some function h that assigns some color code to each center-surround stimulus (a, b)(or, analogously, to other classes of stimuli). Because of extended trichromacy, instead of referring to a pair (a, b) of spectral energy distributions, the corresponding *tristimulus vector* $(\phi(a), \phi(b))$ may be considered, consisting of only $2 \times 3 = 6$ real numbers. And the code assigned to this vector is itself an N-dimensional vector of real numbers:

$$h[\phi(a), \phi(b)] = (h_1[\phi(a), \phi(b)], \ldots, h_N[\phi(a), \phi(b)]).$$

By means of the condition

$$(a, b) \sim (a', b') \text{ iff } h[\phi(a), \phi(b)] = h[\phi(a'), \phi(b')] \tag{4.3}$$

this is made into a model of asymmetric color matching. A function h with this property will be called an *N-dimensional focal color code* (for center-surround stimuli) regarding ϕ.

In the context of isolated patches, a primary color code ϕ may be conceived as an *output code*, in the sense that condition (4.1) implies a one-to-one correspondence between the resulting individual codes $\phi(a)$ and α-colors. In the more complex situation considered now, however, primary color codes merely serve as *input codes*, while—due to (4.3)—the role of an abstract output code for focal colors is taken by h. Suppes et al. (1989, ch. 15) speak of a *complete* color code (or a complete system of color codes) in this connection.

Standard models choose $N = 3$ and propose certain specific types of codes h. These include the idea of taking ratios in the sense that

$$h_i[\phi(a), \phi(b)] = \phi_i(a)/\phi_i(b) \tag{4.4}$$

for $i = 1, 2, 3$, as well as affine or more complex nonlinear functions.

I do not simply take for granted that an N-dimensional focal color code exists for $N = 3$. Rather, I ask in the first place, what is the minimal number N for which such a code exists at all? As such, this is not a substantially meaningful question,[6] however, so some additional criterion needs to be employed, much as the criterion of linearity is invoked in addition to condition (4.1) in the case of primary codes. Since linearity does not seem to generalize in a meaningful way to the present situation, h is required to be *continuous* in the ordinary mathematical sense. This is tantamount to postulating that all its component functions are continuous: intuitively, this means that if one (or more) of the six tristimulus values $\phi_1(a)$, $\phi_2(a)$, $\phi_3(a)$, $\phi_1(b)$, $\phi_2(b)$, $\phi_3(b)$ are varied gradually, then this will never result in a sudden jump of the value $h_i[\phi(a), \phi(b)]$. The functions h_i may become arbitrarily steep, however, and may contain kinks and the like (i.e., they need not be differentiable). Such a code will be called an *N-dimensional continuous focal color code* (regarding ϕ).[7]

So the question pursued here is, what is the smallest integer N for which such a code exists? Note in passing that we cannot derive the mainstream assumption $N = 3$ from extended trichromacy itself. All this principle implies is that if we restrict attention to symmetric matches with a *fixed* surround b_0, then the answer trivially is $N = 3$.[8] But as soon as varying surrounds are allowed for, this is now longer true, since the six values $(\phi_i(a), \phi_j(b))$ (i,j = 1,2,3) might even make an independent contribution to the focal color of the stimuli (a, b), which would mean that $N = 6$ (or even more). On the other hand, extended trichromacy does not imply this, so $N = 3$ remains a possibility.[9]

There are two reasons for choosing the criterion of continuity here, an extrinsic and an intrinsic one. The extrinsic reason is that continuity is a very wide criterion that (implicitly or explicitly) is assumed to be satisfied by virtually all models of asymmetric color matching. So if we rule out the possibility $N = 3$, then all of these models are fundamentally called into question. This is of interest, I think, independently of the second, intrinsic reason.

The intrinsic reason—which may also explain why all these models assume some continuous function h—is the phenomenological observation that the focal color appearance of (a, b) and (a, a) varies "gradually" (even though sometimes quite "rapidly") as a and/or b (and hence $\phi(a)$ and $\phi(b)$) are changed gradually. As noted by Krantz and Simon (1983), who also pursued a topologically motivated continuity-based approach, a similar continuity concept was already expressed in Grassmann's second law, though for the case of single patches: "If one of two mingling lights be continuously altered (whilst the other remains unchanged), the impression of the mixed light also is continuously changed" (Grassmann 1854, 256). In support of that law, he noted, "The principle of continuous transition . . . must be regarded as perfectly established by experience, as a sudden spring in the phenomena would be apparent even in the most crude observations, and such a spring has not yet been discovered" (257). At the mathematical level, this nicely corresponds to the fact that primary color codes, too, are continuous functions.

The above continuity requirement can be understood as expressing the stipulation that this *qualitative continuity* be reflected by the resulting color code. In being related to the subtle mathematical concept of a continuum, this is of course an idealization, though I believe a plausible and fruitful one. Note that it is perfectly compatible with the emergence of discrete color categories. In the space of α-colors, we may for instance (gradually!) pass from the reddish to the greenish domain, that is, from one perceptual category to another. This is generally considered to be perfectly in line with a linear, and hence continuous, concept of primary color coding. (This situation is somewhat analogous to function $f(x) = x^3$ being perfectly continuous at 0, although in changing x gradually from −1 to 1, the values $f(x)$ switch from negative to positive.) In a similar vein, analogous categorical transitions found in center-surround displays (such as the gradual emergence of a blackish color component in certain transitions) do not by themselves speak against a continuity-based approach of color coding. The same goes for threshold phenomena, though this is a subtle point, which will not be discussed here.

4.3 Some Crucial Empirical Observations

The stage has now been set to approach the announced dimensionality results. To this end, I need to invoke additional empirical observations concerning the (im)possibility of certain matches. Keep in mind throughout that (for simplicity's sake) I am referring here to static center-surround displays in dark surround (no light) that are presented for some interval of time. More complex displays would require a somewhat modified approach.[10]

In this section, I present the crucial observations in two ways. On the one hand, I try to convey some idea of how the relevant colors look by using color attributes such as *blackishness*, *dim* vs. *bright*, *achromatic* vs. *maximally saturated* (*colorful*), *white*, *gray*, *brown*, and the like. On the other hand, I look for a (weaker) structural formulation, formally stated as an Observation with a capital O, that avoids this kind of qualitative phenomenal characterization by expressing our observations solely in terms of matches ($\sim$) and mismatches ($\nsim$) regarding the central region of the stimuli; this is in fact all that the dimensionality arguments in section 4.4 require. It provides a clear conceptual basis for the mathematical arguments and circumvents the potential methodological difficulties in connection with qualitative phenomenal characterizations. The latter are often considered somewhat dubious (subjective), or at least less reliable, in part of the color literature, an influential example being Brindley's (1960) distinction between class A and class B observations.

The first relevant observation is well known: among the focal colors evoked by center-surround displays (a, b) are "new ones," such as vivid shades of brown, gray, or black, which cannot be evoked by IPs or LIPs.

This can be stated formally for LIPs as

Observation 4.0 *There are lights a and b such that for all lights c,*

$(a, b) \nsim (c, c)$.

This means that the focal color evoked by such a stimulus (a, b) differs from all α-colors. These colors are called β-colors henceforth. This distinction is *not* meant to imply some substantial dichotomy in the realm of colors; but it is useful for the purposes of this chapter.

To get an idea of what Observation 4.0 is about, it is useful to consider a few examples of β-colors. For each of these, a rough phenomenal description is provided along with a specification of some stimulus by which the corresponding type of color can be evoked. In these stimulus specifications, I sometimes adopt the convenient, but potentially misleading, convention of assigning perceptual attributes of α-colors (white, yellow, bright, etc.) to *lights* themselves in speaking of white (yellow, bright, etc.) lights. This is merely an abbreviation, meaning that normal observers would judge the corresponding α-color evoked by such a light accordingly (under standard viewing conditions).

A prominent example of a β-color is that of a vivid focal impression of brown (which need not look dim at all).[11] To evoke such a color, start out with a (sufficiently bright) yellow light a presented as the central disc (meaning that it looks yellow when presented as an IP stimulus). When this light is surrounded by a white light b whose (psycho)physical intensity (luminance) exceeds that of a by a certain amount, the focal color will turn brown. If instead we start out with a suitable yellowish green light a, we will similarly obtain a focal color that may be described as olive green, which also belongs to the realm of β-colors. Gray or black can be achieved in the same manner by starting out with a white (sufficiently bright) light a. Gray and black can then be evoked, for example, by the stimulus $(t * a, a)$ for suitable factors t, say, $t = 0.4$ or $t = 0.1$, respectively. These "new" β-colors are sometimes termed *contrast colors*, though a closer look reveals that virtually all stable color experiences presuppose some kind of spatial or temporal contrast at the stimulus level, including α-colors (contrast between a light patch and the dark surround). In particular, the limiting case where the eye is exposed to complete darkness results not in the experience of a proper black, but in the impression of a noisy grayish darkness, often called *intrinsic gray*, which can be clearly distinguished perceptually from the shades of black and gray evoked by the just-mentioned stimuli.[12] Note that, unlike a proper black, intrinsic gray by definition is an α-color, even though a limiting case (associated with the stimulus (0, 0)).

Of course, the mere *existence* of β-colors alone as stated in Observation 4.0 does not yet imply higher dimensionality in the sense that $N > 3$, even though certain β-colors exhibit perceptual qualities such as kinds of grayness or blackishness that are not exhibited by α-colors. To see that such an argument would be invalid, assume for a moment that some investigation of α-colors had started by considering only those lights that are not reddish (such as greenish ones). These already make up a three-dimensional subspace of the standard CIE space of α-colors. If the mere existence of "new colors" or "new color qualities" would imply higher dimensionality, then the detection of isolated patches that look reddish—which clearly is a quality not found in the space we started with—would then imply that already CIE space itself would have to be at least four-dimensional, for it comprises greenish as well as reddish colors. This is not the case, of course.

So, to derive the announced result $N > 3$, Observation 4.0 needs to be sufficiently strengthened. To this end, let us have a closer look at how α- and β-colors are related to each other. More specifically, let's look more closely at *homochromatic decrements*—

stimuli of the form $(t * a, a)$ with $t < 1$. Obviously, some β-colors can be generated this way. Think, for instance, of the example of a gray that can be evoked by a suitable stimulus of the form $(t * w, w)$, where w denotes a white light. In homochromatic decrements, a phenomenon is observed that might be called the *induction of blackishness* (sometimes termed *darkness induction*, which I avoid because it is ambiguous): the focal color of $(t * a, a)$ will typically have a blackish component, which, as a rule, gets stronger as t gets smaller, eventually ending up with black. So for a red light a, for instance, this may be described as a blackish red, which is analogous to the case of a white light a, where "blackish whites," or different shades of gray, result. This blackish quality is not encountered in IPs (except perhaps for very dim-looking lights of very low luminance, which look rather different, though). So these *are* examples of β-colors. This means that typically the effect of smaller and smaller factors $t < 1$ in the LIP $(t * a, t * a)$ is different from that of decreasing the factor t in $(t * a, a)$. Simplifying matters, one might say that as a general tendency, the α-color evoked by the former gets less and less bright, while the latter gets more and more blackish.

This suggests the following observation, which indeed is considerably stronger than Observation 4.0.

Observation 4.1 *The focal colors evoked by homochromatic decrements are β-colors. That is, for all lights a and c and all factors $t < 1$,*

$$(t * a, a) \not\sim (c, c).^{13} \tag{4.5}$$

This observation implies that we have to choose $N \geq 4$ (see next section). To derive the result, however, we need only assume that the above mismatch criterion is at least satisfied by some light(s) a (e.g., achromatic ones). But Observation 4.1 assumes the validity of (4.5) for factors t arbitrarily close to 1. This may be deemed a problematic assumption, although I think it is adequate, provided one is willing to accept a little bit of idealization (which has been involved in the preceding discussion just as well). However, being a controversial statement, Observation 4.1, as it stands, does not seem to be the best starting point for a treatment of dimensionality issues. To avoid this difficulty, one might contemplate assuming condition (4.5) to hold only for factors sufficiently smaller than 1, say, for $t < 0.8$. However, this weaker variant of Observation 4.1 would not allow us to derive the desired dimensionality result any more, and thus does not provide a substantial improvement over Observation 4.0. For the method of continuous path analysis to work, it requires some condition that tells us something about the entire range of stimuli $(t * a, a)$ (for suitable lights a) starting from $t = 1$ down to some value $t = t_1$ for which (4.5) holds. The focal colors evoked by such stimuli then constitute a continuous path of colors connecting an α-color at one end with a β-color at the other.

A much weaker variant of Observation 4.1 can be stated along these lines that, to the best of my knowledge, is empirically well established and still suffices to derive the desired result. It is a bit more complicated to formulate and may at first sight look somewhat awkward, but the simple rationale behind it will become clear in the proof of the dimensionality theorems presented in the next section.

To frame this observation properly requires a closer look at the convex cone C (in Re^3) made up by all codes $\phi(a)$. It is sometimes convenient to be a bit sloppy and "identify" these codes in the following discussion with the corresponding α-color evoked by the stimulus (a, a). This cone is unbounded "above," for along with $\phi(a)$ it includes $\phi(t * a) = t\,\phi(a)$ for all real numbers $t > 1$. For the purpose of this analysis, it needs to be cut off by some plane "up high" to make it into a spatially bounded space; lights a outside this space will then be excluded from consideration in what follows. This may be done, for instance, by fixing a luminance value L_{max} and letting the *truncated cone* C^* consist of the ϕ codes of lights with a luminance $\leq L_{max}$ (compare figure 4.2). For the present purpose, we could choose a very high luminance so that lights of this luminance when presented as IPs or LIPs evoke an "extremely bright" color (which need not be made more precise here).[14] The truncated cone C^* has an outer "surface" that completely "covers" it in all directions; the set of codes $\phi(a)$ making up this bounding surface are its *boundary*, denoted by ∂C^*. The corresponding lights a are called *boundary lights*, and the corresponding α-colors (evoked by the LIPs (a, a)), *boundary colors*, which of course is understood as a boundary related to the truncated space of α-colors. The remaining codes, which lie inside the truncated cone, as well as the corresponding lights and colors are called *interior* codes, lights, and colors, respectively, an example being white lights of a luminance below L_{max}.

The class of *boundary lights* (and the corresponding class of evoked *boundary colors*) is composed of the following cases:

(B1) Spectral (i.e., monochromatic) lights and purple lights (additive mixtures of two spectral lights from the violet and the red "ends" of the spectrum), whose chromaticity makes up the outer boundary of the well-known CIE chromaticity diagram (see, e.g., Wyszecki and Stiles 1982). These lights correspond to the maximally saturated (and hence pronouncedly chromatic) α-colors across the entire hue circle.

(B2) At the "bottom," the light $a = 0$ that evokes intrinsic gray.

(B3) At the "upper end," extremely bright-looking lights of luminance L_{max}.

The following empirical observation, although clearly weaker than Observation 4.1, is still sufficiently strong to derive the central dimensionality result of section 4.4.

Observation 4.2 *There are interior lights a_1 and factors t_1, $0 \leq t_1 < 1$, such that the following holds:*

(C1) $(t_1 * a_1, a_1) \not\sim (c, c)$ *for all lights* c; *and*

(C2) $(t * a_1, a_1) \not\sim (b, b)$ *for all factors* $t, t_1 < t < 1$, *and all* BOUNDARY *lights* b.

That is, given some such light a_1 and factor t_1, of interest are the stimuli $(t * a_1, a_1)$ and the resulting continuous path of focal colors as t is varied between t_1 and 1 (compare figure 4.3).

The strict criterion (4.5) of Observation 4.1 is required only to hold for $t = t_1$ here; that is, only the focal color evoked by $(t_1 * a_1, a_1)$ is postulated to be a β-color (C1), so, in particular, Observation 4.0 trivially pops out. For the other stimuli $(t * a_1, a_1)$, it is only required that the focal colors evoked are not *boundary* colors (C2).

The validity of this observation can easily be demonstrated by considering, for example, a white light a_1 of a luminance that is clearly smaller than L_{max} but sufficiently high so that the LIP (a_1, a_1) evokes a "nice" bright white. For the factor t_1, we may, for instance, choose the value 0.4. Then the stimulus $(t_1 * a_1, a_1)$ will clearly evoke a β-color, namely, a "nice" gray (which does not look dim as IPs or LIPs of very low luminance or intrinsic gray). So, (C1) is satisfied. And the stimuli $(t * a_1, a_1)$ will evoke lighter and lighter shades of gray and eventually white as t is varied from $t = t_1$ to 1. So they are clearly different from the boundary colors of all three categories (B1), (B2), and (B3). Hence condition (C2) is also satisfied. Many other examples could be found. Moderately saturated chromatic lights a_1, for instance, would do the job just as well.

A final remark: Observations 4.0, 4.1, and 4.2 indeed refer only to color matching, postulating certain *mismatches*. In this connection it is important to note that the lights a_1 and factors t_1 in Observation 4.2 can be chosen in such a way that the mismatches postulated in (C1) and (C2) are *very pronounced mismatches*. Furthermore, each of these mismatches involves a clear difference in some quality that even in traditional accounts would count as an attribute of "color proper," such as differences in brightness or between various shades of gray and white, or the distinction between chromatic and achromatic colors. So even from a more traditional perspective, these considerations cannot easily be dismissed as referring to some perceptual attribute external to "color."

4.4 Dimensionality Theorems

The core theorems in section 4.4.1 build solely on the Observations 4.0, 4.1, and 4.2. This is done to achieve a maximal degree of rigor and transparency in the corresponding arguments and to free those arguments from any unnecessary conceptual and empirical burden. Some readers may prefer to get a more qualitative understanding of what is going on here, say, in terms of color attributes, which are deliberately excluded from consideration in section 4.4.1. They are invited to read this section in conjunction with the corresponding examples and qualitative descriptions provided in the previous section; further qualitative considerations that could be helpful in this connection are also found later in this chapter.

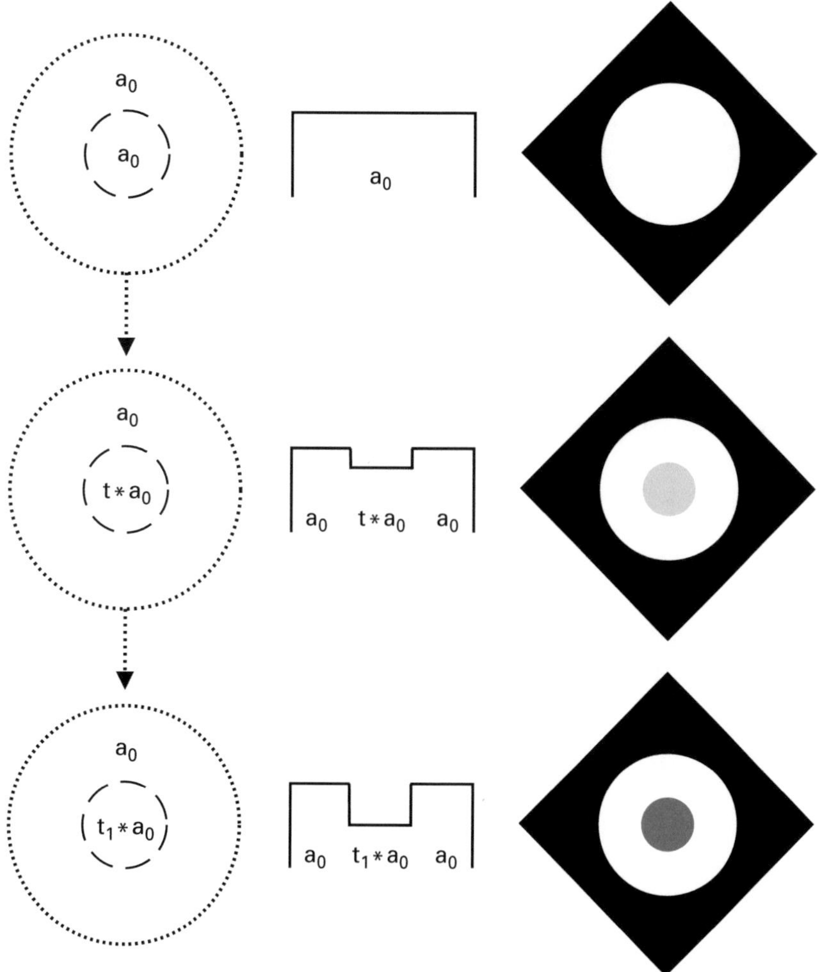

Figure 4.3
An illustration of the path considered as an example for Observation 4.2. Left column: schematic representation of the stimuli; middle column: intensity profiles of these stimuli; right column: schematic representation of what they look like. The right column cannot convey a fully appropriate impression of the colors evoked in a completely dark outer surround, however.

4.4.1 Central Theorems and Proof

Dimensionality Theorem 4.1 *Assume that Observation 4.1 holds along with extended trichromacy, and let h be an N-dimensional continuous focal color code (for center-surround stimuli) regarding some primary color code* ϕ. *Then* $N \geq 4$.

Dimensionality Theorem 4.2 *Assume that Observation 4.2 holds along with extended trichromacy, and let h be an N-dimensional continuous focal color code (for center-surround stimuli) regarding some primary color code* ϕ. *Then* $N \geq 4$.

Proofs If Observation 4.1 is satisfied, then so is Observation 4.2. So Theorem 4.1 follows from Theorem 4.2.[15] So let us turn directly to Theorem 4.2. In this chapter, I sketch only a somewhat simplified version of the proof that conveys the key idea. A two-dimensional illustration of the three-dimensional situation considered in the proof is given in figure 4.4.

The proof is based on an *indirect* argument, a *reductio ad absurdum*. To this end, assume that Observation 4.2 holds and that, contrary to the above claim, a three-dimensional continuous focal color code h exists for for center-surround stimuli. A contradiction will be derived. To keep the argument elementary, I make (with no loss of generality) the simplifying additional assumption that for all lights a, the code h satisfies

$$h[\phi(a), \phi(a)] = \phi(a). \tag{4.6}$$

This may be considered a natural special case anyway, since equation (4.6) already defines a partial three-dimensional continuous focal color code for LIPs (a, a) (due to extended trichromacy). The assumption considered here then means that this partial code can be extended three-dimensionally to the entire gamut of center-surround stimuli.[16] Now, choose some light a and factor t_1 according to Observation 4.2. Let us focus attention on the stimuli $(t * a_1, a_1)$ $(t \leq 1)$ and consider the h codes assigned to these stimuli. These are the vectors:

$$h_t := h[\phi(t * a_1), \phi(a_1)], 0 < t \leq 1. \tag{4.7}$$

Because h is assumed to be continuous, the function that assigns the vector h_t to t is a continuous function defining a *continuous path* in Re^3. Now, condition (C1) of Observation 4.2 along with condition (4.3) of the definition of a focal color code, imply that $h_{t_1} = h[\phi(t_1 * a_1), \phi(a_1)]$ differs from the codes $h[\phi(a), \phi(a)]$ for all vectors $\phi(a)$ in C^*. Because of (4.6), this simply means that h_{t_1} (point *e* in figure 4.4) lies outside of C^*. In contrast, h_1, which because of equation (4.6) is simply the code $\phi(a_1)$, belongs to the interior (point *i* in figure 4.4) of C^*. Hence, the continuous path h_t, $t_1 \leq t \leq 1$, connects the interior and the exterior of C^*. In consequence, this path must some-

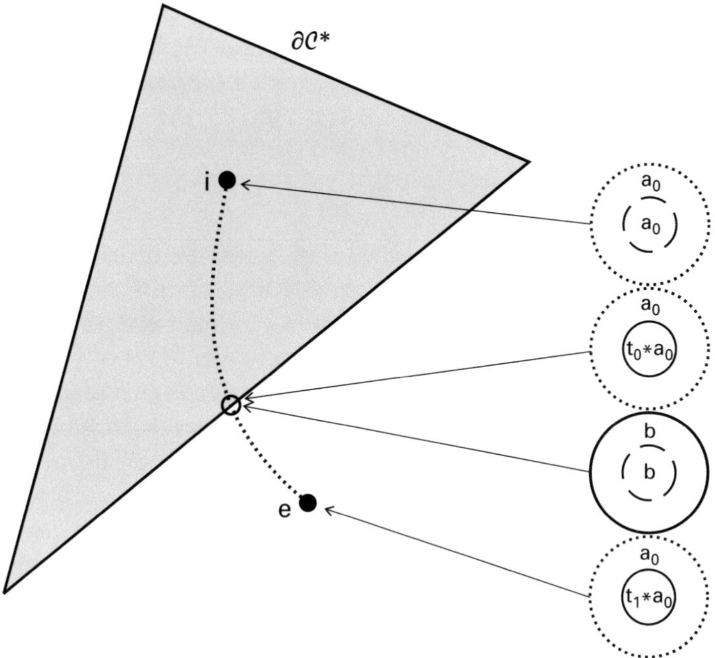

Figure 4.4
Two-dimensional illustration to the proof of Dimensionality Theorem 4.2. To render this illustration three-dimensional, just embed the whole situation into three-dimensional space while replacing the triangle with the truncated cone shown in figure 4.2; the outer surfaces of that cone then jointly constitute the boundary ∂C^*.

where intersect the boundary ∂C^* of C^*. In other words, for some boundary light b and some real number t_2, $t_1 < t_2 < 1$, we must have that

$$\phi(b) = h_{t_2}$$

Because of (4.6) and the definition of h_t given in (4.7), this implies that

$$h[\phi(b), \phi(b)] = h[\phi(t_2 * a_1), \phi(a_1)].$$

Hence, by (4.3),

$$(b, b) \sim (t_2 * a_1, a_1),$$

contradicting the choice of a_1 and t_2 (condition (C2) of Observation 4.2). This contradiction shows that the assumption $N = 3$ is not compatible with Observation 4.2, and hence $N \geq 4$ follows. This completes the proof.

Note that for $N \geq 4$, the problem encountered in the proof for $N = 3$ disappears because, thanks to the additional dimension(s), the corresponding path h_t can then simply bypass ∂C^*. This is analogous to the two-dimensional situation of a square in a plane. Every continuous path in the plane connecting a point i in the interior of the square with a point e outside the square must pass through the boundary of the square. But when embedded in a three-dimensional situation, the path could easily bypass that boundary.

4.4.2 Further Examples of the Method

The above dimensionality theorems are just examples of a general method. Many variants of these theorems are possible when the above definition of a continuous focal color code *mutatis mutandis* is applied to other classes of stimuli and associated "color spaces." May it suffice here to briefly mention two examples.

The first example concerns focal colors evoked by *achromatic center-surround stimuli* $(t * w, w)$ $(t \geq 0)$. Proponents of the mainstream view would postulate that there is a one-dimensional continuous focal color code (brightness) for those stimuli. By means of the above method it can easily be shown that such a code has to be *at least two-dimensional*. All of the relevant empirical observations have already been mentioned. Here is an informal and somewhat sloppy sketch of the argument (where stimuli, evoked focal colors, and the associated color codes are simply lumped together). This sketch can easily be made precise, however, along the lines of Observation 4.2 and Dimensionality Theorem 4.2. To begin with, consider a bounded *one-dimensional* space of achromatic α-colors evoked by LIP stimuli involving achromatic lights with a luminance $\leq L_{\max}$ (evoking white of varying brightness). The boundary here simply consists of the two "endpoints": intrinsic gray (no light) and a very bright white evoked by an achromatic light of luminance $L_{\max}$. As discussed, a continuous path of achromatic colors evoked by stimuli $(t * w, w)$ $(t_1 \leq t \leq 1)$ connects the interior of this bounded one-dimensional space with a "nice" gray outside this space. As can be observed empirically, this path does not pass through the two just-mentioned endpoints serving here as boundary colors. So the extended achromatic color space must be more than one-dimensional.

The second example concerns a *potential* application of the above method. The above observations and arguments refer only to stimuli of the form

$$(t * a, a), \text{for } t \leq 1,$$

that is, homochromatic decrements and LIPs, which jointly will be called *weak homochromatic decrements*. This means that the dimensionality theorems remain true if the phrase "for center-surround displays" is replaced by "for weak homochromatic decrements." Having established this, one might go one step further and propose the hypothesis that for these stimuli, a full-fledged four-dimensional account is appropriate, meaning that both $\phi(a)$ and the factor t are assumed to make an independent contribution to the evoked color. Observation 4.1 would be a simple consequence of this hypothesis. Mathematically speaking, this means the corresponding color space makes up a four-dimensional manifold (in which the achromatic colors constitute a two-dimensional submanifold). This is not yet implied by the dimensionality theorems themselves (see next section), but it seems quite plausible in light of the phenomenon of blackishness induction discussed in section 4.3. So, for the sake of argument, assume there is indeed a four-dimensional continuous focal color code of that kind for weak decrements. If we now extend the scope again to arbitrary center-surround stimuli (a, b), the question arises whether for these stimuli, four dimensions will do, or whether a more than four-dimensional continuous focal color code is needed. In principle, at least, one could then try to show the latter by applying the above method again, that is, by looking for "new colors" (relative to these stimuli/colors) and by studying suitable continuous paths connecting these new colors and the ones evoked by weakly homochromatic decrements, the crucial boundary now being itself three-dimensional. If some such path could be shown not to pass through this boundary, the desired conclusion would follow. Such "new" colors seem to exist indeed, such as focal colors that may be described as simultaneously reddish and greenish, or as bluish and yellowish (see, e.g., Ekroll et al. 2002), as well as, it appears to me, certain shades of brown. But all this is certainly an open issue, which in particular concerns the question of whether a corresponding counterpart to Observation 4.2 can be established empirically. Certainly the corresponding relevant mismatches will be more subtle than those involved in our discussion of Observation 4.2. The same goes all the more for the question of whether dimensionality (in the present sense) goes up even further if more complex displays are considered.

4.5 Discussion

Now, what do the above results tell us above and beyond what has been said so far? I approach this question from two perspectives. First, I raise the question of how the above type of result is related to various concepts of dimensionality. In section 4.5.1 this is briefly discussed from a sensation-oriented perspective. Second, I am concerned with the question of what such results could mean for our understanding of color perception in general. This necessitates a more careful juxtaposition of what has been called here sensation- and perception-oriented approaches. The second question is discussed in section 4.5.2 (which can be read independently of the following discussion of dimensionality concepts).

4.5.1 Dimensions of Dimensionality

The first issue takes into account both the intrinsic aspects of dimensionality mentioned in section 4.3 and the extrinsic ones. Let us begin with the latter. The above theorems simply tell us that all standard three-dimensional models of asymmetric color matching that purport to refer to perfect focal matches are fundamentally flawed empirically, and the same goes for one-dimensional models in the achromatic case. That is, all of these models—provided they are indeed continuous and in line with criterion (4.3)—will be inconsistent with the premises of the theorems. As for Dimensionality Theorem 4.2, this means that any such model must be at odds with extended trichromacy or with Observation 4.2 (or both). The former is the case, for instance, for a ratio-based model as specified by equation (4.4), for this implies that the same code is assigned to arbitrary LIPs (a, a) and (b, b), namely, the code (1, 1, 1). By criterion (4.3), this would imply that these stimuli should look the same. On the other hand, if a model is at odds with Observation 4.2, for some lights a_1 and factors t_1 that satisfy conditions (C1) and (C2), the model implies a match where (C1) or (C2) postulate a mismatch. Since a_1 and t_1 can be chosen in such a way that these mismatches are rather pronounced, this can hardly be counted as a tolerable failure of such a model.

The *intrinsic* aspects mentioned in section 4.3 concern the *topology* of color space, or more precisely, of color spac*es*, proceeding step by step from more restricted color spaces (associated with certain classes of stimuli) to possibly more comprehensive ones. Results like Dimensionality Theorem 4.2 are but a first step in such a project, and only a few, more or less intuitive, remarks along these lines can be made here. For brevity's sake, I continue focusing on colors evoked by center-surround displays (in dark outer surround) unless stated otherwise. The following account easily generalizes to other situations, however.

Let N_0 denote the *least* number N for which there is an N-dimensional continuous focal color code (for center-surround stimuli). From a topological viewpoint, this number may be termed the *global dimensionality* (in the sense of an embedding into Re^N) of the space of focal colors evoked by center-surround stimuli. The dimensionality theorems then state that this global dimensionality is at least four. Now, in comparison, consider the following figures (conceived as topological spaces of points), which are *globally* two-dimensional in this sense: (1) a circle, understood as a closed line; (2) a cross in the sense of two lines intersecting at a point P; and (3), a filled circle (including the enclosed area). In example 1, the situation is *locally* one-dimensional everywhere. In case 2, we may speak of a locally >1-dimensional situation at P, while in case 3, the situation is not only >1-dimensional everywhere, but properly two-dimensional locally, constituting a two-dimensional manifold. Analogously, the conclusion $N \geq 4$ of the above dimensionality theorems does not by itself rule out the possibility illustrated by case 1: *Locally*, the space of focal center-surround colors could perhaps still be three-dimensional everywhere. However, a closer inspection of the situation considered in the proof of the theorem shows that this possibility can indeed be ruled out.[17] Having established this, one might then venture the hypothesis that for the focal colors evoked by center-surround displays, the situation is in fact analogous to case 3, be it for $N =$

4, $N = 5$, or perhaps even $N = 6$ (with possibly a few special regions that call for a separate treatment). Recall that for the color space associated with weak decrements only, a corresponding hypothesis that postulates local four-dimensionality is proposed in section 4.4.2. Strong dimensionality hypotheses of that kind can be sustained by the method discussed before, but they cannot be inferred this way directly. This will always constitute an additional, though often plausible, theoretical step.

One noteworthy consequence of these considerations is that the possibility of obtaining (perfect) asymmetric color matches is far more restricted than often assumed in standard asymmetric color-matching paradigms. For instance, in such an experiment, a subject may be asked to find a light a' such that—for lights a, b, and b' fixed by the experimenter—a (perfect) focal match obtains between the center-surround stimuli (a, b) and (a', b'), with a dark outer surround. This will often be impossible, even if b and b' are rather similar. Problems of that kind have in fact been reported in multiple studies for various types of stimuli (see, e.g., Logvinenko and Maloney 2006, to mention just one recent example). This observation is usually neglected, however, in the mainstream literature. In a way, one problem of standard three-dimensional models of asymmetric color matching is that they imply too many matches.

Not only does a statement concerning "the dimensionality of color space" need to specify the range of colors considered as well as features such as whether the concept is understood locally or globally, but even more conceptual "dimensions" of dimensionality that play a role in the color literature need to be taken into account. So far, this chapter's focus has been on a concept of output dimensionality that is essentially topological in nature. Dimensionality concepts could also be based on different criteria, giving rise to different concepts of dimensionality (none of which is *the* true one; they just capture different, though related, aspects of the situation). One could for instance ask for the least number M for which there is a focal color code $h = (h_1, \ldots, h_M)$ satisfying criterion (4.3) such that one of the following criteria is fulfilled:

- Each component code h_i is a (univariant) *neural code* capturing neural activity at some level of color processing.
- The component codes h_i are associated with *perceptual attributes*.
- The code h represents subjective proximity judgments by some metric on Re^M in the sense of a suitable concept of *multidimensional scaling* (MDS; see, e.g., Suppes et al. 1989, ch. 14).

Usually this will mean that some such criterion is imposed *in addition* to that of continuity, the latter often being assumed tacitly. Provided there is a corresponding code at all, the minimal number M fulfilling such an additional criterion (i.e., the corresponding "dimensionality") will of course satisfy the inequality $M \geq N_0$ (and hence in the present situation, $M \geq 4$). Conversely, if for a class of stimuli, evidence is provided for $M \geq k$ (for some integer k), this does not automatically imply that the corresponding topological global dimensionality N_0 also satisfies this inequality, because possibly $M > N_0$.

This observation is relevant, in particular, to MDS-based criteria, which are quite popular in the color (and brightness) literature.[18] That the MDS-based dimensionality may easily exceed topological dimensionality is illustrated by the following example: A U-shaped line in a plane is one-dimensional from a topological point of view (both globally and locally), but to represent the distances within this line, say, by a Euclidean metric, two dimensions are required. My impression is that this observation applies to various MDS-based studies that, for certain classes of stimuli, suggest a higher-dimensional approach to color or brightness (i.e., $M \geq 4$ or $M \geq 2$, respectively). There are exceptions, however, that also suggest higher dimensionality at a topological level. In a recent MDS-based study, for instance, Logvinenko and Maloney (2006) investigated focal colors in achromatic stimuli (though in a somewhat different setting than the one considered here). They reach the conclusion that in their setting, achromatic colors make up a two-dimensional manifold (implying $N_0 = 2$ for the corresponding color space).

4.5.2 Color Sensation vs. Color Perception Reconsidered

In their study, Logvinenko and Maloney employed stimuli that consisted of central fields of varying surface reflectance embedded in a variegated surround, both of which were exposed to a homogeneous illumination of varying intensity (luminance). According to their two-dimensional model, both the surface reflectance and the intensity of the illumination make an independent contribution to the proximity judgments made by the subjects. This leads us back to issues concerning color *perception* and to issues relating to perceived *surface color* and approximate color constancy.

Restricting attention to matte surfaces, perceived surface color, too, is usually discussed in terms of three dimensions (color solids). Correspondingly, achromatic surface colors are accounted for in terms of one dimension (*albedo*, or *lightness*, ranging qualitatively from black via different shades of gray to white). This *trivariance* of surface color (referring to the perceptual output in complex scenes, i.e., to perceived surface colors) needs to be carefully distinguished from trichromacy as discussed in section 4.2. Trivariance is also not fundamentally called into question by my approach, but it needs to be carefully qualified. If the illumination in a physical scene is changed, then the *overall* appearance of the surfaces will typically change even if the involved perceived surface colors themselves remain (more or less) constant in the sense of color constancy. This difference may be attributed to a difference in the perceived illumination color(s), which is as much a component of the percept as are the perceived surface colors themselves. On this view, the visual system decomposes the locally three-dimensional input into *two* trivariant components (layers, scissions) for each direction in the visual field (*color segmentation*). And the situation further complicates, if (perceived) transparency, gloss, self-luminosity of objects, and so forth are taken into account—what Katz (1911/1935) termed *modes of color appearance* (Erscheinungsweisen der Farben). Similar points have already been made in the early color literature, for example, by Katz (1911/1935), Hering (1920), Bühler (1922), and Gelb (1929), and have been repeatedly brought up again by later authors; see, for example, Adelson (1993), Arend (1994), other papers in Gilchrist (1994), Ekroll et al. (2002; cf. MacLeod 2003), and Mausfeld (1998, 2003). Nevertheless, many accounts of color vision ignore these complexities to the present day.

From this perspective, the above results do not come as much of a surprise. However, I do not think that sensation-oriented aspects of color vision discussed above simply reduce to these complexities of color *perception* proper. It seems, in fact, that both a sensation-oriented and a perception-oriented aspect (level or mode) coexist in the phenomenology of color vision itself, at least to some extent and in a subtly interwoven manner. First, there is the level of a *field of colors*, which are not "yet" attached to perceived objects in perceptual space or perceived illumination (for short: *field colors*). And, second, at the object-centered level, we get surface colors, transparent layer colors, illumination colors, and so forth (for short: *segmented colors*). The latter may sometimes be more or less vague, though. This is particularly true of the reduced stimuli considered in the previous sections, where one might at best speak of an "embryonic" form of segmentation (Mausfeld and Niederée 1993). These phenomenal levels do not kick around independently of each other, of course. Instead, there is some kind of phenomenal binding connecting them. To put it very simply, in each direction of visual space, a field color appears to be associated with the corresponding bundle of segmented colors. Consider, for example, the case where a yellow transparent layer appears to cover part of a blue or a red surface (segmented colors); at the location of the overlapping region, we often simultaneously see a green or an orange, respectively (field colors). This distinction may also help to settle debates such as that between Hering (e.g., 1920, 210) and von Kries (in Helmholtz 1910, 490ff). They had a dispute about whether a gray patch on a wall (in terms of perceived surface colors) should be counted as the same sensation (*Empfindung*) as a suitable gray-looking shadow on a white wall, as advanced by von Kries, or not, as claimed by Hering. Von Kries's position may possibly be adequate for the field level, while Hering clearly refers to the segmented level.

The two levels distinguished here are closely related to the concept of a *double representation* of color in Arend (1994; see also Arend and Reeves 1986), who juxtaposed *unasserted* and *attributed* colors in a similar manner (see also, in a philosophical context, Cohen 2008). In a similar vein, Evans (1974, ch. 11) already pointed out that in color vision not only is there both an "object frame of reference" and an "illumination frame of reference" (which corresponds to our notion of segmented colors), but also that there is a "stimulus frame of reference." In this connection he explicitly warned against an either-or fallacy (193). Sometimes, an analogous *duality of perception* is also invoked in discussions of perception in general (including, e.g., the perception of size and shape). Examples are Helmholtz's distinction of a "Sehfeld" (visual field) and a "Sehraum" (visual space), or Gibson's (1950) distinction between a "proximal mode" and a "world mode" (see also Rock 1983; Todorović 2002). And the picture gets even more complex if one takes into account duality phenomena at the segmented level as involved, for example, in picture perception (see, e.g., Niederée and Heyer 2003).

If one accepts the basic distinction just made, then it is clear that the just-mentioned complexities at the level of segmented colors do not *by themselves* imply a corresponding degree of complexity (or dimensionality) at the level of field colors. It seems that a specific field color (e.g., a certain orange) may in fact be phenomenally associated with different segmentations (such as a yellow transparent layer in front of a red one, an orange surface appearing in neutral illumination, and so forth). This in particular means that color segmentation is *not* per se inconsistent with a three-dimensional account of field colors. In fact, authors entertaining some such distinction of levels or modes often assume three-dimensionality to hold for what is called here the level of field colors.

However, by the above dimensionality results (or related findings reported in the literature, such as, in my view at least, Logvinenko and Maloney 2006), this possibility is ruled out, for these results obviously refer to field colors, thus implying that a higher-dimensional approach is already needed to provide an appropriate account for this seemingly basic level of color vision.[19] So they indeed provide us with an additional piece of information about color vision. As a consequence, the level of field colors should not simply be conceived as a "proximal mode" of color vision. It rather seems to foreshadow some of the complexities showing up at the segmented level, though in a much more rudimentary fashion. And this might in turn play a role in the management of perceptual ambiguity by representing certain possible ambiguities in the precept itself.

In my view, a standard representational or realist account of color vision (among other things) cannot account for the existence of the field level and its relation to the segmented level. Nor can it straightforwardly incorporate simple everyday observations such as that intrinsic color qualities appearing, say, greenish or brownish may pertain both to the field level and to the segmented level, and in turn to colors that are perceived as belonging to surfaces, transparent layers, or (in the case of greenish at least) to self-luminous objects or illumination. For what should all these cases have in common in the represented outer world? These are but elementary examples of a rich *internal* structure governing human color vision, still waiting to be described systematically. Continuous path analysis, that is, investigations into the effect of gradual stimulus variations (which may result in gradual or abrupt changes at the perceptual level) may be expected to yield important insights also in this connection, far beyond the issue of dimensionality. And the same, I believe, goes for accounts of various types of dimensionality, provided they are subsumed under the more general goal of understanding the intricate internal overall logic underlying human color vision.

Acknowledgments

The preparation of this paper was partly supported by the German Federal Department for Education and Science (BMBF-grant 01GWS060 to Rainer Mausfeld) as part of the *Forschungsverbund Interdisziplinäre Anthropologie* (initiated and lead by Wolfgang Welsch). The impetus for this work grew out of many fruitful discussions with Rainer Mausfeld. I am also indebted to Vebjørn Ekroll, Franz Faul, and the editors Jonathan Cohen and Mohan Matthen for their comments on an earlier draft of this paper.

Notes

1. Of course, to the extent that arguments concerning "color out there" also draw on how color is perceived, the considerations in this chapter might have some implications for the corresponding debates, too.

2. As with many physical laws, this is not understood here as a variation in time, but rather as referring to the different causal effects of different (for some time steadily presented) stimuli. Of course, dynamic variations in time are very instructive to study in this connection, but they would require a still more complex formal approach because of the resulting complex temporal dynamics in the evoked colors (successive contrast effects and the like).

3. To keep the exposition simple, the premises of the theorems are formulated in a slightly stronger form than actually needed and a number of standard idealizations and simplifications are tacitly adopted. In particular, the present account generally refers to photopic (i.e., daytime) vision in a so-called human *normal (trichromatic) observer*. A more detailed and comprehensive discussion of most issues dealt with in this chapter is found in Niederée (1998, available at www.psychologie.uni-kiel.de/psychophysik/dimensionality.html).

4. The Grassmann laws essentially state that for all lights a, b, c and positive real numbers t, $a \sim b$ implies that $a \oplus c \sim b \oplus c$ and $t * a \sim t * b$ (and vice versa).

5. That is, for all lights a, there are lights a' and a'' such that $(a, 0) \sim (a', a')$ and $(a, a) \sim (a'', 0)$, where we may or may not adopt the standard assumption that this holds in particular for $a = a' = a''$.

6. In fact, classical set theory (including the axiom of choice) tells us that without some additional criterion, the solution is simply $N = 1$, even though the corresponding functions h may be rather "weird."

7. Strictly speaking, continuity of h is required only within the domain $C \times C$. Outside this domain, the behavior of h is of no interest. Furthermore, all that is required for the present account is that for all vectors $x, y \in C$, the functions that map x to $h[x, y]$ and $h[x, x]$ are continuous in x.

8. Simply let $h[\phi(a), \phi(b_0)] = \phi(a)$. For different fixed surrounds b_0, the focal colors associated with a specific vector $\phi(a)$ may of course be different.

9. A generalized version of the concept of a continuous focal color code is already employed by Krantz and Simon (1983). They attempted to *prove* that a continuous focal color code must exist for $N = 3$, instead of simply taking this for granted as is so often done. Their sophisticated topological argument was based on an assumption concerning the possibility of obtaining color matches at least locally (so-called nonsingularity). This assumption turns out to be incompatible with the present approach.

10. This analysis does in fact apply to several possible experimental settings for such a matching task, as well as to more general situations (e.g., somewhat larger stimuli). All that matters here is that there *are* suitable settings for which the empirical premises employed in the arguments presented in this chapter are satisfied. For some experimental settings found in the literature, these arguments do *not* apply directly to the matching data reported (for those situations, the arguments would have to be stated somewhat differently). A case in point is matching with haploscopically superimposed backgrounds, as employed, for example, by Whittle and Challands (1969); see also the discussions in Whittle (2003) and in Niederée and Mausfeld (1997). These matches are special insofar as in this situation $(a, a) \sim (b, b)$ holds for all lights a and b, which violates condition (4.2) of extended trichromacy (which in this chapter's discussion is presupposed to hold).

11. Of course, there are many shades of brown. The point here is not that *any* color we may tend to categorize as brown or brownish necessarily is a β-color. In fact, by lowering the luminance of an orange light presented as an IP until it looks very dim, eventually a dark-looking α-color will be evoked that one might call brown(ish). But this "peripheral" kind of brown clearly differs from the one I have in mind here. All we need to allow in what follows is that some shades of brown are β-colors. The same goes for other color categories, such as gray.

12. See, for example, Hering (1920, 9), who discussed the role of gray and black along these lines, contradicting Helmholtz, who in some of his writings misleadingly conceived of black as the perceptual response to "no light."

13. Because LIP stimuli are incremental stimuli, this seems compatible with various empirical studies reporting that increments never match decrements, as was reported, for example, for the achromatic case by Whittle and Challands (1969), although they used haploscopically superimposed backgrounds. Recall that I am referring to proper matches here, not to approximate ones. Observation 4.1 is also implied, in principle, by relational, that is, contrast-based accounts (based, for instance, on equation (4.4)). For a discussion of relational approaches, see, for example, Mausfeld and Niederée (1993; also Niederée and Mausfeld 1997) and Whittle (2003). Since these models are typically presented in a three-dimensional framework (but see, e.g., Heggelund 1974),

the results presented in this chapter reveal, in my view, a conceptual tension inherent in these models, calling for a suitable modification. This applies also to Heggelund's higher-dimensional account (which invokes an assumption still stronger than Observation 4.1). See Niederée (1998, sec. 6.5.2) for a more detailed discussion.

14. Similarly, one could, for instance, cut off lights of very low luminance by analogously truncating the cone—by excluding lights a of a luminance below some value L_{min}. This way, very dim-looking patches and intrinsic gray ($a = 0$) could be excluded from consideration. One could also, for instance, restrict C^* to (the primary codes of) those lights that can be realized on suitable color monitors. The following arguments could be easily adapted to these variants and combinations thereof.

15. For an instructive proof of Theorem 4.1 that directly exploits the strong postulates of Observation 4.1, see Niederée (1998, sec. 6.2).

16. Mathematical remark: of course, all that matters for the general case is whether this special case may really be considered "with no loss of generality." In fact, for the general case, one needs to invoke a well-known and deep topological theorem, the *gamut invariance theorem* (which is closely related to Brouwer's Dimensionality Theorem). This allows one to establish, first of all, that because of extended trichromacy, an N-dimensional continuous focal color code must satisfy $N \geq 3$ (i.e., the possibility $N = 1$ or 2 can indeed be ruled out). Now let us assume that h is a three-dimensional continuous focal color code. Observing that, by assumption, the function $h'(\phi(a)) := h[\phi(a), \phi(a)]$ is continuous and one-to-one, and that C^* is a *compact* subset of Re^3, the gamut invariance theorem (plus some standard continuity-based arguments) allows one to establish that $h'[C^*]$ is again a compact subset of Re^3, the boundary and the interior of C^* being mapped into the boundary and the interior of $h'[C^*]$, respectively. Having established this, the proof sketched in this chapter (based on the additional assumption that h' is the identity function) can easily be transferred to the general case. To do so, in certain places of the argument presented here (including the illustration in figure 4.4), C^* has to be replaced with the set $h'[C^*]$ and ∂C^* with its boundary. For a close variant of this theorem, the complete general proof is given in Niederée (1998, sec. 6, particularly 6.1 and 6.4.2). Note that the argument presented hereafter also relies mathematically on the compactness of $h'[C^*]$ along with the consideration of a suitable infimum or supremum.

17. It can be shown that along the path h_t considered in the proof of Theorem 4.2, there must be a point p where the situation becomes locally >3-dimensional. If Observation 4.1 is assumed to hold, this already occurs at the very beginning of this path ($p = i$). For mathematical details and further consideration along these lines, see Niederée (1998, sec. 6.5.4).

18. Note that the inequality $M \geq N_0$ only holds for the "true" dimensionality as defined by the respective MDS model (provided the model is satisfied at all for the situation considered). The *estimated* dimensionality reported in such studies, based on empirical judgments and models that allow for the possibility of judgmental errors, may sometimes be smaller.

19. As pointed out, this already holds for the focal colors in center-surround displays. But the method of continuous path analysis can be brought to bear just as well on other classes of stimuli

and the field colors they evoke, yielding higher dimensionality, such as for the field colors going together with surfaces viewed under different illuminations, or with certain gradual transitions between different modes of color appearance at the segmented level, such as (perceived) self-luminous versus surface colors. See, e.g., Niederée (1998, sec. 6.5 and 7).

References

Adelson, E. H. 1993. Perceptual organization and the judgment of brightness. *Science* 262: 2042–2044.

Arend, L. E. 1994. Surface colors, illumination, and surface geometry: Intrinsic-image models of human color perception. In *Lightness, Brightness, and Transparency*, A. L. Gilchrist, ed., 159–213. Hillsdale, NJ: Erlbaum.

Arend, L. E., and A. Reeves. 1986. Simultaneous color constancy. *J. Opt. Soc. Am.* A (3): 1743–1751.

Brindley, G. S. 1960. *Physiology of the Retina and the Visual Pathway*. London: Arnold.

Bühler, K. 1922. Die Erscheinungsweisen der Farben. In *Handbuch der Psychologie. I. Teil. Die Struktur der Wahrnehmungen*, K. Bühler, ed., 1–201. Jena: Fischer.

Cohen, J. 2008. Color constancy as counterfactuals. *Australasian Journal of Philosophy* 86 (1): 61–92.

Ekroll, V., F. Faul, R. Niederée, and E. Richter. 2002. The natural center of chromaticity space is not always achromatic: A new look at color induction. *Proceedings of the National Academy of Sciences* 99 (20): 13352–13356.

Evans, R. M. 1974. *The Perception of Color*. New York: Wiley.

Gelb, A. 1929. Die 'Farbenkonstanz' der Sehdinge. In *Handbuch der normalen und pathologischen Physiologie. Bd. 12, 1. Hälfte. Receptionsorgane II*, A. Bethe, G. v. Bergmann, G. Embden, and A. Ellinger, eds., 594–678. Berlin: Springer.

Gibson, J. J. 1950. *The Perception of the Visual World*. Boston: Mifflin.

Gilchrist, A. L., ed. 1994. *Lightness, Brightness, and Transparency*. Hillsdale, NJ: Erlbaum.

Grassmann, H. 1853. Zur Theorie der Farbmischung. *Poggendorff's Annalen der Physik* 89: 69–84.

Grassmann, H. 1854. On the theory of compound colours. *Philosophical Magazine*, Series 4, 7: 254–264.

Heggelund, P. 1974. Achromatic color vision. I. Perceptive variables of achromatic colors. *Vision Research* 14: 1071–1079.

Helmholtz, H. von. 1910. *Handbuch der Physiologischen Optik, Bd. 3.* 3rd ed., edited by A. Gullstrand, J. v. Kries, and W. Nagel. Hamburg: Voss.

Hering, E. 1920. *Grundzüge der Lehre vom Lichtsinn*. Berlin: Springer.

Katz, D. 1911. *Die Erscheinungsweisen der Farben und ihre Beeinflussung durch die Individuelle Erfahrung*. Leipzig: Barth. Translated from the 2nd German edition by R. B. MacLeod and C. W. Fox as *The World of Color*, London: Kegan Paul, Trenth, and Trubner, 1935.

Krantz, D. H. 1975. Color measurement and color theory. I. Representation theorem for Grassmann structures. *Journal of Mathematical Psychology* 12: 183–303.

Krantz, D. H., and C. P. Simon. 1983. The topological laws of color matching. Unpublished manuscript.

Logvinenko, A. D., and L. T. Maloney. 2006. The proximity structure of achromatic surface colors and the impossibility of obtaining asymmetric lightness matching. *Perception and Psychophysics* 68: 76–83.

MacLeod, D. I. A. 1986. Receptoral constraints on colour appearance. In, *Central and Peripheral Mechanisms of Colour Vision*, D. Ottoson and S. Zeki, eds., 66–81. London: Macmillan.

MacLeod, D. I. A. 2003. New dimensions in color perception. *Trends in Cognitive Science* 7 (3): 97–99.

Mausfeld, R. 1998. Color perception: From Grassmann codes to a dual code for object and illumination colors. In *Color Vision*, W. G. K. Backhaus, R. Kliegl, and J. S. Werner, eds., 219–250. Berlin: Walter de Gruyter.

Mausfeld, R. 2003. "Colour" as part of the format of different perceptual primitives: The dual coding of color. In *Colour Perception: Mind and the Physical World*, R. Mausfeld and D. Heyer, eds., 381–429. Oxford: Oxford University Press.

Mausfeld, R., and R. Niederée. 1993. An inquiry into relational concepts of colour based on incremental principles of colour coding for minimal relational stimuli. *Perception* 22: 427–462.

Niederée, R. 1998. *Die Erscheinungsweisen der Farben und ihre stetigen Übergangsformen*. Postdoctoral thesis (Habilitation), Philosophische Fakultät der Christian-Albrechts-Universität zu Kiel, Kiel.

Niederée, R., and D. Heyer. 2003. The dual nature of picture perception: A challenge to current accounts of visual perception. In *Looking into Pictures*, H. Hecht, R. Schwartz, and M. Atherton, eds., 77–98. Cambridge, MA: MIT Press/Bradford Books.

Niederée, R., and R. Mausfeld. 1997. Increment-decrement asymmetry in dichoptic matching with haploscopically superimposed displays. *Vision Research* 37: 613–615.

Rock, I. 1983. *The Logic of Perception*. Cambrigde, MA: MIT Press/Bradford Books.

Suppes, P., D. H. Krantz, R. D. Luce, and A. Tversky. 1989. *Foundations of Measurement*. Vol. 2, *Geometrical, Threshold, and Probabilistic Representations*. San Diego: Academic Press.

Todorović, D. 2002. Comparative overview of perception of distal and proximal visual attributes. In *Perception and the Physical World*, D. Heyer and R. Mausfeld, eds., 37–74. New York: Wiley.

Whittle, P. 2003. Contrast colours. In *Colour Perception: Mind and the Physical World*, R. Mausfeld and D. Heyer, eds., 115–138. Oxford: Oxford University Press.

Whittle, P., and P. D. C. Challands. 1969. The effect of background luminance on the brightness of flashes. *Vision Research* 9: 1095–1110.

Wyszecky, G., and W. S. Stiles. 1982. *Color Science: Concepts and Methods, Quantitative Data and Formulae*. 2nd ed. New York: Wiley.

5 Color within an Internalist Framework: The Role of "Color" in the Structure of the Perceptual System

Rainer Mausfeld

Color is, according to prevailing orthodoxy in perceptual psychology, a kind of autonomous and unitary attribute. It is regarded as unitary or homogeneous by assuming that its core properties do not depend on the type of "perceptual object" to which it pertains and that "color per se" constitutes a natural attribute in the functional architecture of the perceptual system. It is regarded as autonomous by assuming that it can be studied in isolation of other perceptual attributes. These assumptions also provide the pillars for the technical field of colorimetry, and have proved very fruitful for neurophysiological investigations into peripheral color coding. They also have become, in a technology-driven cultural process of abstraction, part of our common-sense conception of color. With respect to perception theory, however, both assumptions are grossly inadequate, on empirical and theoretical grounds. Classical authors, such as David Katz, Karl Bühler, Adhémar Gelb, Ludwig Kardos, and Kurt Koffka, were keenly aware of this and insisted that inquiries into color perception cannot be divorced from general inquiries into the structure of the conceptual forms underlying perception. All the same, the idea of an internal homogeneous and autonomous attribute of color per se, mostly taken not as an empirical hypothesis but as a kind of truism, became a guiding idea in perceptual psychology. Here, it has impeded the identification of relevant theoretical issues and consequently has become detrimental for the development of explanatory frameworks for the role of 'color' within the structure of our perceptual system.

The concept of 'color per se' as an abstract attribute that can be dealt with in a decontextualized way has been developed, in the technological context of coloration techniques and dyeing processes, as the basis for standardizations and norms for capturing color appearances.[1] The idea of 'color per se' is, thus, the product of technology-shaped cultural abstractions, including its corollary ideas that color can be characterized by basic color attributes, such as hue, saturation, and brightness, and that color appearances can be represented by a three-dimensional color space.[2] These technology-driven abstractions capture a certain part of our exceedingly complex linguistic usage of color expressions in everyday language and have in turn shaped our ordinary conception of color. However, they do not mirror core properties and principles of the internal organization of color in the perceptual system and, in the context of perception theory, have generated all sorts of spurious questions, such as about the types of "basic color attributes" or the dimensionality of color space. I have dealt in detail with these issues and the relevant empirical evidence elsewhere (Mausfeld 1998, 2003a). Here, it may suffice to point out the gross empirical inadequacy, even in center-surround situations, of the idea that color appearances can be represented by a three-dimensional color space.[3] Its inappropriateness should already be evident from many classical experiments and observations, for instance, Katz's demonstrations of the (at least) bi-dimensionality of achromatic colors. The assumption of three-dimensionality was experimentally tested and shown to be inappropriate in an experiment by Ekroll et al. (2002).[4] Furthermore, Niederée (this volume) rigorously shows that even in center-surround configurations, the dimensionality of color codes must be greater than three if one is willing to accept the topological assumptions that, at least implicitly, underlie almost all models of color coding. This empirical and theoretical evidence against the three-dimensionality of color already indicates that the traditional concept of color is flawed in a fundamental way. Although the entire conceptual framework underlying the idea of a homogeneous and autonomous attribute of color per se has been radically called into question in the earlier literature, it turned into orthodoxy during the first decades of the last century and, since then, is considered a natural and almost compulsory point of departure for dealing with color within perception theory.

Why, then, do corresponding conceptions of color as an autonomous and unitary attribute still pervade perceptual psychology, despite the huge amount of evidence to the contrary? It seems mainly due to the influence of commonsense conceptions of perception, which we illegitimately transfer to scientific inquiry. Of course, all fields of scientific inquiry must inevitably start at their origin from everyday experiences and available commonsense concepts. In the process of their development, however, they have to go beyond commonsense conceptions and develop notions and pursue lines of inquiry that are dictated by the needs of coherence and explanatory width and depth. Perception theory is no exception in this regard. The entire history of the development of the natural sciences, notably physics, is pervaded by a tension between concepts on which successful theories could be based, on the one hand, and deeply entrenched commonsense intuitions and notions about the domains under scrutiny, on the other. Unsurprisingly, a corresponding tension constitutes an incommensurably stronger obstacle with respect to the development of perception theory. In perception theory, dispensing with commonsense intuitions to instead follow the lines of theorizing traced out by successful explanatory accounts will likely prove to be much more difficult than it was in physics. Although we are well aware that commonsense intuitions and concepts are inapt guides for the endeavor to achieve, within the framework of the natural sciences, a theoretical understanding of perceptual phenomena, we are held captive by appearances.[5] Unsurprisingly then, hardly any other domain of rational inquiry is so deeply and almost ineradicably imbued with commonsense intuitions as is perception theory. At the roots of these intuitions is our conviction that perception basically works the way it appears to us. It is, however, an essential part of the functioning of our brain that it does not provide us with mechanisms to observe its own machinery, and this also holds for what may be distinguished as the perceptual system. We therefore can only attempt to better understand the perceptual system's underlying principles by the standard methodological approach of the natural sciences and its characteristic far-reaching abstractions and sharp idealizations. Although in other areas of the natural sciences, we are willing to trade commonsense intuitions and notions for whatever increases the explanatory depth and width of our theories, we are prone—in what Chomsky (e.g., 2000, 77ff) has called "methodological dualism"—to pursue a different path with respect to mental phenomena. Here, it is often claimed that there are, beyond the usual criteria for successful theories, privileged categories of evidence for what we are willing to consider as "psychologically real." Accordingly, in perception theory, the development of explanatory frameworks is often subjected to the unwarranted demand to conform to certain ordinary intuitions of perception. While both philosophy, notably in the seventeenth century, and perceptual psychology, notably Gestaltists and ethologists, have achieved important insights that go beyond commonsense intuitions and reveal the deeper issues involved, current thinking in both disciplines is dominated by presuppositions of commonsense ideas of perception. These conceptions appear to have a particularly unfortunate influence on inquiries into the "nature of color," where they have hampered the right questions being asked and have impeded the development of appropriate explanatory frameworks for color perception. If indeed color cannot be studied in isolation from the type of "perceptual object" to which it pertains, theoretical frameworks appropriate for color perception must be general enough to be appropriate also for dealing with the internal structural form of the "objects" that constitute our perceptual ontology. Detached from an appropriate theoretical account of perception in general, questions regarding the "nature of color" will inevitably remain at the surface of commonsense intuition. I cannot discuss here the vast amount of empirical evidence supporting the view that inquiries into color perception cannot be divorced from general inquiries into the structure of the conceptual forms underlying perception.[6] I will therefore confine myself to argue this contention in the abstract. Before I expound some reflections on what appears to me a fruitful theoretical perspective for dealing with color perception, I briefly, and with an eye on color aspects, deal with some preconceptions that have been illegitimately transferred to perception theory from commonsense intuitions.

5.1 The Impact of Commonsense Intuitions on Scientific Inquiries into Perception

In speaking of commonsense conceptions in the present context, I will understand the term in the broadest possible way, namely as the diversity of modes in which we conceive of perceptual phenomena and the process of perception itself in all contexts other than that of the natural sciences. This usage comprises not only those concepts and ways of world making that underlie, as part of our biological endowment, our ordinary discourse about the world and our acts of perceiving—sometimes referred to as *folk physics* and *folk psychology*—but also derived concepts and notions pertaining to perceptual issues that have been developed for purposes other than those of the natural sciences, whether technological, philosophical, or any other kind.

Common sense tells us that, by and large, perceiving keeps us in direct contact with the world, that it is the external world we perceive and that we perceive it the way it really is. Common sense further tells us that a kind of integral and even immaterial self is in direct contact with the world, and no brain, intermediate substrate, or properties of whatever happens in the body between the sensory stimulation and the percept appear in its ordinary accounts. Common sense discounts, as part of an essential functional achievement of the brain, all the processes that occur between the distal causes and the percept, and thus convince us that we are in direct contact with the world. On this account, perception is an entirely conspicuous process. Of course, common sense is willing to except all sorts of sophistications and, in unusual circumstances, exceptions to this view, but it otherwise regards this account as a kind of truism.

We can therefore distinguish two different, though not unrelated, ideas that characterize, in various guises, commonsense conceptions of perception. The first idea is that perception basically works the way it appears to us, and that therefore explanatory useful categorizations of phenomena are immediately suggested to us. The second idea is a realistic conception of perception in the sense of a (culturally refined) naïve realism. Attempts to provide some kind of philosophical justification for the realism underlying commonsense conceptions of perception tend to go along with the idea of color per se.[7] I will deal in turn with these two ideas and the intuitions ensuing from them, and briefly point out how they have influenced systematic inquiries of perception.

5.1.1 "Perception Basically Works as It Seems to Work"

Hardly any other phenomena of nature appear so plain, self-evident, and intellectually transparent as perception. That we have this impression is itself an essential achievement of the brain, which conveys only the final product of the perceptual system's functioning to our phenomenal experience. It is precisely because this functioning is entirely impenetrable to our phenomenal experience that we are, in our everyday experience, convinced of the integrity of our mental activity. However, the systems of our mind that are involved in perception exhibit the kind of modular structure that is characteristic for all complex biological systems (e.g. Gerhart and Kirschner 1997; Hartwell, Hopfield, Leibler, and Murray 1999). In everyday discourse, no need arises to distinguish the contributions of different subsystems of the mind/brain, in particular between a perceptual system and higher-order intellectual and interpretive capacities. One of the core theoretical concepts of perception theory, namely modularity, is therefore entirely alien to commonsense intuitions and their holistic conceptions of mental activity. In our phenomenal experience, the contributions of the different systems involved are inseparably interwoven, and nothing suggests that the apparent complexity of the percept is reducible to simple principles of separate subsystems.

In an everyday context, it is, for instance, perfectly adequate and useful to distinguish normal perceptions from illusionary ones. Transferred, however, to the context of scientific inquiry, such a distinction amounts to conflating the contributions of subsystems that must be distinguished in explanatory accounts. The insight that explanatory accounts of mental achievements necessitate a distinction into different mental faculties or modular subsystems can be traced back to the pre-Socratic philosophers. The Pythagoreans distinguished *aisthêsis* and *nous*, a perceiving and a reflecting faculty—a distinction that was among the pillars on which Plato built his conception of a differentiated soul, and that constituted a core element in Aristotle's functional conception of the faculties of the soul. The distinction between a perceptual system proper and the higher cognitive and interpretative capacities by which its outputs can be put to various uses has become a central theoretical element in systematic inquiries into the nature of perception.[8] All the same, the pretheoretical notion of a "perceptual illusion" remains a favorite one in traditional approaches to perception, a fact that indicates how deeply perception theory is still imbued with commonsense intuitions. In a more mature science, the idea of classifying phenomena into pretheoretically anomalous or surprising ones, and "normal" ones, would rightly be regarded as rather odd.

Another pretheoretical classification of phenomena suggested by commonsense intuitions is illegitimately transferred to perception theory. If, as we are inclined to believe, perception basically works the way it phenomenally appears to us, it seems natural to group perceptual phenomena according to characteristic phenomenal attributes that they share. Color appears to be a natural candidate for grouping perceptual phenomena. Corresponding classifications according to what are considered elementary perceptual attributes underlie almost all traditional accounts in perceptual psychology. They are based on the hope that such phenomena share distinctive aspects with respect to the functioning of the perceptual system and thus can be subsumed under a common explanatory framework. Classifications in terms of alleged elementary attributes then constitute the starting point for applying a conception of the

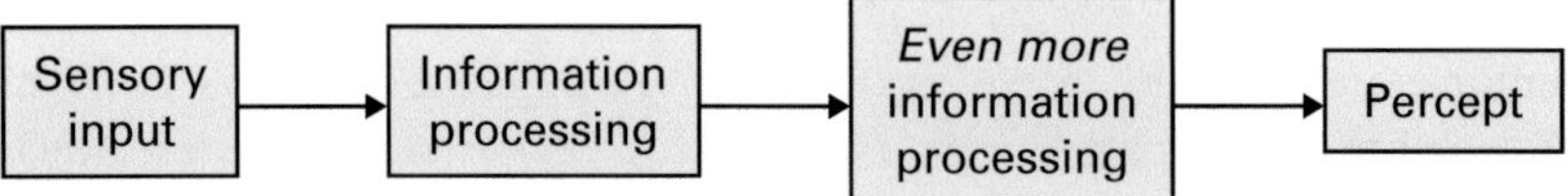

Figure 5.1
Standard model of perception.

nature of perception that has become, implicitly or explicitly, the *standard model of perception* in traditional accounts. The basic form of this model can, following Ulric Neisser's (1976, 17) characterization, be described as depicted in figure 5.1.

Applied to color, a corresponding scheme usually expresses the idea that there are some kinds of "raw colors," or "original colors" that are directly tied to the receptor excitations elicited by the local incoming light stimulus and are transformed and modified in subsequent stages of processing until the percept is yielded.

There are basically two ways open for interpreting this scheme, both equally fatal. On one interpretation, the arrows indicate, in a loose colloquial manner, some temporal sequence, leaving the kind of relations between the boxes, particularly in the last step, entirely unspecified. The second interpretation understands the arrows as indicating consecutive steps of physically definable transformations by which the output of a previous step is transformed to yield the input for the next transformation, by which finally the percept is yielded. While the first interpretation completely bypasses what can be regarded as the *fundamental problem of perception*, namely, to explain how meaningful perceptual categories can arise from a stimulation by physico-geometric energy patterns, the second interpretation amounts to an alleged solution to this problem that is already deeply flawed on conceptual grounds. This has been clearly recognized by Descartes and several others before and after him.

5.1.2 "The World Is Largely as It Appears to Be"

At the core of commonsense intuitions of perception is what is often referred to as naïve realism, that is, the idea that the world, as it really is, independently of an observer, is mirrored in perception. Needless to say, that perception must structurally mirror, or at least not be in conflict with, biologically relevant aspects of the external world. This, however, is hardly an insight but simply rephrases, from a functional point of view, the kind of mental phenomena that have been singled out as an object of inquiry. It does not by any means follow that categories or attributes of perception are categories or attributes of the external world: even if perception would not mirror even in a single case the true manner of being of the external world (whatever that is supposed to be), it still could provide a coupling to biologically relevant structural aspects of it. This has been clearly expressed in Helmholtz's sign theory (and, in different ways, in previous sign conceptions, notably Descartes's).[9]

The explanatory vacuity of the above scheme is, because of our realist convictions about perception, rarely noticed. Our naïve realism regarding perception—which is, needless to say, culturally shaped in complex ways—seduces us to project the categories of the yet-to-be explained output of the perceptual system back to the external world and to use these projections in turn for a description of the external world and the input.[10] By thus conflating the description of the input with that of the output of the perceptual system, core achievements that actually have to be explained become trivialized. According to traditional accounts, the role of the input is to serve as a kind of database from which the distal scene can basically be recovered, yielding the percept. This is, however, as I point out below, a profound misconception of the role of the sensory input.

Part of our ordinary realistic conceptions of perception is that we take it as a matter of course that the meaning of perceptual concepts is largely fixed by reference to categories of the external world. Commonsense conceptions of perception are therefore intimately tied to an *externalist semantics* for perceptual categories and concepts. Because commonsense conceptions have no need to distinguish between the contributions of different subsystems, and thus between the output of a specific subsystem and the potential uses it is put to by other systems, they tend to identify the output of a specific system, namely the perceptual system, with the results of the entire orchestra of mental subsystems, including interpretative ones used for the pragmatics of referring. In ordinary discourse, which has no place for corresponding distinctions, an externalist conception of the meaning of perceptual concepts is thus natural and useful. Scientific inquiries of perception, however, have to pursue a different path, which is dictated by their specific explanatory purposes. Starting with the fact that the output of the perceptual system, namely meaningful categories, is vastly underdetermined, as it were, by the sensory input, namely physico-geometric energy patterns, the core task of perception theory is to understand the conceptual forms with which our perceptual system is endowed and that constitute the basis for meaningful perceptual categories. For this purpose, it is irrelevant whether the sensory input has been causally generated by a real object, by a picture of this object on a computer screen, or by an appropriate stimulation of nerve cells. Fruitful inquiries of perception, notably within ethology and the Gestalt tradition, therefore have pursued an *internalist semantics*, according to which the "meaning" of perceptual categories and concepts is determined by the—yet-to-be-identified—structure of the conceptual forms on which internal information processing is based and by the extremely rich and complex internal interconnections between these conceptual forms. Contrary to our realistic conceptions of perception, meaningful categories are not provided by the external world but by the conceptual endowment of our perceptual system; that is, they are nowhere else than in our head. As Russell (1927, 320) succinctly noted: "Whoever accepts the causal theory of perception is compelled to conclude that percepts are in our heads, for they come at the end of a causal chain of physical events leading, spatially, from the object to the brain of the percipient. We cannot suppose that, at the end of this process, the last effect suddenly jumps back to the starting point, like a stretched rope when it snaps." However, current modes of thinking, both in philosophy and psychology, have remained impervious to this incontrovertible argument. This again bears witness to the disastrous influence which naïve realism exerts on our scientific thinking.

Although naïve realism already founders in the face of the most elementary scientific facts, say, about the properties of our sense organs, it intellectually expresses some of our deepest convictions about the mental activity of perceiving, namely, being in direct touch with a mind-independent world. These convictions are so deeply entrenched in our conception of the world and our interaction with it that their continuous impact on perception research, where they often take the form of a measurement-device (mis)conception of perception, is hardly surprising (cf. Mausfeld 2002).

In philosophy, influential strands have attempted to avoid the obvious problems that even sophisticated and culturally shaped variants of naïve realism are facing, while preserving core elements of realist intuitions about perception. There is a great variety of corresponding philosophical attempts, which go under headings such as "critical realisms," "scientifically informed realism," and so forth. They are generally accompanied by some kind of metaphysical materialism or physicalism, epistemological reductionism, and the idea that the meaning of a percept is determined by its reference to the external world and thus is tied to the truth of a corresponding proposition about the world. Accordingly, to understand the "meaning" of a percept amounts to knowing the conditions under which a corresponding proposition is true. Underlying corresponding philosophical accounts is usually a measurement-device (mis)conception of perception, which takes the form of some kind of local mapping theory of perception. In line with commonsense ideas that perception is a reconstruction of physical world properties and that each perceptual attribute *is* a representation of a corresponding physical aspect, one can then "define" an internal attribute in the following way. First, an external attribute is defined by remapping aspects of the percept back to the world; from this external attribute, then, a "corresponding" internal attribute is established. According to such conceptions, a percept can essentially be reduced to its representational content, which in turn can be identified with a corresponding proposition about an external physical state. Corresponding ideas have fueled a spectrum of sophisticated philosophical discussions. Whatever their philosophical merits might be, they fortunately do not arise within explanatory frameworks of perception theory (as is particularly evident from ethological frameworks).[11] Notions such as "truth conditions," "veridicality," "reference to the world," and "perceptual content" belong to the level of persons and do not enter into explanatory accounts of specific subsystems, such as the perceptual systems. Rather, they describe, mostly in the technical context of philosophical analyses, some aspects of the uses and interpretations to which the outputs of such systems are put by the entire person.

5.1.3 "Colors Out There in the World"

Color has been serving as a paradigmatic study case for philosophical attempts to justify the realism inherent in commonsense conceptions of perception.[12] This is explicitly expressed by McLaughlin (2003, 475): "I persist in the common-sense belief that . . . colours are really 'out there.' Colours are mind-independent properties of things in the physical world: they are objective properties and our visual experience puts us in touch with them." Corresponding presumptions are the basis of a great variety of philosophical accounts of color, whether they conceive of colors as intrinsic properties of external objects (e.g., Byrne and Hilbert 2003), regard them as the physical basis for the disposition to look red to a normal observer under normal conditions (McLaughlin 2003; Cohen 2003), or argue for a "pluralistic realism" (Matthen 1999).

These attempts in particular share with commonsense intuitions of perception the externalist idea that color experiences "represent" an observer-independent property, that is, that colors possess *representational content*, which is given by the external properties (single or composed ones) to which they refer (e.g., Matthen 1988). The primary goal of such *objectivist*, or *relational*, accounts of color is to establish a correspondence between colors and purely physically definable counterparts.[13] These counterparts can be conceived of as, for instance, the "categorical bases of the dispositions to elicit colour experience" (e.g., Jackson and Pargetter 1987; McLaughlin 2003), or as a single physical property that is "truly" represented through color vision (e.g., Byrne and Hilbert 2003).

With respect to corresponding formulations, such as "red things will be disposed to look red because they have the property of redness" (McLaughlin 2003, 480), Maxwell's (1970, 33–34) remark seems still in place:

> A *redefinition* such as: "to be red" means to look red under standard lightning conditions, etc., is absurd since it requires that "red" in the *definiens* have a different meaning from "red" in the *definiendum*—indeed that it have the primary, occurent meaning. If this defect is repaired and a viable causal or dispositional redefinition of color words is produced so that they may be properly predicated of physical entities, then "color" in this new sense will no longer be first order properties, but rather, structural ones. Moreover, we will still need color words that have the primary, occurrent sense to refer to the first order properties that are exemplified in our experience. These cannot be eradicated by defining words. I am sorry to take up space with such obvious matters, but sad experience has indicated that it often is necessary.

In physicalist accounts, as proposed by, for example, Byrne and Hilbert (2003), the goal is to show that under certain assumptions, representational and phenomenal content correspond to each other (except for certain "illusions"). Even if there were such a correspondence, however, no clear-cut physical notion of "color out there in the world" can be justified by it, because "even if there are colored entities—even colored surfaces as we ordinary conceive them—in the physical environment, we never see them and their being colored plays no role in any process whereby we acquire or confirm knowledge. We thus have no more (perhaps less) reason for believing that there are instances of color in the external world than we do for believing in the existence of disembodied spirits," as Maxwell (1968, 170) rightly noticed.[14]

Philosophical attempts to provide a justification for a notion of "colors out there in the world" essentially amount to normatively and prescriptively introducing a kind of philosophically purified language for a discourse about "color out there." Accordingly, what color *really* is, is what it *should* be, given certain philosophical presumptions about the nature of perception. Prominent among these presumptions is the notion that color is an autonomous and unitary attribute (an exception is Matthen 1999), which can be studied as detached from general inquiries into the conceptual bases of our perceptual system. In the context of perception theory, however, no physicalist concept of "color out there" is required, and no issue of subjectivity or objectivity of colors arises.[15] In fact, corresponding notions would, except for metatheoretical discourse, not only be unmotivated but also express an anthropocentric and antibiological attitude.

5.2 "Color" as Part of Different Conceptual Forms of the *Perceptual System*

Prevailing accounts of color in perceptual psychology and in color science have been almost entirely concerned with sensory-based processes and transformations (see figure 5.1). Underlying these accounts, typically, is the concept of "color per se" as a homogeneous and autonomous attribute and a measurement-device (mis)conception of perception. While corresponding conceptions are useful for colorimetrical and certain neurophysiological purposes, they have made the field of color science rather sterile with respect to the explanatory goals of perception theory. Overwhelming empirical evidence indicates that the notion of a homogeneous and autonomous "color per se" is of no explanatory avail for perception theory because the "processes, occurring in acts of perception, instead of being separable into colour-, space- (local sign), and form-processes are processes of field organization; colour, place and form being three interdependent aspects of this general event" (Koffka and Harrower 1931, 215). Color therefore cannot be studied as detached from inquiries into the "processes of field organization" and the structure of the conceptual forms in which it figures as an attribute.

The traditional approach's preoccupation with aspects of processing has been at the expense of inquiries into the structural format of the perceptual "data types," on which computational processes by definition have to be based. Rather, these data types are tacitly borrowed from commonsense conceptions by using the yet-to-be-explained *output categories* of the visual system, such as "surface," "shadow," or "illumination," for a physical description of the input. Conflating perceptual and physical categories in this way trivializes one of the core problems of perception theory, namely, the identification of the conceptual forms with which our perceptual system is biologically endowed and which sharply constrain the perceptual achievements within the class of achievements that are logically compatible with a given type of sensory input. In contrast, orthodox conceptions and approaches attempt to build up from the sensory input—on the basis of thin sets of quite elementary perceptual primitives—the complex categories and concepts that characterize the output of the perceptual system. Only by illegitimately transferring commonsense intuition to perception theory do such conceptions gain some apparent plausibility. However, perception theory is not constrained to preserve our pretheoretical intuitions about perception. In developing explanatory frameworks for perception, we have to be willing to jettison ordinary intuitions about it whenever doing so serves our explanatory needs, and to divest theoretical notions of the distorting residues of commonsense intuitions.[16]

5.2.1 Outline of the Logical Structure of the Perceptual System

Recent decades have brought forth, for the first time in the long history of perception theory, a convergence of quite different fields of inquiry—namely ethology and comparative research, perceptual psychology, and investigations of the perceptual capacities of newborns and very young children—on the contours of a theoretical framework about what appear to be basic principles of perception. The core ideas of this framework have a long history in the *philosophia naturalis* (for some historical aspects, see Mausfeld 2002, appendix). They were (partly) taken up by Helmholtz in his sign theory of perception and became the basis for Gestalt psychology and ethology. Already at the beginning of the last century, the empirical and theoretical evidence in support of corresponding ideas was enormously rich. But only after advances in the computational sciences provided a new conceptual apparatus could these ideas be further explored in a fruitful manner.

Within computational approaches, it is patent that all computational processes require specification of the data format on which they are based. Though prevailing approaches, in line with empiricist conceptions of the mind, exhibit a preference for data formats that can be defined in terms of elementary sensory aspects, pervading evidence has been accumulated, notably in the Gestaltist and ethological tradition, that our perceptual system is biologically endowed with a rich set of conceptual forms. Hence, the core task of perception theory is to identify the structure of these conceptual forms and to better understand how they can combine to produce the kinds of complex perceptual concepts that characterize the output of the perceptual system.

The theoretical picture that is emerging from corresponding inquiries is still very skeletal and inevitably has to be based on considerable theoretical speculation. But even in its currently rudimentary form, it has already yielded intriguing results with respect to a range of significant phenomena and has suggested novel and fruitful questions about the internal architecture of perception. Furthermore, it is consonant with well-supported broader metatheoretical perspectives on the nature of mental phenomena (see, e.g., Strawson 2003; Hinzen 2006). But our confidence in this theoretical picture is, as always in the natural sciences, predominantly due to the fact that it is the result of a theoretical convergence of quite different disciplines.

In its more general aspects, it is rarely spelled out explicitly and therefore has to be extracted and abstracted from the relevant literature. Before putting color into the context of this emerging theoretical picture of the basic principles of perception, I will briefly describe its skeleton in a somewhat bold and, unavoidably, oversimplified way. I have outlined in more detail the general logical structure of the perceptual system on which different disciplines increasingly seem to converge in Mausfeld (2003a).

At core of this theoretical picture is the idea that in more complex organisms, the sensory input serves as a kind of sign for the activation of biologically given conceptual forms, which determine the data format of the computational processes involved. Conceptual forms can be regarded as semantic atoms of the internal semantics of perception, as providing the core semantics of minimal meaning-bearing elements. The conceptual forms, say for *surface, food, enemy,* or *tool,* cannot be reduced to or inductively derived from the sensory input but are part of the specific biological endowment of the organism under scrutiny.[17] To account for the relation between the sensory input and the irreducible and complex perceptual concepts that constitute the output of the perceptual system, a distinction between a *Sensory System* and a *Perceptual System* suggests itself (needless to say, as an idealization).[18] We can categorize as *Sensory System* those computational processes that are closely tied to the physico-geometric data format by which we describe the sensory input. The *Sensory System* thus deals with the transduction of physical energy into neural codes and their subsequent transformations into codes that are "readable" by and fulfill the structural and computational needs of the perceptual system; we can refer to these codes as *cues,* or *signs.* The *Sensory System* preprocesses the sensory input—in a way that is dynamically interlocked with the specific requirements of the perceptual system—in terms of a rich set of essentially input-based concepts that are tailored to the structural and computational demands of the *Perceptual System.* The *Perceptual System,* on the other hand, can be conceived as a self-contained system of perceptual knowledge, which is coded in the structure of its conceptual forms. It contains, as part of our biological endowment, the exceedingly rich set of complex conceptual forms in terms of which we perceive the "external world," such as *surface, physical object, intentional object, event, potential actor, self, other person,* or *event,* with their associated attributes such as *color, shape, depth,* or *emotional state,* and their appropriate relations such as *causation* or *intention.*[19] The *Perceptual System* therefore comprises the rich perceptual vocabulary in terms of which the signs delivered by the *Sensory System* are exploited, and provides the computational means to make its perceptual concepts accessible to higher-order cognitive systems. The sensory codes serve a dual function. They activate appropriate conceptual forms and thus determine the potential data formats in terms of which input properties are to be exploited. Furthermore, they assign concrete values to the free parameters of the activated forms. The conceptual forms that are yielded, in a given input situation, as outputs of the *Perceptual System* are *triggered* by the codes of the *Sensory System,* rather than being computed or inductively inferred from them. We might loosely think of the triggering functions as an interface function that takes specific sensory codes as an argument and activates conceptual forms. The triggering function renders, in principle, the relation between the sensory input and the conceptual forms epistemologically arbitrary.

The abstract functional architecture of perception as suggested by the current convergence of different disciplines can be schematically summarized as depicted in figure 5.2.

We can think of the conceptual forms of the *Perceptual System* as abstract structures, each of which has its own proprietary types of parameters, relations, and transformations that govern its relation to other conceptual forms and to sensory codes.[20] By means of their relational parameters, conceptual forms build systematically connected packages. The *Perceptual System* generates, at its interface to higher-order interpretative systems, packages of activated conceptual forms that consist of legible instructions for these subsequent systems. The values of the free variables of a conceptual form in

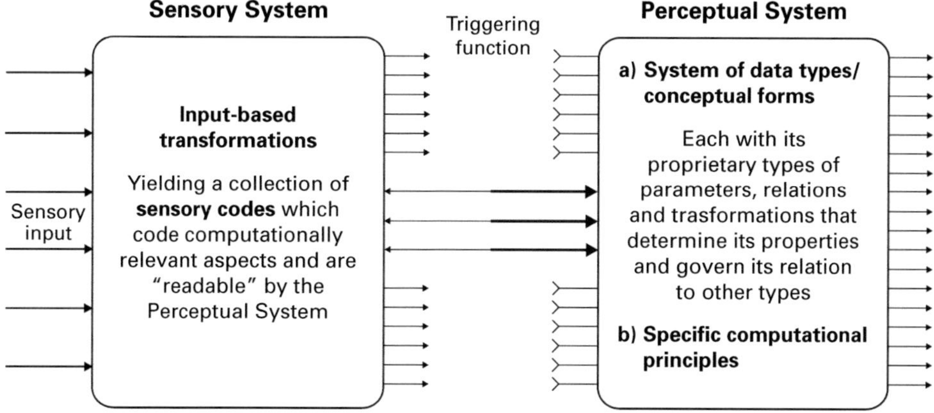

Figure 5.2
Basic structural form of the functional architecture of the perceptual system.

general will not be—and, for subsequent computational processes, need not be—exhaustively specified by the activating input. The examples in figure 5.2 refer to two types of conceptual forms that are of fundamental importance for color perception, namely, types for *surfaces* and types for *ambient illumination* (with some placeholders for the yet-to-be identified relations, transformations, and parameters). Empirical evidence suggests that the *Perceptual System* routinely operates semantically with underdetermined conceptual forms (a structural feature that appears to extend to subsequent systems). Accordingly, the output of the *Perceptual System*, at its interfaces to subsequent systems by which meanings are assigned in terms of 'external world' properties, thus needs not to be semantically determinate or unique but only "good enough" for the semantic needs of the subsequent systems. Underspecification of conceptual forms greatly enhancing the compositional versatility of the *Perceptual System* (Mansfeld, in press). Furthermore, by postponing disambiguation to higher-order interpretative systems, the *Perceptual System* can increase its global stability with respect to the superordinate "interpretations" provided at its interfaces to subsequent systems. This protects the system from settling, under insufficient or "impoverished" input situations, on some definite interpretation that would have to be changed to an entirely different interpretation following a small variation in the input.[21]

The structure of the conceptual forms is only partly visible at the surface of the phenomenal percept. In particular, we do not notice in our phenomenal experience that the conceptual forms involved are underspecified. Rather, the systems that use the outputs of the *Perceptual System* for constructing the phenomenal percept must be furnished with specific computational means to completely specify its phenomenal appearance at each moment.

From an evolutionary point of view, it is usually taken to be a matter of course that the elements of the perceptual system have their specific conceptual and computational structure because they are used to tie the organism to its environment. Conceptual forms by themselves, however, do not refer to the physical world. Rather, their relation is only to other conceptual forms. Their specific form is evolutionarily shaped by the requirement to be functionally adequate, in the sense that they have to fit into the entire perceptual architecture, including its interfaces with the *Sensory System*, the motor system, and the higher cognitive systems. Furthermore, their form is co-determined by physical and architectural constraints[22] as well as by contingent aspects in the course of the evolutionary development of the brain. Because of this, conceptual forms have their own properties, which can be rather surprising when viewed exclusively from the perspective of an adaptive coupling to the external world. Nevertheless, the most complex perceptual achievements—for instance, seeing invisible properties of objects (e.g., pertaining to material qualities), intentional properties of objects (e.g., tools), or mental states of others—were made possible only by decoupling the output of the *Perceptual System* from the given sensory input information and by furnishing the *Perceptual System* with perceptual knowledge, coded in its conceptual forms, which is not derivable from the sensory input or from general sensory-based computations.

Evolutionary observations suggest that the emergence of abstract conceptual forms in the functional architecture of the (evolutionarily younger) *Perceptual System* is the result of an increasing modular differentiation of the underlying neural substrate. Modularity is generally regarded as the basis of the evolvability of biological systems.[23] In the evolution of complex computational systems, an increasing amount of modularity increases the computational need to integrate the outputs of a great variety of subsystems into a common conceptual structure, which has to be on a higher level of abstraction than each of the subsystems that feed into it. In this sense, modularity drives abstraction. The conceptual forms of the *Perceptual System* can be regarded as the result of corresponding processes of abstraction in the evolution of the brain.[24]

These few remarks may suffice in the present context to convey, on a highly abstract level, some core aspects of the theoretical picture that is currently being achieved in the convergence of ethological, developmental, and psychological studies of perception. This conception of perception can, in its basic spirit, be regarded as a fusion of ethological and computational ideas. How radically this conception deviates from commonsense intuitions becomes apparent when one realizes that according to it, there is no difference in principle between, say, the perception of colors and the perception of mental states of others; in both cases, the sensory input serves as a sign to trigger certain conceptual forms, in which *color* and *mental states of others* figure as internal attributes.

5.2.2 The Nonhomogeneity and Nonautonomy of Color

Rich evidence of very different types supports the view that color is not a unitary attribute. Rather, color-type parameters figure, with different properties, in different conceptual forms and computational subsystems.[25] This evidence ranges from comparative studies (e.g., Santos et al. 2001), developmental findings (Leslie et al. 1998), and clinical observations (e.g., Gelb 1920; Stoerig 1998) to experimental findings in perceptual psychology. In Mausfeld (2003a) I deal in detail with the empirical evidence indicating that color-type parameters figure, with different structural properties, in conceptual forms for *surfaces* and for *ambient illumination*, yielding a kind of dual coding of color with intricate interactions and transitions. Here, I will deal with different structural properties (in particular interrelations with conceptual forms for *space* and *ambient illumination*) of color-type parameters in different types and instances of *surfaces*. The natural starting point for this is Katz's distinction of different "modes of color appearance." I briefly discuss this distinction before I turn to the problem of the perception of *material qualities*, which perceptual psychology has, for obvious reasons, notorious difficulties dealing with.

Katz (1911) distinguished, on the basis of phenomenological observations, several types of color, and descriptively classified them into what he called "modes of appearance." Among the ways in which colors appear in space, he in particular distinguished *aperture color* from *surface color*. Aperture colors have no orientation in space and always appear fronto-parallel. Furthermore they appear spatially two dimensional and as having no determinable distance but still render it possible "to visually dive into them to different depths." Surface colors, on the other hand, can have any kind of afrontal orientation and can exhibit a granularity of structure and texture. Only surface colors can appear to have a separate *illumination value* (i.e., as being illuminated). The guiding idea behind this classification is that the appearances of color phenomenally segregate into mutually exclusive categories because they mirror internal processes or states of "essentially different nature." In the wake of Katz's subtle phenomenological observations, which instigated a wealth of further phenomenological explorations (e.g., Martin 1922), controversies arose about whether the different modes indeed mirror internal states of essentially different nature or whether they are merely due to the influences of a "modifying context" on the "raw original colors." The question of whether the modes of appearance are different colors per se or the same colors per se, which are merely modified by "context effects," is, *in nuce*, the question of whether color is or is not a unitary concept with homogeneous coding properties. This question has been intensely debated in the older literature. From the perspective of the colorimetric paradigm, "the mode of appearance does not change colour *per se*. . . . The modes of appearance are simply the various kinds of context or setting in which color is perceived" (Jones 1953, 8). This has been the prevailing view in perceptual psychology since then. The underlying conception of a *raw color, original color*, or *color per se*, which can be indexed by contextual situations to yield colors of transparent, voluminous, glowing, lustrous (and so forth) appearance, has veiled the important theoretical issues involved and brought forth a deeply flawed theoretical picture of color perception. This conception in particular has concealed the intricate way in which color-type parameters are interwoven with other internal attributes within the same conceptual form and other conceptual forms. Among those, *space* (which itself is not a homogeneous concept but comprises various conceptual forms pertaining to spatial aspects) stands out as of unique importance. Katz, following Hering, clearly noticed how intimately color is interwoven with the organization of space, as can be witnessed by his felicitous expression of a "marriage of colour and space." Rich corresponding empirical evidence had been marshaled by Bühler, Kardos, Gelb, and Koffka. These authors also realized that the specific nature of these dependencies cannot, as often has been suggested within traditional accounts, be understood in terms of context-specific modifications regarding rather elementary stimulus aspects, such as form, texture, and so forth (cf. Mausfeld, 2003a, for a more detailed account of relevant findings). Gelb (1929, 672) regarded these dependencies as due to an "expression of a certain structural form of our perceptual visual world," and Cassirer (1929, 155), due to the "very primordial format of organization." In terms of the theoretical framework outlined above, these dependencies mirror the structural nature of the conceptual forms, in which color-type parameters figure. These conceptual forms code our biologically given perceptual knowledge, which comprises a rich internal vocabulary for "material qualities." These internal material qualities go far beyond purely visibly definable attributes and are intrinsically transmodal, which again indicates the high degree of abstractness of the conceptual forms involved. They pertain, for instance, to *stability*, *tenacity*, *ruggedness*, or to attributes such as *lustrous*, *hard*, *juicy*, *dry*, and so forth.[26] The issue of material qualities was almost entirely neglected in traditional accounts because these attributes were erroneously conceived to be obtainable by some associative machinery from elementary sensory attributes. An early exception were lustrous appearances, first studied by Dove (1850), which were considered phenomena of great theoretical importance by Helmholtz, Hering, Kirschmann, Wundt, Bühler, and Katz.

Still, perceptual psychology mostly disregarded these phenomena as mere "side effects" in perceptual psychology until recent years, when attention was drawn to them from an entirely different, technological perspective. The recent interest in the perceptual bases for material appearances emerged from problems of rendering the corresponding qualities on a computer screen.[27] Attempts to identify relevant 2-D image features that yield certain material appearances by mathematically analyzing specific causally responsible physical properties of different types of material proved to be arduous and not very fruitful for perceptual purposes.[28] Rather, these analyses suggested that, as Fleming and Bülthoff (2005, 346) argued in the context of translucency, the physical regularities underlying the interaction of light and surfaces "are too complex for the visual system to estimate intrinsic physical parameters." Instead, as experimental studies in this context have indicated, complex perceptual surface qualities have their specific and, with respect to the underlying physical regularities, often rather simple triggering conditions.[29] These and similar observations, however, cannot be regarded as pertaining merely to problems of "cue" integration. Rather, they point to the inadequacy of all theoretical perspectives that downplay the complexity of the conceptual forms involved.

Classic studies on lustrous appearances already had shown that the types of sensory codes exploited by the internal vocabulary for material qualities are exceedingly variegated and idiosyncratic and cannot be reduced to or understood from physical considerations of physical material properties. This is already obvious from Helmholtz's (1867) well-known stimulus configuration for yielding lustrous appearances, as shown in figure 5.3.

Under stereoscopic viewing conditions, the binocular combination of the two line drawings of inverted luminance contrast yields a vivid lustrous appearance. Similar

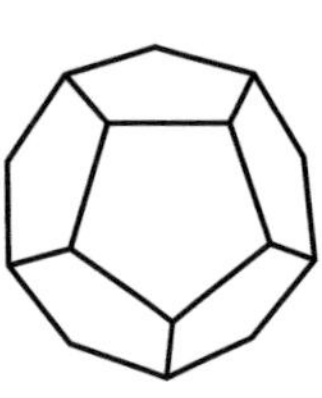
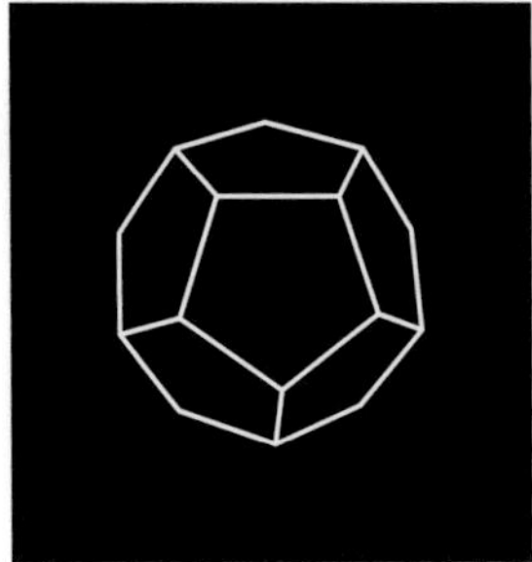

Figure 5.3
Stereoscopic luster.

appearances can be produced by various types of highly reduced stimulus configurations.[30] From corresponding studies, Bixby (1928) obtained instructive phenomenological descriptions of these appearances. Subjects describe them as "light and dark, somehow seen as if in the same place at the same time," "a sort of blending or fusion of light and dark," "a peculiar commingling or sifting-together of dark and light," and "a bulky experience of luminous greyish white." Apparently, the sensory input pattern is sliced into perceptual layers, which pertain to conceptual forms of different types, namely to a surface type and to an illumination type, whose specific interrelations result in the activation of the kind of surface-type attribute that codes a specific internal material quality. The great diversity of highly reduced stimulus configurations that give rise to lustrous appearances shows that material color appearances cannot be derived from an analysis of the sensory-basis level. Rather, the internal logic underlying these appearances can only be revealed on the level of the *Perceptual System* and its conceptual forms. Hering (1879, 576) was guided by an intuition of this kind, which led him to propose what appears to be an essentially correct conjecture: lustrous appearances arise when there is a "surplus of light" with respect to the permissible values of the corresponding free parameters in the conceptual forms for surface and ambient illumination, which yields a "cleavage of sensation" into shallow depth layers of an "essential" and an "accidental" color component of a surface. The shallow depth segmentation[31] involved in most material color appearances again is evidence for the "marriage of colour and space" in the internal coding of our rich vocabulary for material properties.

The neglect of material colors in traditional accounts of color perception again testifies to the theoretical distortions that arose from the "errors of the application of colorimetric thinking to perception" (Evans 1974, 197). The perceptual primacy of material colors is also mirrored in the way we linguistically exploit the output of the *Perceptual System*. For instance, Hochegger (1884, 36) found it "remarkable that etymological investigations on abstract colour names always find the roots in words that mean shiny, glowing, burning, shimmering, dingy, burnt, etc. Even the expressions for colours which seem to be abstract are, in fact, not primordial but rather emerged from paleness, brightness, glossy, matt, dingy etc." In the transition from the ancient Greek's emphasis on forms of light, such as brightness, luster, and the changeability of colors,[32] to the subsequent culturally shaped progression toward an increasingly abstract color vocabulary, we can observe a shift from color appearances as material properties and forms of light to an abstractive notion of color per se as an intrinsic object property.

The notion of color per se as a unitary attribute has been abstracted in the course of a long technology-shaped cultural process. Although such an abstraction is based in perceptual achievements, it shares with a vast class of similar perceptual abstractions that it is essentially a cultural artifact. Of course, the fact that our cognitive capacities provide us with the means to arrive at an abstract notion of color per se is of interest and in need of explanation in itself. However, that we can cognitively attain the concept of color per se implies nothing about whether this concept plays any role in the computational structure of the perceptual system. The available evidence strongly suggests that it does not. In the context of perception theory, a corresponding conception of color has veiled and obfuscated the important theoretical issues with which any account of color perception must cope.

Notes

1. As Wierzbicka (1996, 287) correctly observed, albeit within a different theoretical perspective, "'Colour' is not a universal human concept. It can of course be created in all human societies, just as the concepts 'television,' 'computer,' or 'money' can . . . In all cultures, people are interested in 'seeing' and in describing what they see, but they don't necessarily isolate 'colour' as a separate aspect of their visual experience." See Mausfeld (2003a) for further evidence and corresponding references.

2. This remark does not apply, of course, to the Grassmann space obtained from metameric matches, which, however, does not say anything about color appearances but only represents the equivalence classes on spectral energy distributions yielded by the three types of photoreceptors.

3. For a discussion of theoretical interpretations of center-surround configurations, see Mausfeld and Niederée (1993).

4. All three-dimensional models of color coding imply that in the case of chromatic adaptation, the values of two different operations coincide, namely, the point of achromatic appearance and the convergence point of lines of the same hue. This implication is gravely false, as Ekroll et al. (2002) have shown.

5. Actually, the prevailing conceptions of color go along with an extremely impoverished and oversimplified account of color appearances.

6. See Mausfeld (1998, 2003a) and Mausfeld and Andres (2002) for a more detailed account.

7. The idea of color per se, however, is not tied to these philosophical positions but also prevails in other philosophical perspectives (e.g., Hardin 1988).

8. For instance, Arnauld and Nicole stated in their Port-Royal Logic of 1642 "that there can be no illusion or error" in perception, and that "the whole error solely results from our false judgments" (1685/1996, 75). In the same vein, Kant in his *Anthropologie* (1.10) remarks: "The senses do not deceive us . . . because they do not make judgements at all, and that is the reason why the error always is due to the intellect." And similarly, Helmholtz (1855, 100): "The senses cannot deceive us, they work according to their established immutable laws and cannot do otherwise. It is us who are mistaken in our apprehension of the sensory perception."

9. See Yolton (1996, ch. 8) for an account of Descartes's sign conception of perception.

10. The predisposition to take perceptual concepts for "things in the real world" is the distinguishing mark of all of our mental activity. Kant referred to it as the "transcendental illusion." The transcendental illusion is the propensity to "take a subjective necessity of a connection of our concepts . . . for an objective necessity in the determination of things in themselves" (*Critique of pure reason*, A297/B354). Due to this propensity, whose influence cannot be remedied by intellectual insight into it, we inevitably tend to mistake our own mental categories to hold "objectively" (cf. Grier, 2001).

11. Although I will touch on a few more general presumptions that certain philosophical approaches share with commonsense intuitions about perception, I do not intend to embark on the issues that are, on a high level of technical sophistication, discussed in the field of color philosophy. The reason for this is that corresponding philosophical investigations of color do not seem to me to have any bearing on perception theory.

12. 'Color' as a special domain of philosophical enquiry attains, it appears to me, its particular fascination from tacit commonsense preconceptions of perception. The class of attributes of our perceptual experiences is exceedingly rich. Yet, there is no, say, 'philosophy of motion', 'philosophy of texture', or 'philosophy of timbre'.

13. For comparison, imagine a perceptual system (or an entire organism) that is exactly like ours with two exceptions: It possesses in addition a visual sensitivity for differences in the polarization of light, and comprises furthermore the basic perceptual attribute 'teavy' (to borrow Carnap's famous term). Assume further that this attribute 'teavy' and its qualitative and quantitative instantiations have as triggering conditions equivalence classes (analogous to metameric classes) of the specific polarization effects of certain types of material, say soil and desert sand (e.g., Chen and Rao 1968). Although the perceptual material quality 'teavy' has a physical basis, it would hardly be of any scientific avail to regard 'teaviness' as an aspect of the external world.

14. See also Descartes's remark (*Principles*, 1.70) that, in the context of naturalistic inquiry, "we cannot find an intelligible resemblance between the colour which we suppose to be in objects and that which we experience in our sensation."

15. Also, detached from specific domains of inquiry, no issues of *what colors really are* arise. In science, we consider as real whatever figures in our currently best explanatory theories about a range of phenomena of the natural world. In this sense, we can regard as real an attribute "color" that figures in an explanatory framework for core principles of the perceptual system, whatever the specific properties of this internal attribute will turn out to be. Beyond that, no issues of "realness" arise.

16. As Chomsky (e.g., 2000) has convincingly argued for in the context of linguistic inquiries.

17. It is important to be aware that these linguistic appellations of the conceptual forms of the *Perceptual System* are only makeshift descriptions of non-linguistic entities (which are, of course, further shaped by the properties of subsequent linguistic and interpretative systems).

18. I use the term *perceptual system* to refer loosely to the entire modular system of perception, which includes the *Sensory System* as well as the *Perceptual System* as characterized here.

19. According to the architectural conception proposed here, core aspects of most types of build-in 'knowledge' as assigned to "core knowledge systems" by Spelke (2000) or to "conceptual-intentional systems" in the minimalist program of linguistics have to be imputed to the *Perceptual System* already.

20. In philosophical discourse, corresponding conceptions of concepts are sometimes referred to as Cartesian conceptions (e.g., Fodor 2004). For Descartes, "ideas" are nothing but forms (as expressed in his *Meditations*, in the reply to the fourth set of objections: "ideæ sint formæ quædam"). See Yolton (1984) for Descartes's and Arnauld's conceptions of "idea."

21. See Mausfeld (2003b, 51ff) for a discussion of some empirical evidence related to color.

22. See Stewart (1998) and Carroll (2005) for introductionary expositions from different perspectives.

23. See, e.g., Kirschner and Gerhart (1998); Kitano (2004); Wagner, Mezey, and Calabretta (2005).

24. For some considerations as to the evolutionary emergence and the "biological realness" of abstract conceptual forms, see Mausfeld (in press).

25. Linguistic evidence also appears to provide some indirect support for how intimately color is interwoven with the kind of perceptual object to which it pertains. Interestingly, color-type attributes also figure with different structural and semantic properties in different items of the I-lexicon (Chomsky 2000, 35ff, 125ff). For this and other reasons, it seems to be an intriguing possibility that, from an evolutionary perspective, the conceptual forms of the *Perceptual System* provided the seeds for the development of the items of an I-lexicon.

26. See, e.g., Schapp (1910), who drew attention to the fact that "one directly sees tenacity, brittleness, obdurateness, bluntness and many other attributes for which we lack linguistic descriptions," an ability he ascribed to a given internal vocabulary rather than to sensory-based associative processes.

27. These problems have an interesting counterpart in art history. The simulation of material qualities on a canvas had been regarded as a particular challenge in painting, notably in Dutch renaissance art (Gombrich 1976). Although already Alberti, in his *Trattato della pittura* (1435/1972), had recognized that by a proper juxtaposition of white and black only, the impressions of gold, silver, and glass can be elicited, a realistic impression of material colors in painting turned out to be exceedingly difficult to achieve.

28. As an example, see Koenderink and Pont (2002) for the case of velvet appearances.

29. Corresponding observations can also be made with respect to the auditory perception of material qualities, where relatively elementary features are exploited as cues for complex material properties (e.g., Carello, Wagman, and Turvey (2005).

30. For instance, lustrous appearances can by obtained from spatially homogeneous haploscopically presented half images of different temporal luminance modulations, or from monocularly viewed 3-D objects (e.g., polyeder), the luminance of whose spatially homogeneous faces is independently modulated (Mausfeld and Wendt 2006).

31. This depth segmentation can be obtained without any depth cues, as traditionally conceived, in the sensory input, or, as in the case of Helmholtz's display, with the support of sensory codes for stereoscopic depth.

32. See, e.g., Rowe (1974): "The so-called 'primitiveness' of Greek color-terminology can be seen as a reflection of a greater awareness of changeability of colors in the natural environment; an abstract vocabulary is in a real sense artificial, and in reducing the world of color to a few simple categories, over-simplifies it, and robs it of its subtlety."

References

Alberti, L. B. 1435/1972. *On Painting.* London: Penguin.

Arnauld, A., and P. Nicole. 1685/1996. *Logic or the Art of Thinking.* Cambridge: Cambridge University Press.

Bixby, F. L. 1928. A phenomenological study of luster. *Journal of General Psychology* 1: 136–174.

Byrne, A., and D. Hilbert. 2003. Color realism and color science. *Behavioral and Brain Sciences* 26: 3–21.

Carello, C., Wagman, J. B., and Turvey, M. T. 2005. Acoustic specification of object properties. In *Moving image theory: Ecological considerations,* J. D. Anderson and B. F. Anderson, eds., 79–104. Carbondale, IL: Southern Illinois University Press.

Carroll, S. B. 2005. *Endless Forms Most Beautiful: The New Science of EvoDevo.* New York: Norton.

Cassirer, E. 1929. *Philosophie der symbolischen Formen. Dritter Teil: Phänomenologie der Erkenntnis.* Berlin: Bruno Cassirer.

Chen, H. S., and Rao, C. R. N. 1968. Polarization of light on reflection by some natural surfaces. *Journal of Physics, D: Applied Physics* 1: 1191–1200.

Chomsky, N. 2000. *New Horizons in the Study of Language and Mind.* Cambridge: Cambridge University Press.

Cohen, J. 2003. Color: A functionalist proposal. *Philosophical Studies* 113: 1–42.

Dove, H. W. 1850. Ueber die Ursachen des Glanzes und der Irradiation, abgeleitet aus chromatischen Versuchen mit dem Stereoskop. *Poggendorffs Annalen* 83: 169–183.

Ekroll, V., F. Faul, R. Niederée, and E. Richter. 2002. The natural center of chromaticity space is not always achromatic: A new look on color induction. *Proceedings of the National Academy of Sciences* 99: 13352–13356.

Evans, R. M. 1974. *The Perception of Color.* New York: Wiley.

Fleming, R. W., and H. H. Bülthoff. 2005. Low-level image cues in the perception of translucent materials. *ACM Transactions on Applied Perception* 2: 346–382.

Fodor, J. A. 2004. Having concepts: A brief refutation of the twentieth century. *Mind and Language* 19: 29–47.

Gelb, A. 1920. Über den Wegfall der Wahrnehmung von Oberflächenfarben. *Zeitschrift für Psychologie* 84: 193–257.

Gelb, A. 1929. Die "Farbenkonstanz" der Sehdinge. In *Handbuch der normalen und pathologischen Physiologie.* Bd.12, 1.Hälfte. Receptionsorgane II. A. Bethe, G. v. Bergmann, G. Embden, and A. Ellinger, eds., 594–678. Berlin: Springer.

Gerhart, J., and Kirschner, M. 1997. *Cells, Embryos and Evolution: Toward a Cellular and Developmental Understanding of Phenotypic Variation and Evolutionary Adaptability.* Oxford: Blackwell.

Gombrich, E. H. 1976. *The Heritage of Appelles: Studies in the Art of the Renaissance.* Ithaca, NY: Cornell University Press.

Grier, M. 2001. *Kant's Doctrin of Transcedental Illusion.* Cambridge: Cambridge University Press.

Hardin, C. L. 1988. *Color for Philosophers.* Indianapolis, IN: Hackett.

Hartwell, L. H., Hopfield, J. J., Leibler, S., and Murray, A. W. 1999. From molecular to modular cell biology. *Nature* 402**:** C47–C52.

Helmholtz, H. v. 1855. Über das Sehen des Menschen. In *Vorträge und Reden.* 4. Aufl., Bd.1, 1896. Braunschweig: Vieweg.

Helmholtz, H. v. 1867. *Handbuch der Physiologischen Optik.* Hamburg: Voss.

Hering, E. 1879. Der Raumsinn und die Bewegungen des Auges. In *Handbuch der Physiologie der Sinnesorgane, Bd. 3,* L. Hermann, ed., 343–601. Leipzig: Vogel.

Hinzen, W. 2006. *Mind Design and Minimal Syntax.* Oxford: Oxford University Press.

Hochegger, R. 1884. *Die geschichtliche Entwicklung des Farbensinnes.* Innsbruck: Wagner'sche Universitätsbuchhandlung.

Irwin, E. 1974. *Colour Terms in Greek Poetry.* Toronto: Hakkert.

Jackson, F., and R. Pargetter. 1987. An objectivist's guide to subjectivism about color. *Revue Internationale de Philosphie* 160: 129–141.

Jones, L. A. 1953. The historical background and evolution of the Colorimetry Report. In *The Science of Color,* Commitee on Colorimetry, eds., 3–15. Washington, DC: Optical Society of America.

Katz, D. 1911. *Die Erscheinungsweisen der Farben und ihre Beeinflussung durch die Individuelle Erfahrung.* Leipzig: Barth.

Kirschner, M., and J. Gerhart. 1998. Evolvability. *Proceedings of the National Academy of Sciences* 95: 8420–8427.

Koenderink, J. J., and S. C. Pont. 2002. The secret of velvety skin. *Special Issue on Human Modeling, Analysis and Synthesis of Machine Vision and Applications* 14: 260–268.

Koffka, K., and M. R. Harrower. 1931. Colour and organization. II. *Psychologische Forschung* 15: 193–275.

Leslie, A., F. Xu, P. Tremoulet, and B. Scholl. 1998. Indexing and the object concept: Developing 'what' and 'where' systems. *Trends in Cognitive Sciences* 2: 10–18.

Martin, M. F. 1922. Film, surface, and bulky colors and their intermediates. *American Journal of Psychology* 33: 451–480.

Matthen, M. 1988. Biological functions and perceptual content. *Journal of Philosophy* 85: 5–27.

Matthen, M. 1999. The disunity of color. *Philosophical Review* 108: 47–84.

Mausfeld, R. 1998. Color perception: From Grassmann codes to a dual code for object and illumination colors. In *Color Vision,* W. Backhaus, R. Kliegl, and J. Werner, eds., 219–250. Berlin and New York: De Gruyter.

Mausfeld, R. 2002. The physicalistic trap in perception. In *Perception and the Physical World,* D. Heyer and R. Mausfeld, eds., 75–112. Chichester: Wiley.

Mausfeld, R. 2003a. "Colour" as part of the format of two different perceptual primitives: The dual coding of colour. In *Colour Perception: Mind and the Physical World,* R. Mausfeld and D. Heyer, eds., 381–430. Oxford: Oxford University Press.

Mausfeld, R. 2003b. Competing representations and the mental capacity for conjoint perspectives. In *Inside Pictures: An Interdisciplinary Approach to Picture Perception,* H. Hecht, B. Schwartz, and M. Atherton, eds., 17–60. Cambridge, MA: MIT Press.

Mausfeld, R. in press. Intrinsic multiperspectivity: On the architectural foundations of a distinctive mental capacity. In *Proceedings of the XXIX International Congress of Psychology Berlin 2008,* Frensch, P., ed. London: Psychology Press.

Mausfeld, R., and J. Andres. 2002. Second order statistics of colour codes modulate transformations that effectuate varying degrees of scene invariance and illumination invariance. *Perception* 31: 209–224.

Mausfeld, R., and R. Niederée. 1993. Inquiries into relational concepts of colour based on an incremental principle of colour coding for minimal relational stimuli. *Perception* 22: 427–462.

Mausfeld, R., and G. Wendt. 2006. Material appearances under minimal stimulus conditions: Lustrous and glassy qualities. *Perception* (supplement) 35: 213–214.

Maxwell, G. 1968. Scientific methodology and the causal theory of perception. In *Problems in the Philosophy of Science*, I. Lakatos and A. Musgrave, eds., 148–160. Amsterdam: North-Holland.

Maxwell, G. 1970. Theories, perception, and structural realism. In *The Nature and Function of Scientific Theories*, R. G. Colodny, ed., 3–34. Pittsburgh, PA: University of Pittsburgh Press.

McLaughlin, B. 2003. The place of color in nature. In *Colour Perception: Mind and the Physical World*, R. Mausfeld and D. Heyer, eds., 475–502. Oxford: Oxford University Press.

Neisser, U. 1976. *Cognition and Reality*. New York: Freeman.

Rowe, Ch. 1974. Conceptions of colour and colour symbolism in the ancient world. In *The Realms of Colour*, Eranos Yearbook 1972, vol. 41, A. Portmann and R. Ritsema, eds. Leiden: Brill.

Russell, B. 1927. *The Analysis of Matter*. London: Allen and Unwin.

Santos, L. R., M. D. Hauser, and E. S. Spelke. 2001. Recognition and categorization of biologically significant objects by rhesus monkeys *(Macaca mulatta)*: The domain of food. *Cognition* 82: 127–155.

Schapp, W. 1910. *Beiträge zur Phänomenologie der Wahrnehmung*. Göttingen: Kästner.

Stewart, I. 1998. *Life's Other Secret. The New Mathematics of the Living World*. New York: Wiley.

Stoerig, P. 1998. Wavelength information processing versus color perception: Evidence from blindsight and color-blind sight. In *Color Vision*, W. Backhaus, R. Kliegl, J. Werner, eds.,130–147. Berlin: De Gruyter.

Strawson, G. 2003. Real materialism. In *Chomsky and His Critics*, L. Antony and N. Hornstein, eds., 49–88. Oxford: Blackwell.

Wagner, G. P., Mezey, J., and Calabretta, R. 2005. Natural selection and the origin of modules. In *Modularity. Understanding the Development and Evolution of Complex Natural Systems*, W. Callebaut and D. Rasskin-Gutman, eds., 33–49. Cambridge, Mass.: MIT Press.

Wierzbicka, A. 1996. *Semantics: Primes and Universals*. Oxford: Oxford University Press.

Yolton, J. W. 1984. *Perceptual Acquaintance from Descartes to Reid*. Minneapolis: University of Minnesota Press.

Yolton, J. W. 1996. *Perception and Reality: A History from Descartes to Kant*. Ithaca, NY: Cornell University Press.

II Color Spaces and Explanatory Spaces

6 Into the Neural Maze

Donald I. A. MacLeod

Recent discussions have pointed to many apparently instructive examples of how neurophysiological discoveries advance our understanding of color. But I believe it is equally instructive to consider the difficulties and obscurities that continue to frustrate us in this project. Here I review some of the presumed successes, and some acknowledged and unacknowledged obscurities, in our current physiological understanding of color, including issues that arise in the investigation of trichromacy; psychological primaries; color discrimination; color blindness; and color constancy. My conclusion is that as we trace the flow of information from the object to the retinal image and thence through the retina and visual pathway, it becomes more and more difficult to discern an isomorphism between color as we perceive it and the neural representation of color in the visual system. Perhaps a closer match to phenomenal experience may be found at higher levels of the system, but the empirical and conceptual basis for that is not yet clear.

Research in visual neurophysiology in the past half century has produced a wealth of data on neural representations at various levels in the visual system, and there is a general feeling that progress in relating physiology to perceptual experience has been pretty good. After all, Helmholtz in 1867 could still adopt the simplest possible conception of visual processing: that a single nerve fiber runs from each cone photoreceptor to the brain, without interacting with its neighbors, and there produces its associated sensation, like a single pixel in a full-color Cartesian theater. Now we know better. But how might Descartes or Helmholtz have responded to a revelation of the modern discoveries about the visual system? They would surely have been impressed by the new knowledge, but it also seems plausible that they would have been dismayed and bewildered by the seemingly unhelpful complexity and disorder of the physiological substrate of visual experience as we presently understand it. Perhaps they would have considered current ideas about psycho-neural correspondence unrigorous to the point of glibness, finding that instead of clarifying the relationship between brain events and perception, the detailed physiological knowledge we take such pride in only highlights its inscrutability. Whatever

Descartes or Helmholtz might have thought, that is the position that I mean to defend here.

The accumulated knowledge about neural representations in retina and brain has never put in doubt the general assumption that our experience of color is directly determined by neural processes. But neither has it yielded well-supported specific linking principles that allow particular aspects of color perception to be ascribed to particular kinds of events. Instead, from the very beginning, new knowledge has brought new difficulties to the concept of psychophysical isomorphism. At the outset, without benefit 'of any technical knowledge at all, it is natural to suppose that vision makes contact with external objects. This extramission view of perception has found favor among children, among the ancient Greeks, and—notoriously—among present-day students of psychology (Winer, Rader, et al. 2003). Given this starting point, the early realization that vision depends on light that travels from its objects to the eye comes as a nasty surprise, forcing the realization that perception is internal to the observer. With the limited physical and neuroscientific knowledge available to Descartes, the isomorphism between the objective and subjective worlds could still seem fairly straightforward: the perceptual world simply preserves aspects of the spatial and temporal configuration of what's out there, and specifies information about the values of certain attributes of external things, such as the brightness and color of surfaces. But new problems arose with the recognition that the apparent straightforwardness and immediacy of perception is misleading. Descartes could already trace the beginnings of the causal chain of visual processing in the derivation of the retinal image, the "proximal" stimulus, from the external and distant object. The proximal stimulus is intermediate in the causal chain between the object and the perceptual experience. Yet the perceptual experience is closer to isomorphism with the object than with the proximal stimulus: as perceptual constancies, including color constancy, illustrate, the attributes of perceptual experience are often correlated more closely with properties of their environmental sources than with the proximal signals through which the environment produces its effect (Koffka 1935). This relatively close correspondence is what makes the doctrine of *direct perception* plausible, and it has required sense-datum theorists to delve into the depths of experience, without clear success (Thouless 1931; Arend, Reeves, et al. 1991), in search of something that has a close correspondence to the proximal stimulus.

Since the retinal image is only two dimensional, depth is absent from the proximal stimulus to vision. In closer inspection, the proximal stimulus reveals other disconcerting features. Instead of being continuous, it is fragmented by discrete sampling by the mosaic of retinal photoreceptor cells. So although the proximal stimulus is an intermediate representation of the external world in terms of the causal chain of visual processing, it is not intermediate in form. The stimulus is in fundamental respects less like perception than the original object was.

Newton's demonstration that colored lights generally include all the wavelengths of the spectrum is another nasty surprise, because it disrupts the simplicity of the mapping from object to percept and frustrates realist accounts of color. Some relief comes at the next stage in the causal chain of perception—isomerization of visual pigment—where the infinite spectral degrees of freedom associated with the light stimulus are helpfully reduced to the excitations of only a few types of retinal photoreceptor. But there remains the untidy separation of a single image into three or more different representations that differentially weight the different regions of the spectrum. These must remain segregated, though not completely so, to represent color, but we are left with no answer to the question of how the multiple spectrally selective neural representations are integrated to give a unified field of perceived color. This fragmentation continues at postreceptoral stages. "On-center" and "off center" pathways appear at the bipolar cell. The varieties of amacrine and retinal ganglion cells are numbered in the dozens (Kolb, Linberg, et al. 1992). Further downstream, cells in the primary visual cortex exhibit a practically infinite variety in their functional organization, combining chromatic and spatial selectivity of various sorts, with no indication that any one cell has an independent role in perception, or of what its role would be if it did. After the signals leave primary visual cortex, the progressive complication and fragmentation of the neural representation continues: the signals are dispersed among at least two dozen separate visual areas (Felleman and Van Essen 1991), each forming a woefully distorted map of part or all of the visual field, and none with a very clearly defined role in perception.

How might the sporadic firing of the cells that make up this apparently chaotic tangle form a substrate for perception as we know it subjectively? Surely the honest answer is, we have no idea. We don't know whether or how individual neurons are relevant to perception. An illustrative major advance on this front, following the demonstration that the different visual cortical areas are functionally different, is the finding (e.g., Salzman et al. 1990; review by Parker and Newsome 1998) that monkeys' reports about the motion of a stimulus can be altered by concurrent stimulation of neurons in a region concerned with motion. But even this shows only that the neurons in question have some influence on motion perception, and the influence may only be indirect. It is rather like showing that electrical stimulation of the retina can produce a sensation of light: in both cases the neurons in question may be off-stage members of the supporting crew. Instead of solving a preexisting puzzle, the discovery of cortical functional specialization introduces a new challenge, the "binding problem": how do we correctly integrate the motions, colors, and other attributes of multiple objects? The explanatory gap between perception and neural representation as we currently conceive it is wide, then, and new evidence is not making it seem any narrower.

This chapter surveys a few of the "easy" problems (Chalmers 1996) in understanding color perception. The difficulties encountered in the domain of color will highlight the obscurity of psycho-neural isomorphism in general. While homilies on this theme may be abundant enough already, the exercise will at least provide an occasion to review some interesting facts and ideas about color vision. We consider first trichromacy, since trichromacy provides the most familiar example of physiological explanation in perception—an explanation generally held to be straightforward, simple, and completely satisfactory. But a careful and somewhat lengthy investigation in the following section leads to the conclusion that the neural basis of trichromacy is not yet well understood at all. Later sections discuss, in broader terms, the prospects for physiological explanation of less elementary aspects of color vision. The conclusions are again discouraging for current theoretical perspectives.

6.1 What Makes Us Trichromatic?

Some colors look the same.[1] In trichromatic vision, we can establish an exact color match using a fixed light and three primaries of adjustable intensity additively combined. The trichromacy of color vision is perhaps the clearest case of successful physiological explanation in perception. According to the standard account, two physically different lights will be subjectively indistinguishable if they are equal to one another in their excitation of the three types of cone photoreceptors that mediate color vision (which we may label as L, M, and S for long-wave, mid-spectral, and short-wave sensitive). The infinite-dimensional variation of spectral reflectance functions is thus replaced by a three-dimensional variation in neural effect at the photoreceptor stage. The trivariance of photoreceptor signals is a sufficient explanation for trichromacy, and it requires only that the spectral sensitivities of the operative receptors conform exactly to one or another of three possibilities.

Unfortunately, though, this requirement is not generally fulfilled. For one thing, rods provide a fourth visual pigment and hence a fourth degree of freedom for the neural effects of color stimuli. Rods are important for color vision under a wide range of illumination levels. When rods as well as cones are involved, trichromacy is not established at the photoreceptor level. Yet perceptual color matching under these very generally encountered conditions remains trichromatic (Cao, Pokorny, et al. 2005); this trichromacy presents a challenge that theory has yet to meet. Further, recent research has identified a fifth potential source of color signals—some varieties of retinal ganglion cell (and perhaps some cone photoreceptors; Dkhissi-Benyahya, Rieux, et al. 2006) contain blue-absorbing melanopsin and respond directly to the light that they absorb (as well as conveying signals from the familiar rod and cone photoreceptors) with a spectral sensitivity quite distinct from the rods and cones (Dacey, Liao, et al. 2005; Guler, Ecker, et al. 2008).

A similarly fundamental complication arises in purely photopic (cone-based) vision for many women, who carry a gene for a fourth cone pigment with an abnormal spectral sensitivity in addition to the three normal ones. Male observers with only the abnormal pigment make "anomalous" trichromatic matches that are often strikingly different from those of normal observers. The mothers of these men generally carry one copy of the gene for the anomaly. They can make trichromatic matches, and their matches are within the normal range, with only a slight deviation toward the anomalous match favored by their offspring.[2] They can, for instance, match a monochromatic yellow in the left half of a circular field to a suitable mixture of spectral red and green in the right half of the field to create the impression of a uniform circular disc. In the carrier women, however, unlike most observers, this match can be upset by adding a uniform long- or short-wavelength veiling light to both sides of the disc (Nagy, MacLeod, et al. 1981; MacLeod 1985).

This observation violates Grassmann's third law, which states that matching stimuli can be substituted for one another as constituents of other matches. Since the isomerization of each visual pigment is additive in Grassmann's sense, a match achieved through equality of pigment isomerization rates would satisfy Grassman's third law. It follows that the carrier women's nonadditive trichromatic matches are not matches for the pigments. In the standard genetic model (Nathans, Merbs, et al. 1992), the carriers have at least four visual pigments: one of the X chromosomes specifies the normal L and M pigments, while the X chromosome from the other parent specifies, instead of L or M, its spectrally deviant anomalous counterpart, say L*. The trichromatic matches of these women migrate between the normal match and an anomalous one, depending on whether the added veiling light, red or blue, favors use of the anomalous pigment (L*) or the corresponding normal one (L). The ability to involve the different pigments selectively using adaptation to added red or blue light implies that the four pigments are housed in different receptors. This too is expected: women and other female eutherians are a mosaic of maternal and paternally derived X-linked characteristics, following random inactivation of one or the other X chromosomes on a cell-by-cell basis during development (Mollon 1989).

Since trichromacy in all these cases does not reflect trivariance of the neural response at the photoreceptor level, it must reflect a postreceptoral constraint—a postreceptoral neural trivariance of some kind. But exactly what does this mean? This question is seldom posed, and on close investigation we will discover that there is no well-supported answer.

Conceivably, at some postreceptoral stage of processing, the neural signals might fall into only three chromatic classes, having three distinct spectral sensitivities, with practically complete uniformity in spectral sensitivity within each class (Brindley 1970). But this is not consistent with what is known about the spectral sensitivities of single cells in the visual cortex. These show considerable individual variation, with almost no agreement among investigators on how they should be classified; there is no evidence for a discrete set of cell types each having a precise and invariant pattern of connections to the receptors (Lennie, Krauskopf, et al. 1990; De Valois, Cottaris, et al. 2000). Where in this chaotic diversity of chromatic responsiveness can we find a basis for trichromacy? One could suppose that the untidy complex of color signals observed in the primary visual cortex is superseded by, or perhaps conceals, a cleanly trichromatic organization: the brain might secrete trichromatic colors in the form of three different neurotransmitters (erythrogen, chlorogen, cyanogen?); or there might turn out to be just three chromatic master neurons at each spatial location, which combine in an orderly way the chaotically diverse chromatic signals that they receive as their inputs. But a more plausible alternative to such extravagant ad hoc speculations is that central trichomacy occurs because postreceptoral neural connections treat alike two classes of photoreceptors (for instance, the paternally and maternally derived photoreceptors in the carriers of genes for anomalous trichromacy). An *equivalence* of photoreceptor signals in this sense would occur if the L and L* cones were equivalent in their pattern of neural connections.

A parallel but simpler case is provided by the subjective equivalence of right-eye and left-eye stimulation, which can be made subjectively indistinguishable (Barbeito, Levi, et al. 1985). Newton guessed that optic chiasm unites left and right fibers that serve a common direction in visual space; this could provide a physiological basis for binocular single vision, and also for the subjective equivalence of vision with the left and with the right eye. But electrophysiology has created a serious difficulty for this view. Although binocularly activated single cells exist in the brain, left- and right-eye inputs are relayed to separate alternating thin slabs of tissue in the thalamus and primary visual cortex. If perception made use of the information available in the primary visual cortex, left- and right-eye stimuli should always look very different. Yet these stimuli, so distinct in their cortical representations, are subjectively indistinguishable. One way to resolve this paradox is to suppose that activity in the primary visual cortex has no direct role in perception (Crick and Koch 1995), and that at some later stage of cortical processing, binocular convergence becomes complete. This proposal may not carry conviction, because there is no experimental or theoretical reason to think that the binocular convergence is ever complete enough to account for the subjective identity of left- and right-eye input. And although the proposal requires us to assume that a large proportion of cortical cells—ones with a marked left- or right-eye preference—have no direct influence on perception, it does not come with a principled criterion for deciding which cells have direct influence. Still, it is the best story we have. Perhaps, then, the postreceptoral trivariance that forms the basis for trichromacy when more than three visual pigments are involved could analogously originate from a convergence between the incoming signals that practically obliterates the distinction between L and L* inputs?

It turns out, however, that equivalence of the L and L* photoreceptor signals will not guarantee trichromacy. To see why, denote the cone excitations (the quantum catches in the respective visual pigments) with L, M, S, and L*. On the equivalence assumption, postreceptoral responses must be uniform: it will not be possible for some postreceptoral responses to depend mainly on L, and others mainly on L*. They can, however, depend on both the sum, L + L*, and the product, LL*, of the quantum catches L and L*. For a match to be valid for both a "sum" neuron and a "product" neuron, both L and L* have to be equal (or swapped) for the two stimuli compared, just as if those quantum catches were independently represented by separate postreceptoral neurons. This makes the system dichromatic. Throw in the M and S signals, and it becomes tetrachromatic. The equivalence hypothesis therefore does not guarantee trichromacy.

Besides, there is good evidence against the equivalence hypothesis as a general principle: it fails in certain doubly dichromatic women who have genes for two L pigments from one parent, and for two M pigments from the other. The color vision of these *compound heterozygotes* is fully normal despite the demonstrated red-green dichromacy of both parents or their male descendants (Carroll 2006). Their trichromacy is surprising because it requires the visual system to segregate the photoreceptor signals that originate from the maternally and paternally derived visual pigments. Suppose we classify the photoreceptors as "L" or "M" (in quotes) based on the genetic locus that encodes their pigment (not on the pigment that happens to be encoded there). Each class in the compound heterozygote then has the same proportion of cells with L and M pigment, and if central connections were determined by class membership as we have defined it, dichromacy would result. To explain the compound heterozygote's trichromacy, the fate of her photoreceptor signals must be determined instead by their source pigment. This could happen in two ways: either the photoreceptor cells are labeled, for the purpose of forming connections, by the pigment they contain, or the patterns of connections are contingent on activity—pigment-dependent activity—during development.

But as we have seen, pigment-based segregation does not occur with carriers of anomalous trichromacy, or they would be tetrachromats. So perhaps the number of labels available to classify photoreceptors for the purpose of forming central connections is genetically limited to three? If so, the basis of postreceptoral trichromacy resides in some kind of trivariance of the molecular biological machinery that allows photoreceptor signals to be segregated into three (and only three) classes in the formation of their central connections.

This idea is close to a restatement of the equivalence hypothesis, but by invoking more specific constraints on the processing of the signals, the idea can be put in a form that implies trichromacy without requiring that central color signals fall into just three well-defined classes. A tetrachromatic system will be reduced to a trichromatic one if signals from multiple photoreceptor types are combined *additively* in a fixed manner into a smaller number of postreceptoral signals—a scenario that I will dub *uniform additive combination*. Assume that the *i*th member of a representative array of postreceptoral neurons receives an input

$$E_i = f_i(f_L(\mathrm{L}) + f_{L^*}(\mathrm{L}^*),\, f_M(\mathrm{M}) + f_{M^*}(\mathrm{M}^*),\, f_S(\mathrm{S} + \mathrm{S}^*)). \tag{6.1}$$

Here L and M are the visual pigments from one X chromosome; L* and M* are the pigments from the second X chromosome and apply only to females. Two forms of the S pigment, S and S*, will also be present (even in males), but their excitations are shown as additively combined, since both forms are likely expressed in each S photoreceptor. The nonlinear functions f that relate neural signals to one another, or relate photoreceptor signals to pigment quantum catches, need not be specified here, but f_L and f_{L^*} are the same for all i (though they can be different from one another, thereby relaxing the equivalence condition), as are f_M and f_{M^*}. Only f_i may be different for the different postreceptoral neurons, but this alone allows the different central neurons to have an infinite variety of chromatic spectral sensitivities, in broad agreement with observation. Despite this diversity of spectral sensitivities, the system does satisfy neural trivariance since the three quantities $f_L(\mathrm{L}) + f_{L^*}(\mathrm{L}^*)$, $f_M(\mathrm{M}) + f_{M^*}(\mathrm{M}^*)$, and S determine the inputs to all the neurons at the postreceptoral stage.

Uniform additive combination as embodied in Equation (6.1) guarantees trichromacy for a system with more than three pigments. If the central neurons directly relevant for color perception average across a sufficient number of cones to obtain a fairly representative sample of both maternal and paternal cone signals, this might make trichromatic matches good enough to be acceptable in practice.

These considerations apply not only to carriers of genes for anomalous color vision, but also to many normal women. The normal visual pigments show considerable variation from one person to another (Webster and MacLeod 1986; Neitz and Neitz 2000), so even genetically normal women have both their father's and their mother's versions of the normal L and M pigments. The different versions of the L pigment are often different enough in peak absorption to create quite noticeable differences between Rayleigh matches determined by the maternal versus the paternal L pigments. For most women, therefore, trichromacy is not established at the photoreceptor level: if perception made full use of the information from the photoreceptors, many women would require four or perhaps five primaries to make color matches, depending on whether the maternal or paternal differences between their L pigments, their M pigments, or both were enough to be visually significant. Thus the trichromacy of many women is established postreceptorally, perhaps in the way sketched above. For explaining trichromacy in males when both rods and cones are active, similar considerations apply: trichromacy will hold if rod participation can be modeled by the addition of independent rod terms to each of the three arguments in Equation (6.1). Finally, the melanopsin ganglion cells could be accommodated in an extension of this theoretical scheme; those neurons, however, are thought to have no influence on color perception at all, although there is no obvious reason for their exclusion from perception given their chromatic responsiveness and cortical connectivity (Dacey, Liao, et al. 2005).

A central principle of the naïve but still current interpretation of trichomacy is that each cone has a fixed chromatic signature, characteristic of its spectral class (L, M, or S). This principle has recently been directly challenged, with the remarkable finding of Hofer, Singer, and Williams (2005) that the color sensations elicited by the most elementary possible stimuli—small points of light in a dark field—do not simply depend on the relative stimulation of L, M, and S cones. Instead, the reported colors seem more akin to an appropriate inference from the available data—an inference that takes into account not only the photoreceptor's spectral class but also its place in the retinal array. This observation subverts our cherished simple and definite conceptions of the subjective and postreceptoral neural consequences of excitation of individual cones. It is not yet clear what should supersede them.The general conclusion here is that we have made almost no progress toward establishing a neurophysiological explanation of trichromacy, or even toward clarifying the logical requirements for such an explanation. Even if the present proposal of uniform additive combination were accepted in principle, its empirical status is quite uncertain.

But trichromacy has a phenomenal as well as a behavioral connotation: a three-dimensional space is generally supposed to be enough to characterize the diversity in appearance of colors, at least for uniform fields viewed in a dark surround. I consider next the neural basis for the qualitative characteristics of perceived color under such reduced conditions.

6.2 The Psychological Primaries

Despite the qualifications just noted, the notion that the photoreceptors provide three unipolar signals for representing color is a useful idealization. But at postreceptoral stages, the representation of color undergoes a radical transformation, in which two distinct classes of neurons carry bipolar color-opponent signals, sometimes loosely referred to as a blue-yellow signal and a red-green signal, while a third class is driven by achromatic contrast. At the level of brain processing where input from the optic nerve is received, the three classes are distinct in size and connectivity, and are named konio(cellular), parvo(cellular), and magno(cellular) types (Hendry and Reid 2000).

A recoding of this kind was seen as theoretically attractive as early as the 1890s on the grounds that it would account nicely for the phenomenally unmixed natures of the four psychological primaries, red, green, yellow, and blue (also known as unitary, or unique, colors), and for the fact that other colors appear to be compounds in the sense that orange, for example, is both reddish and yellowish. The electrophysiological confirmation of that conjecture (Derrington, Krauskopf, et al. 1984; De Valois, Cottaris, et al. 2000) has been hailed as an example of successful physiological explanation in perception, perhaps second only to the three-receptor account of trichromacy. But here again, the picture becomes much less clear on close scrutiny. Notably, the "red-green" cells do not correspond well with the psychological primaries (see, for instance, Abramov and Gordon 1994). Whereas ideally they would be unresponsive to colors that are neither reddish nor greenish, they actually respond similarly to green and to blue. They behave in that way because they are driven by a difference between the M and the L cones, and the ratio of the M- to the L-cone sensitivities is in fact even higher for blue than for green. This has been known or strongly suspected since the time of Helmholtz's pupil König, and the relevant cone sensitivities are now known with high precision (Stockman, MacLeod, et al. 1993). Embarrassingly enough, the stimulus that most strongly excites the midspectral ("green") cones, relative to the long-wave cones, and hence maximally polarizes the "red-green" opponent neurons in the "green" direction, is a distinctly reddish spectral violet, with a wavelength near 455 nanometers (nm).

Thus a pure blue, devoid of redness and greenness, stimulates both the parvocellular and koniocellular neurons. The implied code for redness is not a signal confined to the parvocellular cells, but a specific combination of parvocellular and koniocellular activation. In view of this, it is difficult to maintain that the color-opponent neural code has any functional relation at all to the psychological primaries. This leaves us without a known or plausible account of the psychological primaries. The opponent neural recoding may be useful for reasons unrelated to the these primaries. It could be an instance of a frequently encountered principle of functional organization in the visual system, where neural responses are "sharpened," or made more selective, by offsetting excitation with inhibition from nearby inputs. Center-surround antagonism, for instance, is nearly ubiquitous in the spatial receptive fields of sensory neurons, where it enables improvements in spatial resolution. In the color case, it is not the spatial but the spectral sensitivity profile that is sharpened by the arrangement. Perhaps more important, the opponent code improves efficiency by reducing the correlation among the different neural measures of color. And because the opponent signals are minimal for some broad-band gray or near-gray color, the encoding scheme facilitates high sensitivity to small deviations from neutrality. This helps make the relatively desaturated natural colors that are most often encountered in nature highly discriminable, at a small price in reduced discrimination within the seldom frequented extremes of color space (von der Twer and MacLeod 2001).

Some evidence (e.g., He and MacLeod 2001; Shady, MacLeod, et al. 2004; Vul and MacLeod 2006) suggests that the earliest cortical stages, where the color-opponent konio and parvo cells deliver their signals, make no direct contribution to color perception. This leaves open the possibility that somewhere along the neural journey to the unknown seat of consciousness, a color-opponent code could be created that does correspond to the psychological primaries and can account for the phenomenally unitary nature of red, green, yellow, and blue. But alas, there is as yet no sign of such further recoding. On the contrary, the cortical representation of color is so untidy that it has not yet yielded to any simple description (Lennie, Krauskopf, et al. 1990; De Valois, Cottaris, et al. 2000).

6.3 When Is a Neural Substrate Simple?

Many color scientists, acknowledging that the color-opponent signals observed in the pathway to cortex have no relation to the psychological primaries, do nevertheless take it for granted that a color-opponent neural representation capable of accounting for the phenomenally simple or unitary quality of the psychological primaries must exist somewhere in the brain—in a region that is *directly* reflected in phenomenal experience, instead of merely conveying signals from the eye. This tenet was long maintained in the absence of neurophysiological evidence and continues to be maintained even though current neurophysiological evidence does not support it. Its a priori plausibility derives from convictions made explicit as psychophysical axioms by G. E. Müller (in the translation of Boring 1942, 89):

1. The basis of every state of consciousness is a physiological process, to whose occurrence the presence of the conscious state is joined.
2. To an equality, similarity or difference in the sensations . . . there corresponds an equality, similarity or difference in the psychophysical process, and conversely. Moreover, to a greater or lesser similarity of sensations, there also corresponds respectively a greater or lesser similarity of the psychophysical processes, and conversely.
3. If the changes through which a sensation passes have the same direction, or if the differences which exist between series of given sensations are of like direction, then the changes through which the psychophysical process passes, or the differences of the given psychophysical processes, have like direction.

Moreover, if a sensation is variable in *n* directions, then the psychophysical process lying at the basis of it must also be variable in *n* directions, and conversely.

It is the final point, with its reference to multiple directions, that provides the foundation for a physiological account of the psychological primaries. But on close inspection, the meaning and application of the axiom become disturbingly unclear. Some of the difficulties are the following.

First, if, as is generally assumed in the modern story of psycho-neural isomorphism, the different aspects of color are associated with different neural systems each generating a univariant signal, a problem arises in accounting for the subjective integration of the phenomenal dimensions of color. The judgment whether two colors are the same can be made rapidly and with high precision. Deciding whether an orange is redder than a purple is much more problematic. So how can the easy recognition of (multidimensional) color identity depend on the difficult comparison of the individual chromatic dimensions? To answer that those signals are introspectively inaccessible, and are integrated into a unitary percept by some unspecified subsequent processing, would be to give up the desired direct correspondence between perception and the unidimensional physiological signals.

Second, to account for the unmixed simplicity of pure yellow, we want to be able to say that when the addition of a reddish or greenish tint creates a reddish yellow or greenish yellow compound color, this happens because some signal for redness or greenness is introduced into the neural representation—a signal that is absent for yellow. The axiom does not allow this because it does not stipulate any zero point for the neural ("psychophysical") processes. Neurons do have a straightforward and functionally meaningful null signal: zero firing rate. This makes it natural to identify Müller's psychophysical process with the firing rates of *n* sets of neurons. But experimentally, a red-green opponent neuron that fires vigorously for red also fires for yellow, albeit at a more leisurely rate—perhaps close to its *spontaneous* rate, the rate observed without any stimulus—and is inhibited from firing for blue, blue-green or greenish yellow. If such neurons are responsible for phenomenal redness, there should be no uniquely pure and simple yellow: all yellows should be reddish (and by analogous reasoning, greenish as well). A related paradox: when neurons fire at their spontaneous rates, presumably nothing is perceived (since that is the neural state associated with an absence of physical stimulation). This is a simple perceptual outcome, but the corresponding neural state is complex in the sense that we have a full complement of nonzero signals.

Perhaps we might take a different tack by supposing that firing at the spontaneous rate is actually the simplest state of the neural system. But how might one proceed to justify that claim? Perhaps phenomenal simplicity corresponds to a balancing of signals? What, in neural terms, defines balanced signals? Must there be a neural difference signal somewhere in the head, which is zero in the balanced condition? In that case, don't the balanced signals themselves lose their claim to independent and direct phenomenal relevance?

Nevertheless Müller may have been wise to avoid formulating his axioms in terms of explicitly identified signals that range from zero. Although neural firing rates have a straightforward and functionally meaningful zero point, that is not true for many other physiological variables, notably membrane potentials, which are no less plausible than firing rates as contenders for the status of variables in the psychophysical process. This raises the question of the dimensionality *n* of the neural representation. Is there one dimension per neuron? If so, is that the neuron's firing rate? If it is, what is the justification for neglecting innumerable other variable parameters of a neuron's state? Do we have to group neurons in classes, and if so, should their responses just be averaged if the members of the class are not truly homogeneous in their responses?

How could we justify restricting the substrate of color to a circumscribed system of n classes of neuron, given that no subset of neurons operates in isolation from the rest of the brain? And perhaps most intriguing: is the direction of the phenomenal change from gray to green truly the same as the direction from red to gray; and if (as I would contend) it is not, can we reconcile this with the observation that the opponent cells provide two bipolar neural dimensions, or signals, rather than four monopolar ones?

A third, and related, problem in the search for psycho-neural isomorphism in color vision is that the multiplicity of neurons and the diversity of their chromatic responses gives the neural representation a great excess of dimensions beyond the three under discussion. It is important to recognize that the diversity of receptor spectral sensitivities discussed above is only one reason for the postreceptoral diversity. Even if the photoreceptor spectral sensitivities fell into precisely three classes, the responses of postreceptoral neurons conforming to Equation (6.1) could in principle take arbitrary values at any point in the three-dimensional space of receptor excitations, because a suitable choice of f_i, determined by the intervening connections, can shape a neuron's response to any given color independently of its response to similar colors (except perhaps for a continuity constraint), and this can be done independently for different neurons. The potential diversity of response distributions in cone excitation space is therefore limitless. The responses actually observed in neurophysiological experiments are less chaotic than they might be, but they do, as noted, clearly resist description in terms of a small number of classes (Lennie, Krauskopf, et al. 1990; De Valois, Cottaris, et al. 2000). Conceivably, some of this neural diversity may be phenomenally relevant. There could be neurons to indicate whether a color is brown, or belongs to some other particular category, or neurons that represent by their firing how warm or cool a color is felt to be. Still, it would be a daunting challenge to find a phenomenal counterpart for the signal from each of a presumably large class of color-sensitive neurons in, say, the primary visual cortex or later cortical areas specialized for color. But that is what the standard conceptual framework embodied in Müller's axioms leads us to expect.

A final serious difficulty for simple psycho-neural isomorphism is introduced by the mixing of neural signals for color and spatial pattern. Rather than being responsive to a single point in space, visual neurons have a great variety of receptive field profiles. How does the visual system encode both color and form? One possibility is to represent color independently of form; another is to reduplicate the entire apparatus of spatial vision as often as needed to specify color as well as lightness. Neither of these idealizations corresponds to neural reality. There is some functional specialization for color and for form (Livingstone and Hubel 1988; Zeki 1990), yet many neurons are jointly selective for both, as shown by both subjective phenomena (McCollough 1965) and electrophysiology (e.g., Johnson et al. 2001, 2004). It is not clear how in this situation we can identify a limited set of dimensions for a psychophysical process that represents color in the way envisaged by Müller. If there are clean principles underlying the multiplexing of chromatic and spatial information, they remain unknown, although a beginning is being made in investigating this (Engel 2005).

These difficulties are avoided by an entirely different view of the functional role of visual neurons in color perception, a view that provides a more natural account of the diversity of their chromatic responses. We can give up on *isomorphism*, and require only *consistency* of the responses to support perception. This is tantamount to acknowledging that present neurophysiological data do not yet account for phenomenal trivariance, or for the psychological primaries. But it is more than an admission of failure because it opens the door to other theoretical possibilities, discussed in later sections of this chapter.

6.4 Discrimination and Similarity

Recall the second of Müller's psychophysical axioms. "To an equality, similarity or difference in the sensations . . . there corresponds an equality, similarity or difference in the psychophysical process, and conversely." This remains the standard story, and perhaps the best one we have, for the neural basis for similarity and discriminability of colors. But it is problematic almost to the point of incoherence.

The notion of equality between two instances of the psychophysical process is not well defined. A precise identity at the molecular level is a practical impossibility; and if that is not required, what is?

As previously noted, we have no principled way of delimiting Müller's "psychophysical process," the neural substrate *directly* relevant to color perception within the brain. It would be pleasing if the identity principle could be applied piecewise to individual modalities or regions, but there is no reason to think that this is justified, nor is it clear how any suggested partitioning could be acceptable given that every candidate neural subsystem is interpreted by processes from neighboring systems.

Even if we were told which characteristics, or "dimensions," of the psychophysical process need to be considered, we would still have no principled way of determining the similarity of two such processes, because this would require a particular weighting of the differences along each of the given dimensions. A suggestion like "just add the absolute differences in the firing rates of the relevant cells, neuron by neuron" is hardly promising given that the effect of such differences on the behavior of other neurons is dependent on connection strengths that will generally be far from uniform.

The pervasiveness of inhibition in the nervous system complicates any attempt to define similarity of brain states because it implies that metrics of the weighted sum variety are too simple. If an "on-center" and an "off-center" cell, which are generally associated with opposite stimuli, are both more active in brain state A than in brain state B, while in state C the on and off differences are opposite, should this lead us to suppose that from the subjective point of view, A differs more from B than C does? Or the other way around?

The last three objections derive much of their force from the implausibility of the idea that a circumscribed brain system could have phenomenal effects strictly independent of its interaction with the neurons with which it communicates: it makes no sense to apply Müller's axioms to anything outside this circumscribed neural correlate of consciousness. Yet a complete neural model for similarity judgments must provide a causal chain of events underlying decisions of the form "which of these two colors is more like the reference color?" culminating in the binary neuromuscular signal by which the subject indicates his choice. So even if the model does postulate a circumscribed neural system whose activity stands in a one-to-one relation with phenomenal visual experience, differences in that activity can only be the starting point in accounting for discriminative judgments.

A functional perspective can deal with the difficulties raised, if only by bypassing them initially. Perception is subject to random variation. Experimentally, according to a generalization known as Crozier's law (Le Grand 1957; Knoblauch and Maloney 1996), differences that are judged just noticeable are ones that are detected with equal reliability, and small differences are judged equal if they are equal multiples of this threshold (Whittle 1973; Whittle 2003). This encourages hope for reducing the psychoneural gap by measuring the random variation in samples of putatively relevant and accessible neurons: behavioral discrimination can be convincingly modeled on that basis (Shadlen, Britten, et al. 1996), and an account of similarity might follow. In this perspective the meaning of a difference in firing rate is determined by its relevance to behavior, and to the activity of other cells, not by well-defined psychophysical-linking propositions of the Müller sort, where particular phenomenal qualities are inherently associated with particular neural subsystems. This functional view is consistent with an associationist account of similarity, like the one made explicit in Hayek's connectionist manifesto, *The Sensory Order*, conceived in 1920 and elaborated in 1952 (Hayek 1952, 53 and 61). Hayek outlines a mechanistic scheme in which he tries to account for everything consequential about sensation while abandoning "the 'absolute' qualities of sensation"—with no obvious regret—as "a phantom-problem." He proposes

> that the sensory qualities are not in some manner originally attached to, or an original attribute of, the individual physiological impulses, but that the whole of these qualities is determined by the system of connexions by which the impulses can be transmitted from neuron to neuron; that it is thus the position of the individual impulse or group of impulses in the whole system of such connexions which gives it its distinctive quality; that this system of connexions is acquired in the course of the development of the species and the individual by a kind of "experience" or "learning"; and that it reproduces therefore at every stage of its development

certain relationships existing in the physical environment between the stimuli evoking the impulses. . . . The connexions . . . are thus the primary phenomenon that creates the mental phenomena.

6.5 What Do the Color-Blind See . . . and Why?

Does the meaning of neural signals reside in their relation to other parts of a developing neural system (Hayek), or in phenomenal qualities with which they are inherently associated (Müller)? The sensations of the color-blind provide an opportunity to investigate this point. Red-green color blindness is nearly always traceable to a straightforward and simple structural change: a swapping of the L-cone pigment for an M pigment, or vice versa (Nathans, Merbs, et al. 1992), makes the visual system dichromatic rather than trichromatic. But can this model, together with otherwise unchanged neural processing, account for the sensations of the color-blind? It turns out that it cannot.

Any attempt to compare the experiences of different subjects encounters a familiar obstacle. Through learning, people with normal color discrimination will achieve some degree of consensus in their use of color names, even if the phenomenal experiences on which they base their judgments differ greatly. Two approaches are available to surmount this difficulty. First, as Palmer (1999) suggests, perhaps we can trust others to distinguish their phenomenally unitary psychological primaries from the other, compound colors where the phenomenal properties of primary pairs are mixed. Even if you and Franz both call grass green, the two of you may do so in virtue of phenomenally different experiences. But if you merely agree about what it means to be composite and what it means to be pure, or unitary, that is sufficient grounds for you to trust Franz's claims of the form "what I am now experiencing is an unmixed or unitary color." This doesn't require any precarious psycho-neural-linking hypotheses, or theories of neural coding, just a shared understanding of what it means to be unmixed as opposed to composite.

This limited trust can be extended not just to normal observers but to the color deficient. Many red-green–blind observers insist that they see red (and greens of sufficiently long wavelength) as a pure and unitary yellow. This claim is credible enough if other colors are reported as compound but perhaps less convincing if all colors are reported as unitary. Fortunately, a second approach is available, thanks to the rare appearance of individuals with one trichromatic and one color-blind eye. For any stimulus viewed with the red-green–blind eye, a unilaterally color-blind subject can select a color viewed with the normal eye to match it. When this is done, the claims of the bilaterally red-green blind are generally supported: for most of few such observers known to science, the sensations from the red-green–blind eye range from pure yellow to pure blue. But one clear exception has come to light. MacLeod and Lennie

(1976) found a case where red and green appeared orange (610 nm, which he considered far redder than pure yellow); saturated blues and violets appeared slightly greenish blue; and spectrally neutral colors appeared gray or greenish. Thus when represented in the color space of normal vision, the locus of perceived colors derived from the red-green–blind eye follows an arc rather than a straight line. In its avoidance of purples and its termination in the orange-red, the arc is directed through a set of colors that are environmentally likely given the information available from the red-green blind eye, and avoids those less likely. This suggests a more complex basis for color appearance than a simple deletion of the red-green opponent signal: the colors seen are roughly ones that minimize the average disagreement between the eyes (Alpern, Kitahara, et al. 1983).

Neither the curvilinear locus of our observer, nor the yellow-blue locus of most others, is consistent with a model where only the visual pigments are affected in the color-deficient eye. The prediction from that model is quite precise and very different (MacLeod and Lennie 1976): someone with a normal nervous system, but with an L/M pigment swap that makes the pigments in the L and M cones of the red-green–blind eye identical, will match all colors viewed by that eye to a set of colors that stimulate the L and M cones of the normal eye in a constant ratio and are differentiated only by the S cones of the normal eye. This is because if there is bilateral symmetry in postreceptoral processing, the binocular match must be a match at the photoreceptor level. At that level, the identity of visual pigment between the L and M cones in the red-green–blind eye will make their relative stimulation the same for all color stimuli. So the color stimuli selected as matches using the trichromatic eye must also be equal to one another in their excitation of the L and M cones. The predicted locus of the matching colors is called a *tritanopic confusion line*, because these colors are confused by tritanopes, who lack S cones and have only the normal L and M cones; the tritanopic confusion line through white connects deep violet at short wavelengths to greenish yellow for long wavelengths.[3] Yet no unilaterally red-green–blind subject has reported colors that lie along a tritanopic confusion line. It follows that all known unilateral color-blind individuals have atypical neural processing of color, in addition to their pigment swap. The neural change is most likely a reorganization elicited by the altered input from the photoreceptors (MacLeod and Lennie 1976). For the cases where reported colors range from blue to yellow, it might be tempting to invoke atrophy of a red-green opponent system, as Byrne and Hilbert assume (present volume). This does not seem very likely, though, since the parvo (red-green) system is thought to be important for visual acuity, and acuity is unimpaired in the red-green blind. Moreover, loss of input from the electrophysiologically documented "red-green" system (which is driven by L and M cones only), with otherwise normal processing, would lead to perception of violet and lime, not blue and yellow, since violet and lime are the colors that in normal observers fail to excite that system. And for the

MacLeod and Lennie case, no such simple account suffices, since redness and greenness are present but are a function of S-cone input, in a way that reduces the average discrepancy between the perceptions of the two eyes.

A seldom recognized theoretical point is that in principle, even ordinary (bilateral) red-green–blind observers may experience both unitary and compound colors, as MacLeod and Lennie's unilateral case did. And by Palmer's argument above, they should be able to recognize and reliably report this.

A priori, the sensations of the red-green blind could well include a huge diversity of hues. Even a straight line locus in the chromaticity diagram of normal trichromats—such as a tritanopic confusion locus that skirts the subjectively achromatic point—traverses nearly half the hue circle, proceeding (for instance) from violet, through (desaturated) greenish colors, to a yellow nearly complementary to the initial violet. The variety of hues traversed by a curved locus can be even greater, as with the violet-blue-green-yellow orange trajectory suggested by MacLeod and Lennie's observer. In these cases a unitary green and a unitary blue should be identifiable among the binary colors of the dichromatic spectrum. Only one paper has raised, let alone investigated, these interesting possibilities. Cicerone, Nagy, and Nerger (1987) found that red-green–blind protanopes can not only identify a neutral point in the spectrum, but also can locate a distinct spectral point in the blue that they described as bluish but (uniquely) lacking in greenness or redness, an observation parallel to those of MacLeod and Lennie's subject.

Thus the sensations of the color-blind are quite different from the receptor-based prediction. The mapping from receptors to sensations appears to be highly plastic, in ways that are not yet well documented or well understood.[4]

Plasticity of color perception has also been shown experimentally in people with normal color vision. Long-term exposure to saturated red light over many days caused a persistent (though not completely permanent) redward shift of "pure yellow," as if an adaptive control system adjusted the neural red-green null point toward the centroid of recently encountered color stimuli (Neitz, Carroll, et al. 2002). McCollough's observation of orientation-contingent color aftereffects (McCollough 1965) suggests a similar recalibration, with the complication that here, spatial and chromatic perceptions interact. It has been proposed (Barlow 1990) that in the induction of these aftereffects, individual neurons change their chromatic signature, so that after viewing a diet of reddish verticals, neurons selectively sensitive to vertical form change their color preference from achromatic to reddish. Neurons do change their stimulus "trigger features" in this way (Kohn and Movshon 2004). By the same token, it is likely, though not yet demonstrated, that the perceptual consequences of activity in early cortical neurons can undergo adaptive modification. Cortical plasticity is a problem for explanatory frameworks that attribute particular aspects of color to neurons with definite identities, and associated connectivities, via well-defined psychophysical-linking propositions of the Müller sort. This type of scheme has worked well in correlating color with the earliest neural stage of processing, at the photoreceptors, but it may be inappropriate for the more plastic brain. An alternative can be found in connectionist systems like those of Hebb (1949) and Hayek (1952), where the stimulus requirements and perceptual sequelae of neural activity are modified as connections are strengthened during development by correlated excitation of the input and output neurons. Such a coupling from monocular inputs to binocular cells would naturally lead to a discrepancy-minimizing pattern of connection for at least some unilaterally color-blind observers.

6.6 Constant Colors: Compensated or Constructed?

In tracing the flow of information from its external source through the photoreceptors and subsequent neural representations, we have encountered more and more fragmentation, and less and less straightforward isomorphism with either the external environment or the phenomenal world that mirrors it. We are in danger of getting lost in the maze. Where to find a way out? What we are looking for is a neural representation more closely isomorphic with phenomenal experience, and hence with the three-dimensional external world that the phenomenal world mirrors. This suggests the possibility that the neural maze has a kind of symmetry, so that the jumble of signals in the primary visual cortex gives way to representation that is in some sense more orderly at higher levels of the cortex. Many mazes have a symmetrical construction, so that once you are halfway in, you can find an onward path to the exit that is a mirror image of the inward one. Perhaps the neural maze is like those? Not much is known from direct neurophysiological observations to support such a conjecture, but visual phenomena give some clues to the kind of organization that might be involved in bringing high-level neural representations into closer correspondence with perception.

As noted in the introduction, color constancy provides one example of the closer relation of perception to the proximal than to the distal stimulus. How does perception recover colors and lightnesses that depend closely on the reflectances of external objects, while the retinal stimulus varies due to changing illumination? Fundamental though this question is, our ignorance about its neural basis is remarkably complete.

There is strong support for some elements of Land's early retinex (retina + cortex) model (Land and McCann 1971). Signals from the L, M and S photoreceptors are independently subject to rapid local adaptation that takes the form of a nearly reciprocal adjustment of sensitivity (He and Macleod 1998; Lee, Dacey, et al. 2003). Because of this, the signals that pass from retina to cortex when a new stimulus replaces a prior one depend mainly on the ratio of the cone excitations, rather than on their absolute values (Enroth-Cugell and Shapley 1973); this has the effect of compensating fairly well for coloration of the illumination on a scene, as Monge (Mollon 2006), Helmholtz (1925), and Cornsweet (1970), for example, pointed out. If there is no new stimulus, vision simply fails: without motion of the retinal image, objects fade to invisibility over a period of several seconds, much as afterimages do. In normal vision, the fading is prevented by small involuntary eye movements that modulate the excitations of the photoreceptors lying sufficiently close to each boundary in the visual field by sweeping the boundary to and fro (Desbordes and Rucci 2007). Although these eye movements critically affect the neural representation, they play no role in neurophysiological recordings, where the animal is paralyzed and the test stimuli have to be flashed, and perhaps for this reason, the neural basis of the fading has not been clearly elucidated despite its obviously fundamental importance. One might hope that retinal output signals monitored electrophysiologically using flashed stimuli would have the same dynamics as the subjectively observed fading, but they never do. Instead, each new stimulus gives rise to a brief transient burst of impulses (or a comparably transient suppression of firing) that lasts only a fraction of a second and is followed by a weak and variable maintained discharge (Marrocco 1972). It is not clear whether the progressive fading of static images occurs because some cortical circuit responds only to change in the retinal input, or whether it is the brief transient signals that sustain vision, by triggering a persisting change of cortical state appropriate to the new stimulus (as happens in other contexts; Ferber, Humphrey, et al. 2003). In either case, the signals arriving at the cortex are very different from the stimulus in their dependence on temporal and spatial as well as spectral parameters. In their strong dependence on chromatic or achromatic *contrast*, they differ radically from the "wavelength-dependent" caricature sometimes drawn by authors who wish to highlight the distinctive contribution of cortical processing to color constancy. But they are also far from being isomorphic to perception, resembling instead (from the point of view of a cortical homunculus) something like a line drawing where lines indicate the boundaries of high contrast. In the retinex model, the cortex resurrects brightness and color at each point by integrating these contrast-dependent signals at nearby and more remote boundaries in the visual field. For the neural mechanism of this tricky accomplishment, there are no well-supported conjectures. One interesting class of proposal, consistent with a symmetric scheme where sensory analysis is followed by perceptual synthesis, is the idea (Hurlbert 1986) that the reconstruction could be made by an inverse transform. A high-level representation potentially isomorphic with the stimulus is relayed back to the primary visual cortex by the brain's plentiful feedback connections. If the descending pathway duplicates the spatiotemporal filtering applied to the afferent signals en route to the primary visual cortex, the high-level representation that matches the descending signals to the incoming ones is the correct one. This *predictive coding* scheme is one way to close the loop from the stimulus to the perceptual model, allowing the centrally generated model to preserve isomorphism while the intervening data (the afferent signals) have a quite different form and are used only to test and adjust the model.

However it is managed, the integration of local contrast signals is an incomplete model for color constancy. It fails to explain, for instance, how we are able to distinguish a red scene under neutral light from a neutral scene under red light. Our recognition of that distinction appears to exploit subtle cues, based on the statistical distribution of colors in the image (MacLeod 2003; Brainard, Longere, et al. 2006). In view of this, the ideas of a passive correction of the early cortical representation to compensate for changing illumination, and the idea of inverse filtering in a feedback loop, are both too simple. Helmholtz's conception that our estimate of the illuminant represents an "unconscious inference" can accommodate a much wider range of processes. But the neural basis of these inferences is still completely obscure, since in this view the signals monitored in current neurophysiological experiments could be as different from the centrally generated model as scientific data are from models used to account for them. By the same token, the neural embodiment of the perceptual model is not clarified by investigating the encoding of the sensory data.

It has been tempting to think of color as a kind of bedrock of experience, inherent in sense data and requiring no overlay of interpretation. Yet it is clear that the perceptual specification of color and lightness must include a fairly complete model of the scene, including much information about three-dimensional shape and arrangement as well as conditions of observation, such as the nature and location of illumination sources. Phenomena such as the Mach card show that with constant sensory input, perception of lightness and color can be abruptly modified by a revised decision about the surface orientations in a bistable stimulus (Bloj, Kersten, et al. 1999; Bloj and Hurlbert 2002). The decision to classify an edge as a shadow or surface property has similar repercussions. And the estimation of the lightness profile on a curved surface is inextricably intertwined with the estimation of its shape. There is thus no good reason to consider color as "given" while three-dimensional scene geometry is constructed: both are constructed! How their construction is implemented in neural machinery is far from clear, but it seems likely that the cortical input serves only to trigger, guide, or direct the perceptual construction of color as well as of other aspects of the environment (Yuille and Kersten 2006).

A recently investigated case of vision recovered in adulthood (Fine, Wade, et al. 2003) provides a dramatic and instructive contrast with the normally sighted. Uniquely, as far as we know, this subject (MM) has "the eye of the artist" in the sense that he accurately matches the retinal illuminances associated with light and shadowed regions in a scene. (This ability is desirable for artists but never achieved by them: "phenomenal regression to the real object" [Thouless 1931] seems inescapable.) MM does not automatically perceive the visual world as a three-dimensional arrangement, and his computation of lightness and color is correspondingly simplified. He does, however, compensate fairly well for changes in the intensity or color of the illumination on the scene as a whole; this may be a result of retinal adaptation processes.

If we make a distinction between sensory processes that deliver a progressively diversifying set of signals along an array of ramifying afferent chains, and perceptual processes where centrally generated hypotheses are tested and modified using feedback-based comparison with the incoming stream of sensory data, constancy seems to owe something to both of these broad stages of processing. But even the relevant sensory processes are only partially elucidated at the neural level as yet, and the perceptual ones hardly at all. Although the schemes described, involving "generative models" or "predictive coding," feature prominently in current speculation about perception, little or no research in the physiology of color vision has been conducted with such possibilities in mind. If the generative scheme is correct, we would expect some cortical signals to represent the discrepancy between perceptual prediction and retinal input rather than being directly determined by the input. Some beginnings have been made in testing this in conscious humans, for instance, in a study of perceptual organization using brain imaging (Murray, Kersten, et al. 2002; Murray, Schrater, et al. 2004). Transient visual impairment can be induced by transcranial magnetic stimulation, and in some experiments, this has been attributed to interference with neural feedback loops (Pascual-Leone and Walsh 2001), but that observation is not clearly supportive of a predictive coding scheme. Testing the proposal with microelectrodes in animals will be difficult.

6.7 Concluding Summary

As we have seen, the project of correlating color experience with neural events is hampered not only by the irreducible "explanatory gap" of the metaphysicians, with its attendant "hard problem," but by major additional gaps that reflect our limited current knowledge and understanding of neural processing. The neural representations of color revealed by neurophysiology, particularly in the cortex, are untidy, and new discoveries in neuroscience are only making them more complex. I hope my readers will concur with Hardin (1999) that even the "easy" problems we have considered are hard enough—and that they are far harder than is generally acknowledged. If we are ever to discern an isomorphism between brain events and the experience of color, it will surely require further discoveries and conceptual advances that we can't presently anticipate.

Although receptoral processes are fundamental for trichromacy—the first, and easiest, problem encountered—its basis has not been completely elucidated. And as more central processes become relevant, the significance of neural events becomes increasingly obscure. The postreceptoral recoding of color in terms of opponent signals has long been thought to be a basis for psychological primaries, but the known opponent signals cannot serve as such a basis, and could have quite a different functional significance.

The colors seen by the color-blind are not well predicted by their light-sensing abnormality alone; in unilateral cases, the colors may be selected, through some kind of neural adaptation, to minimize the discrepancy with what is seen by the trichromatic eye. Although color appearance and color similarity are usually modeled with the assumption that neurons are associated with fixed subjective qualities, like pixels in a Cartesian theater, the plasticity of the relation between afferent signals and perception encourages an associative account instead.

The basis of the "unconscious inferences" on which color vision depends in all but the simplest situations remains obscure. Perhaps feedback to the primary visual cortex allows a centrally generated perceptual model to be checked and revised in the light of incoming sensory messages. Such schemes give hope for the possibility of some kind of isomorphism between color perception and as yet unknown central neural events, but they discourage the search for a simple correspondence between subcortical or early cortical neural events and perception.

Acknowledgments

The editors, Rolf Kuehni, and Steve Shevell provided useful comments on an earlier draft. Research supported by NIH grant EY01711.

Notes

1. Here "color" refers specifically to the spectral energy distribution that constitutes the color stimulus, whereas elsewhere I mostly intend other meanings when I use the word. This is the linguistic practice for which Humpty Dumpty has been strongly criticized, but it seems to work perfectly well as long as the shared understanding of context removes serious ambiguity. Many philosophers evidently feel a need for a univocal definition of color—physicalist, subjectivist, dispositional, or whatever—but Whittle (2003) encourages us to abandon that dream, and I find his pleading persuasive.

2. Part of the fascination of research with these subjects is the intriguing possibility that they might be tetrachromatic and have an added dimension of color experience. Jordan and Mollon (1993) consider this an open question, and some of their experiments do suggest an ability to distinguish between Rayleigh-matched color pairs. The most economical explanation of that finding, however, is that the two fields appear to differ in texture, as would be expected where a trichromatic system has nonuniform spatial properties owing to the interleaving of cones specified by maternally and paternally derived genes. Informal reports of our subjects are consistent with this interpretation. Jameson, Highnote, and Wasserman (2001) found that their sample of

such women divided the spectrum into more bands than normal; this is consistent with tetrachromacy but is not strong evidence for it.

3. Although tritanopes are sometimes called blue-blind or blue-yellow–blind, they do *not* confuse blue with yellow, since these are quite distinct in their effects on the L and M cones.

4. One little noted fact about unilaterally color-blind people is startling enough to warrant mention. At least three of the eight or so putatively congenital unilaterally red-green–blind subjects gave no indication of knowing, prior to investigation, that their two eyes were different! Our subject noticed the difference only when it led to binocular rivalry on one occasion, which led us to bring him into the lab for testing. Only following controlled experiments with binocular matching was he slowly persuaded that his two eyes were different. As a biologist, he had considerable experience viewing things in a monocular microscope. That a person can be red-green–blind in one eye without noticing goes beyond oddness. It seems to me to meet the requirements for a philosophically useful datum in that it nibbles at the edges of conceivability; I for one would not have considered it conceivable until I encountered it. Such cases provide an empirical answer to the question, Can an observer be unaware of his own qualia? On the evidence from unilateral congenital color blindness, the answer, in at least a limited sense, would appear to be yes.

References

Abramov, I., and J. Gordon. 1994. Color appearance: On seeing red—or yellow, or green, or blue. *Annual Review of Psychology* 45: 451–485.

Alpern, M., K. Kitahara, et al. 1983. Perception of colour in unilateral tritanopia. *Journal of Physiology* 335: 683–697.

Arend, L. E., Jr., A. Reeves, et al. 1991. Simultaneous color constancy: Paper with diverse Munsell values. *Journal of the Optical Society of America A* 8 (4): 661–672.

Barbeito, R., D. Levi, et al. 1985. Stereo-deficients and stereoblinds cannot make utrocular discriminations. *Vision Research* 25 (9): 1345–1348.

Barlow, H. B. 1990. A theory about the functional role and synaptic mechanism of visual aftereffects. In *Vision: Coding and Efficiency*, edited by C. Blakemore. New York: Cambridge University Press.

Bloj, M. G., and A. C. Hurlbert. 2002. An empirical study of the traditional Mach card effect. *Perception* 31 (2): 233–246.

Bloj, M. G., D. Kersten, et al. 1999. Perception of three-dimensional shape influences colour perception through mutual illumination. *Nature* 402 (6764): 877–879.

Boring, EG. 1942. *Sensation and Perception in the History of Experimental Psychology*. New York: Appleton-Century-Crofts.

Brainard, D. H., P. Longere, et al. 2006. Bayesian model of human color constancy. *Journal of Vision* 6 (11): 1267–1281.

Brindley, G. S. 1970. *Physiology of the Retina and Visual Pathway*. London: Edward Arnold.

Cao, D., J. Pokorny, et al. 2005. Matching rod percepts with cone stimuli. *Vision Research* 45 (16): 2119–2128.

Carroll, J. 2006. Colour-blindness detective story not so simple. *Clinical Experimental Optometry* 89 (3): 184–185; author reply, 185–186.

Chalmers, D. J. 1996. *The Conscious Mind*. New York: Oxford University Press.

Cicerone, C. M., A. L. Nagy, et al. 1987. Equilibrium hue judgements of dichromats. *Vision Research* 27 (6): 983–991.

Cornsweet, T. N. 1970. *Visual Perception*. New York: Academic Press.

Crick, F., and C. Koch 1995. Are we aware of neural activity in primary visual cortex? *Nature* 375 (6527): 121–123.

Dacey, D. M., H. W. Liao, et al. 2005. Melanopsin-expressing ganglion cells in primate retina signal colour and irradiance and project to the LGN. *Nature* 433 (7027): 749–754.

De Valois, R. L., N. P. Cottaris, et al. 2000. Some transformations of color information from lateral geniculate nucleus to striate cortex. *Proceedings of the National Academy of Sciences USA* 97 (9): 4997–5002.

Derrington, A. M., J. Krauskopf, et al. 1984. Chromatic mechanisms in lateral geniculate nucleus of macaque. *Journal of Physiology* 357: 241–265.

Desbordes, G., and M. Rucci. 2007. A model of the dynamics of retinal activity during natural visual fixation. *Vision Neuroscience* 24 (2): 217–230.

Dkhissi-Benyahya, O., C. Rieux, et al. 2006. Immunohistochemical evidence of a melanopsin cone in human retina. *Investigative Ophthalmology and Vision Science* 47 (4): 1636–1641.

Engel, S. A. 2005. Adaptation of oriented and unoriented color-selective neurons in human visual areas. *Neuron* 45 (4):613–23.

Enroth-Cugell, C., and R. M. Shapley. 1973. Flux, not retinal illumination, is what cat retinal ganglion cells really care about. *Journal of Physiology* 233 (2): 311–326.

Felleman, D. J., and D. C. Van Essen. 1991. Distributed hierarchical processing in the primate cerebral cortex. *Cerebral Cortex* 1 (1): 1–47.

Ferber, S., G. K. Humphrey, et al. 2003. The lateral occipital complex subserves the perceptual persistence of motion-defined groupings. *Cerebral Cortex* 13 (7): 716–721.

Fine, I., A. R. Wade, et al. 2003. Long-term deprivation affects visual perception and cortex. *Nature Neuroscience* 6 (9): 915–916.

Guler, A. D., J. L. Ecker, et al. 2008. Melanopsin cells are the principal conduits for rod-cone input to non-image-forming vision. *Nature* 453 (7191): 102–105.

Hardin, C. L. 1999. Color Quality and Color Structure. In *Toward a Science of Consciousness III, The Third Tucson Discussions and Debates,* edited by S. R. Hameroff, A. W. Kaszniak and D. J. Chalmers: MIT Press.

Hayek, F. A. 1952. *The Sensory Order: An Inquiry into the Foundations of Theoretical Psychology.* London: Routledge.

He, S., and D. I. Macleod. 1998. Contrast-modulation flicker: Dynamics and spatial resolution of the light adaptation process. *Vision Research* 38 (7): 985–1000.

He, S., and D. I. Macleod. 2001. Orientation-selective adaptation and tilt after-effect from invisible patterns. *Nature* 411 (6836): 473–476.

Hebb, D. O. 1949. *The Organization of Behavior.* New York, John Wiley.

Helmholtz, H. von. 1925. *Helmholtz's treatise on physiological optics, Vol. Ill: The Perceptions of Vision,* J. P. C. Southall, trans., J. P. C. Southall, ed. Rochester, New York: Optical Society of America. Original edition, 1867.

Hendry, S. H., and R. C. Reid. 2000. The koniocellular pathway in primate vision. *Annual Review of Neuroscience* 23: 127–153.

Hofer, H., B. Singer, and D. R. Williams. 2005. Different sensations from cones with the same photopigment. *Journal of Vision* 5 (5):444–454.

Hurlbert, A. 1986. Formal connections between lightness algorithms. *Journal of the Optical Society of America A* 3 (10): 1684–1693.

Jameson, K. A., S. M. Highnote, and L. M. Wasserman. 2001. Richer color experience in observers with multiple photopigment opsin genes. *Psychonometric Bulletin Review* 8 (2):244–461.

Johnson, E. N., M. J. Hawken, and R. Shapley. 2001. The spatial transformation of color in the primary visual cortex of the macaque monkey. *Nature Neuroscience* 4 (4): 409–416.

Johnson, E. N., M. J. Hawken, and R. Shapley. 2004. Cone inputs in macaque primary visual cortex. *Journal of Neurophysiology* 91 (6):2501–2514.

Jordan, G., and J. D. Mollon. 1993. A study of women heterozygous for colour deficiencies. *Vision Research* 33 (11):1495–508.

Knoblauch, K., and L. T. Maloney. 1996. Testing the indeterminacy of linear color mechanisms from color discrimination data. *Vision Research* 36 (2): 295–306.

Koffka, K. 1935. *Principles of Gestalt Psychology.* New York: Harcourt, Brace.

Kohn, A., and J. A. Movshon. 2004. Adaptation changes the direction tuning of macaque MT neurons. *Nature Neuroscience* 7 (7): 764–772.

Kolb, H., K. A. Linberg, et al. 1992. Neurons of the human retina: A Golgi study. *Journal of Comparative Neurology* 318 (2): 147–187.

Land, E. H., and J. J. McCann. 1971. Lightness and retinex theory. *Journal of the Optical Society of America* 61 (1): 1–11.

Le Grand, Y. 1957. *Light, Color and Vision*. London: Chapman and Hall.

Lee, B. B., D. M. Dacey, et al. 2003. Dynamics of sensitivity regulation in primate outer retina: The horizontal cell network. *Journal of Vision* 3 (7): 513–526.

Lennie, P., J. Krauskopf, et al. 1990. Chromatic mechanisms in striate cortex of macaque. *Journal of Neuroscience* 10 (2): 649–669.

Livingstone, M., and D. Hubel. 1988. Segregation of form, color, movement, and depth: Anatomy, physiology, and perception. *Science* 240 (4853): 740–749.

MacLeod, D. I., and P. Lennie. 1976. Red-green blindness confined to one eye. *Vision Research* 16 (7): 691–702.

MacLeod, D. I. A. 1985. Receptoral constraints on color appearance. In *Central and Peripheral Mechanisms of Color Vision*, D. Ottoson and S. Zeki, eds., 103–116 London: MacMillan.

MacLeod, D. I. A. 2003. Colour discrimination, colour constancy and natural scene statistics. In *Normal and Defective Colour Vision*, J. D. Mollon, J. Pokorny, and K. Knoblauch, eds. London: Oxford University Press.

Marrocco, R. T. 1972. Maintained activity of monkey optic tract fibers and lateral geniculate nucleus cells. *Vision Research* 12 (6): 1175–1181.

McCollough, C. 1965. Color adaptation of edge-detectors in the human visual system. *Science* 149 (3688): 1115–1116.

Mollon, J. 2006. Monge: The Verriest lecture, Lyon, July 2005. *Vision Neuroscience* 23 (3–4): 297–309.

Mollon, J. D. 1989. "Tho' she kneel'd in that place where they grew ..." The uses and origins of primate colour vision. *Journal of Experimental Biology* 146: 21–38.

Murray, S. O., D. Kersten, et al. 2002. Shape perception reduces activity in human primary visual cortex. *Proceedings of the National Academy of Sciences USA* 99 (23): 15164–15169.

Murray, S. O., P. Schrater, and D. Kersten. 2004. Perceptual grouping and the interactions between visual cortical areas. *Neural Networks* 17 (5–6): 695–705.

Nagy, A. L., D. I. MacLeod, et al. 1981. Four cone pigments in women heterozygous for color deficiency. *Journal of the Optical Society of America* 71 (6): 719–722.

Nathans, J., S. L. Merbs, et al. 1992. Molecular genetics of human visual pigments. *Annual Review of Genetics* 26: 403–424.

Neitz, M., and J. Neitz. 2000. Molecular genetics of color vision and color vision defects. *Archives of Ophthalmology* 118 (5):691–700.

Neitz, J., J. Carroll, et al. 2002. Color perception is mediated by a plastic neural mechanism that is adjustable in adults. *Neuron* 35 (4): 783–792.

Palmer, S. E. 1999. Color, consciousness, and the isomorphism constraint. *Behavior Brain Science* 22 (6): 923–943; discussion 944–989.

Parker, A. J., and W. T. Newsome. 1998. Sense and the single neuron: probing the physiology of perception. *Annual Review of Neuroscience* 21: 227–77.

Pascual-Leone, A., and V. Walsh. 2001. Fast backprojections from the motion to the primary visual area necessary for visual awareness. *Science* 292 (5516): 510–512.

Salzman, C. D., K. H. Britten, and W. T. Newsome. 1990. Cortical microstimulation influences perceptual judgments of motion direction. *Nature* 346: 174–177.

Shadlen, M. N., K. H. Britten, et al. 1996. A computational analysis of the relationship between neuronal and behavioral responses to visual motion. *Journal of Neuroscience* 16 (4): 1486–1510.

Shady, S., D. I. MacLeod, et al. 2004. Adaptation from invisible flicker. *Proceedings of the National Academy of Sciences USA* 101 (14): 5170–5173.

Stockman, A., D. I. MacLeod, et al. 1993. Spectral sensitivities of the human cones. *Journal of the Optical Society of America A Opt Image Sci Vis* 10 (12): 2491–2521.

Thouless, R. H. 1931. Phenomenal regression to the real object. II. *British Journal of Psychology* 22 (1): 1–30.

von der Twer, T., and D. I. MacLeod. 2001. Optimal nonlinear codes for the perception of natural colours. *Network* 12 (3): 395–407.

Vul, E., and D. I. MacLeod. 2006. Contingent aftereffects distinguish conscious and preconscious color processing. *Naure Neuroscience* 9 (7): 873–874.

Webster, M. A., and D. I. A. MacLeod. 1988. Factors underlying individual differences in the color matches of normal observers. *J the Optical Society of America A* 5 (10):1722–1735.

Whittle, P. 1973. The brightness of coloured flashes on backgrounds of various colours and luminances. *Vision Research* 13 (3): 621–638.

Whittle, P. 2003. Contrast colours. In *Colour Perception: Mind and the Physical World,* R. Mausfeld and D. Heyer, eds., 115–138. London: Oxford University Press.

Winer, G. A., A. W. Rader, et al. 2003. Testing different interpretations for the mistaken belief that rays exit the eyes during vision. *Journal of Psychology* 137 (3): 243–261.

Yuille, A., and D. Kersten. 2006. Vision as Bayesian inference: Analysis by synthesis? *Trends in the Cognitive Sciences* 10 (7): 301–308.

Zeki, S. 1990. Parallelism and functional specialization in human visual cortex. *Cold Spring Harbor Symposium Quantitative Biology* 55: 651–661.

7 Where in the World Color Survey Is the Support for Color Categorization Based on the Hering Primaries?

Kimberly A. Jameson

There is considerable debate in the study of human color categorization and naming regarding (1) the degree to which universal tendencies exist in the ways different linguistic societies categorize and name perceptual color experiences, and (2) the possible basis for such universal tendencies. Regarding this controversy, the most popular view in the empirical literature is that a panhuman regularity in human visual processing, specifically features related to the *Hering opponent-color* construct, gives rise to a standard, universally shared phenomenological color experience, and that this in turn is the basis for the empirically demonstrated similarity in color categorization and naming across cultures (see Hardin 2005; Kuehni 2005b; Kay 2005; Regier, Kay, and Cook 2005; or Philipona and O'Regan 2006 for an extreme variant of this approach). This view is widely held and is referred to here as the *standard view* of the area.[1]

Expressions of the standard view can be found throughout the color categorization literature. For example, in recent analyses of the World Color Survey (WCS) data, Malkoc, Kay, and Webster (2005) report:

> The centroids of the stimuli labeled by basic color terms in these [WCS] languages cluster strongly around similar points in color space, showing that respondents view the spectrum in very similar ways regardless of the varying number of categories into which their lexicons partition it. While counterexamples have been noted [Davidoff 2001] . . . the similar clustering across languages suggests that the special and shared status of basic color terms may reflect special and shared properties of the human visual system or of the visual environment. (2154)

From time to time in the empirical literature, supporters of this standard view mention influences from culturally relative factors on color-naming and categorization behaviors (Kay and Kempton 1984; Kuehni 2005a, 2005b).[2] But, in general, explanatory mechanisms beyond those expressed as the standard view have not figured prominently in the mainstream theories of the area.[3]

The present chapter focuses on one factor widely considered by the standard view to be the basis for color-naming phenomena and explores some plausible, comparatively uninvestigated factors that might underlie color naming. These are illustrated, in part, through a reexamination of WCS data as it has been presented by Kuehni (2005b).

The aim of this chapter is to examine the appropriateness of Hering opponent-color salience as a theoretical foundation for explaining patterns of color naming in datasets like the WCS, which include many languages that do not use Hering color terms.[4] The main conclusion reached is that a proper explanation for cross-cultural color naming and categorization should not depend on the Hering opponent-color construct.

In section 7.1, Hering opponent-color theory is described; section 7.2 suggests the extent to which color categorization literature relies on classical Hering color theory; section 7.3 examines some evidence regarding whether unique hue experiences are shared across individuals—a key assumption in color-naming research; section 7.4 reexamines Kuehni's (2005b) analysis of WCS data in light of sections 7.1–7.3; section 7.5 briefly discusses the circumstances under which the construct of *hue salience* is an appropriate modeling construct in color categorization theory, discusses the appropriateness of alternative modeling constructs to unique hue salience, and provides some empirical support for the suggested alternatives; and section 7.6 reviews the main points discussed and offers some conclusions.

7.1 What Are the Hering Colors?

The Hering opponent-color theory (Hering 1920/1964) was a prominent psychological processing component in the original color-naming theory of Berlin and Kay (1969), which continues to permeate contemporary theories as an important factor underlying color-naming regularity across individuals (e.g., Kay 2005; Regier, Kay, and Cook 2005; Griffin 2006; Lindsey and Brown 2006). It was originally proposed by Hering largely as a model of individual color phenomenology and was subsequently developed as a model of physiological processing by Jameson and Hurvich (1955, 1968).

In its standard form, this three-channel model of color opponency uses salient points in color space, or primaries, based on three opponent axes in color space: *black* versus *white*, *red* versus *green*, and *yellow* versus *blue* (as in figure 7.1). This color-opponent model assumes that the higher-order structure of individual color appearance, and an individual's color-similarity judgments, are directly based on these Hering (black-white, red-green, and blue-yellow) color appearance dimensions.

In addition, the psychological literature assumes that cognitively the black-white, red-green, and blue-yellow axes are largely linear and independent, which is implicit in the widespread use of the *unique hue construct* (as discussed below) in color catego-

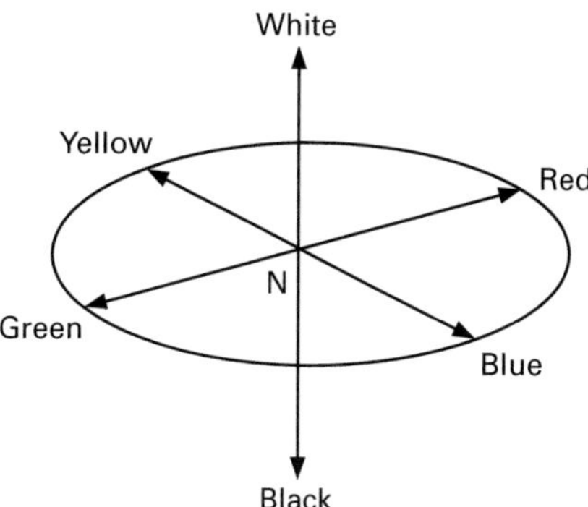

Figure 7.1
Hering (1920) opponent-color axes.

rization research (e.g., Sivik 1997; Hård and Sivik 2001). Thus, as in figure 7.1, Hering opponent-color theory classically positions the polar endpoints of a red-green axis, perpendicular to a polar yellow-blue axis, and this conceptualization remains influential in color appearance theory (e.g., Hardin 2000, 2005; Nayatani 2004). Largely due to the work of Hurvich and Jameson,[5] the Hering opponent-color model was long considered an appropriate description of both early visual processing (i.e., chromatic response mechanisms in the lateral geniculate nucleus) and higher order (cortical or phenomenological) color representation.

This view changed when the empirical findings of Krauskopf, Lennie, and colleagues prompted a reanalysis of the issue in the 1980s (Krauskopf, Williams, and Heeley 1982; Derrington et al. 1984; Krauskopf, Williams, Mandler, et al. 1986). Since that time, opponent-color axes, and the relationships between the axes, were found to depend on the level of cortical processing considered. For example, representation at the level of postreceptoral excitation involves color space angles shifted off the classic Hering axes—suggesting a different picture from that shown in figure 7.1—such that red is opposed, or nulled, by a blue-green,[6] which is near-orthogonal to an axis formed by a green-yellow nulled by violet.[7]

Classic opponent-color theory suggests that the axes in figure 7.1 should yield unique hue relationships (Hering's *Urfarben*) shown in figure 7.2.[8] Instead, when unique hue settings are displayed in an approximately perceptually uniform space, it becomes apparent that average Hering unique hue settings (Kuehni 2004) are displaced off the axis endpoints defining CIELAB space (shown in figure 7.3).[9] These unique hue results are discussed again below in detail, in the context of other relevant findings (e.g., Kuehni 2003, 2004).

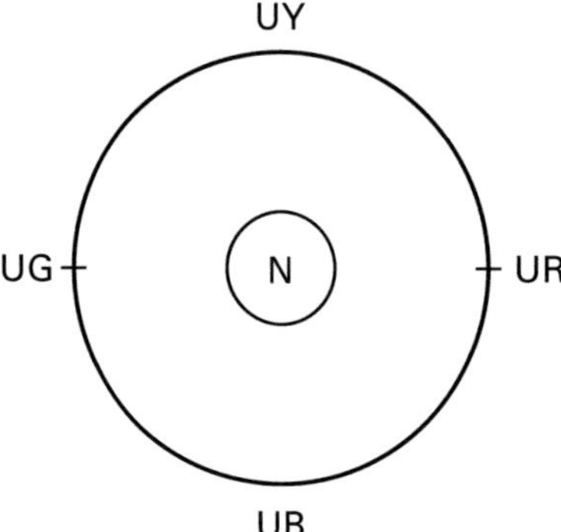

Figure 7.2
Classical relations among the Hering unique hues, unique yellow (UY), unique red (UR), unique blue (UB) and unique green (UG), on the hue circle plane of figure 7.1.

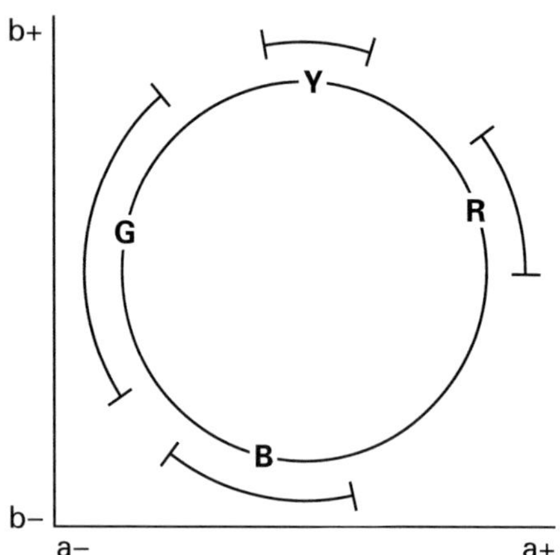

Figure 7.3
Unique hue settings and ranges in CIELAB (1976) *a* and *b* dimensions. Uppercase letters show the average of empirical unique hue settings for red (R), yellow (Y), green (G) and blue (B). Segments adjacent to each setting represent the ranges of unique hue setting averages. Data included were originally discussed by Kuehni (2004) and by Bruce MacEvoy at www.handprint.com.

7.2 Hering Colors in Color Categorization Research

In an important departure from early philosophical theories of color appearance realism and categorization, Hardin (1988) advanced an empirically motivated view of subjective color, making a move toward solidifying the psychological reality of color experience as constructed by an observer. And even for the much loved and historically important Hering colors, Hardin argued for color as subjective and mainly in the heads of perceivers.

The color categorization literature has generally resisted critical examination of the Hering opponent colors (and opponent-process physiology) theory and it's central role in the explanation for human color-naming phenomena. One exception was Jameson and D'Andrade (1997) who (1) questioned the physiological reality of classic opponent-color theory (as developed by Hurvich and Jameson 1957), and (2) argued against the special status of the Hering colors as plausible explanatory factors for color naming and categorization findings. Jameson and D'Andrade (1997) based their arguments on a clear disconnect between the physiological data and the phenomenological data. However, at the time these suggestions were considered extreme, as dissenting from the mainstream, and contrary to the well-accepted theory stating that unique hue phenomena were (1) linked to highly specific, early visual-processing mechanisms, and (2) responsible for the universal structure of human color categorization and naming.

In current research the Hering primaries construct remains robust and figures prominently in mainstream theory as the basis for universal tendencies in color categorization. For example: "The Kay and Maffi model takes universal constraints on color naming to be based on presumed universals of color appearance for example, on opponent red/green and yellow/blue phenomenal channels" (Kay 2005, 52); and "the six Hering primaries: white, black, red, yellow, green, and blue [suggest] that these points in color space may constitute a universal foundation for color naming. These foci in color space have also appeared to be cognitively privileged, in non-linguistic tasks with speakers of languages that have dissimilar color naming systems" (Regier, Kay, and Cook 2005, 8386).[10]

C. L. Hardin also argues strongly for the psychological reality of individual unique hue appearances: "Given a particular observer in a particular state of adaptation and a particular set of observational conditions, there is a way to . . . [empirically assign colors to stimuli]. The names of just four perceptually basic hues—red, yellow, green, and blue—are both necessary and sufficient to describe every hue" (Hardin 2004, 32). See also Hardin (2005) for a more extensive presentation of this view.

The research cited above underscores the prominent status opponent-color theory continues to receive in color categorization and naming research.

7.3 Do Shared Perceptual Experiences Underlie Similar Color-Naming Behaviors?

The foregoing discussion raises some questions about the continued use of the Hering opponent-color construct as the explanatory basis for similar patterns of color categorization and naming observed across cultures. To clarify this issue, we can examine results on empirically measured Hering primaries in the form of unique hue settings across individual observers, and compare such results with data on best-exemplar choices for Hering colors, as done in section 7.4.

Evidence suggests that across subjects with normal color vision, there is wide variation in the stimuli they select for the unique hues (e.g., Kuehni 2001, 2003, 2004, 2005a, 2005b; Webster et al. 2000, 2002; Otake and Cicerone 2000; Jordan and Mollon 1995; Boynton and Olson 1990). Figure 7.3 presented earlier shows CIELAB hue angle ranges of individual unique hue settings across several different studies (Kuehni 2004), illustrating, for example, that the variation for the unique green setting spans a considerable portion on the hue circle,[11] compared with unique yellow, which shows relatively less variation. Such results exemplify how group unique hue ranges can be rather large as a result of unique hue settings that vary considerably across individuals. (See also discussion in Kuehni 2005a.)

Aside from this large individual variation across observers' unique hue locations, there seems to be effectively no correlation (r = –0.02) between subjects' Rayleigh matches and settings of unique green (Jordan and Mollon 1995, 616), as well as a lack of correlation among individuals' unique hue settings—results contrary to what classic opponent-color theory might predict. Regarding the latter, Webster, Miyahara, Malkoc, and Raker (2000) expressed surprise at observing no correlation among the stimuli individual observers selected for unique hues. That is, participants' unique hue variations did not arise from idiosyncratic individual biases that could shift all of a given individual's unique hue settings to systematically differ from the average settings observed. Thus, Malkoc, Kay and Webster (2005) report, "A subject whose unique yellow is more reddish than average is not more likely to choose a unique blue that is more reddish (or more greenish) than average. The independence of the unique hues is surprising given that many factors that affect visual sensitivity (such as differences in screening pigments or in the relative numbers of different cone types) should influence different hues in similar ways and thus predict strong correlations between them" (2155).

The results of Malkoc et al. (2005) give reason to doubt the shared uniformity of Hering color experiences, because (a) variation across individuals is substantial and reliable, and (b) there is no explanation for (a) based on idiosyncratic biases. Regarding the presumed unique hue basis for color categorization and naming, Malkoc et al. (2005) report, "The range of variation in the hue settings is pronounced, to the extent that the range of focal choices for neighboring color terms often overlap . . . [and] . . .

some subjects chose as their best example of orange a stimulus that other subjects selected as the best example of red, while others selected for orange a stimulus that some individuals chose for yellow" (2156).

Thus, existing research does not give an affirming answer to the question posed at the outset of this section (namely, do shared perceptual experiences underlie similar color-naming behaviors?), and suggests that a widely shared perceptual experience is not a likely explanation for existing empirical color-naming results. To summarize, the reasons for this are (1) a significant lack of evidence for shared unique hue settings across individuals either within or across ethnolinguistic groups,[12] (2) no evidence to support the assertion that enough similarity exists in unique hue settings to explain the amount of color-naming agreement observed either within a given ethnolinguistic group or across ethnolinguistic groups (elaborated in the next section), and (3) no clear demonstration that robust, congruent unique hue settings give rise to equivalent internal experiences in any two individuals (as discussed in Jameson et al. 2006).[13] From a historical perspective, then, one might also consider that color categorization theory has overgeneralized Hering's unique hue construct because, unexpectedly, unique hues are not shared phenomenologically, they are not linked to well-defined ranges of *focal chip* reflectances, and (perhaps more important) individual idiosyncratic category variation cannot be accounted for by systematic shifts in personal "landmark color" settings,[14] or by some identifiable bias across individuals that systematically affects unique hue settings. For these reasons, unique hues seem deficient as a basis for explaining the specific physical stimuli that are individually or collectively identified as color category focal exemplars.

7.4 How Does Individual Perceptual Variation Relate to Cross-Cultural Color-Naming Results?

When similarities between two color categorization systems are found empirically, the usual assumption is that it is largely a consequence of similar perceptual experiences across observers. This applies equally to cases of observed similarities across individuals in the same ethnoliguistic society, and cases of observed similarities across individuals from different societies, using different color lexicons. Usually such explanations anchor similar perceptual experiences to the Hering primaries. This long-standing historical practice is widely employed and accepted in mainstream color categorization research (e.g., Kay 2005; Kay and Regier 2003; Regier, Kay, and Cook 2005; Hardin 2000, 2004, 2005).

However, given the discussion in section 7.3, the question we now need to consider is whether, in general, perceptual experience is the largest component of the explanation behind color categorization similarities, or if, perhaps, other overlooked factors can be identified that play equally important roles in color-naming phenomena.

The earlier suggestion that unique hue settings are not shared in a way that identifies specific physical stimuli does not rule out the possibility that other identifiable mechanisms might be responsible for the shared *focal regions* found in some cross-cultural color-naming research.[15] For example, cross-cultural focal region similarity could be the product of restricted *ranges* of observed unique hue settings (viz., figure 7.3's relatively narrow "Y" setting range compared to that shown for "R," "B," or "G") or a nonphysically based individual cognitive construct of shared phenomenal color salience for the Hering colors.

To examine the possibility that other mechanisms are at play, and to explore the relationship between section 7.3's unique hue variation and color-naming patterns, a reexamination of results discussed by Kuehni (2005b) is now presented.

7.4.1 Cross-Cultural Evidence Presented by Kuehni (2005b)

Empirical studies that permit a proper assessment of color-naming patterns across cultures—especially when done in the field—are costly and demanding undertakings; and much cross-cultural color-naming evidence shows that subjects with normal color vision vary widely in the stimuli they select as the focal stimuli for basic color terms (e.g., Berlin and Kay 1969; Davidoff 2001; Roberson, Davies, and Davidoff 2000; Kay and Regier 2003; Jameson and Alvarado 2003a, 2003b; Lindsey and Brown 2006).

The most recently available database (WCS; Kay and Regier 2003; Cook, Kay, and Regier 2005) is unique and extremely valuable considering the vanishing opportunities to observe color-naming behaviors undisturbed by direct contact with outside cultures (MacLaury 2005). The WCS consists of color-naming data for 110 unwritten languages from nonindustrialized societies with minimal exposure to external industrialized influences. The WCS field investigations (detailed in Cook, Kay, and Regier 2005) were conducted using fewer empirical controls compared with laboratory studies, but because of this, they assess more naturalistic naming behaviors than one might encounter under the typical circumstances of a controlled laboratory experiment. The WCS used, in part, a Mercator projection stimulus, shown here as figure 7.4 and plate 12 (with the hue columns reordered as reported by Kuehni 2005b) from the *Munsell Book of Color* (Munsell 1966; Newhall, Nickerson, and Judd 1943).[16]

Using the publically available WCS database,[17] Kuehni (2005b) examined a universal perceptual categories hypothesis. Esssentially, Kuehni asks, what if across WCS languages, the ranges of stimuli from figure 7.4 (plate 12) that are chosen as *focal* exemplars are found to resemble ranges of empirically observed unique hue settings? If such a correspondence is seen, then, Kuehni suggests, a panhuman universal perceptual basis for color category focals can be argued.

Figure 7.3 presents unique hue ranges in a CIELAB approximation of perceptual color space, and figure 7.5 schematically depicts how such unique hue ranges might be used to predict focal exemplar ranges for English speakers relative to the vertical

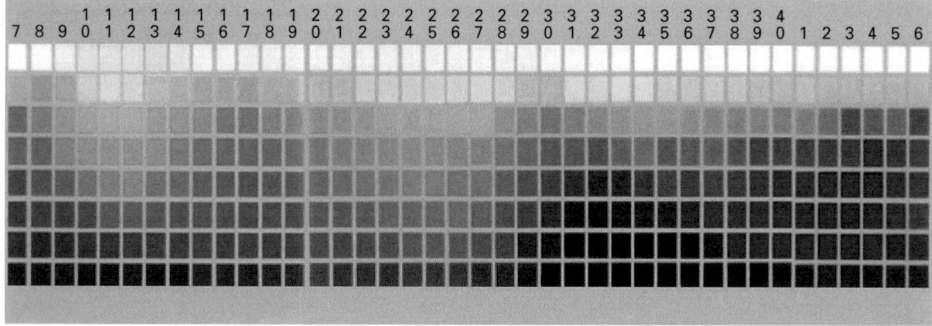

Figure 7.4
The World Color Survey stimulus reordered to produce a continuous reddish region, with columns 1–6 appearing after column 40. See plate 12 for a color version of this figure.

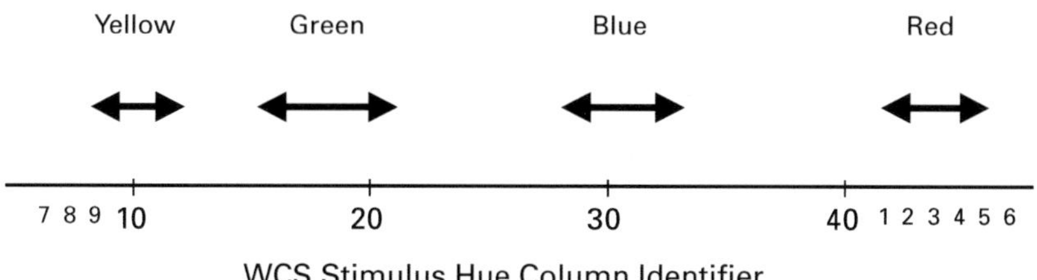

Figure 7.5
Hypothetical range variation for English focal color terms in the WCS stimulus as might be predicted by unique hue settings (e.g., figure 7.3) relative to the WCS stimulus. Depicted ranges are schematic, do not capture brightness variation, and are drawn to illustrate how observed trends in scaled unique hue data might be compared to the WCS stimulus array.

hue columns of figure 7.4's WCS stimulus. With regard to figure 7.4, Kuehni's (2005b) empirical question asks, do observed unique hue ranges resemble the focal exemplar ranges seen in WCS languages? The representation in figure 7.6 (plate 13), derived from data presented in Kuehni's (2005b) article, permits a renewed examination of this empirical question.

7.4.2 Considering the WCS Data

Kuehni (2005b) asked whether observed unique hue ranges resembled the focal exemplar ranges found in all 110 of the WCS languages. However, as he reports, 65 percent of WCS languages do not have linguistic glosses for all four unique hue terms (i.e., for yellow, green, blue, and red). In fact, Kuehni found that only thirty-eight WCS languages had the linguistic glosses needed for a comparison against the four unique hue ranges.[18] Also, in some of these thirty-eight languages, more than one distinct linguistic gloss was found for unique hue categories. Thus, across thirty-eight languages Kuehni identified thirty-nine distinct glosses for yellow, thirty-nine glosses for green, forty-four glosses for blue, and forty-five glosses for red categories. Moreover, in the process of identifying the focal term ranges from each language, Kuehni also encountered a need to prune focal color choice outliers. That is, in 76 percent of the thirty-eight languages he considered, some participants "had distinctly different interpretations of a given color name, as demonstrated with their choice of focal color" (Kuehni 2005b, 412). Using the same data refinements Kuehni employed, figure 7.6 (plate 13) provides an alternative representation of the data found in Kuehni's (2005b) figure 3.

Figure 7.6 provides two types of data relative to the WCS hue scale (x-axis): (1) median focal color term ranges from 38 languages, and (2) unique hue setting

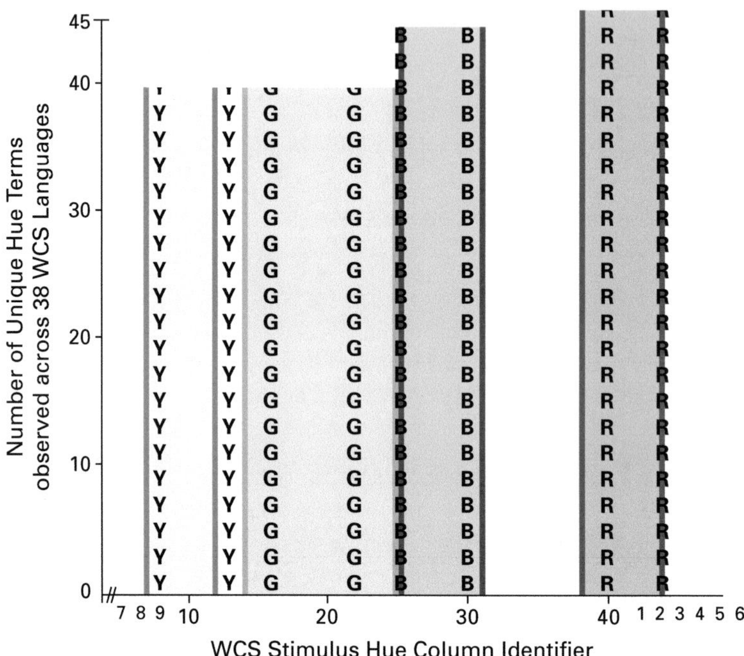

Figure 7.6

Ranges for unique hue settings and corresponding focal color term ranges in the WCS stimulus array. Two types of data are shown relative to the WCS hue scale on the x-axis: (1) Median focal color term ranges for thirty-eight languages (i.e., columns delimited by congruent alpha-character lines), and (2) unique hue ranges (i.e., shaded columns delimited by solid lines). Figure 7.6 is a reanalysis of the data presented by Kuehni (2005b, 417, figure 3). See plate 13 for a color version of this figure.

ranges. Thus, figure 7.6's data differs from that of Kuehni (2005b, figure 3) by showing the median range for each focal term gloss instead of presenting separate ranges for each language examined. Median focal term ranges are denoted by alpha-character columns. Thus, the pair of vertical "Y" lines show the median denotative range observed for *yellow* focal terms, followed by that for the *green* focal range ("G" lines), the *blue* focal range ("B" lines), and the *red* focal range ("R" lines).

In contrast to Kuehni's presentation of these data, figure 7.6's (plate 13) median focal ranges emphasize each term's modal denotative range, or the central extent of signification across the thirty-eight languages examined. This is both a rigorous and a fair alternative presentation of the data, that does not capitalize on the most (or least) variable ranges seen across languages and instead concentrates on the core meaning of the term given by the modal amount of range variation for each category gloss across the thirty-eight languages considered.[19] This use of modal ranges can be interpreted as the average (or universal) meaning of these four color terms across all thirty-eight languages examined.

Thus, the x-axis width of a given focal term column shows the median denotative range (across thirty-eight languages) of that focal term relative to the WCS stimulus array. Y-axis height of alpha-character lines gives the total number of terms observed across the thirty-eight languages for each unique hue category. The second type of data seen in figure 7.6 (plate 13) is graphed as shaded columns that show unique hue setting ranges based on the data of some three hundred observers.[20] Pairs of solid lines show unique hue setting ranges relative to WCS stimulus array. From left to right, shaded columns show the hue ranges for *unique yellow* settings, *unique green* settings, *unique blue* settings, *unique red* settings.

An important caveat is needed when interpreting the results in figure 7.6: Focal color term ranges are properly interpreted as category *best-exemplar ranges* (i.e., "universally shared focal points, or prototypes, in color space" Regier, Kay, and Cook 2005, 8386) as opposed to illustrating full category ranges denoted by a given color term.

Now let us consider how figure 7.6's alternative representation of the WCS data facilitates the examination of the empirical question: *do unique hue ranges resemble the focal exemplar ranges seen in WCS languages?*

First, note that relative to the WCS stimulus array, both types of data ranges span such a large extent of figure 7.6's horizontal axis that they essentially imply that only the unassessed "purplish" stimulus region gets excluded from the so-called highly salient Hering-color regions.[21] Quantitatively, figure 7.6's unique hue ranges cover 68 percent of WCS stimulus hue columns (see figure 7.4), and the four median focal ranges span 49 percent of the WCS stimulus hue columns (making "unique" a misnomer).[22]

With regard to the empirical question posed by Kuehni's (2005b) analysis, such large variation in both focal and unique hue ranges makes the chance for a failed correspondence between these two types of data ranges exceedingly unlikely. Thus, although Kuehni presents a fairly exacting analysis regarding his universal perceptual categories hypothesis, of these same data he, somewhat surprisingly, concludes "The results from the 38 languages provide support for the perceptual salience of the Hering UHs" (Kuehni 2005b, 423).

Here a different interpretation of these data is offered because although range correspondences do exist, it seems that a criterion of *being anywhere in the color stimulus ballpark* is a poor measure of correspondence in support of a panhuman universal basis for color category focals. In addition, this is not what one would expect if color-naming behaviors were actually based on Hering-color appearance universals (Kay 2005, 52) and actually does not accord with the spirit of the construct as expressed in the original formulation of Hering's opponent-color theory (more on this later).

Finally, consider that for 72 of the 110 languages contained in the WCS database, the above comparisons between focal and unique hue ranges is not even possible because those languages do not have glosses for the presumed panhuman salient Hering categories (despite many having lexicons that include color terms robustly used to denote non-Hering colors).

In view of these WCS data, the standard view's use of the Hering unique hue construct as a panhuman-shared phenomenal basis for color category focals and category naming seems much less compelling; and the way that individual variation in perceptual experience relates to cross-cultural color-naming results seems much less dependent on the Hering unique hue construct.

7.5 When Is Unique Hue Salience Appropriate for Color-Naming Modeling?

Section 7.4 implies that historically, Hering's unique hue construct has perhaps been over extended as an explanation for color category focal exemplar results.[23] Still, due in part to the prominence of Hering's opponent-color theory, color salience is considered an important shared color-processing feature throughout the color-naming literature. Thus, in view of the above-mentioned variability of the highly salient unique hues, one might wonder what general role *hue salience* plays in color categorization and naming. The following analysis of hue salience addresses this issue.

7.5.1 Revisiting the Definition of Hering's Unique Hues

As classically formulated, in Hering's opponent-colors theory (Hering 1920/1964), "Unique hues are defined as those hues that are phenomenologically pure or unmixed in quality: thus unique green is that green that appears neither bluish nor yellowish. The four unique hues (blue, yellow, red and green) are central to classical Opponent Process Theory and are held to be those colours for which one of the putative opponent processes . . . is in balance" (Jordan and Mollon 1995, 614).

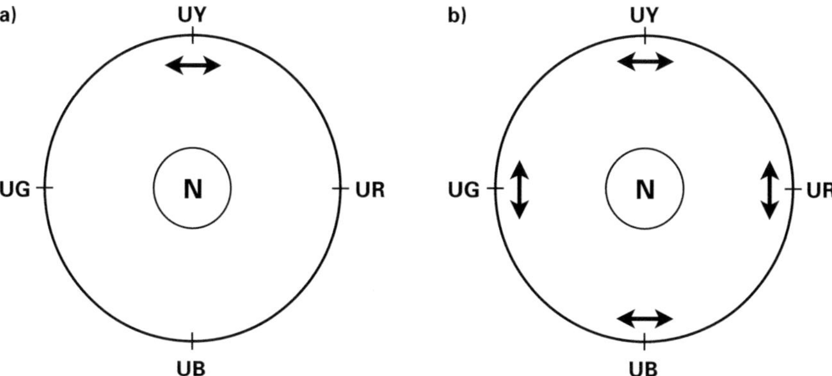

Figure 7.7
Unique hues defined by a color exclusion operation.

Historically, this unique hue construct has presented two important components relevant for color categorization and naming: (1) Unique hues are theoretically construed as color purity relations (or privileged appearances obtained by color mixture procedures) found in psychophysical experiments; and (2) They are phenomenologically defined as (i) of high subjective salience, and (ii) are necessary and sufficient descriptors of all visible colors (see Hardin 2004, 32).

Figure 7.7 presents a schematic of the standard color-mixing construct implicit in unique hue settings (shown in figure 7.2). Figure 7.7a illustrates a process of *narrowing in* on an individual's unique yellow (UY) setting by additively mixing proportions of red primary and green primary lights with the aim of canceling any visible red or green tinge in the mixture, and producing a unique, pure, yellow appearance. Thus, determining unique yellow settings requires a person to successively adjust primary ratios until a yellow appearance that is neither reddish nor greenish is achieved. The two-headed arrow in figure 7.7a illustrates this operation of canceling, involving the exclusion or repelling of neighboring primaries to achieve a pure unique yellow.

Figure 7.7b shows similar operations for all unique hues: UY, UR, UG, and UB. The two-headed arrows placed at these four Hering hue points diverge from the points on the hue circle at each unique hue location to represent the canceling, or exclusion, of the depicted flanking primaries as required by the instructions of the empirical procedure. The identification of all Hering primaries via analogous unique hue settings is a defining feature of the unique hue construct (Jameson and Hurvich 1955).[24]

However, one might question whether the cancellation procedure just described, with its ability to isolate an individual's psychologically compelling, pure hue settings,

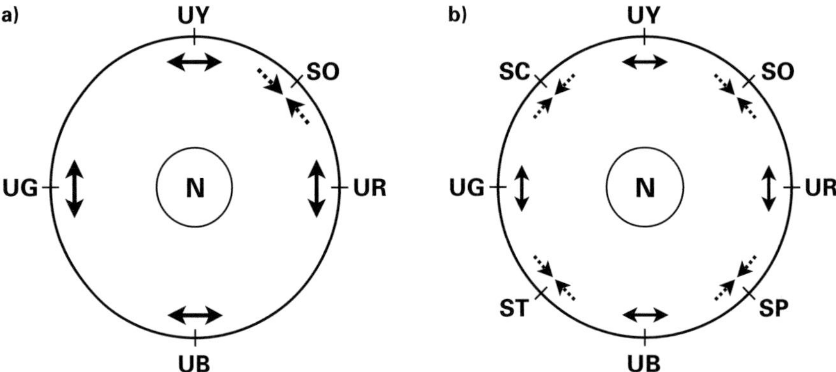

Figure 7.8
Alternative salient hues defined by varying color-mixing instructions: (a) an alternative salient hue setting SO (salient orange) intermediate to the UY and UR positions; (b) four proposed alternative salient hue settings on the hue circle. Hering's classic opponent colors are denoted by UY, UR, UB, and UG. Four new salient hue points are SO (salient orange), SC (salient chartreuse), ST (salient turquoise), and SP (salient purple). The relationships represented here are not suggested as a physiological processing model underlying any of the colors represented. Figure adapted from Jameson et al. (2007).

is necessarily the most appropriate procedure for objectively identifying sets of *privileged perceptual fundamentals* (as the Hering colors are commonly known).

In point of fact, much evidence supports the existence of other highly salient color appearances representing opposing colors that are as empirically robust and compelling as the Hering opponent colors (e.g., Malkoc et al. 2005; Webster, Miyahara, et al. 2000; D'Zmura and Knoblauch 1998; Webster and Mollon 1994; Krauskopf, Williams, and Heeley 1982; and others). In light of this, what aspect of the empirical procedures used to define the Hering unique hues distinguishes those colors from other opposing hues paired on an axis across the color space? The answer may hinge on a detail as simple as a modification in the unique hue task instructions.

Figure 7.8 illustrates a modification in task instructions that initially presumes the classic Hering colors as mixture primaries (similar to the figure 7.7 procedure), but that permits the identification of robust alternative salient hue settings. For example, figure 7.8a depicts (in addition to figure 7.7's four unique hue settings) an alternative salient hue setting SO (denoting salient orange) intermediate to the UY and UR positions. The dashed arrowheads converging near figure 7.8's SO setting illustrate a mixing operation on adjacent primaries (i.e., UY and UR are combined to reach an equilibrium color with subjectively equal proportions of each), thereby differing from the usual cancellation operation shown in figure 7.7. This simple change in instruction, then, is a plausible basis for achieving an individually compelling and robust SO hue setting. The point is that *alternative* salient hue settings are just as easily obtained from the classic color-mixing procedures described above, given a very minor modification of task instructions. Thus, on one hand (using usual Hering unique hue settings), the instruction to "mix two classic primaries until no classic primary is apparent" should confirm the privileged salience of Herings unique hues. Whereas, on the other hand, a slightly different instruction (namely, "mix two classic primaries until both are equally apparent"), should produce settings with similarly privilaged salience for the proposed alternative salient hue settings shown in figure 7.8.[25]

Figure 7.8b illustrates additional salient points: SO (salient orange), SC (salient chartreuse), ST (salient turquoise), and SP (salient purple). The dashed arrows at these four additional points converge at each of the four alternative hue settings to represent four equally apparent mixtures of adjacent classical primaries. These alternative salient hue settings become possible through an uncomplicated, natural variation on the unique hue empirical task, consisting only of a slight rewording of the task instructions. Clearly, based alone on the earlier empirical definition of color salience, there seems no good reason not to accept all eight of the subjectively compelling hues shown in figure 7.8b as equally compelling privileged perceptual fundamentals.[26]

The analysis above questions the accepted theoretical idea that the unique hues alone connote privileged perceptual salience. It suggests that classic Hering unique hue salience could be tied to empirical task instructions, and that with sensible, minimal variations to the instructions, other alternative hue settings with comparable salience (i.e., empirical robustness, compelling subjective salience, etc.) may be established. This raises some doubt concerning the usual assumption of unique hue special status that is implied by part 1 of the two-part definition stated at the outset of this section.

Part 2 of the definition given earlier was, historically, the unique hue construct is phenomenologically defined as *(i) of high subjective salience,* and *(ii) are necessary and sufficient descriptors of all visible colors*. The analysis just described suggests that all eight hue settings in figure 7.8b meet the *high subjective salience* requirement. The *necessary and sufficient descriptor* criterion seems to this author to be (a) culturally relative, since languages that do not have glosses for all four unique hue categories (but do have glosses for non-Hering-color categories) are apparently able to sufficiently capture all the color experiences of their speakers; and (b) unrealistic as a criterion in that it does not apply in 65 percent of WCS languages that do not have a full complement of Hering-color glosses. In the best of worlds, a *necessary and sufficient descriptor* criterion that is used to argue universal color salience should employ constructs that are similarly manifest across all groups of individuals assessed. While the Hering color category descriptors may be seen in a subset of the world's languages, they are not found in many languages, and this limits their utility as universally necessary and sufficient descriptors. Finally, in my opinion the *necessary and sufficient descriptor* criterion is really misplaced in the discussion of salient hues identified using perceptual color space considerations, and a better explanation is needed than a linguistically based descriptor rationale if color-processing fundamentals are underlying color-naming phenomena.[27]

7.5.2 Empirical Support for a More Broadly Defined Notion of Hue Salience

Recent empirical support exists for section 7.5.1's suggestion that the alternative hue points described may be similar in phenomenological salience to the Hering unique hues. Malkoc, Kay and Webster (2005) used hue cancellation and focal-naming tasks to compare individual differences in stimuli selected for unique hues (e.g., pure blue or green) and binary hues (e.g., blue-green, or turquoise as described above).[28] They did not find any distinction between unique and binary hues in terms of variability in the settings, and, like the unique hues, the binary hue settings were surprisingly uncorrelated with other hues. They also state, "the degree of consensus among observers did not clearly distinguish unique from binary hues, nor basic terms from nonbasic terms" (2165).

Their study shows that there is little to differentiate binary from unique hues, and "no clear tendency for unique and binary hues to behave differently" (Malkoc et al. 2005, 2158). They conclude that "the processes underlying subjective color experience, and how they are derived from the opponent organization at early postreceptoral stages of the visual system, remain very poorly understood" (2164), and that "the unique hues do not emerge as special and do not alone fully anchor the structure of color appearance for an individual" (2155).

Malkoc et al.'s (2005) findings also accord with results from simulated color category solutions (Jameson and Komarova 2009a, 2009b). That is, simulations involving realistic populations of observers find that all optimal population color category solutions on a hue circle exhibit rotational invariance (i.e., randomly rotated yet structurally similar solutions) as long as some symmetry-breaking influence is not introduced to the color space by an environmental or perceptual processing nonuniformity. Generalized to humans, this suggests that in the absence of a marked nonuniformity in perceptual space (or in environmental color utility and so forth), color categories are less likely to be fixed by Hering-type color appearance saliences and are more likely to depend on relational structure inherent in categorization pragmatics and features of the stimulus domain (Jameson and Komarova 2009a, 2009b). This observation of rotational invariance in color category solutions suggests that the sharing of unique hue privileged–processing salience is (a) not necessary for the development of optimal color categorization population solutions, and (b) that an optimal color category solution need not realize Hering-color categories, even when the simulated observers in such populations are based on color discrimination data from actual human observers.

7.5.3 Summary

Are Hering opponent-color experiences and the unique hue construct the exclusively appropriate perceptual basis for color categorization and naming theory? The answer to this question could be "yes" if cross-cultural color-naming theory aimed primarily to model the classic color-mixing results for the subset of languages that have linguistic glosses appropriate for such paradigms.[29] Otherwise, if cross-cultural color-naming theory aims to capture commonalities across perceptual color space that are shared, communicated, and represented across many additional languages, then a more inclusive and comprehensive theoretical basis is needed to capture all the factors that contribute to the hue saliences that shape color categorization across cultures.

7.6 Conclusions

The aim of this chapter was to consider a central assumption inherent in the standard view explanation for color-naming behaviors both within and across cultures. I explored the possibility that the empirical results showing individual differences in perceptual processing undermine the argument that a shared phenomenal salience for the Hering unique hues is the sole explanatory factor of shared color category structures within a given society. Other analyses presented suggest that primary mixture settings for Hering's unique hues do not provide a robust basis for explaining the prevalence of WCS color term ranges for the corresponding glosses.

It is important to note that although the present analyses do not find that naming patterns in the WCS data are explained by individual Hering-color salience, and suggest that a different explanation be sought for the basis of human color-naming similarities, it is not implied that no support exists for shared patterns of color naming in the WCS data (e.g., Lindsey and Brown 2006). In general, the present critique of the Hering-color construct does not bear on uninterpreted statistical demonstrations of shared patterns in the WCS data (e.g., Regier, Kay, and Khetarpal 2007).

The view expressed here goes beyond the issue of what relevance the Hering colors might have to color categorization results. In general, color perception, environmental colors, and pragmatic constraints must all place clear structure on color categorization phenomena. For example, color in the environment frequently signals important information, and primate color perception has almost certainly evolved in ways that allow recognition of such signals (Regan et al. 1998; Osorio and Vorobyev 1996). The ability to effectively communicate about valuable color signals, even when individual variation in visual processing exists, seems like a desirable capacity (and many aspects of human evolution underscore the value of within-species variation and the evolutionary ability for specialization and adaptation), as does the ability to maintain communication about such information under circumstances where environmental colors vary seasonally and geographically. It therefore seems reasonable to seek explanations for color categorization similarities across human societies that do not strictly depend on fixed visual environments or fixed attributes of perceptual processing. This chapter simply expresses the perspective that beyond the typically considered explanatory features, there are clearly panhuman cognitive and communication universals that are arguably more plausible sources for explaining similarities seen in cross-cultural color categorization and naming (see Komarova et al. 2007; Komarova and Jameson 2008; Jameson and Komarova 2009a, 2009b). If the standard view were to incorporate as a portion of its emphasis the serious investigation of such plausible factors, it seems likely that color-naming research efforts might build a more accurate description of the phenomena.

The main conclusion of this chapter is that although perceptual processing is an important constraint on color categorization and naming, a more extensive set of factors is needed to account for (a) the process by which cultures similarly categorize and name color experience, and (b) the factors that lead such systems to differ across cultures. Explicitly recognizing the limited utility of the Hering color salience argument in color categorization research is an important step toward establishing a proper explanatory model of cross-cultural color-naming phenomena. With this in mind, future empirical tests of color categorization hypotheses should take into account perceptual color organization as well as more complicated, pragmatic, social interactions that also play a role in categorization and naming phenomena.

Acknowledgments

Thanks to Angela Brown, Louis Narens, Wayne Wright, and Ragnar Steingrimsson for helpful comments. This article extends ideas introduced by Jameson et al. (2007). Partial support was provided by NSF#07724228 grant from the SES Methodology, Measurement, and Statistics Board.

Notes

1. Alternative perspectives that emphasize culturally relative influences on color-naming phenomena—for example, that supported by D. Roberson, J. Davidoff, and colleagues (e.g., Roberson et al. 2000, Roberson and Hanley 2007, etc.)—are also frequently seen in the literature but are not featured in the analyses presented here.

2. Especially regarding possible linguistic influences.

3. A noteworthy exception to this is the recent shift in theoretical emphasis of Regier, Kay, and Khetarpal (2007).

4. Especially *blue* and *green* glosses in the WCS languages.

5. Beginning around the time of Hurvich and Jameson (1957), and including their subsequent opponent-color articles.

6. Or the (L – M) mechanism—which is somewhat misleadingly referred to as a "red-green" mechanism.

7. Or the "tritan" S – (L + M) mechanism, see Gunther and Dobkins' (2003; description of both mechanisms).

8. Hering's Urfarben are six fundamental perceptions, or pure color perceptions that, as defined, when proportionally mixed can produce all color experiences that humans observe. Figure 7.2 shows Hering's theoretical relationships among four chromatic Urfarben on a hue circle plane from a schematic color sphere. According to this theory of phenomenology, unique red (UR) is nulled (N), or mixed to a neutral appearance, by some proportion of unique green (UG).

9. CIE (Commission Internationale de l'Eclairage, or the International Commission on Illumination) is the original organization responsible for setting standards for color and color measurement; it developed the CIE XYZ (1931) model, which was the first of a series of mathematical models that describe color in terms of synthetic primaries based on human perception. The primaries are imaginary mathematical constructs that model our eyes' response to different wavelengths of light. The CIELAB (1976) system aims to approximate a perceptually uniform color metric, and is often used in color appearance applications and in modeling the structure of psychological color relations. Here it is used for evaluating Hering opponent-color theory predictions because it adequately models color appearance structural relations, and the just-noticeable-difference similarity orderings, most likely present in a given individual's phenomenological color space.

10. After the present chapter was submitted for publication, Regier, Kay, and Khetarpal (2007) published an alternative to the Hering-colors explanation using the WCS database. They suggest adoption of an interpoint distance model explanation (Jameson and D'Andrade 1997; Jameson 2005) in future color-naming research.

11. Jordan and Mollon (1995) observed that although intersubject variation was substantial (demonstrated by a Gaussian-shaped frequency distribution of the unique green settings of 97 observers, with a mean at 511 nanometers (nm) and a standard deviation of 13 nm), the separate estimates of each subject's individual unique green setting showed good agreement (with the average within-subject standard deviation being only 1.63 nm).

12. The suggestion that there is an insufficient basis to support the idea that individuals share similar perceptual experiences when viewing physically identical stimuli (regardless of whether they are of the same or different cultural affiliation) is supported by the large individual variation seen in measured color-matching settings, seen over a wide range of empirical studies, stimulus formats, and investigators (see Hardin 2004).

13. Essentially, there is no way to prove equality of perceptual experiences across the internal states of two individuals because they are subjective class B observations as described by Brindley (1960). Also discussed by Mollon and Jordan (1997).

14. Or as regularized rotational shifts across individuals' categorization results.

15. Throughout this article, *focals, focal ranges*, and *focal regions* are used as defined originally by Berlin and Kay (1969), as empirically identified stimulus regions for category best-exemplars that are "universally shared focal points, or prototypes, in color space" (Regier, Kay, and Cook 2005, 8386).

16. Note that the WCS stimuli reproduced here as figure 7.4 (plate 12) omits the achromatic stimuli from the center of the Munsell color solid—chips A to J—used in the WCS investigations to represent the ten levels of Munsell value from white through gray to black.

17. See WCS Data Archives, www.icsi.berkeley.edu/wcs/data.html.

18. Note that of the WCS languages that lacked linguistic glosses for one or more of the Hering primaries, many exhibit distinct terms for color categories considered by the standard theory as noncore, or non-elemental, based on a visual-processing emphasis. Observing languages that adopt glosses for theoretically low-salience, non-elemental colors, prior to the naming of so-called highly salient, elemental colors, raises additional concern for the standard view's Hering-color assumptions.

19. Thus, unlike the cases in Kuehni's figure 3 where the denotative range of one focal term (say, yellow) invaded the territory of an adjacent focal term (say, green) for a given language, in figure 7.6 (plate 13), these overlapping semantic ranges are limited to a range consistent with a core meaning across the thirty-eight languages. Note that this alternative does not, in principle, pose a hindrance to confirmation of Kuehni's empirical question.

20. These unique hue setting data are identical to those shown in Kuehni's figure 3 representing three hundred observers who were not part of the WCS sample (see Kuehni 2005b, 415).

21. Purplish stimuli occupy the region between blue and red columns, or the stimuli flanking the hue at column 35.

22. Compared to median focal ranges observed, focal range extent for the four minimum observed ranges cover 25 percent of the WCS stimulus, whereas the four maximum focal ranges observed span 114 percent of the WCS stimulus (exceeding 100 percent of the stimulus because some of the observed maximum ranges overlap considerably).

23. The issue here is not the use of the unique hue construct in general, but its specific use in color categorization theory. There is a long-standing practice of connecting unique hue experiences with neurophysiological mechanisms, and this has been an important theoretical and modeling emphasis in vision science. For example, while Parry, McKeefry, and Murray (2006) show that the red-green and yellow-blue opponent-color relations are not the same for foveally versus peripherally presented stimuli. They suggest that some unique hues may serve as anchor points in color space, with appearances that stay comparatively stable and do not shift with retinal eccentricity of the stimulus. If this were shown to be exclusively a unique hue characteristic, such a finding would warrant a special processing and phenomenological salience for some unique hue color appearances, but such findings are not among those that have historically motivated the use of unique hues in color categorization research and theory.

24. Where a pure yellow is neither reddish nor greenish (UY); a pure red is neither yellowish nor bluish (UR); a pure blue is neither reddish nor greenish (UB); and a pure green is neither bluish nor yellowish (UG).

25. Neither form of these instructions requires that the "classic primaries" adjusted be the Hering primaries—they could just as well be any other similarly distributed points along the hue circle that were hypothesized to invoke privileged phenomenological salience or processing.

26. Note that although figure 7.8b schematically depicts both unique hue points and alternative salient hue points on orthogonal axes and uniformly distributed on the hue circle, the current argument does not rely on such color space orthogonality, and there is evidence that such regularity is not empirically seen in individual unique hue settings (Kuehni, personal communication, October 2006; Malkoc et al. 2005).

27. Although necessity and sufficiency may prove to be important factors when cultural considerations are more prominently figured into color categorization investigations.

28. Malkoc et al. (2005) assessed hues comparable to the eight shown here in figure 7.8b.

29. Here "classic color mixing" refers to the monocular-view type experiments used to establish color mixture ratios (which really are not subject to shared communication pragmatics) for Hering primary colors (e.g., determination of yellow settings that are neither reddish nor greenish, and so on), whether employing light mixtures or color papers (see Kuehni 2004).

References

Berlin, B., and P. Kay. 1969. *Basic Color Terms: Their Universality and Evolution.* Berkeley: University of California Press.

Boynton, R. M., and C. X. Olson. 1990. Salience of chromatic basic color terms confirmed by three measures. *Vision Research* 30: 1311–1317.

Brindley, G. S. 1960. *Physiology of the Retina and Visual Pathway.* London: Arnold.

Cook, R. S., P. Kay, and T. Regier. 2005. The World Color Survey database: History and use. In *Handbook of Categorization in the Cognitive Sciences,* H. Cohen and C. Lefebvre, eds., 223–242. Amsterdam and London: Elsevier.

Davidoff, J. 2001. Language and perceptual categorization. *Trends in Cognitive Science* 5: 382–387.

Derrington, A. M., J. Krauskopf, and P. Lennie. 1984. Chromatic mechanisms in lateral geniculate nucleus of macaque. *Journal of Physiology* 357: 241–265.

D'Zmura, M., and K. Knoblauch. 1998. Spectral bandwidths for the detection of color. *Vision Research* 38: 3117–3128.

Griffin, L. D. 2006. The basic colour categories are optimal for classification. *Journal of the Royal Society: Interface* 3: 71–85.

Gunther, K. L., and K. R. Dobkins. 2003. Independence of mechanisms tuned along cardinal and non-cardinal axes of color space: Evidence from factor analysis. *Vision Research* 43: 683–696.

Hård, A., and L. Sivik. 2001. A theory of colors in combination—A descriptive model related to the NCS color-order system. *Color Research and Application* 26: 4–28.

Hardin, C. L. 1988. *Color for Philosophers: Unweaving the Rainbow.* Indianapolis, IN: Hackett.

Hardin, C. L. 2000. Explaining color appearance. In *Special Issue on Art and the Brain,* Part II, J. A. Gougen, and E. Myin, eds. *Journal of Consciousness Studies* 7 (8–9): 21–27.

Hardin, C. L. 2004. A green thought in a green shade. *Harvard Review of Philosophy* 12: 29–39.

Hardin, C. L. 2005. Explaining basic color categories. *Cross-Cultural Research: The Journal of Comparative Social Science* 39 (1): 72–87.

Hering, E. 1964. Outlines of a theory of the light sense. Translated by L. M. Hurvich and D. Jameson. Harvard University Press. Originally published in 1920 by Springer, Berlin.

Hurvich, L. M., and D. Jameson. 1957. An opponent-process theory of color vision. *Psychological Review* 64: 384–404.

Jameson, D., and L. M. Hurvich. 1955. Some quantitative aspects of an opponent-colors theory. I. Chromatic responses and spectral saturation. *Journal of the Optical Society of America* 45: 54–52.

Jameson, D., and L. M. Hurvich. 1968. Opponent-response functions related to measured cone photopigments. *Journal of the Optical Society of America* 58: 429–430.

Jameson, K., and R. G. D'Andrade. 1997. It's not really red, green, yellow, blue: An inquiry into cognitive color space. In *Color Categories in Thought and Language,* C. L. Hardin and L. Maffi, eds., 295–319. Cambridge: Cambridge University Press.

Jameson, K. A. 2005. Culture and cognition: What is universal about the representation of color experience? *Journal of Cognition and Culture* 5 (3–4): 293–347.

Jameson, K. A., and N. Alvarado. 2003a. The relational correspondence between category exemplars and naming. *Philosophical Psychology* 16(1): 26–49.

Jameson, K. A., and N. Alvarado. 2003b. Differences in color naming and color salience in Vietnamese and English. *COLOR Research and Application* 28(2): 113–138.

Jameson, K. A., D. Bimler, D. Dedrick, and D. Roberson. 2007. Considering the prevalence of the "stimulus error" in color naming research. *Journal of Cognition and Culture* 7: 119–142.

Jameson, K. A., and N. L. Komarova. 2009a. Evolutionary models of color categorization. I. Population categorization systems based on normal and dichromat observers. *Journal of the Optical Society of America A* 26 (6): 1414–1423.

Jameson, K. A., and N. L. Komarova. 2009b. Evolutionary models of color categorization. II. Realistic observer models and population heterogeneity. *Journal of the Optical Society of America A* 26 (6): 1424–1436.

Jordan, G., and J. D. Mollon. 1995. Rayleigh matches and unique green. *Vision Research* 35: 613–620.

Kay, P. 2005. Color categories are not arbitrary. *Cross Cultural Research* 39: 39–55.

Kay, P., and W. Kempton. 1984. What is the Sapir-Whorf hypothesis? *American Anthropologist* 86: 65–79.

Kay, P., and T. Regier. 2003. Resolving the question of color naming universals. *Proceedings of the National Academy of Sciences* 100: 9085–9089.

Komarova, N. L., and K. A. Jameson. 2008. Population heterogeneity and color stimulus heterogeneity in agent-based color categorization. *Journal of Theoretical Biology* 253: 680–700.

Komarova, N. L., and K. A. Jameson, and L. Narens. 2007. Evolutionary models of color categorization based on discrimination. *Journal of Mathematical Psychology* 51: 359–382.

Krauskopf, J., D. R. Williams, and D. W. Heeley. 1982. Cardinal directions of color space. *Vision Research* 22: 1123–1131.

Krauskopf, J., D. R. Williams, M. B. Mandler, and A. M. Brown. 1986. Higher order color mechanisms. *Vision Research* 26: 23–32.

Kuehni, R. G. 2001. Determination of unique hues using Munsell color chips. *COLOR Research and Application* 26: 61–66.

Kuehni, R. G. 2003. *Color Space and Its Divisions*. Hoboken, NJ: Wiley.

Kuehni, R. G. 2004. Variability in unique hue selection: A surprising phenomenon. *COLOR Research and Application* 29: 158–162.

Kuehni, R. G. 2005a. Unique hue stimulus choice: A constraint on hue category formation. *Journal of Cognition and Culture* 5 (3–4): 387–408.

Kuehni, R. G. 2005b. Focal color variability and unique hue stimulus variability. *Journal of Cognition and Culture* 5 (3–4): 409–426.

Krauskopf, J., D. R. Williams, and D. W. Heeley. 1982. Cardinal directions of color space. *Vision Research* 22: 1123–1131.

Krauskopf, J., D. R. Williams, M. B. Mandler, and A. M. Brown. 1986. Higher order color mechanisms. *Vision Research* 26: 23–32.

Lindsey, D. T., and A. M. Brown. 2006. Universality of color names. *Proceedings of the National Academy of Sciences* 102: 16608–16613.

MacLaury, R. E. 2005. So-called brightness in color ethnography: Potentials for LCD technology in fieldwork and categorization research. *Cross-Cultural Research* 39 (2): 205–227.

Malkoc, G., P. Kay, and M. A. Webster. 2005. Variations in normal color vision. IV. Binary hues and hue scaling. *Journal of the Optical Society of America* 22: 2154–2168.

Munsell. 1966. *Munsell Book of Color (Matte-Finished)*. Baltimore, MD: Munsell Color Company.

Mollon, J. D., and G. Jordan. 1997. On the nature of unique hues. In *John Dalton's Color Vision Legacy*, C. Dickinson, I. Murray, and D. Carden, eds., 381–392. London: Taylor and Francis.

Nayatani, Y. 2004. Proposal of an opponent-colors system based on color appearance and color-vision studies. *COLOR Research and Application* 29: 135–150.

Newhall, S., D. Nickerson, and D. Judd. 1943. Final report of the OSA subcommittee on spacing of the Munsell colors. *Journal of the Optical Society of America* 33: 385–418.

Osorio, D., and M. Vorobyev. 1996. Colour-vision as an adaptation to frugivory in primates. *Proceedings of the Royal Society of London B*, 263: 593–599.

Otake, S., and C. M. Cicerone. 2000. L and M cone relative numerosity and red-green opponency from fovea to midperiphery in the human retina. *Journal of the Optical Society of America A* 17: 615–627.

Parry, N. R. A., D. J. McKeefry, and I. J. Murray. 2006. Variant and invariant color perception in the near peripheral retina. *Journal of the Optical Society of America A* 23: 1586–1597.

Philipona, D. L., and J. K. O'Regan. 2006. Color naming, unique hues, and hue cancellation predicted from singularities in reflection properties. *Visual Neuroscience* 23: 331–339.

Regan B. C., C. Julliot, B. Simmen, F. Vienot, P. Charles-Dominique, J. D. Mollon. 1998. Frugivory and colour vision in *Alouatta seniculus*, a trichromatic platyrrhine monkey. *Vision Research* 38: 3321–3327.

Regier, T., P. Kay, and R. S. Cook. 2005. Focal colors are universal after all. *Proceedings of the National Academy of Sciences* 102: 8386–8391.

Regier, T., P. Kay, and N. Khetarpal. 2007. Color naming reflects optimal partitions of color space. *Proceedings of the National Academy of Sciences* 104: 1436–1441.

Roberson, D., and J. R. Hanley. 2007. Color vision: Color categories vary with language after all. *Current Biology* 17: R605–R607.

Roberson, D., I. R. L. Davies, and J. Davidoff. 2000. Color categories are not universal: Replications and new evidence from a stone-age culture. *Journal of Experimental Psychology: General* 129: 369–398.

Sivik, L. 1997. Color systems for cognitive research. In *Color Categories in Thought and Language*, C. L. Hardin and L. Maffi, eds., 163–193. Cambridge: Cambridge University Press.

Webster, M. A., and J. D. Mollon. 1994. The influence of contrast adaptation on color appearance. *Vision Research* 34: 1993–2020.

Webster, M. A., E. Miyahara, G. Malkoc, and V. E. Raker. 2000. Variations in normal color vision. II. Unique hues. *Journal of the Optical Society of America A* 17: 1545–1555.

Webster, M. A., S. M. Webster, S. Bharadwaj, R. Verma, J. Jaikurnar, G. Madan, and E. Vaithilingham. 2002. Variations in normal color vision. III. Unique hues in Indian and United States observers. *Journal of the Optical Society of America A* 19: 1951–1962.

Plate 1

View of a Uniform Color Scales model showing different kinds of cleavage planes, which form scales of approximately equally distant colors. Horizontal planes represent colors of equal lightness, adjusted for the Helmholtz-Kohlrausch effect (Image courtesy D. L. MacAdam).

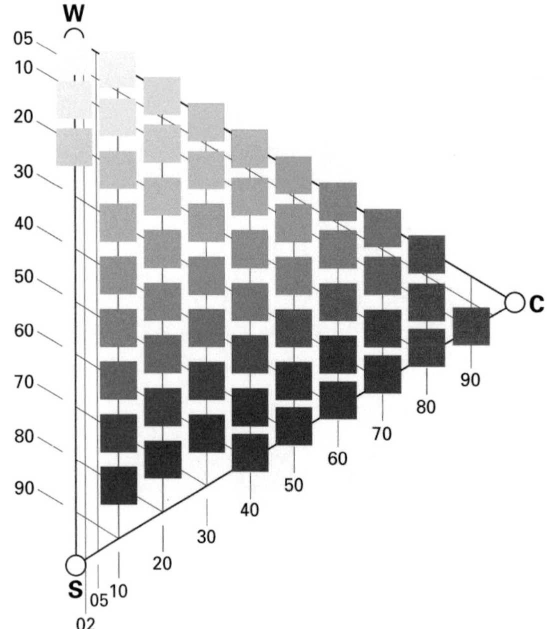

a)

Reflectance efficiency
100%
50%
0%
.40μm .45μm .50μm .55μm .60μm .65μm .70μm
Wavelength
Single-wavelength color spectrum

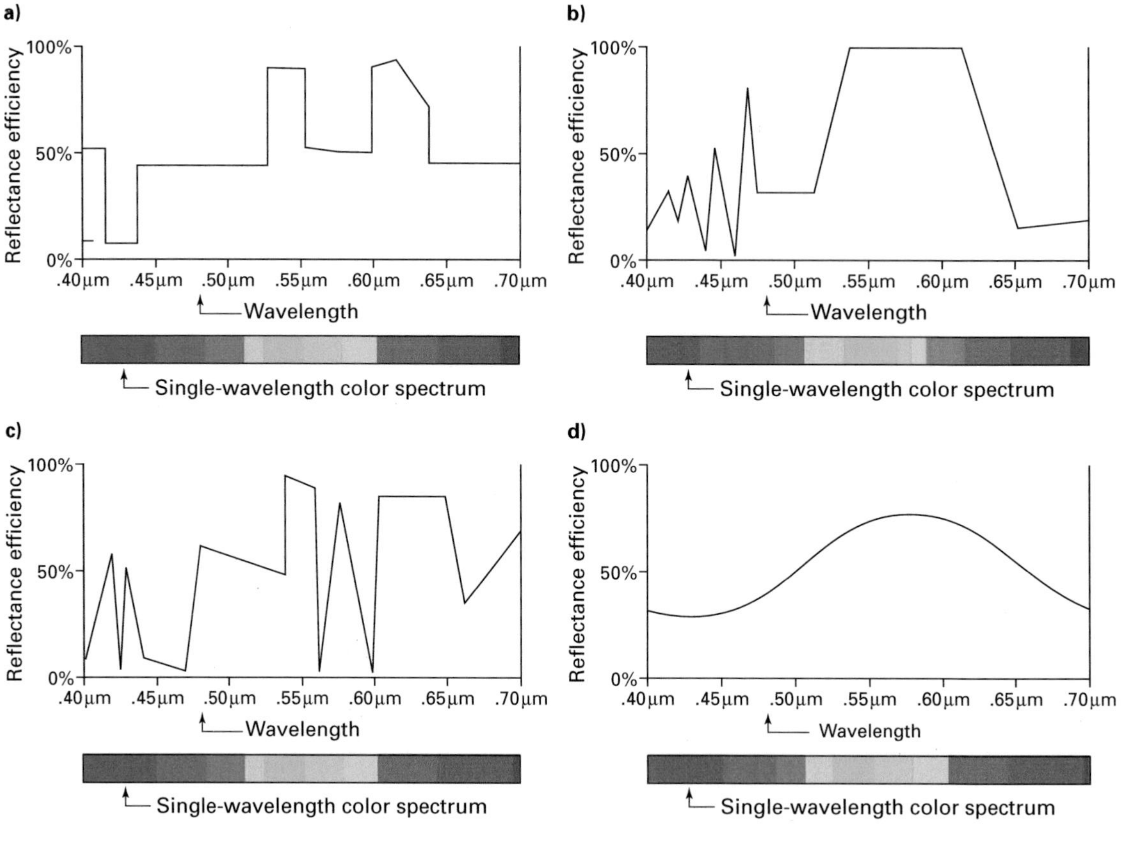

b)

Reflectance efficiency
100%
50%
0%
.40μm .45μm .50μm .55μm .60μm .65μm .70μm
Wavelength
Single-wavelength color spectrum

c)

Reflectance efficiency
100%
50%
0%
.40μm .45μm .50μm .55μm .60μm .65μm .70μm
Wavelength
Single-wavelength color spectrum

d)

Reflectance efficiency
100%
50%
0%
.40μm .45μm .50μm .55μm .60μm .65μm .70μm
Wavelength
Single-wavelength color spectrum

◀ **Plate 2**

Image of the constant hue plane Y90R of the Swedish Natural Color System in the form of an equilateral triangle with white (W), black (S), and the full color (C) in the corners. Lines of constant blackness *s* and of constant chromaticness *c* are indicated (image courtesy NCS Color Center/USA).

◀ **Plate 3**

Reflectance profiles for four distinct material objects.

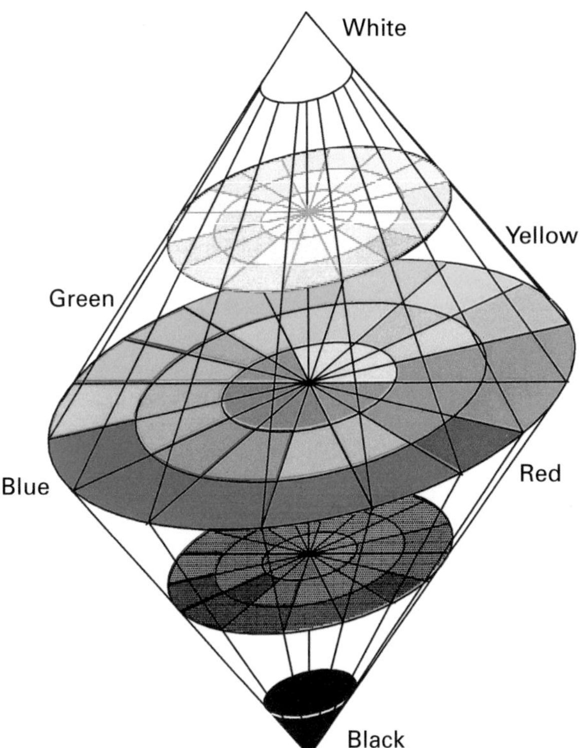

Plate 4

Three-dimensional spindle representing the human phenomenological color space.

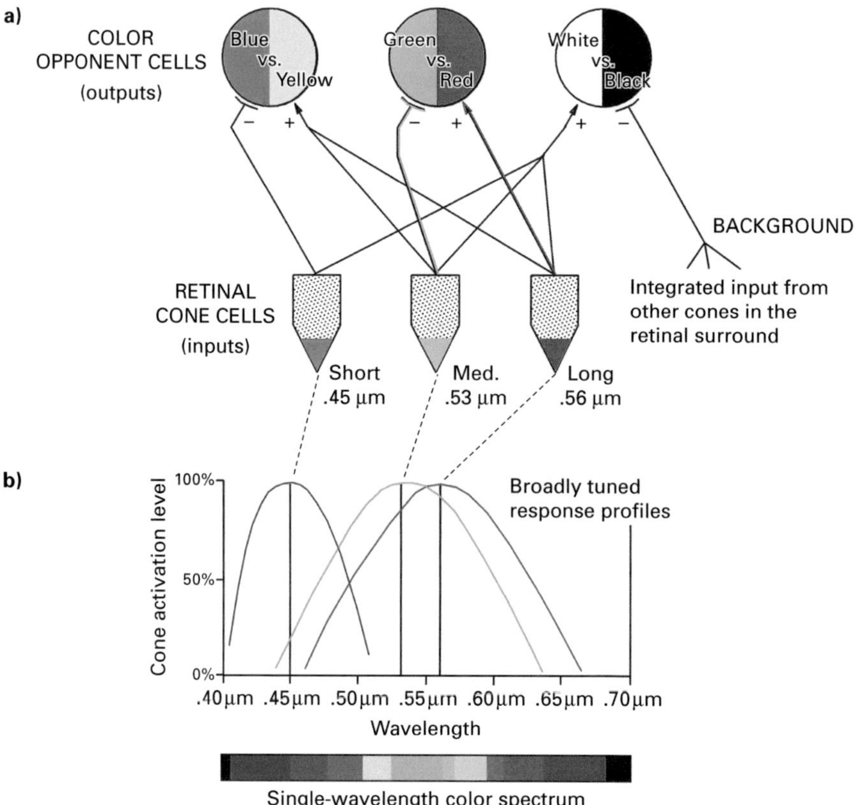

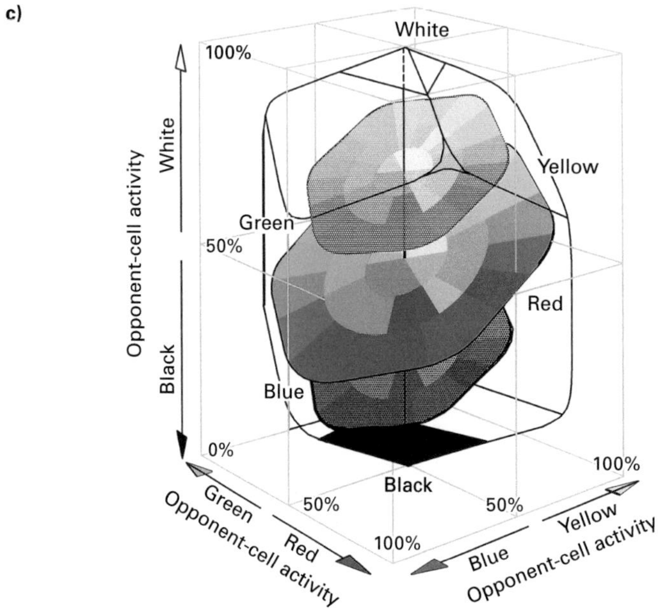

The Hurvich activation-triplet spindle

◀ **Plate 5**

Empirical details of the Hurvich-Jameson model of human color coding: (a) connectivity of the neural network; (b) wavelength-sensitivity of input cones; and (c) color-coding space of possible activation patterns.

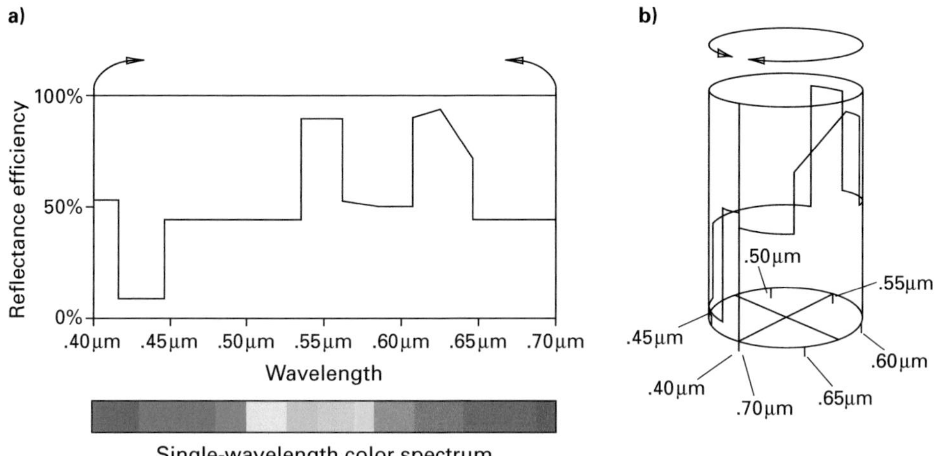

Plate 6

Reflectance profile within the visible spectrum: (a) planar and (b) cylindrical representations.

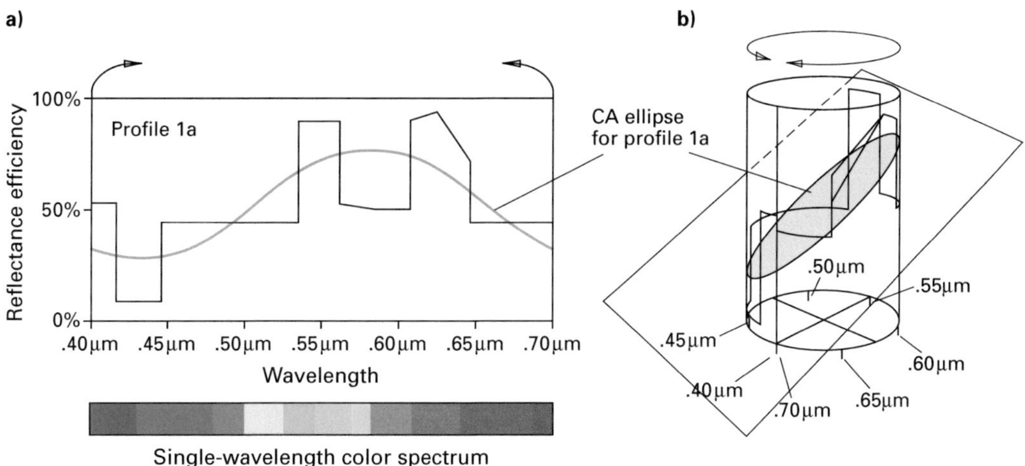

Plate 7

The visible spectrum reflectance profile (a), and the elliptical locus of a planar cut (b) through that profile's cylindrical representation.

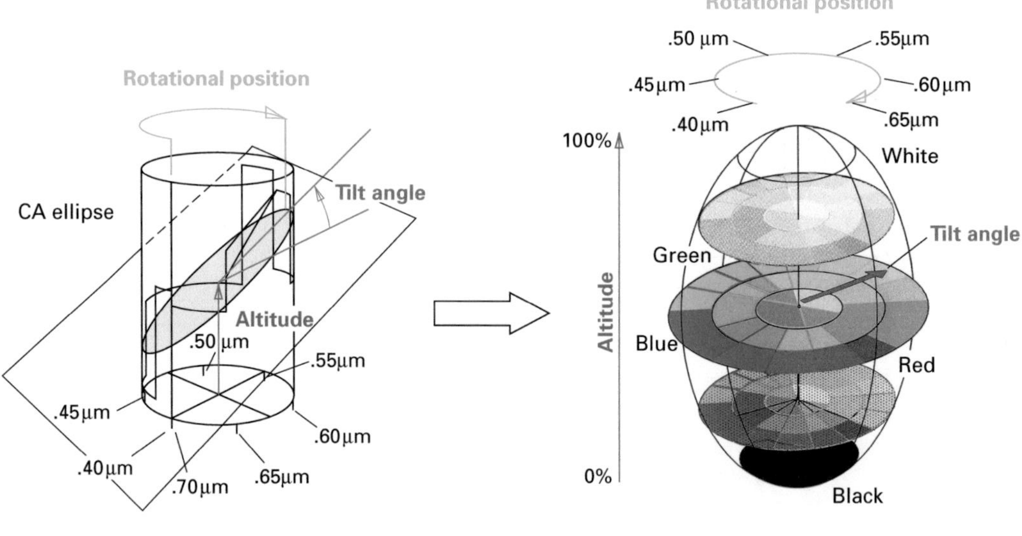

Plate 8
Three dimensions of variation in a space of possible CA ellipses.

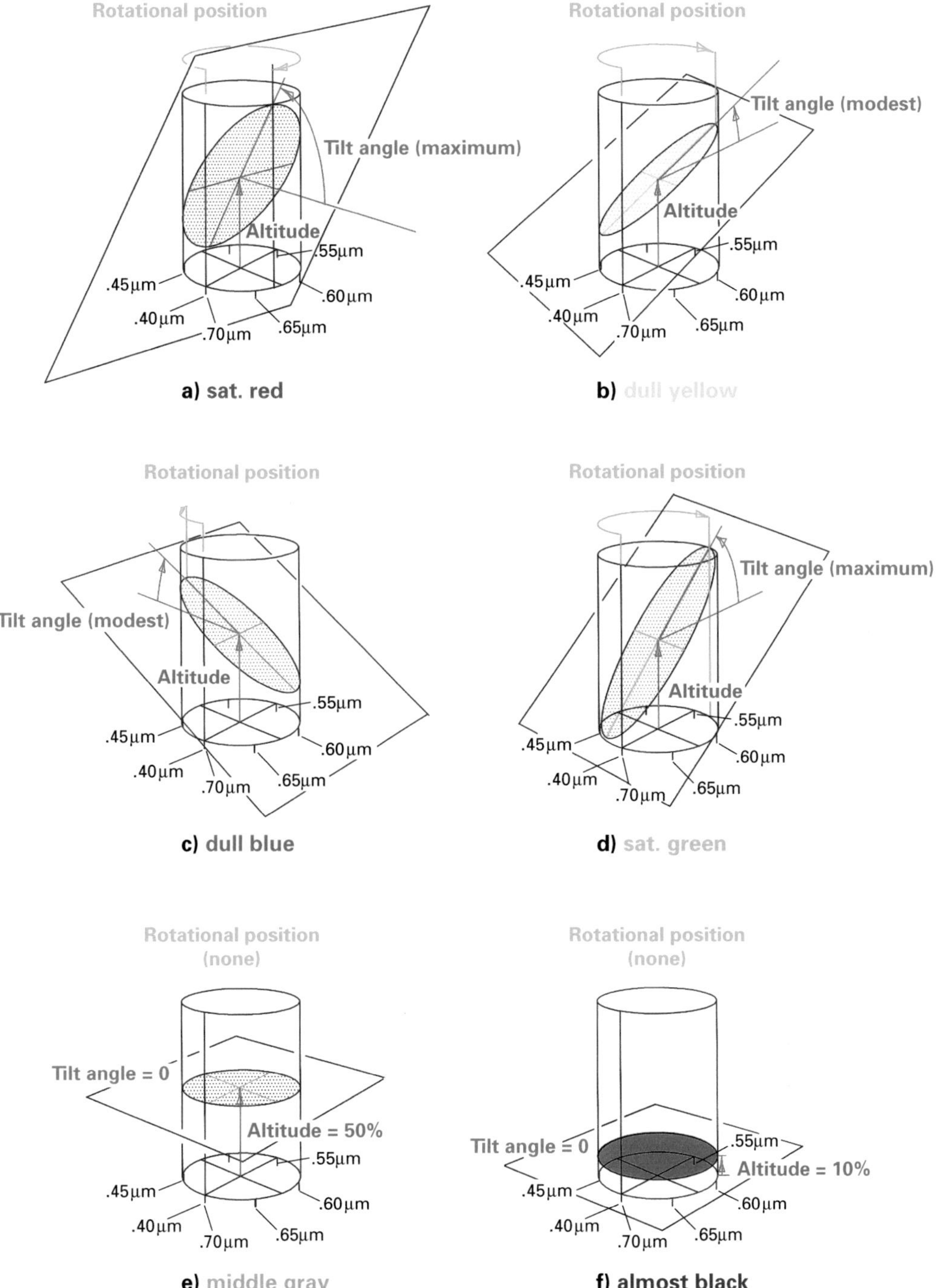

Plate 9

CA ellipse, tilt, and rotation position for sample color reflectance profiles.

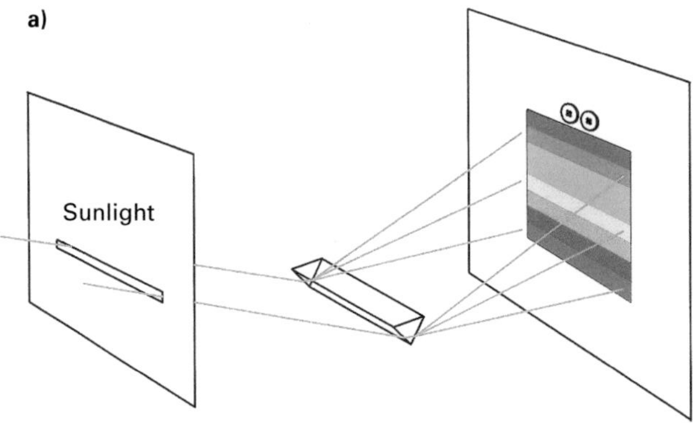

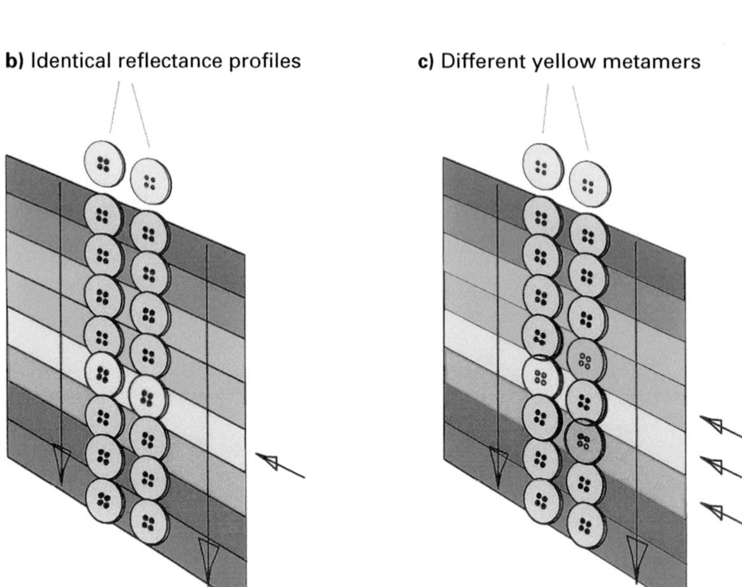

Plate 10

Using (a) a rainbow projected on a wall, we can compare two buttons of apparently the same color to determine if they are (b) identical or (c) different metamers.

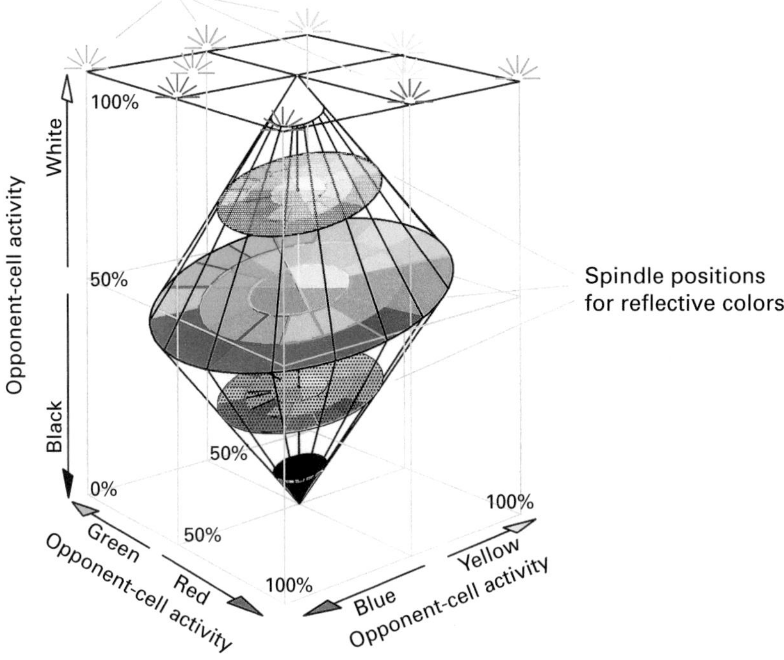

Plate 11

The different representation points of self-luminous colors and reflective colors within the human phenomenological color space.

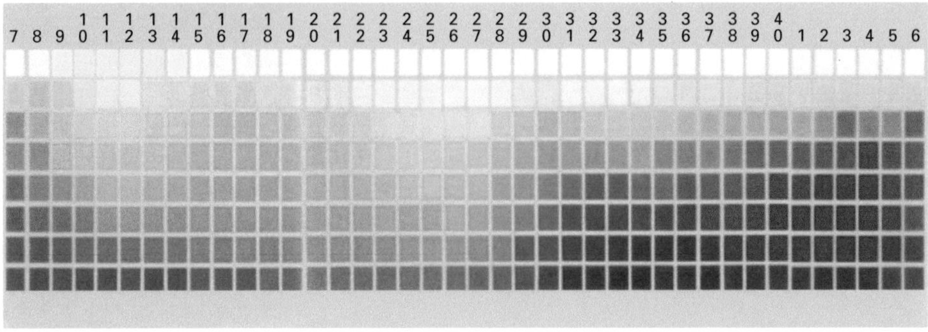

Plate 12

The World Color Survey stimulus reordered to produce a continuous reddish region, with columns 1–6 appearing after column 40.

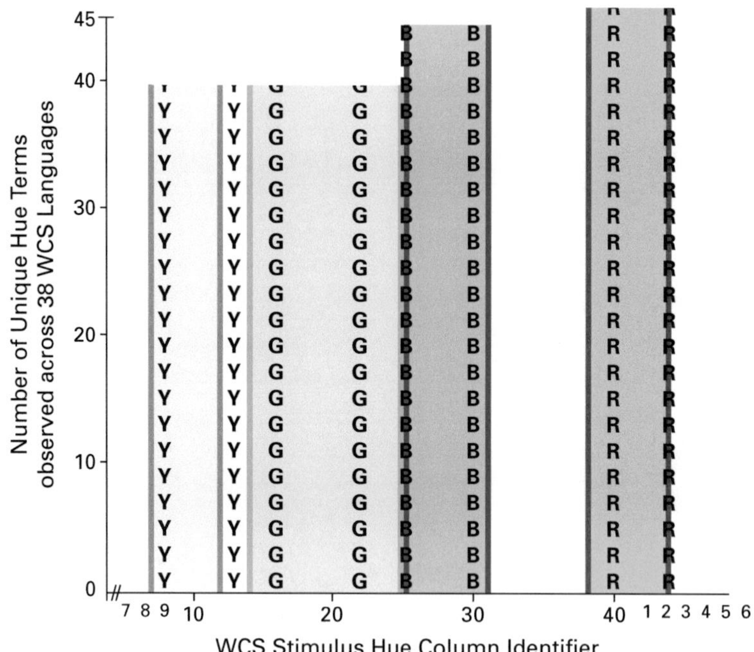

Plate 13

Ranges for unique hue settings and corresponding focal color term ranges in the WCS stimulus array. Two types of data are shown relative to the WCS hue scale on the x-axis: (1) Median focal color term ranges for thirty-eight languages (i.e., columns delimited by congruent alpha-character lines), and (2) unique hue ranges (i.e., shaded columns delimited by solid lines). This figure is a reanalysis of the data presented by Kuehni (2005b, 417, figure 3).

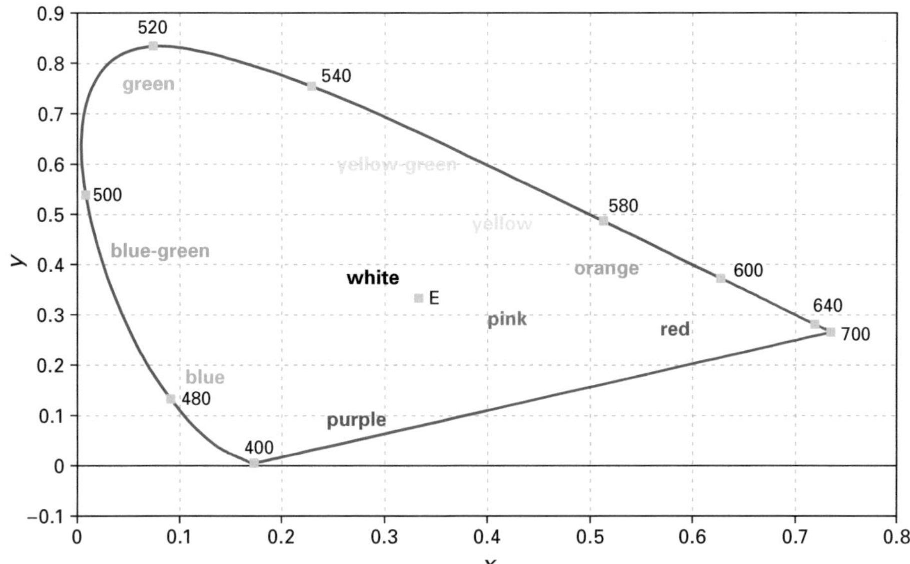

Plate 14
1931 CIE chromaticity diagram.

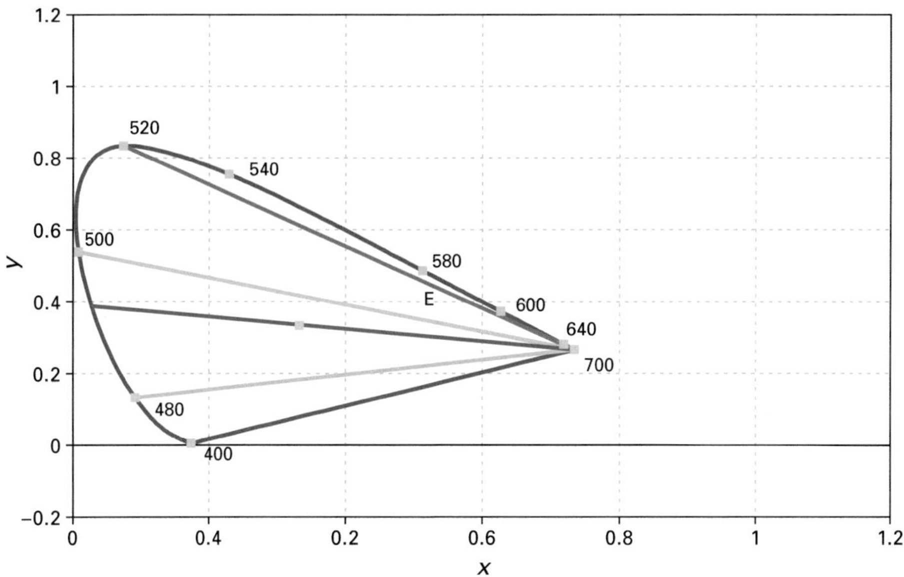

Plate 15
Four protanope confusion lines.

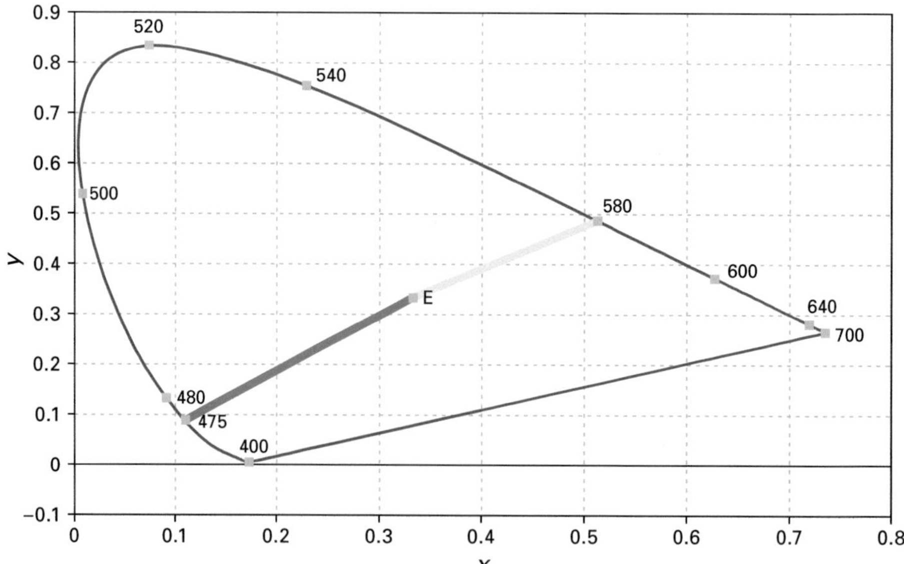

Plate 16

Gamut for protanopes and deuteranopes (on standard reduction view).

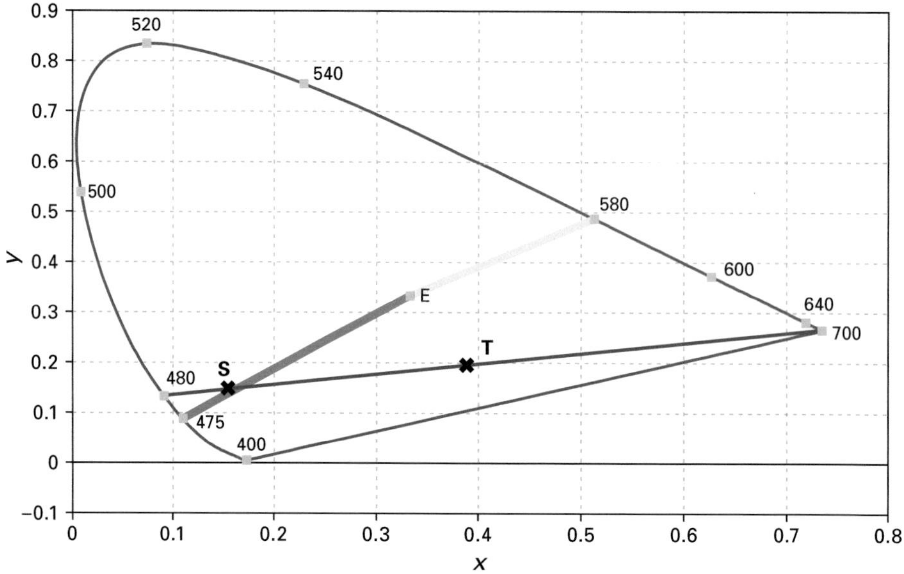

Plate 17

S looks to a normal as *T* looks to a protanope.

8 Color, Qualia, and Attention: A Nonstandard Interpretation

Austen Clark

A state of *phenomenal consciousness* is, minimally, a state of consciousness that has some kind of phenomenal character. Both clauses are open to multiple interpretations, the discussion of which can grow quite heated. For the purposes of this chapter, I propose to avoid as many of those debates as possible by confining the discussion to a simple kind of mental state that all sides agree belongs to the genus *phenomenal consciousness*, no matter what other species the latter might include. This simple kind might be called *perceptual awareness* or *perceptual experience*: episodes in which a subject both perceives something and is aware of what is perceived or of some aspect of what is perceived. Its perceptual origin gives the episode its distinctive phenomenal character, and often that phenomenal character is the feature of which one is aware. Examples include seeing and thereby becoming aware of the red glow of the sky at sunset, or feeling the cool breeze that springs up as the sun goes down. One might also call these *sensory experiences*: one experiences—becomes conscious of—some sensible feature of something that one senses.

In these states one both senses something and is aware of some aspect of what one senses. The latter condition is what qualifies the members of this species as *states of consciousness*. In virtue of having them, one is aware of something: here, specifically, one is aware of some aspect of what one senses. One may, or may not, also be conscious of being in that state. It is possible that one is so absorbed by the sunset—the red glow of the sky, the cool breeze—that one is not also simultaneously aware of *seeing* the red glow, or of *feeling* the cool breeze. We talk of "losing ourselves" in various experiences, aesthetic and otherwise, and in such cases we seem to be aware simply of what we see, or what we feel, not also of the seeing of it, or of our internal state of sensation. So a state of perceptual awareness (as used here) is not necessarily a state *of which* one is conscious. Instead it is a state in which one is conscious of what one perceives.

Now phenomenal consciousness is a puzzle, and intent study of neuropsychology makes it, if anything, more puzzling rather than less. Among the most startling of incongruities are the phenomena found in the *disconnection* syndromes of blindsight and hemineglect. Patients with various kinds of brain damage can pass many of the normal tests for seeing things and having states with phenomenal character, even though they sincerely deny seeing those things or being aware of any features of those things. The blindsight subject DB can pass many of the normal tests for seeing shape, orientation, and location, and for discriminating different shapes, orientations, and locations; though if the stimuli in question are presented within his scotoma, DB sincerely denies seeing the stimulus and denies being aware of any feature of anything located within the scotoma. In pure "unawareness" mode, DB is not aware of any sensible feature of a stimulus presented within the blind field, yet he successfully "guesses" its location, shape, and orientation.

These results seem to drive a wedge between the "phenomenal" and the "consciousness" parts of phenomenal consciousness. They challenge the widely held assumption that states with phenomenal character are necessarily states of consciousness. More broadly, they pose a puzzle for anyone interested in the relations between perceptual appearance and the awareness thereof. In the first of the two conferences that prompted this volume, Larry Hardin revealed his longstanding interest in this problem. He posed (as an agenda item for conference two) the question, "What does the awareness of color add if one can already make chromatic discriminations?"[1] It is a wonderful question, and I propose to address it by examining some of the disconnection phenomena. While this chapter will certainly not answer the question, I hope the energetic beating of bushes will give us some sense of the surrounding undergrowth and perhaps flush out some of its interesting and reclusive inhabitants.

8.1 Blindsight

Blindsight is well known, if not well understood, so I can provide a brief summary and point the reader to fuller descriptions found elsewhere (see Weiskrantz 1986, 1997). Certain kinds of brain damage produce a scotoma, or blind spot, within some portion of visual perimetry. That portion is delimited retinotopically, by azimuth and elevation relative to the point on which the eyes are focused. If the eyes are focused on a fixed point, and stimuli are presented within the scotoma, the patient will, for many classes of stimuli, deny seeing any stimulus, and indeed deny being aware of anything at all in that region. Nevertheless, if the patient is prompted to hazard a guess about various attributes of the stimulus, or somehow to indicate a value, sometimes the resulting indicators correlate very highly with the actual values of the stimulus presented. Attributes can include location (indicated by pointing), shape (X vs. O), motion (vertical vs. horizontal), and, most surprising of all, color.

The latter deserves some elaboration (see Stoerig and Cowey 1992). Color discrimination is tested through a long series of forced-choice discrimination tasks. Any given series of trials involves just two spectral colors, matched for luminance, that are first named for the subject (*orange* and *yellow*, for example) and presented for examination. Then in a given trial, the subject fixates on a central point, and one of the two colors, picked randomly, is presented in that subject's scotoma. Since the subject cannot tell whether the stimulus is present or absent, a click indicates when it is time to respond; the task is to "guess" the name of the color that was just presented. But with a long enough series of such guesses, one can assess the ability of the subject to discriminate different wavelengths. Other tests allow comparison of the spectral sensitivity curves (specifically, threshold sensitivity at different wavelengths) of the blindsight subject with that of normals (see Stoerig and Cowey 1991). The absolutely stunning result is that some blindsight subjects can indeed discriminate some wavelength differences independently of luminance, at statistically significant levels, even for relatively small wavelength differences (20 nm, the difference between the yellow and the orange stimuli). As Weiskrantz puts it,

> The latter capacity—discrimination of colors—presses credulity to the limit, because in those tests—which by their nature were very time-consuming and lasted for several days—the subjects uniformly and consistently denied seeing color at all, and yet performed reliably above chance, even between wavelengths falling relatively close together. Moreover, the fine-grained features of the spectral sensitivity curves of these subjects (carried out, again, by forced-choice guessing) suggested that wavelength opponency, that is, color contrast, was intact (Stoerig and Cowey 1989, 1991, 1992). The subjects seemed to be able to respond to the stimuli that would normally generate the philosophers' favorite species of "qualia," namely colors, but in the absence of the very qualia themselves! (Weiskrantz 1997, 23).

In this last sentence, Weiskrantz puts his finger on the problem. Behaviorally, these subjects pass some of the tests that would normally be taken to indicate that they see the differences between different colors. They pass some of the tests that would indicate that they are sensitive to differences among chromatic appearances. Yet they resolutely and sincerely deny being aware of any aspect of the stimulus presented within the scotoma. So do they have chromatic qualia, or not?

8.2 The Standard Interpretation

If you think the answer to this question is "obviously not!" then there is a nonnegligible probability that you share an interpretation of the experimental and theoretical literature that I will call the *standard* interpretation. I admit the label is tendentious; it is "standard" only in the sense that it seems to be a common view in both philosophy and psychology. On this view, qualia and phenomenal properties require consciousness. More precisely, if at time *t* a subject is not aware of any color at all, then it follows necessarily that at time *t* that subject does not have any chromatic qualia. For any property to be a quale *Q*, some subject must be aware of, or conscious of, *Q*. So showing that DB is not aware of any colors of stimuli in his scotoma settles definitively the question of whether the perceptual states responsible for his discriminations of stimuli therein sport any qualia. They do not and cannot. Similarly, characterizing a property as a *phenomenal* property, or as a property of appearance, implies (according to this view) that some subject is aware of that property. Barring awareness is a way of barring the applicability of all the phenomenal locutions—all the being-appeared-to's, the looking-like *F*'s and seeming-to-be *G*'s.

It is easy to produce examples of philosophers promulgating this view, though no such list can suffice to show it is the "standard" view, or even a widely held view. Nevertheless, a few such examples will have to suffice. One classic source is Thomas Nagel. What he calls *phenomenological qualities* are subjective: "A feature of experience is subjective if it can in principle be fully understood only from one type of point of view: that of a being like the one having the experience, or at least like it in the relevant modality. The phenomenological qualities of our own experience are subjective in this way" (Nagel 1979b, 188).

This subjective, or *perspectival*, character is, according to Nagel, what we are talking about when we say the state is a *conscious* state (Nagel 1979a, 166). John Searle (1997) is a recent advocate of both these claims. Consciousness is a subjective phenomena; it has a "first person ontology" (Searle 1997, 98, 114, 120). Qualia, or characteristics of appearance, cannot be divorced from consciousness: "I myself am hesitant to use the word 'qualia' and its singular 'quale' because they give the impression that there are two separate phenomena, consciousness and qualia. But of course, all conscious phenomena are qualitative, subjective experiences, and hence are qualia. There are not two types of phenomena, consciousness and qualia. There is just consciousness, which is a series of qualitative states" (Searle 1997, 8–9).

Finally, David Chalmers provides some classic statements of the view that "the problem of explaining these phenomenal properties is just the problem of explaining consciousness" (Chalmers 1996, 4). He says that the words "qualia" or "phenomenal" pick out approximately the same class of phenomena as "consciousness," and that "'To be conscious' in this sense is roughly synonymous with 'to have qualia'" (Chalmers 1996, 6). Color sensations invariably present qualia, so color sensations are the "paradigm examples of conscious experience" (Chalmers 1996, 6). There is indeed a robust sense in which to characterize something phenomenally is to characterize how that something is experienced. What I am calling the standard interpretation leaves no space at all between the concepts of being-appeared-to and being aware-of; the former locutions apply only where the latter are ensconced.

Among psychologists, the influence of the standard interpretation is often seen in a reluctance to apply certain locutions to the states involved in early, preattentive, perceptual processing. Although DB can (with good though not perfect reliability) discriminate the orientations and shapes of stimuli in his scotoma, it is widely thought that all such discriminations can be explained by the brute mechanical registration and processing of information in early visual channels. Provided DB operates in pure "unawareness" mode, there is no call to say that DB "senses" any such features, that he is presented any sort of appearance by stimuli in the scotoma, that those stimuli have any sort of phenomenal property, or that any of the states that subserve his successful discriminations have any sort of qualitative character. All of these locutions, it is thought, imply that DB is aware of the features, appearances, or characters involved. DB manifestly is not. Ergo any account of the success of those discriminations must avoid any mention of sensation, appearance, or phenomenology. The words are banned. We have instead mere information processing, of the sort that occurs in an utterly unconscious automaton—a telephone switch, a thermostat, or a laptop computer. Note that this is the contrapositive of the philosophers' version of the standard view. It reads: if one is not conscious of property *P*, then *P* is not a quale, not a property of appearance, not a phenomenal property. We see this view at work in the Weiskrantz quote above: since DB is not aware of the features he successfully discriminates, those discriminations proceed in the absence of the "very qualia themselves."

Now of course everyone has Humpty Dumpty's right to use a word however they please, but I will argue that insisting on this usage erects a permanent roadblock in the way of our understanding phenomenal consciousness. If the argument works, then it will be worth looking at some alternatives that might allow us to skirt around that roadblock.

8.3 Phenomenal Properties

Terms such as *qualia* and *phenomenal* have become loaded and overloaded with multiple and conflicting theoretical connotations, so it is a good idea to start with the simplest and most traditional sense of the word, and confine ourselves to that. *Phainomenon* is Greek for "appearance," and a *phenomenal property,* on one traditional reading, is a manner of appearance, a characteristic of how things appear. The "things" in question are, for the purposes of this chapter, resolutely confined to the things one perceives; their "appearances" are characteristics of how they (for example) look, or feel, or taste, or (generally) seem. If the mountains look blue in the distance, then "blue" is used in that context to attribute a phenomenal property, characterizing how those mountains appear. We have other "verbs of appearance" that are used, like "looks," to characterize such appearances while remaining neutral on the question of whether those appearances are sustained by reality (see Chisholm 1957). For example, before one adapts to a newly filled eyeglass prescription, one might experience a brief period when rectangles *look* trapezoidal; straight lines *appear* slightly bowed; objects in the periphery *seem* to move when one turns one's head; distant objects *appear* distorted; and the floor *feels* as if it undulates when one walks down the hall. It can make one seasick. Thankfully, adaptation proceeds quickly, and soon one can abandon all these verbs of appearance. Thereafter, the floor feels, and is, motionless and flat; rectangles look, and are, rectangular; and so on.

To use C. D. Broad's (1927) phrase, phenomenal properties characterize the "facts of sensible appearance": how things (in the broadest sense of the word) appear to the senses. One can attribute such a phenomenal property without committing oneself on the question of whether its target is as it appears. Something that (merely) looks blue can look exactly like something that is blue. Judging from how it looks, a witness unaware of any peculiarities of the circumstances of perception would judge the thing to *be* blue. But the appearance is not merely a mistaken judgement; even after one learns of the conditions that make this nonblue thing (in these conditions) look blue, it will for all that continue to *look* blue.

This gives us a clue for unpacking the traditional notion of phenomenal properties (see Clark 2008). There is a similarity in appearance between this thing, seen in these particular circumstances, and some other, avowedly blue thing, seen in other circumstances. At the limit this similarity yields a match, or even phenomenal identity. If two things appear the same (in some respect), then they share a property of appearance. Episodes of perceiving those things present the same phenomenal property. For example, if stimulus x looks exactly the same as stimulus y, then x and y share visual phenomenal properties. If y feels exactly the same as z, then y and z share tactile phenomenal properties.

The entities to which the properties of appearance are ascribed are, paradigmatically, things one perceives (or more broadly, portions of the perceived world), but a person's use of a verb of appearance, instead of the standard copula, flags some oddity in the circumstances of perception, which might be responsible for some aspects of that appearance, and which are such as to call for caution in what would otherwise be a straightforward perceptual judgement. Two portions of the world, presented under different circumstances, might present the same appearance, but in one of them that appearance is more a function of the circumstances of perception than of the thing perceived. It looks funny because of this light, or it appears oddly shaped because of my new eyeglass prescription. Change the light or the prescription, and the appearance changes as well. A speaker's awareness of such circumstances is marked by the choice of verb.

The limit of sensed similarity is absolute indiscernibility, but many variants of similarity relations are pressed into service: matching, discriminability, discernibility, relative similarity, recollection of similarity, and all their negations and contraries. Traditional phenomenal properties are individuated by appeal to the discriminations that a subject can make. Stimuli that are absolutely indiscernible present the same appearance; stimuli that can be discriminated from one another present at least slightly different ones. In this way, a traditional phenomenal property—an attribute of sensible appearance—is tied to capacities to perceive similarities and differences among stimuli.

Capacities to discriminate can provide tests sufficient to attribute or deny a (traditional) phenomenal property. If subject S can routinely and reliably discriminate between occasions in classes *P* and *Q*, then stimuli in class *P* do not present the same appearance as those in *Q*—and S can perceive the difference. If, for example, she can reliably discriminate line orientations that differ by just one-tenth of a degree, then to her the orientation of one line in such a pair must not appear the same as that of the other. If she can reliably discriminate spectral light of 580 nanometers (nm) from light of 584 nm, then however they look, we know those two wavelength packets do not look the same to her. Whether some human females have a four-dimensional color space is found by testing whether they can reliably discriminate differences (and relative similarities among) a class of stimuli that to normal trichromats are all mutually indiscriminable.

One way to summarize what seems paradoxical about blindsight is that subjects can pass many of the tests that would typically be taken to show that they are perceiving something, and can discriminate it from other stimuli, even though they sincerely deny being aware of any aspect of the stimulus in question. Consider the evidence mentioned in section 8.1. DB can with surprising reliability discriminate location, X's from O's, and the direction of motion, even though the stimuli are presented in such a manner that he is not aware of any feature of them. The subjects in the Stoerig and Cowey experiments could with surprising reliability "guess" whether the light was yellow or orange, even though they were not aware when the stimulus was present, and had to be prompted when to "guess." In fact, I will argue, they pass some of the tests that indicate they register the chromatic appearances of stimuli within the scotoma. An alternative phrase: they pass some tests that indicate they have states with qualitative character. They pass such tests even though they are not aware of the objects (or stimuli) presenting those appearances.

8.4 Mysteries inside Enigmas

Clearly DB is discriminating some differences between stimuli presented in his scotoma, even though (if we take him at his word) he is unaware of those stimuli and must be prompted to guess. The simplest hypothesis, which I will favor, is that the difference he discriminates between stimuli *is* a phenomenal difference. It is simply one of which the subject is unaware. Otherwise, it has all the characteristics of your run-of-the-mill difference in phenomenal properties. To stand the Weiskrantz quote on its head: DB has the very qualia themselves but is simply not aware of them. Why not?

The question is apt to be met with incredulity, sputters of outrage, and then finally an argument: "But that's not what the word means! Qualia must be conscious. The proposal that there exist unconscious phenomenal differences, or that DB is being appeared-to in a matter of which he is unaware, is speculative, useless metaphysics. It distorts language and can only sow confusion."

I will repeat that one can, if one likes, reserve the words "qualia," "phenomenal character," and "appearance" to just the ones of which someone is conscious. But the critical question is whether or not there can exist a family of other properties, which differ from these "qualia," "phenomenal characters," and "appearances" in *only one* respect: that no one is aware of them. Could there be differences that are exactly like those differences you call "phenomenal," excepting only the fact that no one is conscious of them?

The question needs some tightening. Of course there will be many consequences—logical consequences, psychological consequences, and other nomological consequences—of the fact that no one is conscious of these differences. For example, the particular instances will not be reported by the subject in question, or by any subject; they will not be described in English, or in any natural language; they will not appear in that subject's autobiography, if such an autobiography is ever produced. They will inspire neither sonnets nor self-recriminations. Some philosophers will urge that it follows as well that they are not in any sense phenomenal. Lump all these consequences together and call them *consequences of nonconsciousness*. Could there be a family of properties *P* such that *all* the differences between *P* and the normal, conscious, phenomenal properties *Q* are found among the consequences of nonconsciousness? Could there exist such creatures of darkness? If you think the answer to that question might be yes—that a "yes" is conceivable—then you are open to the possibility of a "nonstandard" interpretation of the relations of chromatic qualia to awareness. If you think the answer must be no, then I fear you have erected an a priori roadblock against the possibility of our ever understanding phenomenal consciousness.

Consider the implications of a "no." To say no is to say that it is not the case that there exists a family of properties whose differences from the normal qualia (or normal phenomenal properties) are all found among the consequences of nonconsciousness. So for any quale *Q*, any such candidate property *P* must differ from *Q* in some other respect besides those consequent upon the fact that someone is conscious of *Q* and no one is conscious of *P*. In other words, quite apart from the fact that *Q* characterizes something of which someone is conscious, and *P* does not, and apart from all the consequences of that fact, there is also some *other* difference, or differences, between them.

This implies in turn that that of which one is aware when one is aware of a quale cannot be identical to any property that can exist even if no one is aware of it. All the candidates must differ, not only in not being conscious, but also in some other respect besides those consequent upon nonconsciousness. One might think that one could give a characterization of that of which one is conscious that is independent of the fact that one is conscious of it. In the typical two-term relation, one can name the terms independently of their relationship to one another. But the claim that qualia *must* be conscious implies that here the rule fails: qualia cannot be identified with any properties for which such an independent characterization is possible. Not only is consciousness of them a mystery, but that of which one is conscious when one is conscious of them is a mystery as well. A solution to either requires the other already be solved. This is not a mystery inside an enigma, but something worse: two mysteries, each of which is *wholly* inside the other.

In contrast, a "yes" answer allows that there might be candidate properties that differ from qualia in (essentially) just one respect: that no one is conscious of them. The very same difference of which a subject is aware when a subject is aware of the difference between the appearance of yellow and the appearance of orange can exist even if no subject is aware of it. If so, it is possible—conceivable—to analyze a phenomenal property of which a subject is conscious independently of coming to understand what it is to be conscious of it. This mimics a strategy applied to understanding many perceptual capacities, such as what it is to see a color. Eventually, after many centuries, our kind came to understand some of the physical characteristics of the stimulus—specifically, of the electromagnetic radiation impinging on the retina. That understanding at last allowed investigators to explore how a physical nervous system might detect and discriminate variations in those physical characteristics. It unlocked the doors. Without that characterization of the stimulus for color vision, it is hard to see how human color scientists ever would or could have made what progress they have made in understanding human color vision.

A similar divide-and-conquer approach applied to awareness of phenomenal similarities and differences suggests that we first try to understand that of which one is aware, independently of the awareness of it, and then try to understand what makes it true that some subject S is aware of some such thing. Blindsight, hemineglect, and other disconnection syndromes strongly suggest that differences in sensible appearance can be registered in various preconscious, preattentive perceptual processing mechanisms. This registration is not yet "phenomenal consciousness," but perhaps all that is missing is awareness and what follows therefrom. If so, we could characterize that of which one is aware independently of the awareness of it. To such registrations. all we need add are capacities that suffice sometimes to make one conscious of what is registered.

Can this latter step ever be completed? The audience (including me) at last confronts the question (Larry Hardin's question) squarely in the face: what must we add to preattentive perceptual information processing in order to instantiate a subject who is aware of a quale of color?

8.5 Selective Attention

Surprisingly enough, I think we can start to provide parts of an answer to this question. These parts derive from a philosophically neglected psychological capacity: the capacity for selective attention. How does contemporary cognitive psychology treat the phenomena of selective attention? The first surprise (for philosophers) is that the emphasis is very much on the word "selective." It is thought that there exist many independent, peripheral, preattentive perceptual processes that operate in a parallel, bottom-up fashion for many different modalities, and different channels within modalities, simultaneously. It is also thought that there exists relatively more central, lower-bandwidth, general-purpose processing, which can absorb inputs from all the different peripheral processes and integrate them to form one action plan for the organism as a whole. Somewhere between peripheral and central processes, it is thought that there must be some process that selects which of the peripheral processes are going to receive the benefits of further, central processing. Psychologists fiercely debate the details of the characterization of peripheral and central processes. They debate where, exactly, the selection process occurs. They debate why the selection occurs. But it is (almost) universally acknowledged that there is some selective process, somewhere. That selection is the job of selective attention.

One old and somewhat discredited explanation for why selection is required appeals to the sheer differences in bandwidth—in bits per second—of all the perceptual data coming in per second, compared to what a central processor might be able to manage. During our every waking moment, all the sensory modalities are firing away all the time, transmitting impulses up their many tracts into the central nervous system. These transmissions are initiated peripherally, as independent and bottom-up processes, and while a certain amount of central (and peripheral) gating is possible, there is no central way to turn off receptors. The resulting volume, even in a single modality, beggars the imagination. For example, the optic nerve in humans contains about a million fibers for each eye, and each nerve can (theoretically) fire up to a thousand times per second. If we simply want to record for each millisecond whether the nerve has fired or not, we require a recording capability for each optic nerve of a gigabit per second.[2] And what does one do with all the data? The typical scholarly monograph requires roughly a million typed characters (a megabyte), which with endnotes and references takes roughly ten megabits to store; and so the optic nerve from one eye could theoretically transmit a small library, of one hundred such volumes, each second. Over the years book storage becomes problematic even at very low annual rates of acquisition, and here we are adding a hundred per second.

And remember, these are transmissions from just one eye; we have the other eye, and all the other senses, to manage as well. Humans have about a million muscle spindles, each with its own neural control. The auditory nerve may contain fewer than a hundred thousand neurons, but it almost certainly uses the full temporal resolution that is possible, so even a mere hundred thousand neurons might yield a bandwidth of a hundred megabits per second. No one quite knows how to assess the information-bearing capabilities of dynamic touch and the somesthetic modalities, much less taste or smell. But in any case, it is clear that if one considers the transmission lines in a pure information-theoretic way, their sum remains many orders of magnitude beyond the capacities of any computer available today.

The necessity, then, is to select. It is hard enough to keep up with the current rates of academic book publication; it is humanly impossible to keep up with a hundred books per second. The trick lies in selecting what to ignore, to pick the relevant bits out of the stream and focus one's efforts on them. This selection is the key job of what cognitive psychologists think of as selective attention. One set of analogies likens it to a gateway or filter through which only selected streams of peripheral information are allowed to pass.[3] The sources of these streams are conceived as parallel and autonomous sensory processes, which mostly proceed in data-driven, independent, and bottom-up fashion. The destination for this information is "central processing." It is "central" in various senses: it is not scattered about in multiple, independent, parallel-processing modules; it is a common destination for all perceptual representation; it makes its results available to all portions that are "central"; and, relative to the amount of processing power that would be needed to analyze fully all that perceptual information, it has a limited capacity. This limited capacity may give it the appearance of proceeding serially, one step at a time, rather than in multiple, parallel streams. Furthermore, it is "central" in that its resources can be allocated and reallocated under some degree of central control; more or less all of them can be assigned to or focused on one or another task. The function of selective attention is to select which streams will receive this further, central, processing.

This picture of the landscape lying between perception and attention has undergone elaborations and amendments since it was first proposed over fifty years ago (see Broadbent 1958; Neisser 1967; Fodor 1986), but in broad outline it is a well-established part of the conventional wisdom of cognitive psychology. In particular, the need for *some* sort of selection *somewhere* between peripheral sensory processes and central cognitive ones is almost universally acknowledged, and that selection is acknowledged to be the job of selective attention. This job specification can help us to understand some of the powers that folk psychology ascribes to attention. For example, it is a truism that if you focus your attention on something you perceive, you can often perceive more of its details, perceive it more quickly, and perceive it more reliably. A good reason to advise novice drivers to pay attention to the traffic is that doing so helps them to anticipate and respond more quickly, and thereby avoid some accidents. Cognitive science provides an explanation: by "paying attention" to the traffic, the novice driver opens certain switches, gates, and channels within the nervous system, and inhibits others, so that visual information (and not other kinds) gets the benefit of central processing. This further processing allows finer details to be resolved within the flow of visual data and allows for a quicker, more accurate, and centrally coordinated response to exigencies that are detected visually. In folk terms: if you pay attention to what you're looking at, you can see things better.

We can also demystify some of the persistent optical metaphors applied to selective attention. It is often compared to a spotlight: the stimuli to which one attends are "lit up" by a spotlight, which is under some degree of voluntary control. Similarly, consciousness is often described as an inner light, which lights up certain processes, leaving others in the dark (see Rey 1997, 472). The light picks certain items out from the background and enables one to see them better. Now, it is true that more illumination can help one see better, but additional processing power devoted to the task can do the same thing. As anyone who has used a digital camera knows, the more processing power one can throw at a picture, the quicker its details can be resolved. So there is more than one way to "see" better. The spotlight analogy is probably as old as the human capacity to direct artificial illumination upon things, but wider use of computers may someday provide it with a rival. "Focus" itself provides a second optical analogy: only when it is held at the correct distance will the magnifying glass bring the object of interest into focus, but when it does, the view provides more detail and finer structure. Again the analogy points to a way of receiving more information from the object of interest, but better processing of the same information could yield the same effect.

8.6 The Act/Object Divide

Psychologists, then, emphasize the selective functions of selective attention. The reason this is something of a surprise to philosophers is that philosophers tend to pay attention to other aspects of attention: in particular, to its intentional and quasiperceptual attributes. In ordinary language, attention is described as picking out an "object," in something like the way an intentional state represents or is about an intentional object. We speak of focusing attention on an object, or directing attention at an object. The object is picked out by the attentional state in something like the way an object one perceives is picked out by a perceptual state. Such an object need not be a physical object or a "thing" in any ordinary sense of the word "thing." It should be read as an intentional object: as whatever that state represents or is about. So the object to which one pays attention might be a physical thing, but it might also be a feature or property, a set of relations among features, a state of affairs, a region of space, an event, a process, or whatever. This intentional or quasiperceptual aspect of attention is what grabs the spotlight in philosophical discussions.

Some theorists think of attention as functioning something like a demonstrative: it is a way of picking out some of the objects one perceives without requiring a description true of just those objects (see Pylyshyn 2001; Campbell 2002). This "picking out" is done directly, in something like the way the "this" is assigned a referent when a speaker makes a successful demonstrative use of the word. The referent must be among the objects currently perceived, but that condition does not suffice, since many things are perceived to which it does not refer. Typically, an additional demonstrative gesture helps to pick out the referent. Likewise, many objects might be currently perceived, but attention somehow picks out from among them some proper subset, whose members compose the current "focus" of attention. The optical analogies suggest a model: more light is shone on the objects to which one attends, or they are in better focus. In this way, the objects to which one attends would be perceptibly different from ones to which one does not attend, and the difference between attending and not attending would yield a difference in appearance. Some are "lit up" or "in focus."

Now if the psychological consensus described in the previous section is correct, then the real nature of selective attention—its scientifically discovered real essence—is that certain channels are opened or enhanced, and others are closed or inhibited, so as to allow a select subset of sensory representations to enter central processing. To pay attention to an object *x* is to open certain gates and switches in the nervous system to allow representations of *x* to be further processed in "central" portions of one's cognitive architecture.

This explains an interesting feature of the locutions we use when we talk about attention. Those locutions seem to speak of objects, but when one talks of "attending to an object" one is certainly not attributing any kind of property to that object. Instead one is talking about relations among psychological states that are directed at, or that are about, that object. For example, when we say, "Sam is paying attention to the fishing line," we are not saying that some ghostly illumination shines on the fishing line, or that it shines somewhere inside Sam's head; instead, ostensible reference to the fishing line is used to characterize relations among some of Sam's perceptual representations. In particular, representations of that object are the ones selected to receive the benefits of Sam's central processing. So Sam is more likely to react quickly if the line twitches.

Our ordinary locutions for "attention" employ this handy shorthand. We mention the "object" of attention as a way of picking out a set of representations whose intentional object is the one mentioned. To "pay attention to" *x* is to alter the configuration of gates and channels inside one's head so that representations of *x*, and not of other things, receive the benefit of further central processing. That processing in turn allows one to resolve more sensory and structural detail in *x*, perceive it more quickly, remember it better, and so on. But how one actually manages to open the right gates and channels, and close or inhibit others, is something of a mystery. How one *finds* the appropriate ones to control is a mystery as well. But that is what the exercise of selective attention *does.*

8.7 Perceptual Awareness Demystified

The fact that selection by selective attention is competitive imposes some significant structure on what might otherwise be a gaggle, a horde, an unruly mob of ongoing mental processes. One implication, for example, is that some of those processes are *not* selected. Put in folk terms, at any given moment you may be paying attention to something you perceive—the twitches on the fishing line, for example—but there are certainly plenty of other things, features, processes, or events you perceive at the same time to which you are not paying attention: the sound of your breathing, the pull of gravity, the taste in your mouth, and so on. One feels the pull of gravity, and one could focus attention on it, but attention is focused elsewhere. It would be handy to have a label for all these ongoing sensory processes that could have been selected by selective attention, and would have been, but for the selection of something else. I suggest that we *do* have a label for sensory processes that have exactly this status: we say that those are phenomena of which we are *aware.* These are processes that are in the running for selection but happen not to be first at the moment.

The idea is that selective attention picks out a subset of all the streams of perceptual processes currently under way, and the objects of those streams are the ones in the "spotlight" or in the "focus" of attention. But one also perceives objects not currently in the spotlight. One must be able to gather some information about such objects, or attention could never shift away from its current focus in an adaptive fashion. One sense of the ordinary term "aware" applies precisely to all those objects not in the spotlight, but in the penumbra: all the ones perceived, and potentially selectable, but currently not winning the competition for selective attention. You are "aware" of them even though you are not "attending" to them.[4] Often the only thing precluding such a win was the fact that something else won instead.

There are various bits of evidence for this account of the distinction between "attending to" and "aware of." Attention is directed at some "focus," some selected subset of the objects currently perceived; and the directing of it is under some degree of voluntary control. Otherwise a teacher who raps the table and says to the class, "Pay attention!" would be issuing an imperative that is absurd. Indeed the class may in time learn how to better focus its attention, though such outcomes seem rare these days. Whereas "being aware of" applies indifferently to a much larger field; it is not directed at anything in particular. One cannot "focus" one's awareness, unless this just means focusing one's attention. The spotlight metaphor gets no grip on our locutions for awareness: phenomenologically we do not find a foreground of things of which one is aware surrounded by a background of things of which one is unaware. Phenomenologically, such a background is equivalent to no background at all. Nor is awareness under the same degree of voluntary control as selective attention; a teacher who raps the table and says "Be aware!" would be issuing either an absurd imperative or a joke (unless it means simply "Beware!" or "Pay attention!").

The suggestion is that "aware of" applies to all the objects of representations that are in the running for selection by selective attention but that are not the current winners.[5] This can be a large field, because it includes all the also-rans. But one does have to be in the running. Precisely because awareness includes all those that are *not* winning, it has no foreground-background organization; it has no particular focus or direction, so it cannot be directed deliberately at anything. There is other evidence for this analysis. Etymologically, the word "aware" comes from the same root as "wary," "warn," and "beware," denoting the exercise of at least some caution, care, preparedness, wariness, vigilance, or responsiveness. (OED: OE *wær, gewær* corresp. to OS *war*, OHG *giwar* [G *gewahr*], f. Gmc base meaning "observe, take care.") One is aware of all the things that could have attracted one's attention in the prior moment or might do so in the next. A tripwire is in place, which if tripped would attract attention, which in turn could call up concerted action of some sort—a sort to be determined by central cognitive resources. In contrast, a guard caught unawares is exercising no such wariness; the bad guys can sneak in without the guard noticing, without catching the guard's attention, without alerting the guard. I suggest that the level of vigilance described by the former locutions is that events *could* attract selective attention; the only impediment to their success in doing so is that something else is, at the moment, dominating the selection. The tripwire in place (if one is "aware" of one's surroundings) is one that simply draws attention; it does not lead directly to any actions except those of noticing, alerting, and attending to.

Whereas if one proceeds unawares, then even those actions do not occur. In the guard caught unawares, the channels between the receipt of sensory information and the engagement of selective attention are, unfortunately, closed, or at least attenuated.[6] To say "the guard was not aware of the sound" implies that the guard's auditory representation of the sound was one that could not draw attention, and so was not a candidate for entry into central processing. If one is completely unaware of the sound, one is incapable of shifting attention to it. The cost is that no central planning can be aroused.

Now according to the argument in the previous section, our talk about the objects of attention is a way of referring, indirectly, to a status enjoyed by representations *of* that object: that those representations are selected for further, central, processing. If this is so, then our talk about the objects of which we are aware is likewise a kind of shorthand. To say "Sam is aware of the slosh of water in his boots" implies that at any moment Sam *might* focus his attention on the water sloshing in his boots. He could do so. But mention of the putative object is a means of indirect reference; the ultimate translation is of the form "Sam's somesthetic sensations of water-sloshing-in-boots are in the competition, and could be selected for further, central, processing but at the moment are not in first place." Talking about the "awareness" of something is a way of talking about the status of perceptual representations of that thing; in particular, their status in the competition for selective attention.

An analogy from economics might help. Such talk characterizes what might be called the informational "liquidity" of perceptual representations. If the representations have high liquidity, the information therein is transmissible, conveyable, mutable; it can be deployed and employed elsewhere. Channels leading from the representations are open and ready and could, with but a twitch in the direction of attention, send them downstream to engage central capacities, alter responsiveness, and so on. One can speak of the downstream subpersonal modules as "consumers," except one must keep in mind that information, like capital, is not so much "consumed" as distributed, manipulated, analyzed, invested, stored, reinvested, saved, and recycled again. Here and there it is wasted, squandered, or lost. So we have not just downstream "consumers" but investors, analysts, loan agents, manufacturers, profiteers, distributors, and so on. Some minds contain wise guys and loan sharks. In terms of this analogy, awareness does not alter the value or even the currency of the capital received; instead it changes the routes through which that same capital can be deployed. It changes the liquidity of that capital: its potentialities for deployment, investment, reinvestment, and so on. Some types of capital are more liquid than others; and given a fixed type, different economic arrangements can make that same type more or less liquid. One implication of perceptual awareness is that some perceptual representations have high liquidity. They are perched in such a place that they could, with just a flick of the switch, be transmitted and invested elsewhere. That is, dropping the analogy, they could engage selective attention.

8.8 Preattentive Registration

The winners of the competition to attract selective attention are the ones we attend, scrutinize, notice, or heed. We are "aware" of the ones that are competing but not winning. But notice that there is a third category. What of all the ones who do not even enter the race? These would be sensory representations to which one could not shift selective attention; for one reason or another, they are not even competitors. What of them? What is their status?

I will argue that this category is indeed a robust one, and it is filled with states that occur during any episode in which perceptual information about something is preattentively registered even though the subject is not aware of the thing in question. Various circumstances can produce such episodes, such as those found in masking studies, priming studies, search under extreme time pressure or high cognitive load, covert recognition, and perhaps hypnotic suggestion. And this at last is the category that can help us to understand what is happening in the disconnection syndromes of hemineglect and blindsight. What we need for them is precisely a category of what might be called "mere" sensing: specifically, of sensory registration without awareness.[7]

Such a category is, first, logically possible. To revert to an earlier example, suppose we find some specific set of conditions under which it would be impossible (it would contradict the laws of psychology) for Sam (or any human in the same situation) to focus attention on the feeling of water sloshing into boots. Selective attention has operating parameters, and they can be exceeded, jammed, or abused. Selection is not purely random; it is based in part on information about current goings-on. The mechanisms that determine where attention will go next must also have access to such information. Suppose that our conditions are carefully designed to overflow the information buffers, or cause some other glitch in the operations, so that at that moment the selecting process literally cannot select the representations of water sloshing into boots. Our conditions are somewhat similar to finding a "bug" in the operating system.[8] Perhaps a task overload or "denial of service" attack will suffice: Sam is simultaneously landing a difficult fish and trying to complete some nearly impossible cognitive tasks designed by a fiendish psychologist. With tasks sufficiently well tuned to exploit particular weaknesses in the mechanisms of selection, the sensing of wet feet might get knocked right out the race. Perhaps, given that task load, it is not even in the running, and no normal human in that situation could at that moment shift attention to feeling the slosh.

If in this sense Sam *cannot* turn his attention to the water sloshing into his boots, then common sense would suggest he is as good as insensible to that state of affairs. He certainly cannot notice it, or observe it deliberately, or scrutinize it, since all those require success at attracting attention. For the same reason, the event is not one that Sam could report or describe; nor could he use a demonstrative to refer to it. He could not use "*That* water," even in thought, to secure reference to the water in question. He could not satisfy any of these ordinary indices of being *conscious of* water sloshing into his boots. His overt actions would not differ from those of someone who does not feel the water at all. In fact, he could not do anything deliberately—could not perform any voluntary intentional action—that would indicate that he had felt the influx.[9]

Common sense makes no distinction between Sam being insensible to, and Sam being unaware of, the water sloshing into his boots. There is indeed a sense of "insensible" that is, simply, "unconscious." It is not surprising that these two concepts are often conflated. Clearly enough, one cannot be aware of a sensible feature (like the feeling of a slosh) unless one senses it. But common sense perforce relies on overt behavior and verbal report; the slow progress of experimental science in this domain has also gradually provided covert and indirect indicators, which can tease apart the two notions. So for example, skin conductance can indicate *covert* recognition in someone whose prosopagnosia eliminates all overt or verbal indicators of recognition (such as naming the person). Similarly, one can get indirect evidence (or covert measures) indicating that Sam is not strictly insensible to the water (he does indeed *sense* it: the information registers somesthetically), but simply that he cannot focus his attention on it. He "senses" it perfectly well—the preattentive registration of somesthetic information proceeds normally—but here there is a hitch in the attentional mechanisms. Such a case would be one in which he "senses" the water but is not aware of it.

If you are persuaded that this state of affairs is possible, one more step will complete my argument. Within that envisioned state of affairs occur psychological states of preattentive registration of perceptual information. Suppose that given normal background conditions, some of those states are necessary and sufficient to subserve the discriminability of, and relative similarities among, various sensible features. Discriminations hang on the presence or absence of these particular mental states. For example, they carry the information sufficient to render the feeling of water sloshing into boots discriminable from the feeling of (say) mud, or motor oil, or applesauce, or some other liquid sloshing into boots; and likewise they carry information necessary and sufficient to discriminate water sloshing into boots from water sloshing into trousers, or sleeves, or hat. Now, as argued above (in section 8.4), if we are ever to get a nonmysterious account of phenomenal consciousness, at some point we will find a family of states and properties that differ in (essentially) just one way from the normal phenomenal properties of which one is aware, that difference being that no one is aware of them (and all that follows from that). Suppose in the fullness of time we identify the preattentive states that subserve the discriminabilities and relative similarities of the sensible features in question. We would then have grounds to conclude that those states, even though they are preattentive, register *phenomenal* differences; all they lack is some standing in the competition for selective attention, which (per hypothesis) would make one aware of what they represent.

A terminological variant on this conclusion is that these states of preattentive registration of perceptual information are states with *qualitative character*.[10] They have all the properties of the normal ones, except those that follow from the fact that no one is aware of their objects. DB might have "the very qualia themselves," although he is not conscious of them. To use Chisholm's terminology, DB is in a state of being "appeared-to," even though he is not at all aware of the appearances.

The suggestion is that states of being appeared-to, states that have a qualitative character, need not be states that make someone conscious of something. The concepts *being appeared to* and *being conscious of* are distinct and independent. If so, something can look red to someone even though that someone is not aware of the thing that looks red, of the reddish look of it, or of seeing something that looks red. The only sense in which such states are "states of consciousness" is simply that a creature who has them is not an entirely unconscious creature, since some mental activity is going on.[11] Likewise, states of "being appeared to" are in that minimal sense states of consciousness: they indicate that the animal in question is not comatose or asleep. But almost any mental activity suffices to show that.

A likely response to this suggestion is: how can one be appeared to without being conscious of the appearances? I admit the suggestion is weird, but that is all it is: weird, uncommon, improbable. By hypothesis, your own states of unconscious sensation are not accessible to you. The proposal is logically analogous to Freud's proposal (1900/1913) that unconscious wishes exist: it is not refuted by angry members of the audience declaring that they are aware of no such wishes in themselves. Freud delighted in presenting the hypothesis to the public, and took the many expressions of anger as evidence *for* his hypothesis, though this latter inference was not sound. One can simply turn the question on its head: why must every appearance be an appearance of which someone is conscious? Being appeared to and being conscious of are distinct concepts; what is the necessary link between them?

8.9 Two Objections

One response to this question might be put as follows: if these really are episodes of being appeared to, instances of mental states that have some qualitative character, then we ought to be able to answer the question, *who* is being appeared to? How can an appearance be "presented" if there is no one *to whom* it is presented? As noted above (section 8.2), Nagel himself argued that phenomenological properties, like consciousness, are subjective: they can only be understood from the "point of view" of the subject who has them (Nagel 1974). So perhaps what links "being appeared to" and "being conscious of" is that both of them are subjective phenomena. They can only be understood if one understands the "point of view" of the subject who has them.

Now there is a sense in which it is quite clearly true that to understand phenomenal properties, we have to understand the point of view of the creature whose perceptions instantiate those properties. Namely, we have to understand the similarities and differences which *that creature* senses to obtain among the stimuli that confront it. These are not determined by the stimuli themselves. As earlier emphasized, the similarities and differences of phenomenal properties need not and do not map in any direct way onto similarities and differences of the physical properties of stimuli. Instead one has to understand how the creature's sensory systems work, so that one can understand the similarities and differences *it* perceives among those stimuli. Part of the objection is hence well founded: we do, in a certain sense, need to enter into the point of view of the creature who has these sensations.

But the sense in which we need to do this does not sustain the rest of the objection. In particular this *perspectival* character of phenomenal appearance is not sufficient to yield "subjective facts," or facts of a different ontological or metaphysical kind than ordinary facts. Nor does it license the inference to the conclusion that there is no appearance unless there is someone or something *to whom* the appearance is "presented." The "presentation" terminology was initially introduced as a handy way of stating the converse of the relation "being acquainted with," and so it does sound odd to say that DB is "presented the appearance" of something even though he roundly denies being aware of any such thing. It sounds less odd to say he simply "has" a sensory state that underwrites the successful discrimination. That state stands in the same place in the sensing of similarities and differences—it has the same qualitative character—as the states that normally make one conscious of seeing those things. But in DB that very state does not make anyone conscious of anything.

A second objection might be put as follows. Clearly states of preattentive registration of perceptual information are not states of phenomenal consciousness. I have urged that adding the capacities of selective attention yields an architecture that is closer to one to which we are willing to ascribe states of phenomenal consciousness. But how do the additions do that, exactly? Let us imagine this in some detail. We might construct a special purpose wine-tasting machine, for example. One pours a sample of some vintage into the input funnel. The machine takes various measurements, computes some numbers, and eventually types out "a chemical assay, along with commentary: 'a flamboyant and velvety Pinot, though lacking in stamina'—or words to such effect" (Dennett 1988, 46). This machine might have enological discriminative powers that far exceed any human's, but at best it has the functional equivalent of preattentive sensory registration. It is not aware of anything it "tastes," nor is there anything it is like to *be* that wine-tasting machine. All the processing proceeds in the dark.

Now to this machine we add some additional circuitry, which adds the functional equivalent of the mechanisms of competition and selection, and of the releasing, shifting, and maintaining of attention on some selected subset of the representations coursing through its innards. How does any such addition make it any more reasonable to say that the machine is now aware of what it "tastes"? Couldn't all the latter proceed in the dark as well?

There is room for but the barest sketch of an answer. Notice first that if a psychological model of perceptual experience is thought to be possible at all, then if our search for it goes well, at some point it will be reasonable to ascribe such mental states to an instantiation of such a model. After all, per hypothesis, *we* are an instantiation of such a model. The real question is what one must put into such a model to make it a likely candidate for satisfying such ascriptions. The ones found early in the search are absurdly simple, but that does not show that all the later ones will be as well.

Second, it would be no mean feat to add to any artifact the functional equivalent of the mechanisms of competition and selection, and of the releasing, shifting, and maintaining of attention. This would add much more complexity than one might think. For example, to have competition, there must be competitors, implying a range of possible alternative activities to which the relatively limited resources of something that it is fair to call "central" processing could be applied. So a single-purpose, one-track device cannot qualify. There must be a plethora of distinct representational streams to which one might "pay attention." This in turn requires a genuine architectural distinction between peripheral and central processing; a multitude of the former processes must compete for the benefits of the latter. The system must also be able to select one such stream, shift resources to it, keep others in abeyance, yet allow itself to respond to emergencies, or shift to others as needs or priorities change. A system that can pay attention must also be distractable; and to say a system is "distractable" is to say something rather complicated about it. It has a finite attention span. Even during that span, competitors beckon. It can be heedless or focused. It can become enthralled, absorbed, or bored.

If one did duplicate the architecture of selective attention, then competition and selection, the allocation of central processing, the runner-up status of inputs not selected, the monitoring of those runner-ups, and all the rest would be duplicated as well. And I suggest that it is then not quite so easy to imagine such a device as totally lacking awareness. Awareness becomes a live question only if our wine-tasting machine could become distracted or bored, or if it took some effort to get it to focus on the task we would like it to complete—but then, I think, it really *is* a live question. So, oddly enough, it is the (temporary) absence of focused attention that betokens the presence of awareness. The short attention span of some undergraduates is regrettable, but not because it indicates lack of consciousness. In fact, it indicates precisely the opposite: the undergraduate is attending to something *else*. Even an attention span of zero fails to indicate lack of consciousness. If the idioms of attentiveness apply at all, then so do the idioms of awareness. To get to the pure repose of no consciousness, one must entirely eliminate the applicability of our idioms of attention: not an attention span of zero, but no attention span at all.

Acknowledgments

I thank Jonathan Cohen and Mohan Matthen for all their work organizing the color science conferences and the resulting volume of papers, and for providing very useful comments to me on this particular paper as well. Thanks also go to Valtteri Arstila, Murat Aydede, Steven Biggs, and Alex Byrne, who provided comments on versions of this paper, as well as to audiences at the University of Toronto, Harvard University, the University of British Columbia at Vancouver, Simon Fraser University, and Rutgers University, where versions of these ideas were presented.

Notes

1. My notes indicate that the first question Larry Hardin posed was "Why should there be any color experience at all, rather than just information processing?" The question listed above came second, as a follow-up and elucidation of this first one.

2. The problem with this sort of calculation is that no one knows what the relevant ensembles are at the two ends of the channel: how the information might be coded or "chunked." It is almost certainly not the case that the lateral geniculate nucleus needs to record data at the rate of one thousand bits per second for each nerve in the optic tract. Visual information is probably coded and chunked in ways that does not require such high fidelity. But, as mentioned later, auditory information may in fact require such high fidelity, since minute differences in the timing of wave crests arriving at the two ears are used to determine the direction of the sound.

3. Other analogical models include gain control, amplifier, zoom lens, telephoto lens, and channel. Some of these analogies emphasize amplification of the selected signal; others, the inhibition of the nonselected signals. See LaBerge (1995, 39).

4. Or at least not *focally* attending to them. In the end, probably both sets of phenomena are products of "divided" attention, of the allocation of attentional resources. The ones in the spotlight win the plurality of those resources, while those in the penumbra get a tiny share. See Dehaene and Naccache (2001).

5. Again, recent models suggest that such selection is not all or none; the "winner" just wins a plurality of the available attentional activation, while those that remain "in the running" receive just enough to keep them active.

6. Since selection is not all or none, one should not think of representations passing through the "gate" of selective attention as analogous to a cow passing through a gate in the fence. The gate might simply weaken the signal strength of the cow. (See the other analogies mentioned in note 3.)

7. Psychologists these days are rightly leery of the words "sensation" and "sense," because they suggest to some ears an appeal to an old and discredited theory according to which each perceptual modality starts with elemental states, accessible to consciousness, called *sensations*, and from them build up *perceptions*. I do not mean to endorse that theory by using the verb "sense." Nor do I mean to imply that those processes are either elemental or accessible to consciousness. The verb has not yet been banned from philosophical discourse, and there I think it sometimes means something close to what is suggested here: the preattentive registration of perceptual information. Note that showing sensation without awareness is something less than showing *perception* without awareness; in particular, we are accustomed to the idea that information extracted in early sensory systems is not invariably available for the guidance and control of

action. However, many of the problems that Dretske (2006) found in experimental tests for the *lack* of awareness infect the sensory domain as well.

8. The so-called attentional blink provides a real example of pushing the envelope of attentional operating parameters. There is an attentional "dwell time" of several hundred milliseconds (which may reflect the time it takes to disengage and move attention); if a second stimulus that needs attention is presented during that interval, error rates go up enormously. See Raymond, Shapiro, and Arnell (1992, 1995).

9. Such actions are precluded because they require attention to the somesthetic event. The impossibility in question would also rule out an *exogenous* shift of attention—one in which the subject does not deliberately shift attention, but has attention drawn involuntarily to some stimulus (for example, by movement or change).

10. I follow the terminological convention of treating *qualitative properties* as properties of mental states, and *phenomenal properties* as characteristics of sensible appearance, which are attributed rightly or wrongly to the things one perceives. So how the water feels when it sloshes into one's boots is a phenomenal property, characterizing how a portion of the world appears. Whatever properties of somesthetic sensory states subserve the discriminability of that feeling from others constitute the qualitative character of those sensory states. Unfortunately, the term "phenomenal character" is often used in the literature as a synonym for "qualitative character"; I use it that way myself at the beginning of this chapter.

11. One might say that these states have *phenomenality,* even if they are not cognitively accessible (see Block 2002, 394.) Or one might say that their "phenomenology overflows accessibility" (Block 2007, 487). But the terminology is treacherous. I do say that bottom-up preattentive perceptual systems have a far greater bandwidth than central processing (section 8.5), and that some states within those preattentive systems suffice to sustain the presentation of phenomenal properties (section 8.8). These statements imply that some systems sufficient to sustain the presentation of phenomenal properties overflow (have greater bandwidth than) that which is cognitively accessible (e.g., that which can be processed in the global workspace). In places Block's conclusion is put in convivial terms: "If we assume that the strong but still losing coalitions in the back of the head are the neural basis of phenomenal states (so long as they involve recurrent activity), then we have a neural mechanism which explains why phenomenology has a higher capacity than the global workspace." (Block 2007, 498). But there is at least one disagreement: I see no reason to believe that those preattentive states that have qualitative character are necessarily "experiential" or that the subject is in any sense "aware" (or "Aware") of them. Even though Block is now "abandoning the term 'phenomenal consciousness'" (2007, 485), for him states that have "phenomenology" still seem to be "experiential" states. He describes subjects in the Sperling experiment as having "experiences as of specific alphanumeric shapes" (2007, 487) and says, "The argument of this paper depends on the claim that subjects in the Sperling and Landman experiments have phenomenal experiences of all or almost all of the shapes in the presented array" (Block 2007, 491). Finally, activation of the fusiform face area in patient G. K. is described as possibly yielding "phenomenal experience that the subject not only does not know about but in these circumstances cannot know about." (Block 2007, 498). If these three

citations were amended, and the assumption that qualitative states are necessarily experiential were dropped, I think there would be very little difference between Block's conclusion and the one here.

References

Block, N. 2002. The harder problem of consciousness. *Journal of Philosophy* 99 (8): 391–425.

Block, N. 2007. Consciousness, accessibility, and the mesh between psychology and neuroscience. *Behavioral and Brain Sciences* 30 (5/6): 481–548.

Block, N., O. Flanagan, and G. Güzeldere, eds. 1997. *The Nature of Consciousness: Philosophical Debates*. Cambridge, MA: MIT Press.

Broad, C. D. 1927. *Scientific Thought*. London: Routledge and Kegan Paul.

Broadbent, D. E. 1958. *Perception and Communication*. New York: Pergamon Press.

Campbell, J. 2002. *Reference and Consciousness*. Oxford: Clarendon Press.

Chalmers, D. 1996. *The Conscious Mind*. New York: Oxford University Press.

Chisholm, R. 1957. *Perceiving: A Philosophical Study*. Ithaca, NY: Cornell University Press.

Clark, A. 2008. Phenomenal properties: Some models from psychology and philosophy. *Philosophical Issues* 18: 406–425.

Dehaene, S., and L. Naccache. 2001. Towards a cognitive neuroscience of consciousness: Basic evidence and a workspace framework. *Cognition* 79: 1–37.

Dennett, D. 1988. Quining qualia. In *Consciousness in Contemporary Science*, A. J. Marcel and E. Bisiach, eds., 42–77. Oxford: Clarendon Press. (Reprinted in *The Nature of Consciousness: Philosophical Debates,* N. Block, O. Flanagan, and G. Güzeldere, eds., 619–642, Cambridge, MA: MIT Press, 1997.)

Dretske, F. 2006. Perception without awareness. In *Perceptual Experience,* T. S. Gendler and J. Hawthorne, eds., 147–180. Oxford: Oxford University Press.

Fodor, J. A. 1986. *Modularity of Mind.* Cambridge, MA: MIT Press.

Freud, S. 1900/1913. *The Interpretation of Dreams.* Translated by A. Brill. New York: MacMillan.

LaBerge, D. 1995. *Attentional Processing.* Cambridge, MA: Harvard University Press.

Nagel, T. 1974. What is it like to be a bat? *Philosophical Review* 83: 435–450. (Reprinted in Nagel 1979a, *Mortal Questions*, 165–180, Cambridge: Cambridge University Press; and in *The Nature of Consciousness: Philosophical Debates,* N. Block, O. Flanagan, and G. Güzeldere, eds., 519–527. Cambridge, MA: MIT Press, 1997.)

Nagel, T. 1979a. *Mortal Questions*. Cambridge: Cambridge University Press.

Nagel, T. 1979b. Subjective and objective. In *Mortal Questions*, 196–214. Cambridge: Cambridge University Press.

Neisser, U. 1967. *Cognitive Psychology*. New York: Appleton-Century Crofts.

Pylyshyn, Z. 2001. Visual indexes, preconceptual objects, and situated vision. *Cognition* 80: 127–158.

Raymond, J. E., K. L. Shapiro, and K. M Arnell. 1992. Temporary suppression of visual processing in an RSVP task: An attentional blink? *Journal of Experimental Psychology: Human Perception and Performance* 18: 849–860.

Raymond, J. E., K. L. Shapiro, and K. M Arnell. 1995. Similarity determines the attentional blink. *Journal of Experimental Psychology: Human Perception and Performance* 21: 653–662.

Rey, G. 1997. A question about consciousness. In *The Nature of Consciousness: Philosophical Debates*, N. Block, O. Flanagan, and G. Güzeldere, eds., 461–482. Cambridge, MA: MIT Press. (Originally published in *Perspectives on Mind,* H. Otto and T. Tueidio, eds., 5–24. Norwell: Kluwer Academic Publishers, 1988.)

Searle, J. R. 1997. *The Mystery of Consciousness*. New York: New York Review of Books.

Stoerig, P., and A. Cowey. 1989. Wavelength sensitivity in blindsight. *Nature* 342*:* 916–918.

Stoerig, P., and A. Cowey. 1991. Increment threshold spectral sensitivity in blindsight: Evidence for color opponency. *Brain* 114: 1487–1512.

Stoerig, P., and A. Cowey. 1992. Wavelength sensitivity in blindsight. *Brain* 115: 425–444.

Weiskrantz, L. 1986. *Blindsight: A Case Study and Implications*. Oxford: Oxford University Press.

Weiskrantz, L. 1997. *Consciousness Lost and Found*. Oxford: Oxford University Press.

9 It's Not Easy Being Green: Hardin and Color Relationalism

Jonathan Cohen

It's not that easy being green;
Having to spend each day the color of the leaves.
When I think it could be nicer being red, or yellow or gold . . .
or something much more colorful like that.
—Kermit the Frog

C. L. Hardin is that rarest of things—a philosopher who has changed the world. Prior to the publication of *Color for Philosophers: Unweaving the Rainbow* (Hardin 1988), philosophical work on color had been conducted in roughly the same terms in which it was carried out by the famous moderns—Galileo, Boyle, Locke, et al. But once Hardin pointed out that a vast field of empirical research had developed since the modern period, and showed convincingly that these developments impose serious constraints on ontological and epistemological disputes about color, the philosophical landscape was forever changed. Subsequently, philosophical work on color has increased dramatically in both sophistication and interest (as shown by its growth in recent years), and we have Hardin to thank for these salutary developments.

But Hardin hasn't contented himself with reframing traditional philosophical issues about color in a way that is sensitive to relevant empirical constraints. In addition, he has been a staunch defender of color irrealism—the view that there are no colors, *qua* properties of tables, chairs, and other mind-external objects—and a vociferous critic of several varieties of realism about color that have been defended by others (e.g., Hardin 2003, 2004). Among the realist views that Hardin has targeted are the so-called color physicalism of Hilbert (1987), Byrne and Hilbert (1997a, 2003), and Tye (2000),[1] and, inconveniently for me, the relationalist view of Cohen (2003a, 2003b, 2004, 2009).

My main purpose in this chapter is to defend relationalism against Hardin's most recent criticisms of it (Hardin 2004, 2006, 2007). However, to motivate this discussion, it will be helpful first to consider Hardin's recent objections against color physicalism, and why existing physicalist defenses against such objections are unsatisfactory (section 9.1). This will put us in a position to see some of the virtues of color relationalism (section 9.2), and then to evaluate Hardin's complaints against it (section 9.3). I'll argue that relationalism survives Hardin's criticisms, and that it is ultimately preferable to both color physicalism and Hardin's own irrealist view.

9.1 Hardin on Color Physicalism

Many philosophers inclined toward realism about color have proposed that (surface) colors are some sort of mind-independent properties of object surfaces. Recent color physicalists have defended the more specific view that (surface) colors are classes of surface spectral reflectance distributions.[2]

Hardin and others have pointed out that this form of color physicalism faces significant hurdles stemming from the remarkably widespread interpersonal and intrapersonal (and interspecies) perceptual variation with respect to color: a given stimulus produces a variety of effects in different perceptual systems, and produces a variety of effects on a single perceptual system when viewed under different perceptual circumstances. On standard assumptions that are common ground in the current dispute, each of these different effects is a representation of the color of the stimulus.[3] But if colors are mind-independent and circumstance-independent properties of surfaces, as are spectral reflectance distributions (or classes thereof), then physicalists are committed to saying that at most one of these varying effects represents the color of the stimulus veridically. However, the objection goes, it is hard to see that anything could (metaphysically) make it the case that one of the variants is veridical at the expense of the others: it seems that any considerations that could be brought forward in support of the veridicality of one of the variants could be matched by considerations of equal force in favor of some other variant.

As Hardin (2004) hammers home, the sort of variation of color vision at issue is no mere imagined possibility. For one thing, there is overwhelming and unambiguous evidence of actual variation of color vision in normal human subjects (and even more when we look at nonhuman visual systems), holding all viewing conditions (e.g., illumination, background, viewing distance, angular subtense of visual field, etc.) fixed. One among many clear instances of this sort of interpersonal variation, discussed at length by Hardin (2004), is the variation in the spectral wavelength (alternatively, in the Munsell chip) selected by subjects as *unique green* (i.e., as looking greenish without looking at all bluish or at all yellowish). When two normal trichromatic observers view chip *C* under identical perceptual conditions, *C* looks unique green to one of them but bluish green (hence not unique green) to the other. For that matter, as described by Cohen (2004) and Hardin (1988), there is equally overwhelming and unambiguous evidence of intrapersonal variation of color vision as a function of many parameters of the viewing condition. Consequently, there are cases of intrapersonal variation that mirror the interpersonal one just discussed. For example, a chip C can look greenish to you when presented against one background and bluish when presented to you against another background. If color physicalism is true, then in this intrapersonal case as well, it must be that at most one of the competing representations veridically represents C's color. But, once again, it is extremely hard to imagine what could (metaphysically) make it the case that one of the representational variants is veridical at the expense of the other.[4]

Physicalists have sometimes suggested in response to this concern (e.g., Byrne and Hilbert 2003, 2004; Tye 2000) that the difficulty here is merely epistemic—that while facts about variation might prevent us from *knowing* which of the competing variants is veridical at the expense of the others, there is nonetheless a (possibly unknown) fact of this matter, just as color physicalism requires. But, as far as I can see, there is literally no reason to believe that that is true. On the contrary, there is substantial (but defeasible) reason to believe that it is false—namely, the failure of several hundred years of systematic efforts directed at uncovering such facts of the matter establishes a presumptive case against their existence. As such, the physicalist's insistence that there is an epistemically unavailable fact of the matter strikes me as a piece of unwarranted optimism.[5,6]

9.2 Color Relationalism

As we saw in section 9.1, color physicalists are committed to claiming that at most one among conflicting perceptual variants veridically represents the color of a single stimulus. The difficulty with this commitment, I suggested, is that there is no reason for believing that any one of the variants is uniquely distinguished from the others in the way that the physicalist maintains. In the face of these difficulties, the relationalist suggests that we should avoid the trouble by refusing to choose between the variants. That is, she suggests, the way out of the trouble is to hold that the conflict between the variants is only apparent, insofar as the single stimulus is genuinely both unique green to observer S_1 and not unique green to S_2. Likewise, since a single stimulus can look unique green to a single observer S under one viewing condition C_1 but fail to look unique green to S under another viewing condition C_2, and since there seems not to be any fact of the matter that makes one of these two representations of the stimulus's color veridical at the expense of the other, the relationalist will refuse to choose between the two: instead, she'll insist, the stimulus is genuinely both unique green to S under C_1 and not unique green to S under C_2.

What the relationalist proposes, then, is that colors are not (as the physicalist maintains) subject- and condition-independent properties of their bearers, but relational properties constituted in terms of relations to subjects and viewing conditions. Since, on this view, colors are relational properties, the perceptual variants that we initially characterized as conflicting are in fact not conflicting. This view does justice to the facts about perceptual variation, and it does so without requiring either unmotivated choices between variants or unjustified optimism that there is some unknown (or unknowable) fact that could motivate such a choice. This, it seems to me, is an important virtue of the view, and one that makes it worth taking seriously for those who aspire to realism about color.

That said, it is worth emphasizing that relationalism is not, by itself, a theory of the nature of color. It is a theory about what sorts of properties colors are, namely, that they are relational properties; but it does not say which properties of that sort—which relational properties, in particular—colors are. In fact, there are several species of relationalism: these include the dispositionalist view of McGinn (1983), Peacocke (1984), and Johnston (1992); the so-called enactive view of Thompson et al. (1992) and Thompson (1995); the selectionist view of Matthen (2005); the role functionalist position of Cohen (2003a, 2009); and others. The question of whether colors are relational puts substantive constraints on what counts as an adequate color ontology, but answering that question will leave plenty of room for disagreement about what colors are. Consequently, relationalism might aptly be regarded as a family of accounts of color ontology—a family whose members share some commitments but not others, rather than as an account in its own right. (I'll return to this theme in section 9.3.2.)

9.3 Hardin on Relationalism

As we have seen, relationalist views avoid at least one important class of difficulties that plague color physicalism. However, Hardin presses further worries against relationalism. First, he urges that the view doesn't deliver the kind of objectivity that a color realist should want, and so doesn't amount to an acceptable form of color realism after all. Second, he worries that color relationalism is unacceptably liberal about the attribution of colors. I'll take these worries in turn (sections 9.3.1–9.3.2), and then turn to a conjecture about what might ultimately underlie Hardin's other reasons for being worried about relationalism (section 9.3.3).

9.3.1 Objectivity Lost?

Hardin's first concern about color relationalism is that the view makes colors insufficiently objective—no more objective, in any case, than properties like *being beautiful* and *being ugly*. If so, he suggests, the kind of objectivity relationalism can bestow to color properties is too attenuated for a realism worthy of the name:

> To me, this woman's face is the very Form of Beauty incarnate. To Jonathan, it is a face only a blind mother could love. Beauty is in the eye of the beholder, you say? No, it is an objective property of the woman's face, for I need only to relativize it to Larry's gaze, if only at time T. Ugliness is also an objective property of her face, provided of course we understand it as being

relative to Jonathan's eye at time T. And so this same woman has the possibility of being all things to all men (Hardin 2004, 36).[7]

Is Hardin right that relationalism threatens to make *being beautiful* an objective property also? Obviously, the answer to that question depends rather a lot on what one means by 'objective,' and unfortunately there are a variety of candidate understandings with a historically salient claim on the word.

It does seem that there are several relevant understandings of 'objectivity' on which *being green* and *being beautiful* (as understood by the relationalist) turn out to be objective. For example, suppose a property *P* is objective just in case individuals in the world genuinely exemplify *P*; for the relationalist, about *being green* and *being beautiful*, both properties are objective in this sense, insofar as things genuinely are green or beautiful (viz., by virtue of standing in certain relations to subjects). Alternatively, suppose we count *P* as objective just in case *P* would continue to have its instances if (counterfactually) there were no perceiving subjects; here, again, a relationalist can hold that both properties are objective in the sense at issue in that the relations in virtue of which things count as green or beautiful can be spelled out in terms of how those things *would* (possibly counterfactually) strike subjects.

In addition, there are other historically salient understandings of objective that underwrite the opposite verdict about both properties. For example, objectivity is sometimes understood in terms of interpersonal convergence; for example, one might think of *P* as an objective property just in case undistracted and knowledgeable subjects agree in judging whether *x* is *P* for a wide range of *x*. Of course, one of the lessons of section 9.1 is that color properties fail this criterion; and, of course, many (e.g., Hardin, in the quotation above) have thought the same is true of *being beautiful* (but see, e.g., Mothersill 1984; Zangwill 2001). Another criterion for property objectivity, built on the traditional subject-object contrast, might hold that property *P* is objective if and only if it is constituted independently of subjects. Of course, relationalism about *P* just amounts to denying that this criterion holds of *P*, so the relationalist about *being green* and *being beautiful* will hold that both properties are nonobjective on the present understanding.

So there are several criteria for objectivity that agree in treating *being green* and *being beautiful* as objective properties, and there are several others that agree in treating them both as nonobjective properties. On the other hand, it is not inevitable that the color relationalist will classify the two properties on the same side of the objective-nonobjective distinction.

First, suppose that, as before, we accept relationalism about both *being green* and *being beautiful*, but that we understand property objectivity in terms of the subject-independence of its causal effects; in particular, suppose a property *P* counts as objective just in case the event of its exemplification by *x* has causal effects that are independent of the effects *x* has on subjects. As far as I can see, this further (and perfectly good) sense of property objectivity leaves it entirely open whether distinct relational properties (*being green* and *being beautiful*) are alike or different in respect of their objectivity, depending on the contingencies of causal pathways in the world. Moreover, and independently of this last point, it is important to recall that a color relationalist is not per se committed to a relationalist treatment of *being beautiful*; consequently, if she favors an alternative view about the latter, many of the criteria for property objectivity listed above may count color properties as objective and *being beautiful* as nonobjective, or vice versa.

It seems, then, that contrary to what Hardin suggests, the color relationalist is forced to conclude neither that *being green* is nonobjective, nor that it is only as objective as *being beautiful*. But suppose (contrary to fact) that a relationalist were committed to regarding colors as no more objective than *being beautiful*. Why would that be an objectionable conclusion? Of course, some understandings of realism would have the effect of undermining the relationalist's realist aspirations—namely, those that make criterial one of the senses of objectivity that are unavailable to the relationalist. On the other hand, this point itself seems a reason for being dissatisfied by those construals of 'realism,' rather than taking it as an indictment of relationalism.[8] After all, insofar as relationalism is precisely a theory of how colors are constituted, relationalists are committed to saying that colors are *bona fide* properties that are (sometimes) exemplified by *bona fide* instances—otherwise there would be no target for relationalism to be a theory about in the first place (cf. Sayre-McCord 1988; Sober 1982; McDowell 1985). So far as I can see, then, there is no reason that a relationalist need fear the conclusion that *being green* is only as objective as *being beautiful*.

9.3.2 A Chromatic Swarm?

Hardin has a second reason for being unhappy with color relationalism that turns on the plurality of color properties recognized by the view. In this connection, he employs a venerable philosophical strategy—the argument by quotation. In this case, he quotes Meno's explication of Gorgias's definition of virtue:

> First of all, if it is manly virtue you are after, it is easy to see that the virtue of a man consists in managing the city's affairs capably, and so that he will help his friends and injure his foes while taking care to come to no harm himself. Or if you want a woman's virtue, that is easily described. She must be a good housewife, careful with her stores and obedient to her husband. Then there is another virtue for a child, male or female, and another for an old man, free or slave as you like; and a great many more kinds of virtue, so that no one need be at a loss to say what it is. For every act and every time of life, with reference to each separate function, there is a virtue for each one of us, and similarly, I should say, a vice (Plato, *Meno*, 71E).

The argument finishes off by quoting Socrates's sarcastic reaction to Meno's disquisition: "How fortunate I am, Meno! I wanted one virtue and I find that you have a whole swarm of virtues to offer."

Despite the exegetical perils attendant on the attempt to understand arguments by quotation (especially those quoting such vexed texts as the *Meno*), I think I can discern two possible antirelationalist objections that Hardin might have in mind here. First, one might worry that relationalism is objectionable on strictly numerical grounds: the thought would be that relationalism sustains too many color attributions, that it results in what Hardin at one point terms "ungainly pluralities" (2004, 36). A second possible concern, which I take to be closer to what underlies Socrates's objection to Gorgias's account of virtue, is that the distinct colors recognized by the relationalist fail to be appropriately unified. Once again, I'll take these concerns one at a time.

Ungainly Pluralities It seems hard to deny that there is something *prima facie* counterintuitive in the relationalist's claim that a given stimulus can exemplify (at the same place, at the same time) multiple colors, each relativized to different subjects and perceptual conditions. Most significantly, this relationalist claim seems to fly in the face of our ordinary conversational practice of attributing a single color to a given object at a given time: we say that the ripe lime is green, and leave it at that, and speak without hesitation about *the* color of the lime. There is, I think, no serious question about these facts about the way we attribute colors. The only serious question is how damaging these facts are to color relationalism.

My reason for thinking that they are not so damaging after all is that data about the property attributions occurring in ordinary conversation are the joint result of (1) facts about the properties figuring in the attributions, and (2) the communicative (and other) presuppositions we bring to the linguistic exchanges in which the attributions occur. I suggest that the apparently troublesome data about color attributions are to be explained in terms of factors of type (2) rather than type (1), and so are not ultimately at odds with color relationalism *qua* account of the nature of color properties.[9]

That is, the reason we typically attribute in conversation only a single color to a particular ripe lime at time t is that, while the lime exemplifies many colors at t (it is green to S_1 in C_1, red to S_2 in C_2, orange to S_3 in C_3, and so on), we care only about a very small finite minority of those colors—in many cases, only one of them—in typical conversational settings. For one thing, in such settings we are generally unconcerned with the colors that the lime exemplifies in virtue of its relations to visual systems very much unlike our own (e.g., octopus visual systems, imagined Martian visual systems, and even human dichromatic and anomalous trichromatic visual systems), and so don't mention these colors that (according to the relationalist) the lime genuinely exemplifies. And although there will be nontrivial variation between the colors the lime has relative to visual systems we do care about (see section 9.1) that is discernible through systematic testing in the psychophysics lab, this variation won't reveal itself in ordinary conversational settings outside the lab; consequently, ignoring this variation by the use of a more general color term—e.g., green *simpliciter*—will serve well enough for ordinary conversational purposes. Likewise, in such settings we are generally unconcerned with the colors that the lime exemplifies in virtue of relation to viewing circumstances substantially unlike those we typically encounter, or in which our interests in making color classifications are not in play (e.g., circumstances of extremely low illumination, or viewed through a chromatic filter, or where the stimulus subtends 0.01° of visual angle, or against a background of diagonal achromatic lines and presented for 20 milliseconds), and so don't mention these colors that (according to the relationalist) the lime genuinely exemplifies. And although there will be nontrivial variation between the colors the lime has relative to visual circumstances we do care about (see section 9.1) that is discernible through systematic testing in the psychophysics lab, this variation won't reveal itself in ordinary conversational settings outside the lab; consequently, ignoring this variation by the use of a more general color term—e.g., green *simpliciter*—will serve well enough for ordinary conversational purposes.

My proposal, then, is that the presuppositions we bring to ordinary conversational settings in which we make color attributions include general ways of filling in the parameters to which (according to relationalism) colors are always relativized. In effect, when we say that the lime is green, we mean that the lime is green to visual systems of the kind that are contextually salient, under visual circumstances of the kind that are contextually salient. These parameters go unmentioned in ordinary conversation not because they are not there (a conclusion that would be damning to color relationalism), but because the presuppositions in force in these settings allows us to leave them tacit (a conclusion that is consonant with color relationalism). Of course, if this strategy serves ordinary purposes, this leaves it open that there are some non-ordinary purposes (e.g., those in the psychophysics lab) in which the simplifications break down. This is, I suggest, exactly what we find: when systematic variation of the relevant parameters reveals the contribution they make to the color of a fixed stimulus, we make explicit the relationality of colors that had heretofore remained tacit—for example, we say that the chip is unique green to *this* observer, but not *that* one, in *this* visual circumstance, but not *that* one.

By way of analogy, consider the weather. I take it that meteorological status is obviously constituted in terms of a relation to locations, which means that there really is a swarm of meteorological conditions: it is cold and raining in location L_1, it is hot and dry in location L_2, and so forth. But as long as I don't investigate the systematic effects on weather of considering different locations, conversational presuppositions constrain the way we fill in the relevant relativizations; consequently, my neighbors and I can get by asking each other (not explicitly relativized) questions like "what is the weather today?" This strategy fails, however, as soon as more wide-ranging investigation begins: when I talk to friends across town, I notice that what they say about the weather is very different from what I would say, and this effect increases as I vary the value of the locational parameter more dramatically. For example, Hardin lives in Syracuse, and I live in San Diego, so he'll always say that it is raining and overcast, and I'll always say that it is sunny and 72°F. When we notice that this occurs, we make explicit the relationality of the meteorological situation that had heretofore remained tacit—e.g., we say that the weather is sunny and 72°F in San Diego, but not in Syracuse.

It seems, then, that dramatic meteorological variation can often be ruled out effectively by conversational presuppositions (e.g., those shared by my neighbors and me) that restrict attention to a small range of locations. This does not mean that there is no meteorological variation left in play even when such presuppositions are in force—there really can be meteorological variation over small changes in location. But it does mean that the variation in question is much less dramatic than it would be were they not in force, and that the residual variation is often small enough that it is unlikely to manifest itself to casual observation (as opposed to detailed meteorological investigation).

So, too, I suggest, dramatic perceptual variation with respect to color can often be ruled out by conversational presuppositions that restrict attention to a small range of contextually salient visual systems and viewing circumstances. This does not mean that there is no perceptual variation with respect to color left in play even when such presuppositions are in force—there really can be perceptual variation with respect to color between contextually salient visual systems and in contextually salient viewing conditions. But it does mean that the variation in question is much less dramatic than it would be were they not in force, and that the residual variation is often small enough that it is unlikely to manifest itself to causal observation (as opposed to detailed psychophysical investigation).

For these reasons, I think color relationalism can reasonably be reconciled with our ordinary practices of color attribution, and that this defuses the numerical worry we have been considering.

Unity of the Colors Let us return, then, to the other worry discernible in the passage Hardin quotes, which concerns the unity of the colors. As I read the text, the heart of Socrates's objection to Gorgias's view of virtue is not that it results in a mere plurality of virtues, but that there is no principled criterion for inclusion in the plurality. As Socrates puts the point a few lines down from the quoted passage, "Even if they are many and various, yet at least they all have some common character which makes them virtues. That is what ought to be kept in view by anyone who answers the question, What is virtue?" (72d). If this is the right way to read Socrates's concern about virtue, then the analogous worry for the color relationalist is not merely that the view proliferates colors, but that there should be something—and something that it is the central task of a theory of color to elucidate—shared by the several colors recognized by the theory in virtue of which they do (and other properties do not) count as colors.

I am sympathetic to Socrates's objection to Gorgias (at least as I have construed it); I think it is fair to ask that a theory of virtue should amount to more than a motley list—that it should provide a principled and projectible criterion that unites those things it recognizes as virtues and distinguishes them from those it does not. Moreover, I share Socrates's concern that Gorgias's account, as related by Meno, does not meet this desideratum. If this is right, it seems only fair to insist that an adequate account of color should, likewise, provide a principled and projectible criterion that unites those things it recognizes as colors and distinguishes them from those it does not. I want to argue that the color relationalist is, unlike Gorgias, in a position to meet this desideratum.

Unfortunately, saying why requires some caution, insofar as the relationalist's answer will depend on commitments that go beyond the strict bounds of relationalism—namely, on details of the specific form of relationalism that she holds. Recall that, as emphasized in section 9.2, relationalism is not by itself an account of the nature of color. Relationalism has it that colors are relational properties constituted in terms of relations to subjects and viewing conditions; however, saying this leaves open the crucial question of exactly which relations to subjects and viewing conditions are relevant—and the latter question is one on which different relationalists hold sharply divergent views (along a number of different dimensions). What I want to insist on for present purposes is that (i) each relationalist has an answer to this question, and (ii) each answer to this question will provide a principled and projectible criterion that unites those things that the theory recognizes as colors and distinguishes them from those things it does not. It follows from these two points that each relationalist will have an answer to the challenge under consideration, even if relationalism per se is agnostic between these answers.

To make this so-far-schematic point clearer, I'll illustrate by adverting to the example of my own preferred version of relationalism—the role functionalist account I have defended elsewhere (Cohen 2003a, esp. section 1.3, 2009). Suppose, therefore, that you are a role functionalist about color: you hold that *being green for subject S in condition C* is the (multiply realizable) functional role of disposing its bearers to look green to *S* in *C*, that *being red for subject S in condition C* is the (multiply realizable) functional role of disposing its bearers to look red to *S* in *C*, and so on for the other colors. Then you will say that *being green for S in C* and *being red for S in C* are united in being colors because of this fact: they are both (multiply realizable) functional roles of disposing their bearers to look a certain way to certain subjects in certain circumstances. But this still isn't sufficient for answering Socrates's worry about unity, because we still owe a story about what it is for something to look one of those certain ways to certain subjects in certain conditions—we need an account of *looks green to S in C*, *looks red to S in C*, and so on. To answer this question is to go beyond relationalism once again; indeed, it is to go beyond functionalism *qua* theory of the nature of color and to defend a theory about the nature of color experience. As before, I take it that a number of options are open at this point and will provide only one way of fleshing out what is needed for the sake of illustration. Say that *x* looks green to *S* in *C* just in case by visually attending to *x* in *C*, *S* is appropriately caused (in *C*) to have an experience of green; and say that *experience of green* is a type of state of the opponent-process system (viz., a neurocomputationally individuated state type; *mutatis mutandis* for *looks red to S in C* and its ilk).

On the functionalist flavor of relationalism, then, the unity of colors comes from the unity of the set of color experiences; and (on the neurocomputational identity-theoretic strategy sketched above) the latter gets its unity—that is, the set is distinguished from other things in a principled and projectible way—by the type individuations made available by a well-known neurocomputational theory (viz., opponent-process theory; Hurvich and Jameson 1957; Boynton 1979; Hurvich 1981; Hubel 1988). We are therefore in a position to meet Socrates's challenge, for we have a principled and projectible criterion for being a color experience, and (derivatively) we have a principled and projectible criterion for being a color.

That said, I want to emphasize that my point here is more general than the particular relationalist-friendly proposals sketched above, and so should not be taken to stand or fall with those particular proposals. Relationalists are a varied lot. They agree that colors are constituted in terms of relations to perceivers and viewing conditions, but they disagree among themselves about *which* relations to perceivers and viewing conditions constitute colors. Still, every relationalist (or, at any rate, every relationalist who has worked out a full account of the nature of color) is committed to some or other answer to this question. But since any answer to the question will amount to a principled and projectible criterion for being a color, this means that every relationalist (who has worked out a full account of the nature of color) can provide for the sort of unity that Socrates complains is lacking from Gorgias's account of the virtues. Even if there is no single response to Socrates that would be acceptable to every relationalist, no relationalist (who has worked out a full account of the nature of color) will be without an answer to Socrates.

9.3.3 Relationalism and Reality

Though I have addressed the complaints about relationalism that Hardin (2004) made explicit, he has revealed in comments on an earlier draft of this chapter a more fundamental worry that underlies the other criticisms discussed, and that, I think, warrants discussion. Namely, Hardin worries that color relationalism is an insufficiently robust form of realism about color, and so would be more forthrightly cast as a species of irrealism.

Unsurprisingly, given his views, Hardin endorses the relationalist's strategy of explaining what unites the instances of colors ultimately in terms of facts about perceiving subjects rather than (exclusively) in terms of perceived objects. Similarly, he commends the relationalist for recognizing the need to appeal to perceiving subjects in explaining central (and arguably essential) properties of colors, such as the similarity relations that obtain between them and their unique-binary structure. From his point of view, however, accepting that colors are constituted in subject-involving ways (and thereby making available these explanatory strategies that he agrees are needed) is tantamount to giving up on realism about color. If colors and their most important features are explained in terms of what is in heads rather than what is in objects, this is for Hardin reason to regard our ascription of colors to objects as a false projection onto those objects of what is not (or is not wholly) there. Hardin urges this point in the following characteristically eloquent passage: "The color structure of external physical objects is, on the relationalist view, parasitic on the color structure that manifests itself in perception. If external objects were the bearers of color, one would suppose that it is they who would bequeath color structure to perception rather than the other way around. Indeed, if that were so, I might be prepared to gesture outwardly and say, 'Lo, the colors!' But it is not so, and thus my gesture must be inward" (Hardin 2007, 4).

But if relationalism is, in fact, a closet irrealism, then Hardin's advice to relationalists is to join him in embracing that irrealism with a clear conscience. Thus, Hardin concludes by extending an olive branch: "Come home, Jonathan. All is forgiven" (Hardin 2007, 5).

Although I am grateful for the forgiveness Hardin offers, and flattered to be invited to join his ranks, I must demur. For I continue to believe that there is an important sense in which color relationalism is a form of realism (and in which Hardin's view is not). Of course, it is not the word 'realism' that matters; I am happy to surrender the word. The more substantive issue is that, according to color relationalism (and not according to Hardin's irrealism), colors *qua* properties that objects bear are among the true inventory of what there is. Indeed, one way to think about the lesson of relationalism about colors (or about *being beautiful*, *being humorous*, and so on) is that, while what there is contains some elements constituted independently of subjects, there are other and equally good elements of what there is that are constituted in a way that depends essentially on subjects. In the particular case at hand, color relationalism teaches us that, contrary to what Hardin supposes in the quoted passage, external objects can be bearers of colors even if the unity and structure of colors is bequeathed to them by (subject-involving) perception.

Hardin's invitation depends on jettisoning from our ontology whatever is constituted in a way that depends on subjects. Therefore, it turns on what will seem to the relationalist an inappropriately restricted accounting of what there is. For this reason, I cannot accept his gracious invitation in good faith.

9.4 Conclusion

I think Hardin is right to reject color physicalism. That view makes commitments that we have empirically motivated reasons for doubting and can react to these doubts only by an unwarranted optimism (section 9.1). On the other hand, I think Hardin is too quick to reject relationalism about color. Relationalism has the virtue that it avoids the unmotivated choices and unwarranted optimism of color physicalism and does justice to the full range of observed perceptual variation with respect to color (section 9.2). Moreover, I have argued (section 9.3) that relationalists can answer Hardin's concerns about the objectivity of the view, its commitment to ungainly pluralities, and its lack of unity, while continuing to recognize colors as *bona fide* elements of what there is. While relationalism might involve a revision of some of our naïve views about color, I propose that the view allows us to hold onto more of what we take to be central, while accommodating the empirical facts, than competing alternatives. In particular, it is a much less radical revision than Hardin's own irrealist view that does away with colors *qua* properties of extramental entities entirely. As a famous philosopher has pointed out, it's not easy *being green*; however, the viability of relationalism shows that it is not impossible.

Acknowledgments

I am grateful to Larry Hardin, John Jacobson, Dom Lopes, Mohan Matthen, and Dana Nelkin for helpful comments on earlier drafts, and to an audience at the 2007 Society for Philosophy and Psychology who heard an earlier version of this paper.

Notes

1. 'Physicalism' strikes me as an inapt label for the view of Hilbert et al., insofar as (1) alternative accounts also allow that colors are physical; (2) the view presently under consideration comes without any substantive characterization of the physical; and (3) the issue about what counts as physical is, in any case, orthogonal to the main lines of contrast concerning the nature of color. That said, I'll adhere to the label preferred by the defenders of the view for present purposes.

2. A surface will reflect some percentage of the light of wavelength λ that falls on it. If we collect the percentages of reflected to incident light for each visible wavelength, we get a function (from visible wavelengths to numbers in the interval [0,1]) that characterizes the disposition of a surface to affect light in the visible range in a certain way. This function is the *surface spectral reflectance distribution*.

Physicalists typically prefer to identify colors with classes of such functions (rather than with the functions themselves) because of the phenomenon of *metamerism*: under any given illumination, an infinite number of surfaces (distinct in their surface spectral reflectance distributions) will be visually indistinguishable for a given observer. What this suggests is that identifying colors with reflectance functions yields an excessively fine-grained individuation of the colors. The move to identify colors with classes of reflectance functions is intended to get around this problem. (An exception to this generalization is Churchland [this volume], who identifies colors with reflectances and is prepared to live with the resultant extremely fine-grained individuation of colors.)

3. These standard assumptions are rejected by, for example, Smith (2002) and Travis (2004), who deny that vision and color experience represent the stimulus and its properties in the relevant way. Since, as noted in the main text, these assumptions are not disputed by the main parties to the current dispute, I'll continue to ignore this issue in what follows.

4. Significantly, the similar structure of the problem that arises in both interpersonal and intrapersonal cases of perceptual variation suggests that we should be seeking a single solution that applies to both sorts of cases.

5. Cohen (2007) elaborates this response and provides some additional reasons for doubting that there is some unknown fact of the matter that would sustain the physicalist's commitment.

6. Hardin, (2004, 35) offers a further, if less-developed, complaint against the color physicalist's insistence that there is an unknown fact of the matter about which perceptual variant veridically represents the color of a given stimulus. Namely, he suggests, if the color physicalist is driven to posit unknown (or even unknowable) facts about colors, it restores just the sort of veil of perception between us and the world that, one might have hoped, a physicalist ontology held the promise of sweeping away, and thereby undercuts one of the initial attractions of the view. I'll put this point aside, since I am not convinced that such epistemological motivations are as essential in motivating color physicalism as Hardin maintains, and since I think physicalism faces other serious problems (as discussed in the main text).

7. Hardin (2006, 342) makes a similar complaint against color relationalism.

8. I'll return to this point in section 9.3.3.

9. Cohen (2004, sec. 4) develops this idea in greater detail.

References

Boynton, R. M. 1979. *Human Color Vision.* New York: Holt, Rinehart and Winston.

Byrne, A., and D. R. Hilbert. 1997a. Colors and reflectances. In *Readings on Color,* vol. 1, *The Philosophy of Color,* A. Byrne and D. R. Hilbert, eds., 263–288. Cambridge, MA: MIT Press.

Byrne, A., and D. R. Hilbert, eds. 1997b. *Readings on Color.* Vol. 1, *The Philosophy of Color.* Cambridge, MA: MIT Press.

Byrne, A., and D. R. Hilbert. 2003. Color realism and color science. *Behavioral and Brain Sciences* 26 (1): 3–64.

Byrne, A., and D. R. Hilbert. 2004. Hardin, Tye, and color physicalism. *Journal of Philosophy* CI (1): 37–43.

Cohen, J. 2003a. Color: A functionalist proposal. *Philosophical Studies* 112 (3): 1–42.

Cohen, J. 2003b. Perceptual variation, realism, and relativization, or: How I learned to stop worrying and love variations in color vision. *Behavioral and Brain Sciences* 26 (1): 25–26.

Cohen, J. 2004. Color properties and color ascriptions: A relationalist manifesto. *Philosophical Review* 113 (4): 451–506.

Cohen, J. 2007. Color, variation, and the appeal to essences: Impasse and resolution. *Philosophical Studies* 133 (3): 425–438.

Cohen, J. 2009. *The Red and The Real: An Essay on Color Ontology*. Oxford: Oxford University Press.

Hardin, C. L. 1988. *Color for Philosophers: Unweaving the Rainbow*. Indianapolis, IN: Hackett.

Hardin, C. L. 2003. A spectral reflectance doth not a color make. *Journal of Philosophy* 100 (4): 191–202.

Hardin, C. L. 2004. A green thought in a green shade. *Harvard Review of Philosophy* 12: 29–39.

Hardin, C. L. 2006. Comments. *dialectica* 60 (3): 341–345.

Hardin, C. L. 2007. Comments on Cohen's "It's Not Easy Being Green." 2007 Society for Philosophy and Psychology, Toronto.

Hilbert, D. R. 1987. *Color and Color Perception: A Study in Anthropocentric Realism*. Stanford, CA: CSLI.

Hubel, D. H. 1988. *Eye, Brain, and Vision*. New York: Scientific American Library.

Hurvich, L. M. 1981. *Color Vision*. Sunderland, MA: Sinauer.

Hurvich, L. M., and D. Jameson. 1957. An opponent-process theory of color vision. *Psychological Review* 64: 384–403.

Johnston, M. 1992. How to speak of the colors. *Philosophical Studies* 68: 221–263. Reprinted in *Readings on Color,* vol. 1, *The Philosophy of Color,* A. Byrne and D. R. Hilbert, eds., 137–176. Cambridge, MA: MIT Press, 1997.

Matthen, M. 2005. *Seeing, Doing, and Knowing: A Philosophical Theory of Sense Perception*. Oxford: Oxford University Press.

McDowell, J. 1985. Values and secondary qualities. In *Morality and Objectivity: A Tribute to J. L. Mackie*, T. Honderich, ed., 110–129. London: Routledge and Kegan Paul.

McGinn, C. 1983. *The Subjective View: Secondary Qualities and Indexical Thoughts*. Oxford: Oxford University Press.

Mothersill, M. 1984. *Beauty Restored.* Oxford: Oxford University Press.

Peacocke, C. 1984. Colour concepts and colour experiences. *Synthese* 58 (3): 365–381. Reprinted in D. Rosenthal, *The Nature of Mind,* 408–416. New York: Oxford University Press, 1991.

Rosenthal, D. 1991. *The Nature of Mind.* New York: Oxford University Press.

Sayre-McCord, G. 1988. The many moral realisms. In *Essays on Moral Realism,* G. Sayre-McCord, ed., 1–23. Ithaca, NY: Cornell University Press.

Smith, A. D. 2002. *The Problem of Perception.* Cambridge, MA: Harvard University Press.

Sober, E. 1982. Realism and independence. *Noûs* 16: 369–385.

Thompson, E. 1995. *Colour Vision: A Study in Cognitive Science and the Philosophy of Perception.* New York: Routledge.

Thompson, E., A. Palacios, and F. Varela. 1992. Ways of coloring: Comparative color vision as a case study for cognitive science. *Behavioral and Brain Sciences* 15: 1–74.

Travis, C. 2004. The silence of the senses. *Mind* 113 (449): 57–94.

Tye, M. 2000. *Consciousness, Color, and Content.* Cambridge, MA: MIT Press.

Zangwill, N. 2001. *The Metaphysics of Beauty.* Ithaca, NY: Cornell University Press.

10 How Can the Logic of Color Concepts Apply to Afterimage Colors?

Jonathan Westphal

10.1 Personal

It is often remarked that since the 1980s, there has been a dramatic increase in the amount of work being done in analytic philosophy on the philosophy of color. One of the causes was the publication of C. L. (Larry) Hardin's groundbreaking work in the area, *Color for Philosophers: Unweaving the Rainbow* (1988, 1993), but apart from causing lots of other people to get into the field, Larry has been the main contributor to it. In addition to that book, I count nineteen very substantial articles about color in the ten years from 1983 to 1993, and there are probably more that I don't know about. I was introduced to Larry at an Eastern Division APA meeting in the mid-1980s by Hilary Putnam, who thought we should get together because we both seemed obsessed with color. I remember that Larry and I had dinner and talked about our plans for the subject. He wanted to develop the philosophy of color in close connection with the science of color, and although I knew that this was very important, and wanted to contribute as much as I was able, my own more primitive interests seemed to point toward Wittgenstein and toward a limit to the science of color in its received form. "OK," I said, "You do the science, and I'll do Wittgenstein and some metaphysics," and instantly regretted it. But for the most part we have stuck to the agreement. I am enormously proud to say that we reviewed one another's books in the same number of *Mind*, his on color and science and mine on color and Wittgenstein. I had no idea, on that evening in Washington, that I was talking to someone who would become the doyen of color philosophy and a model practitioner of the new art, and I would have been even more shocked to have been told that within twenty years, the vague hope I had for the subject would become a reality. The vague hope was that there would be a detailed, responsible, and scientifically well-informed philosophy of color that was entirely sensitive to logical and conceptual issues. To some extent this already existed, for example, in the work of the Australian School (Armstrong, Smart and Campbell), and in the still earlier debate about the persistent problem of color incompatibility, though this debate tended not to include science. But Hardin's work brought the different strands together, and it was the focus of what became in effect an academic specialty. He showed the way to other talented people who are now working in the field so effectively, such as Justin Broackes, Alex Byrne, Jonathan Cohen, Austen Clarke, Don Dedrick, David Hilbert, and Evan Thompson. So I am very happy about the state of the art.

On the other hand, I am not as happy about the way in which color science has been deployed to sort out the problems of color incompatibility that were so important in the history of twentieth-century metaphysics and epistemology, starting with the *Tractatus*. Many quasilogical, semantic, and metaphysical questions having to do with incompatibility come up in color theory, and the problem is so tricky and delicate that I do not believe we have got it quite sorted out yet, despite some marvelous work on the topic. The trouble is especially acute in Larry's favorite haunt, *the subjective*, as it is called, which has flying around in it, among other things, what are known as *afterimages*.

My topic in this chapter is the incompatibility of afterimage colors. Every naïve subject I have talked to who encounters afterimages without prejudice has agreed that they have color; I mention this because I think it is the initial and also the commonsense view. But if a physical account of color incompatibility is right for physical colors, how can we account for the incompatibility of afterimage colors? One can of course give a physiological account based on opponent processes. But this will deliver at best a contingent truth, or a necessity relative only to a particular retinal structure and postretinal neural coding.

10.2 Wittgenstein's Puzzle Propositions

10.2.1 Introduction and the Puzzle Propositions and Questions

In *Remarks on Colour*, Wittgenstein (1978) wrestled with certain questions, very important ones in the development of analytic philosophy, about the impossibility of certain colors and color effects. I call these the *puzzle questions*. I call the negative existential and modal propositions corresponding to the puzzle questions the *puzzle propositions*. Some characteristic examples of the puzzle questions, with their corresponding puzzle propositions, are

(10.1) Why can there be nothing transparent and white?
There can be nothing transparent and white.

(10.2) Why can there be no reddish green?
There cannot be a reddish green.

(10.3) Why can there be no brown light?
There can be no brown light.

(10.4) Why can there be nothing glowing and gray?
Nothing can be glowing and gray.[1]

It is not always entirely easy to see what Wittgenstein's own answers to the puzzle questions were, and it may be that they were all variants of the thought that we should somehow learn not to ask the puzzle questions and concentrate on keeping our own conceptual houses in order. Even if this is the right sort of reading of *Remarks on Colour*, the project of keeping one's concepts straight is a partially positive and at least an intermittently systematic one, and I have felt justified, to some extent, in searching for the most straightforward possible physical solutions, while remaining sensitive to alternative possibilities of therapy. In this paper I want to begin to extend these physical solutions to what had seemed to me, up till now, to have almost the force of a counterexample: afterimage colors.

10.2.2 Proposed Solutions and Their Logic

In *Colour: A Philosophical Introduction*, I gave answers to Wittgenstein's puzzle questions, which had the advantage of being all the same sort. They all involved the relation between the illuminating light that is incident on the colored object and the light reflected and absorbed. For (10.1) and (10.2), the answers are roughly as follows.

1. (10.1) A surface cannot be both white and transparent, because if it is white, then it reflects all or almost all of the incident light, diffusely, and transmits none, and if it is transparent, it transmits all or almost all of the incident light, and reflects none, or relatively little. If it transmits all or almost all of the incident light, then it does not reflect all or most of it. It cannot be both white and transparent, because it cannot both transmit (and not reflect) all or most of the incident light (it is transparent), and reflect (and not transmit) all or almost all of the incident light (it is white).[2] The logic of the analysis is illustrated in figures 10.1 and 10.2, with the arrows representing implications.
2. (10.2) For the impossibility of anything red and green, the analysis is similar, but restricted to ranges of red and green. If a surface is red, then
(a) there is a red light of which it reflects all or most, and
(b) there is a light, having a color complementary to the color of the red light of which it reflects all or most, of which it absorbs all or most (figure 10.3).
3. If a surface is green, and complementary to the red of the light described above in (a), then
(a) there is a green light, of which it reflects all or most, and
(b) there is a light, having a color complementary to the color of the green light of which it reflects all or most, of which it absorbs all or most.

But the color of the light complementary to the color of the green light described in 3(b) is the color of the red light described in 1(a; figure 10.4). So if a surface is the

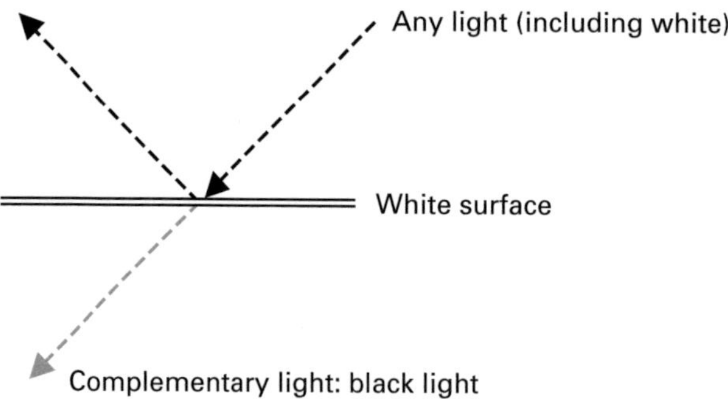

Figure 10.1
Perfectly white surface.

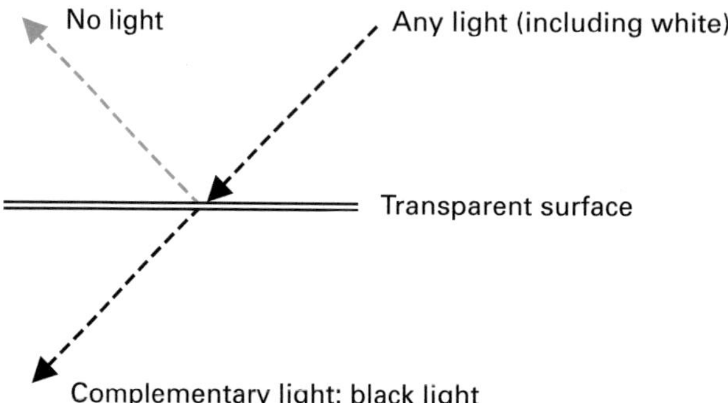

Figure 10.2
Transparent surface.

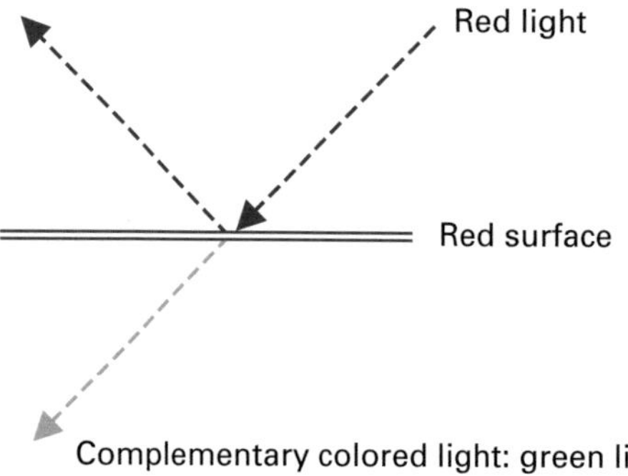

Figure 10.3
Red surface.

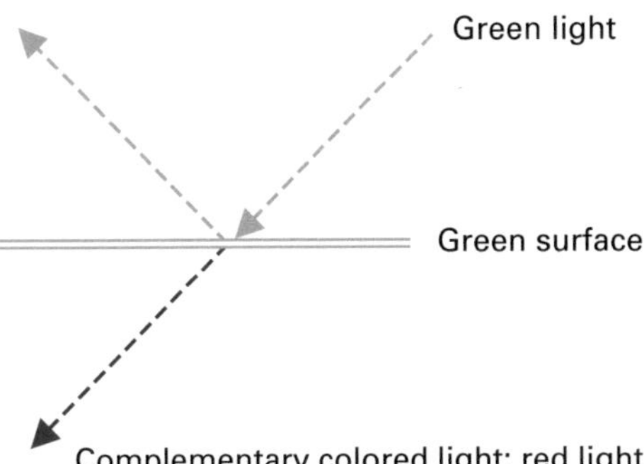

Figure 10.4
Green surface.

complementary red and green described, then there is a red light of which it reflects all or most (from 1[a]), and there is a red light (the same red light) of which it absorbs all or most (from 2[b]).

If we add the simplifying principle (ignoring transparency and other factors) that if a surface reflects some given light, then it does not absorb that light, we can derive the contradiction that, if a surface is red and green of the two kinds described in 1 and 2, then
(a) there is a red light of which it reflects all or most;
(b) there is a red light (the very same red light) of which it does not reflect all or most.

I must point out that here I am giving an analysis that is directed toward the impossibility of "Nothing can be red and green *colored*, all over, at the same time," rather than Wittgenstein's actual puzzle proposition, "There cannot be a reddish green." I do not feel that I understand the relationship between the two propositions at all well, and so, reluctantly, I will leave the topic hanging.

10.3 Reflectance Physicalism

Is the account I have just given a form of physicalism, specifically *reflectance physicalism*, as it has been called (Byrne and Hilbert 2003, section 3, 3ff)? Reflectance physicalism is the view that colors should be identified with spectral reflectance curves. In *Color: a Philosophical Introduction*, I objected strenuously, for a variety of empirical reasons, to physicalism in the Australian form, which identifies color with the wavelength of reflected light (Westphal 1991, ch. 6). I also anticipated and declined to accept reflectance physicalism, arguing that the x-axis for spectrophotometric curves for colored surfaces and materials should be a spectrum labeled with the *psychological* color terms, not with wavelength values, and that this is the only way to preserve Wittgenstein's necessities (Westphal 1991, 36).[3]

One of the reasons I chose not to accept reflectance physicalism in its entirety is that Wittgenstein's puzzle propositions are necessary, and the identification of colors with reflectance and absorption spectra is contingent. If the reflection-absorption spectrum (RAS) of red is what explains why red cannot be greenish, and it is possible for red to have the RAS that in fact belongs to green, then it is possible not only for the red we see—phenomenal red, assuming there is some other kind—to be greenish, but also for red to be green. All we have to do to make that happen is to keep red visibly the color it is and change its reflection-absorption (RA) curve. As I see it, the case parallels the identification of water with H_2O. It is a necessary truth that pure water is clear. Suppose that being H_2O is what explains the clearness of water. But then it turns out that water is only contingently H_2O. It could also quite possibly have had a structure of D_2O or Z_2O or whatever. And suppose that it does not follow from water being Z_2O that it is a clear liquid. What then? Or again, if it is a necessary truth that triangles have straight sides, and suppose it is only *contingently* true that a triangle is a three-sided plane figure made by joining straight-line segments, then the necessary will be after all contingent. It is certainly a tautology, and therefore necessary, that red is whatever it is, but it is not necessary that red, or rather being red, is having a particular set of RA curves, that is, those which happen to peak at the long-wave end of the spectrum. Or is it?

10.4 A Problem for the Necessities

10.4.1 The Existence of Afterimage Color

The analyses given above (in section 10.2.2) face a very interesting and important difficulty. Suppose they are correct. Color incompatibility of the relevant sort (the impossibility of anything red and green, and, with a bit of luck, of reddish green) is explained by RA spectra, suitably "phenomenalized" with a psychological x-axis.[4] (I call the account given in Westphal 1991 and in figures 10.1–10.4 the P-RAS theory, for phenomenalized reflection-absorption spectra.) Now we can begin to form a link between the RA spectra and the color terms in the analysis, which allows us to make good the claim that the puzzle propositions are necessary, as in the figures above. But there is a problem. Suppose that afterimages have color. What explains incompatibility among colors is phenomenalized RA spectra. But afterimages do not have RA spectra. They do not literally reflect or absorb electromagnetic radiation. So there are no logical relations between afterimage colors, and afterimages can be incompatible colors; the same is true of all psychological colors. Yet the necessities of color logic do extend to the psychological colors, *pace* Crane and Piantanida.[5]

The central difficulty is that with afterimages, there is no light in the sense in which "light" is used in physics, and there are no reflecting and absorbing surfaces. Though psychophysics tells us that afterimage color does vary in the standard psychological

HSB (hue, saturation, brightness) dimensions, afterimages do not reflect, absorb, emit, or transmit light. Nor, apparently, is there any illumination by which we see them. The P-RAS explanations of the necessities depend on relations between light (the word taken in an objective sense) and surfaces (also taken in an objective sense). But light (the word taken in a subjective sense) does not exist, and nor do surfaces (taken in the same sort of sense) in the world of afterimages.

10.4.2 A Solution to the Problem: *As If*

Consider, by way of the beginning of an argument for a psychological P-RAS solution, an alternative. One common reaction to the question of whether afterimages have color is very well described by J. J. C. Smart (1963): "What is going on in me is like what is going on in me when my eyes are open, the lighting is normal, etc. etc., and there really is a yellowish orange patch on the wall" (94). When I see a yellowish orange afterimage, in this view, on one reading, it seems to me *as if* I were seeing a yellow-orange patch on the wall.

In Westphal (1991, 109), I offered a fierce and perhaps unfair criticism of this suggestion. I took it to mean that what it seems to me is going on when there is an orange afterimage before me is *phenomenologically* just like what it seems to me is going on when there really is an orange patch on a wall in front of me. I still think that I was right about the phenomenological differences between the two cases, but now it seems worth saying that on the P-RAS view, what is going on in me when I see the afterimage is in the crucial respect like what is going on in me when I see the orange patch on the wall. There is just the same sort of relationship between the perceived illumination environment and the spectral profile of the image. As I shall try to show in a minute, from this point of view, Smart was right after all.

Part of the criticism I made of Smart's view (that when I see a yellowish orange afterimage it is *as if* I see a yellowish orange patch on the wall) was mistaken. But for the purpose of explaining the various color impossibilities with respect to afterimage color, there remains a difficulty. Suppose there is a strange kind of imaginary or hallucinated dagger with which one is actually able to stab people, unlike Macbeth's, and that is how it comes to have "gouts of blood" on it. How can we explain the action of the dagger? It will not do to say that what Macbeth hallucinates is an *as if* dagger, that *as if* daggers are daggers, and that daggers can stab, so that what Macbeth hallucinates can stab. (Macbeth knew enough of the philosophy of hallucinations to ask, "Art thou not, fatal vision, sensible to feeling as to sight? Or art thou but a dagger of the mind, a false creation, proceeding from the heat-oppressed brain?" (II.1)

The plain difficulty is that the "as if" operator does not preserve logical relations. The "as if" yellowish orange patch will not inherit the physical impossibilities offered in a physical explanation from the real yellowish orange patch it resembles, unless there is some further reason to suppose that the "as if," or psychological color, and the physical color exist in some sort of common structure of colors.

10.4.3 A Better Solution to the Problem?

Are afterimages, then, a counterexample to P-RAS explanations? My answer is that they are not.

The first and one of the more important things to note is that what is impossible in the puzzle propositions is typically a color, and *color*, along with *hue*, *saturation*, and *brightness*, is a purely psychological term, "purely" here intended to exclude psychophysical terms. So paradoxically, what makes the problem for afterimages (that they lack RA spectra) is *also* true of the colors as they appear in cases of regular external color perception. When we see an external red surface illuminated with greenish light, say, and looking a bit black, that is just as much an image in a surround, which is to be described in purely psychological terms as any afterimage is.

If we say that light is electromagnetic radiation, then the physics of light is as irrelevant to the incompatibility of regular physical colors as it is to afterimages! The dialectic goes somewhat as follows.

We start by noting that afterimages are not associated with physical light, whereas external perception and external images are.

Stage 1: Problem

Color afterimages	External color images
x	Light

The next thing to note is that external color images are not associated with light either, because *color* is a psychological term. So, paradoxically, we get the situation in which both kinds of images (remembering that it is these images that are the carriers of color)[6] lack the concepts we need for the logical explanations: *light, reflection, absorption* and so forth.

Stage 2: Negative Solution

Color afterimages	External color images
x	*x*

Clearly, however, there is a difference between the relation of the afterimage colors to physical light and the relation of the external image colors to light. The difference is that the external color images are caused and explained by the presence of RA spectra.

Stage 3: Primitive Physicalism

Color afterimages	External color images
x	RA spectra

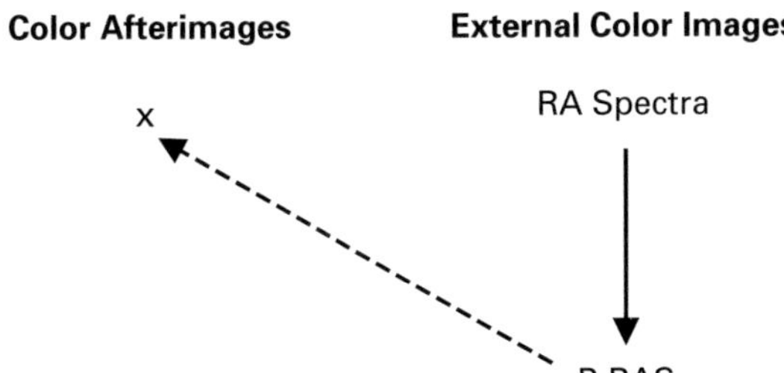

Figure 10.5
Stage 4: Enlightenment.

The next step is to realize that the logic of the color relations of the external image cannot be imported into the required explanations, except by a correlation (see figure 10.5). And this destroys the modality of the puzzle propositions. The presence of RA spectra is irrelevant.

The final step I want to take is to phenomenalize the physical RA spectra. This will keep the puzzle propositions necessary for physical, or external, colors. But now we can look for P-RA spectra even in the case of afterimages as well. For the P-RAS explanations used only psychological terms.

And it turns out that there are analogues for physical light and physical surfaces, and for the concepts of absorption and reflection in the case of the afterimages. What happens in a case of regular external color perception is that the eye is aware of the illumination and compares it to the surface in which it is interested. From this it determines the color of the surface. So from the relevant point of view—color perception—*light is illumination*, and illumination is a sampled standard. But from this point of view, there *is* a surrounding illumination that bathes the perceived afterimage. An afterimage that is seen against this illumination will be some particular color, whereas seen within or under another illumination, it will assume a different color. There is an exact analogy here with the change of perceived size of afterimages against backgrounds at apparently different distances. Suppose I see an afterimage, with my eyes closed, that is reddish in color, and surrounded by a light green illumination as I am calling it. But now the illumination changes; it turns a deeper blue-green, say. The result is that the red image becomes an even sharper red, because it darkens the given illumination even more. So the phenomenology of light and darkening (the analogues of reflection and absorption) do exist in the domain of afterimages.

The transparency of an image is a harder problem, but important work has been done here in psychophysics by and others (beginning with Metelli 1974; see also Westland et al. 2002).

What I have wanted to do is show the existence and availability of a set of phenomenological concepts that can be used to describe afterimage color. These phenomenological concepts are the very ones that are relevant to the physical explanations that can be given of Wittgenstein's puzzle propositions. And Larry Hardin was certainly not wrong in thinking that the subjective world has something important to teach us about color. I think that he would agree that colors can be subjective (because he thinks that they *are* subjective) and yet conform to the logical "laws" that make the incompatibilities between colors exist. I would like to say that I hope that I have given an explanation of subjective color incompatibilies that he can accept. I know that in fact his subjectivism is so strong as not to allow any logical relations of necessity between colours—if subjectivism is true, then there are no such relations. To me that is proof of the falsity of subjectivism, as there obviously are logical relations. To him it will appear as a proof that there are no logical relations, since subjectivism is true. We still seem to be more or less where we were when Putnam introduced us.

Notes

1. Wittgenstein (1978, 1.19 [10.1]; 1.9–1.14 [10.2]; 3.60 [10.3]; 1.36 [10.4]).

2. For simplicity of exposition, I have ignored *absorption* in the above.

3. The development of *productance physicalism* is important because, as Byrne and Hilbert have shown, it meets some of the worries raised about the narrower reflectance physicalism.

4. David Wiggins has said that this use of the "phenomenalized," "phenomenal" etc., is "truly terrible," I think because of the confusion with phenomenalism in its usual sense (that every categorical proposition about a material object can be translated without loss into a conditional proposition about phenomena). I don't disagree, but what I intended was a comparison between (a) the logical construction of colors (or concepts of colors, rather) out of concepts embodying descriptions of reflection and absorption spectra with psychological terms, and (b) the logical construction of material objects out of psychological phenomena.

5. I merely assume this. If an argument is wanted, here is one from Justin Broackes:

> We might instead have been built in such a way that after looking at something red, we actually had an image somehow of red and green mixed or superposed, quite puzzlingly (as in the Crane and Piantanida reports). But in actual fact we aren't built like that, so the after-image of red is simply green. (Or actually, more accurately, green-blue.) When people say "What about the Crane and Piantanida cases?" I say "I'm not claiming that nothing can ever in some sense *look* red and green together. That may well in fact happen. I'm claiming that we work with concepts of red and green that are such that they seem to exclude something's *being* red and green. Think of a Max Escher print: two places x and y can be such that x *looks* to be higher than y, and that x *looks* (when you consider other aspects of the picture) to be lower than y. But our concepts of higher and lower are such that they seem to exclude x actually *being* ("in the world," so to speak) both higher and lower than y.

So I'm not convinced that we should expect the relations among properties to be perfectly captured in the relations among the *appearances* of properties.

6. I mean almost nothing by this remark, using *image* only in the sense in which we speak of an image being preserved, in specular reflection, or an image as part of a photograph, in the *OED* sense of "An optical appearance or counterpart of an object, such as is produced by rays of light either reflected as from a mirror, refracted as through a lens, or falling on a surface after passing through a small aperture." In this sense, a shadow is an image. And by *carrier*, I mean the following. If a square shape, for example, is a "carrier" of color, then the shape has the color, just as the color, one might argue, has the shape.

References

Byrne, A., and D. R. Hilbert. 2003. Color realism and color science. *Behavioral and Brain Sciences* 26: 3–21.

Crane, H. D., and T. P. Piantanida. 1983. On seeing reddish green and yellowish blue. *Science* 221: 1078–1080.

Hardin, L. 1988. *Color for Philosophers: Unweaving the Rainbow*. Indianapolis, IN: Hackett.

Hardin, L. 1993. *Color for Philosophers: Unweaving the Rainbow*. Expanded ed. Indianapolis, IN: Hackett.

Metelli, F. 1974. The perception of transparency. *Scientific American* 230 (4): 90–98.

Smart, J. J. C. 1963. *Philosophy and Scientific Realism*. London: Routledge.

Westland, S., O. da Pos, and C. Ripamonti. 2002. Conditions for perceptual transparency. In *Human Vision and Electronic Imaging. VII. Proceedings of the SPIE*, B. E. Rogowitz and T. N. Poppas. ed.

Westphal, J. 1991. *Colour: A Philosophical Introduction*. Oxford, Blackwell.

Wittgenstein, L. 1978. *Remarks on Colour*. G. E. M. Anscombe, ed. Oxford: Blackwell.

III Color Blindness

11 How Do Things Look to the Color-Blind?

Alex Byrne and David R. Hilbert

GENTLEMEN,—Colour-blindness is not a good name for the condition to which it is applied
—F. W. Edridge-Green, 1911

Our question is, how do things look to the color-blind? But what does that mean?

Who are the "color-blind"? Approximately 7 percent of males and fewer than 1 percent of females (of European descent[1]) have some form of inherited defect of color vision and as a result are unable to discriminate some colored stimuli that most of us can tell apart. (*Color defective* is an alternative term that is often used; we will continue to speak with the vulgar.) Color-vision defects constitute a spectrum of disorders with varying degrees and types of departure from normal human color vision. One form of color-vision defect is *dichromacy*: by mixing together only two lights, the dichromat can match any light, unlike normal trichromatic humans, who need to mix three. The most common form of dichromacy (afflicting about 2 percent of males) is red-green color blindness, or red-green dichromacy, which itself comes in two varieties. A red-green dichromat will not be able to distinguish some pairs of stimuli that respectively appear red and green to those with normal color vision. For simplicity, we will concentrate almost exclusively on red-green color blindness.[2]

In a philosophical context, our question is liable to be taken in two ways. First, it can be straightforwardly taken as a question about visible properties of external objects, like tomatoes. Do tomatoes look colored to dichromats? If so, what color? Second, it may be interpreted as the more elusive—although closely related—question of "what it's like" to be color-blind. Forget about tomatoes—what is a dichromat's *experience* like?[3] This chapter addresses the first question and, for reasons of space, does not explicitly address the second. Put another (arguably equivalent) way, we are asking what colors are *represented by* a dichromat's experience.

Having now identified the "color-blind," and the straightforward way in which our question should be taken, a further point of clarification might be helpful. Imagine a bright blue car parked in an underground garage illuminated by orange lighting. Bright blue objects under this lighting look quite distinctive. Those accustomed to the garage can tell by looking whether something is bright blue. They may even say, pointing to the car, "That looks bright blue." Those with more linguistic scruples will perhaps prefer instead to say, "That looks *to be* bright blue," or "That looks *as if* it is bright blue." This distinction, of course, is sometimes explained in terms of various "senses" of "looks." In the alleged *phenomenal* sense of "looks," the car looks bright blue in sunlight, but bluish black in the garage.[4] Whether or not "looks" in fact has a special phenomenal sense, there is clearly an important difference between viewing the car in sunlight, and viewing it in the garage, even if one is inclined to say that it looks bright blue both times. "Looks" in our question is to be stipulatively interpreted so that the following is true: in the garage, the car looks bluish-black and does not look bright blue.

With our question clarified, we can now briefly outline the two main candidate answers. On what we will call the *reduction view*, a red-green dichromat enjoys a reduced range of normal color appearances. On the *standard* version of the reduction view (which is orthodoxy, if anything is), things look yellow and blue to dichromats, but not red or green. On the alternative *alien view*, a red-green dichromat does not see yellow or blue (or, for that matter, red or green), but sees some other colors entirely; as Hardin puts it, "what he sees is incommensurable with what we see" (1993, 146). Deciding between these views turns out to be no simple matter.

The dispute between the reduction and alien views, we should emphasize, turns on *hue*, not on saturation or lightness. There is no reason to believe dichromats are blind to saturation or lightness; in particular, we will assume throughout that dichromats see completely desaturated colors—white, black, and gray.[5] (Following common practice, we sometimes use "color" to mean *hue*: the context should make this clear.)

Not surprisingly, color scientists have addressed the question of this chapter, and at least some are skeptical about the prospects of answering it. A (slightly dated) example is provided by Boynton's excellent 1979 text, *Human Color Vision*, which includes a section titled, "What Do Red-Green Defective Observers Really See?" After a couple of pages of discussion, Boynton ends by saying that "the issue of what dichromats 'really' see probably can never be fully resolved" (1979, 382; see also Kaiser and Boynton 1996, 456).[6] In the optimistic camp, a more recent paper in *Nature*, "What Do Color Blind People See?" (Viénot et al. 1995; see also Brettel et al. 1997), contains color illustrations purporting to show to normal subjects what a picture of flowers would look like to dichromats. Similar illustrations can be found on many websites.[7]

The reduction and alien views lend support to different answers to the question of veridicality: do red-green dichromats see the true colors of things, or do they suffer from many color illusions? John Dalton, the great British chemist, produced the first systematic investigation of color blindness and was red-green color-blind himself (hence "Daltonism"). In the first published account of red-green dichromacy, Dalton reported that the pink flower of *Geranium zonale* "appeared to me almost an exact sky-blue by day" (1977, 520).[8] On the face of it, the flower is simply pink, and not also sky blue. So, if we take his words at face value, Dalton was suffering from a color illusion. Alternatively, perhaps the flower appeared to be some *other* ("alien") color to Dalton, and moreover one that is not a contrary of pink. On this view, there need be no illusion, although Dalton did make a (perhaps understandable) error in using an ordinary color word to describe the flower's appearance. (We will return to Dalton later, in section 11.1.3.)

In the next section, we supply some background on color vision and color blindness and examine four pieces of evidence bearing on our question: similarity judgments, the use of color language, the opponent-process theory of color vision, and (rare cases of) unilateral and acquired dichromacy. Similarity judgments and color language are of little help; the opponent-process theory and the two unusual forms of dichromacy apparently support the standard reduction view. In section 11.2, we examine the reduction view in more detail, including the issue of veridicality just mentioned. In section 11.3 we turn to the alien view and evaluate two arguments for it. Section 11.4 returns to the reduction view, which we argue needs revision. Rather surprisingly, when the reduction view is appropriately amended, it turns out to *be* a version of the alien view!

11.1 Background: Color Vision and Color Blindness

11.1.1 Trichromacy and the CIE Chromaticity Diagram

The normal human eye contains three kinds of *cones*, photoreceptors used for color vision. (The rods, photoreceptors used for vision in dim light, play no significant role.[9]) Cones contain *photopigments* that enable the cone to respond to light. The three cone types are distinguished by their respective photopigments, which are sensitive to different parts of the spectrum. The L cones are maximally sensitive to long-wavelength (yellowish green) light, the M cones to middle-wavelength (green) light, and the S cones to short-wavelength (bluish violet) light. Although cones are more likely to respond to light of certain wavelengths than others, this difference can be compensated by a difference in intensity. For example, if one M cone is stimulated by low-intensity light of wavelength 530 nm (close to its peak sensitivity), and another M cone is stimulated by an appropriately selected high-intensity light of wavelength 450 nm, the two cones will respond identically. The individual cone responses, then, do not contain any information about wavelengths (other than information about very broad bands). Wavelength information, and therefore information about the colors of things, is obtained by comparing the outputs of the different cones.

Consider a color-matching experiment: the observer views an illuminated disc divided in two horizontally. The *test light* appears in the upper semicircle; the appearance of the lower semicircle is the product of three *primary* lights. The observer's task is to adjust the individual intensities of the three primary lights so that the two semicircles match. If the three primaries have been chosen so that no two can match the third, then the observer will always be able to match the test light. Put more precisely: sometimes a match will not be achieved simply by adjusting the three primaries, but will require transferring one primary so that it mixes with the test light, not with the other two primary lights. Since two primary lights will not suffice for a match, normal human color vision is *trichromatic*.

This empirical result about matching is a consequence of there being three cone types (*retinal* trichromacy, as opposed to the just-mentioned *functional* trichromacy), together with some other simple assumptions.[10] It allows us to represent any test light using three coordinates, specifying the intensity of the primaries required to match the test light. A test light might be represented by (–2, 1, 3)—1 unit of the second primary, 3 of the third, and –2 of the first (that is, 2 units of the first primary added to the test light). It is convenient—or was in 1928 when the matching data were first obtained—to get rid of the negative numbers by a linear transformation. Any test light is then represented by three positive coordinates, which we can think of as specifying the intensities of three "imaginary" primaries required for a match.

Since every test light can be assigned three coordinates, it has a location in a three-dimensional space, with the axes representing the intensity of the corresponding primary. If the (imaginary) primaries are those chosen by the Commission Internationale de l'Eclairage (CIE), and if the matching data is taken from a certain small group of Englishmen, viewing a stimulus of 2° (about the size of a quarter at arm's length) against a dark neutral background, then the resulting space is the *1931 CIE tristimulus space*, with every test light assigned three *tristimulus coordinates*. Because wavelength information is lost at the retina, due to the broad-band response of the cones, numerous physically different lights will have the same tristimulus values—the phenomenon of *metamerism*.

The tristimulus space is sometimes called a *color* space, which can be misleading. It is a space of *lights* (or stimuli more generally[11]), not colors. (The Natural Color System space is an example of true color space.[12]) Further, there isn't a natural way of transforming the tristimulus space into a color space. With the assumption that every light with the same tristimulus values has the same color, the tristimulus space could be converted into a literal color space by taking the items in the space to be the colors of the lights, whatever they are. But this assumption is not at all plausible. More important, the tristimulus space is not any kind of color *appearance* space—the space does not encode how a light would *look*. What is encoded is whether two lights will *match*—at least for the majority of us who approximate the "standard observer": the lights will match, or appear the same, just in case they have the same (or, more exactly, very close) tristimulus coordinates.

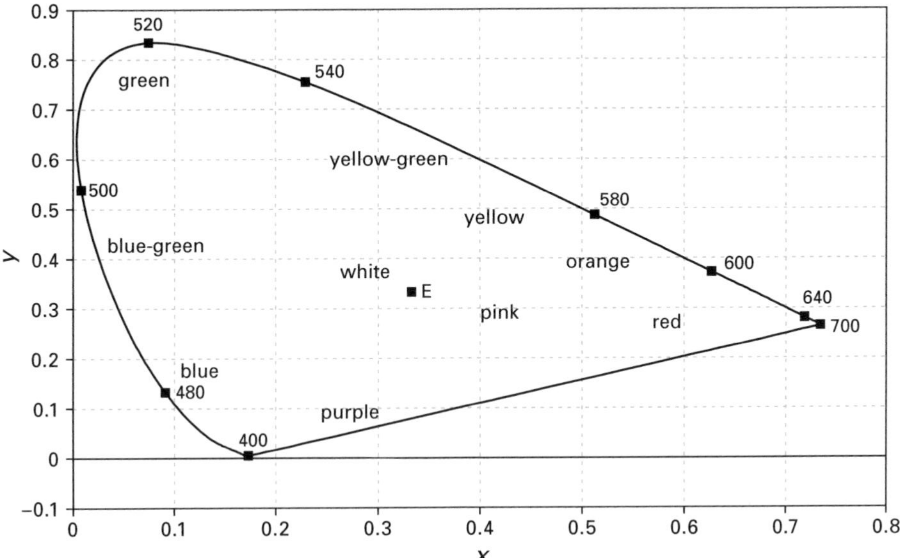

Figure 11.1
1931 CIE chromaticity diagram. See plate 14 for a color version of this figure.

Consider a point (X, Y, Z) in the tristimulus space, and another point that lies *n* times further out on the same ray from the origin, (*n*X, *n*Y, *n*Z). The corresponding second light needs primaries in the same ratio for a match, but *n* times more of each. Bearing in mind the three dimensions of color appearance (hue, saturation, and lightness), one might expect that the second light would appear brighter (lighter) than the first, with the same hue and saturation. This is indeed the case, to a fair approximation. Hence, if the tristimulus space is reduced by one dimension by mapping each ray to a single point, we obtain a two-dimensional space in which stimuli with different locations will appear to differ in hue, saturation, or both. With a particular choice of axes, this space is displayed on the CIE *chromaticity diagram*, which is often annotated with the approximate color appearance of the stimuli, as in figure 11.1 (plate 14).

The *chromaticity coordinates* (x, y) are derived from the tristimulus coordinates (X, Y, Z) as follows: $x = X/(X + Y + Z)$, $y = Y/(X + Y + Z)$; (x, y) specifies the *chromaticity* of the corresponding light. The spectral (single-wavelength) lights are arranged around the horseshoe-shaped line, from spectral red on the bottom right (starting at 700 nm, the approximate long-wavelength end of the visible spectrum), green on the top, and violet on the bottom left (ending at 400 nm, the short-wavelength end). White is roughly in the middle: the chromaticity coordinates of the "equal energy white" light (marked *E*) are (1/3, 1/3).[13] The lights become progressively less saturated as they approach white, as illustrated by the location of pink; the spectral lights are therefore maximally saturated. (In addition to the compression due to ignoring brightness, metamerism ensures a many-one relation between stimuli and chromaticity.) The chromaticity diagram has the nice property that the chromaticity of any mixture of two lights lies on the line connecting the chromaticities of the two lights. Since any light is a mixture of spectral lights, this means that all stimuli lie in the region spanned by the horseshoe and a line (the "purple" line) connecting its two endpoints. The fact that many chromaticity coordinates have no corresponding physical stimulus is due to the use of imaginary primaries. Bearing in mind the dangers of labeling the chromaticity diagram with color names, it is helpful to think of the imaginary colors that lie outside the spectrum locus and purple line as more saturated instances of the hues that are found within them.

11.1.2 Confusion Lines

As we mentioned at the beginning, there are two forms of red-green dichromacy. These are *protanopia* and *deuteranopia*: protanopes lack the L photopigment, while deuteranopes lack the M photopigment. (Tritanopes, the third kind of dichromat, lack the S photopigment.) Consider a protanope. As one might expect, 700 nm light at the far (red) end of the spectrum stimulates the L cones and negligibly stimulates the M or S cones; 700 nm light is (near enough) the *L-cone primary* (or *fundamental*). Suppose we take a spectral light (say, one of 520 nm) and mix it with 700 nm light. The protanope will not be able to distinguish the second light from the first (provided the second contains the same intensity of 520 nm light as the first), because the 700 nm light will not affect the M and S cones, and the protanope has no L photopigment. So, given the way mixtures are represented on the chromaticity diagram, a line drawn from 700 nm to 520 nm specifies the chromaticities of stimuli that a normal observer can distinguish, but will match for a protanope—one of the protanope's *confusion lines*. Repeating this procedure for other spectral lights results in an array of confusion lines all converging on 700 nm, the *copunctal* (or *confusion*) *point*, as illustrated in figure 11.2 (plate 15).

One of the four confusion lines shown is of particular interest. It passes through the white point *E* and ends in the vicinity of 495 nm, marked *N* on the figure. Therefore a protanope will not be able to distinguish this region of the spectrum from a neutral white stimulus—495 nm (or thereabouts) is the protanope's *neutral point*.

If two stimuli lie on the same confusion line, they will match (equating for brightness). The converse is also true. Suppose that two lights 1_1 and 1_2 lie on different confusion lines: 1_1 from 700 nm to λ_1, and 1_2 from 700 nm to λ_2. Then the first light has the same chromaticity as a mixture of light of wavelength 700 nm and λ_1, and the other as a mixture of 700 nm and λ_2. Now, λ_1 and λ_2 will affect the M and S cones differently; otherwise they would lie on the same confusion line. Adding an amount

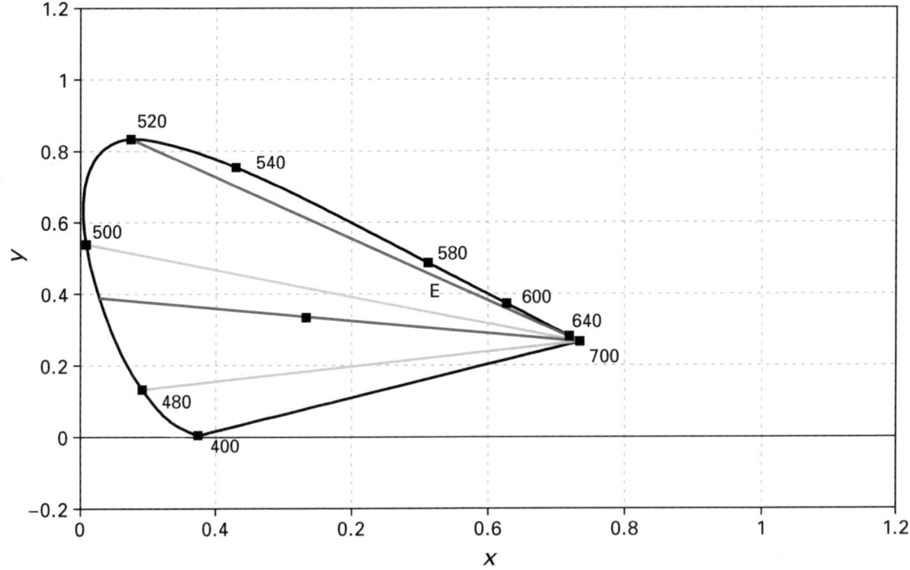

Figure 11.2
Four protanope confusion lines. See plate 15 for a color version of this figure.

of 700 nm light will make no difference, since that will not affect either the M or the S cones. So the protanope will be able to distinguish 1_1 and 1_2. Hence, two stimuli lie on the same confusion line if and only if they match.

Similarly, deuteranopes and tritanopes will have their own distinctive confusion lines and neutral points. Since light of 400 nm stimulates the S cones but not the M or L cones, the copunctal point for a tritanope will be (near-enough) at 400 nm. However, because there is no light that stimulates just the M cones, the copunctal point for a deuteranope lies outside the region of real colors, at about (0.9, 0): the M-cone primary is, like the CIE primaries, "imaginary." A deuteranope's neutral point is a little greener than a protanope's, at about 500 nm.

The theoretical basis for the discrimination data just presented is entirely at the receptor level, with the following very simple assumption connecting receptor activity and discrimination. In the sort of matching experiments described, two stimuli are indiscriminable just in case they have the same effects on the receptors. (For simplicity, we are ignoring the phenomenon of "just noticeable differences.") No assumptions were made about the way color information is processed in the brain; in fact, even the dimensions of color appearance were ignored. Admittedly, the CIE chromaticity diagram has some labeled color regions to aid understanding, and it is intended to differentiate between stimuli that differ in hue or saturation, but none of this information is necessary to predict matches.

Clearly, then, discrimination data cannot predict how things *look* to a dichromat. For all the discrimination data say, a green 510 nm light looks to a dichromat exactly as an orange 620 nm light looks to a normal trichromat, or looks gray, or even looks to have a hue that normals never see. To make progress, different kinds of data are needed.

11.1.3 Similarity

Perceptual tasks involving arranging stimuli by similarity can help, but only up to a point. The Farnsworth-Munsell 15-D color blindness test consists of fifteen differently colored caps of equal lightness and saturation, plus one reference cap. If asked to arrange the caps in order according to color, starting with the reference cap, a red-green dichromat will choose a sequence that interleaves some red and green caps, and that approximates their order along the yellow-blue dimension (Jameson and Hurvich 1978, figure 2). Shepard and Cooper (1992) asked subjects to arrange cards displaying pairs of colors in order of the similarity between the pairs. Multidimensional scaling produced the familiar color circle for normal subjects and (approximately) a line for dichromats.[14] These sorts of result suggest that the dichromat's color space has two opposing hues, and that *if* these hues are familiar ones, they are yellow and blue. But that does not decide between the reduction and alien views.

11.1.4 Color Language

One obvious way of getting information about perceived hue is simply to ask the color-blind how things look. If, as the reduction view has it, satsumas look yellow to deuteranopes, not orange, won't they admit this when questioned? Perhaps surprisingly, the matter is not this simple.

The color-blind, like the blind and those with other sensory deficits, learn native languages in their entirety, not some subset with the terms corresponding to their deficit subtracted. In the case of color blindness, where many individuals with the disorder are unaware of their condition, not only are a full set of color terms learned, but they are applied to objects on the basis of how those objects look. That is, color terms like "red" and "blue" function for the color-blind much as they do for those with normal trichromatic vision. It isn't true, as Hacker claims, that the red-green color-blind "cannot use ["red" and "green"] correctly in the way that we do, and will characteristically eschew their use" (1987, 152). Of course, the color-blind will sometimes mislabel the colors of things, but they will call tomatoes "red" and grass "green." And neither do their everyday mistakes with color terms expose them as especially noteworthy linguistic deviants, contrary to Hacker's suggestion: the normally sighted also misapply terms for shades, especially relatively unusual ones like "teal" and "puce."[15]

We mentioned Dalton in our introductory section; his description of his experience with English color language is instructive:

> In the course of my application to the sciences, that of optics necessarily claimed attention; and I became pretty well acquainted with the theory of light and colours before I was apprized of any peculiarity in my vision. I had not, however, attended much to the practical discrimination of colours, owing, in some degree, to what I conceived to be a perplexity in their nomenclature . . . With respect to colours that were *white, yellow,* or *green,* I readily assented to the appropriate term. *Blue, purple, pink,* and *crimson* appeared rather less distinguishable; being according to my idea, all referable to *blue.* I have often seriously asked a person whether a flower was blue or pink, but was generally considered to be in jest. Notwithstanding this, I was never convinced of a peculiarity in my vision, till I accidentally observed the colour of the flower of the *Geranium Zonale* by candle-light, in the autumn of 1792. The flower was pink, but it appeared to me almost an exact sky-blue by day; in candle-light, however, it was astonishingly changed, not having then any blue in it, but being what I called red, a colour which forms a striking contrast to blue. (Dalton 1977, 520)

Notice that Dalton refers to a broad range of colors, and that evidently his application of color terms to ordinary objects was not so eccentric as to convince him of the "peculartiy" in his vision. Even his expertise in "the theory of light and colours" was not enough, with the discovery of his color blindness happening by chance.

Still, the color-blind face difficulties in using color terms, as Dalton recounts. The perceived similarities among colored objects differ markedly between dichromats and trichromats (see again figure 11.2, plate 15, in section 11.1.2). Consequently the perceptible difference between, for example, objects correctly labeled *cobalt* and those correctly labeled *violet* is not at all obvious to dichromats. As Dalton saw it, there were just too many words marking overly subtle distinctions in certain regions of color space. However, despite his "perplexity" about nomenclature, he appears to have used color language with tolerable accuracy.

Dalton's linguistic behavior in the greenhouse and potting shed suggests, if anything, that his color vision was near-enough normal. In fact, more careful investigation of his linguistic behavior would probably have reinforced this incorrect conclusion. Jameson and Hurvich (1978) asked dichromatic subjects to name the Farnsworth-Munsell caps. The chosen names were quite similar to those of normal subjects. One deuteranope in particular behaved exactly like someone with normal color vision. Bonnardel (2006) required subjects to name 140 colored chips, presented individually against a gray background in random order, using eight basic color terms. The agreement of the two deuteranopic subjects with a typical normal classification was 66 percent and 72 percent.

Further, dichromats associate the correct similarity relations with color terms. In addition to the perceptual similarity task involving pairs of colors (see the previous section), Shepard and Cooper (1992) also elicited judgments of similarity using just pairs of color names. They found a circular similarity space that strongly resembled the similarity space of normal subjects.

So if dichromats can tell us how things look to them, the circumstances must be carefully chosen. In particular, the subjects must be able to set aside years of trying, and largely succeeding, to speak with the trichromatic majority. Consider the hue circle; in particular, the hue circle devised by the nineteenth-century French chemist Michel Chevreul, divided into seventy-two sectors of equal angle: clockwise from top, green-olive-yellow-orange-red-purple-violet-blue-turquoise and back to green.[16] Suppose we showed Chevreul's circle to a red-green dichromat who was aware of his deficit and explained to him the question of this chapter. How would he describe the circle's colors?

William Pole, a professor of civil engineering at University College London, tried this experiment on himself, shortly after Chevreul had published his "Cercle Chromatique":

> Now if I follow the Chevreul circle, starting from red [at the bottom], and going round, in the direction of a watch-hand, toward blue, in every division which I pass, the sensation of yellow becomes fainter and fainter . . . until very soon the yellow disappears altogether, and nothing but a dark grey or perfectly colourless hue remains . . . the blue I see perfectly, but the various tints of violet are to me only a darkened blue . . . at about the second or third division beyond "bleu vert," the blue has entirely disappeared, and nothing is left but a neutral gray. Beyond this the illumination begins to increase again, and at the same time a sensation of *yellow* begins to enter; the light and the colour both gradually heightening as I advance, until at the division "jaune" the darkening influence has entirely disappeared, and the full normal yellow hue is obtained. (Pole 1859, 329–328)

The hue circle, as Pole reports it, consists of two opposing chromatically colored sectors, blue and yellow, joined by two small achromatic regions centered around red and green. Other red-green dichromats have also succeeded in convincing themselves that they see things as either yellow, blue, or achromatic, just as the reduction view predicts.

There are two problems, however. First, dichromats are not in agreement, and there are only a handful of published cases like Pole's.[17] Second, and more important, even if we accept that Pole sees only two hues, we have not yet seen any compelling reason to think that these are yellow and blue. Granted, Pole sees lemons and satsumas as having the same hue *C*, but why think that is yellow? After all, satsumas *aren't* yellow. Pole *calls C* "yellow" but—assuming that lemons and bananas, but not satsumas, are the best exemplars of *C*—what else is he expected to call it? (Cf. Judd 1948, 248.)

11.1.5 Opponent-Process Theory

One argument that Pole was right, albeit for uncompelling reasons, is based on the widely accepted opponent-process theory of color vision (see, e.g., Hardin 1993, 26–58; Kaiser and Boynton 1996, ch. 7). According to this theory, color information is processed in three independent channels: red-green, yellow-blue, and white-black. The channels are antagonistic in the following sense: the output of the red-green channel, for example, is either biased toward red, or biased toward green, or at the balance point that is biased neither toward red or green.[18] The output of the red-green channel depends on the difference of the output of the L and M cones (which we can label L and M). Then (ignoring units, scaling, and other complications), if $L - M > 0$, then the red-green channel is biased toward red; if $L - M < 0$, it is biased toward green; and if $L = M$, then it is at the balance point. Similarly for the yellow-blue channel, which takes inputs from all three cone types: if $(L + M) - S > 0$, it is biased toward yellow; if $(L + M) - S < 0$, it is biased toward blue; and if $(L + M) - S = 0$, the yellow-blue channel is at the balance point. The S cones make no contribution to the white-black channel, which becomes more biased toward white as $L + M$ increases. (All this is at best considerably oversimplified, but it will do for our purposes.)

Because deuteranopes and protanopes have, respectively, no M or L cones, one would expect them either to have nonfunctioning red-green channels, or to lack them altogether. (This is discussed further in section 11.4.4.) Similarly, the yellow-blue channel should be either inoperative or absent in tritanopes.

Plausibly, however, a deuteranope or protanope *will* have functioning yellow-blue and black-white channels. Hence opponent-process theory leads naturally, if not inevitably, to a vindication of Pole and the standard reduction view: the colored world of deuteranopes and protanopes is much like ours, but with red and green missing. Things look blue, yellow, white, black, and gray, but nothing looks red, purple, green, or turquoise. (See, e.g., Hurvich 1981, 242–243; Hardin 1993, 145; Kaiser and Boynton 1996, 145.)

11.1.6 Unilateral and Acquired Dichromacy

The most direct source of data on color appearance for dichromats would be the testimony of a normal trichromat who became a dichromat for a day. This would require importing the experiences of a dichromat into the sensorium of a trichromat, which might sound medically impossible, if not conceptually confused. However, there are two actual conditions that resemble this procedure.

First, there are rare individuals who have (relatively) normal color vision in one eye and who exhibit one of the three types of congenital dichromacy in the other. These individuals would appear to be able to make the required direct comparison. They see the usual range of colors with their normal eye, and so they are able to compare this range with the colors they see with their dichromatic eye.

The results of this research are suggestive, although not conclusive. Two basic techniques account for most of the data. Unilateral dichromats can be asked to describe stimuli presented to the dichromatic eye or they can be asked to match stimuli between the two eyes. For unilateral protanopes and deuteranopes, the evidence is generally in agreement with the standard reduction view. They describe stimuli presented to the dichromatic eye as yellowish and bluish, and the only stimuli that match across the two eyes are those that are unique yellow and unique blue as seen with the normal eye (see Judd 1948; Sloan and Wollach 1948; Jackson 1982; Ruddock 1991, but compare MacLeod and Lennie 1976). There are some complications here, sometimes with the data, but also with the exact nature of the subjects' color vision. In a significant fraction of the small number of subjects studied, the "normal" eye itself is anomalous to varying degrees. In addition, there are cortical interactions between the two eyes relevant to perceived color, and consequently the central processing of color signals from the normal eye may reflect the influence of a history of stimulation by the abnormal eye as well.[19]

Second, there are trichomatic individuals who become dichromats as the result of injury or disease. Acquired dichromacy, as opposed to some more complicated disturbance of color vision, is quite rare, and identified cases tend to involve loss of S-cone function, rather than M- or L-cone function (Sperling 1991). It is quite common in acquired color-vision defects for both red-green and yellow-blue systems to be affected to varying degrees, and they are usually associated with other disorders of vision. Acquired defects are typically unstable and often spatially varied, with one eye more affected than the other, or regions within the eye varying in the degree and nature of the impairment. The endpoint is often complete blindness rather than color blindness. The impairments also need not be the straightforward loss of color discrimination seen in the inherited dichromacies. For example, some acquired color-vision defects take the form of *chromatopsia,* in which an endogenous color is added to the perceived object color (Krastel and Moreland 1991). Although the development of impairment is often rapid enough to be noticeable to subjects, there are no useful published data on color change deriving from acquired dichromacy. This is presumably due to the rarity of the condition and the difficulty of making color comparisons over periods of weeks to months.

Acquired dichromacy is of little help, then. However, the evidence from unilateral protanopia and deuteranopia supports the standard reduction view.[20]

11.2 The Standard Reduction View

So far, we have seen that the standard reduction view has a lot to be said for it (we will now drop "standard" when the context makes it clear). This section elaborates the reduction view further, ending with a discussion of three problems. The first two are only apparent; the third is more serious.[21]

11.2.1 The Reduction View Elaborated

Let us stick to protanopes for ease of illustration. The reduction view tells us what the range of perceivable colors is for a protanope—the protanope's *gamut.* If we take a familiar three-dimensional color space, with the colors arranged along two dimensions of hue (red-green and yellow-blue), and one achromatic dimension (black-white), then the protanope's color space is just a two-dimensional plane spanned by the yellow-blue and white-black axes. However, what this *doesn't* tell us is how particular stimuli look to a protanope. For all that's been said, a protanope might see a banana as *blue*.

However, once the reduction view is granted, this further issue is reasonably tractable. Certain stimuli in a normal will keep the red-green channel in balance; plotted on a chromaticity diagram these stimuli will be all and only those that are mixtures of spectral unique blue (approximately 475 nm) with white, and spectral unique yellow (approximately 580 nm) with white—the gamut of the protanope, consisting of the variously saturated shades of unique blue and unique yellow (ignoring lightness). Since mixtures lie along straight lines, the stimuli that keep the red-green channel in balance will lie on the curve consisting of the line connecting 475 nm with the white point *E* (or thereabouts), and the line connecting *E* with 580 nm. As far as the appearance of these stimuli go, the normal trichromat's red-green system seems to be irrelevant. Hence these stimuli, as displayed in figure 11.3 (plate 16), should (at least approximately) appear the same to a protanope (and, by the same token, to a deuteranope) as they do to a normal.

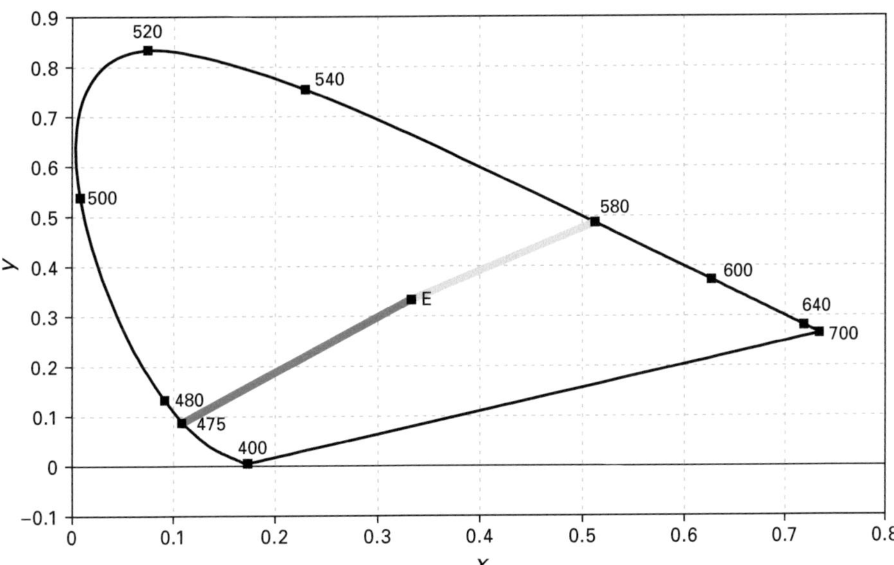

Figure 11.3
Gamut for protanopes and deuteranopes (on standard reduction view). See plate 16 for a color version of this figure.

With the usual rule of thumb connecting color appearances (in the viewing conditions of the color-matching experiment), figure 11.3 (plate 16) may also be taken approximately to represent how stimuli lying on the two lines meeting at E look to a protanope.

But that's not all the stimuli, of course—what about the rest, the vast majority? The facts about confusion lines mentioned in section 11.1.2 provide the answer. Take some stimulus T, say, one in the purple region of the diagram. The protanope will not be able to distinguish between T and a stimulus S whose chromaticity coordinates are at the intersection of the confusion line passing through T, and the curve in figure 11.3 (plate 16), as illustrated in figure 11.4 (plate 17).

Since S looks the same to a normal as it does to the protanope, and since S looks to a normal to be a fairly saturated blue, that is how T will look to the protanope. Purple and turquoise stimuli will (tend to) look blue, and orange and olive stimuli will (tend to) look yellow.

Using these sorts of assumptions (with a number of sophisticated variations), a full-color image I_T can be converted into a reduced color image I_S that appears to a normal the way I_T appears to a protanope, and similarly for the other two forms of dichromacy.[22]

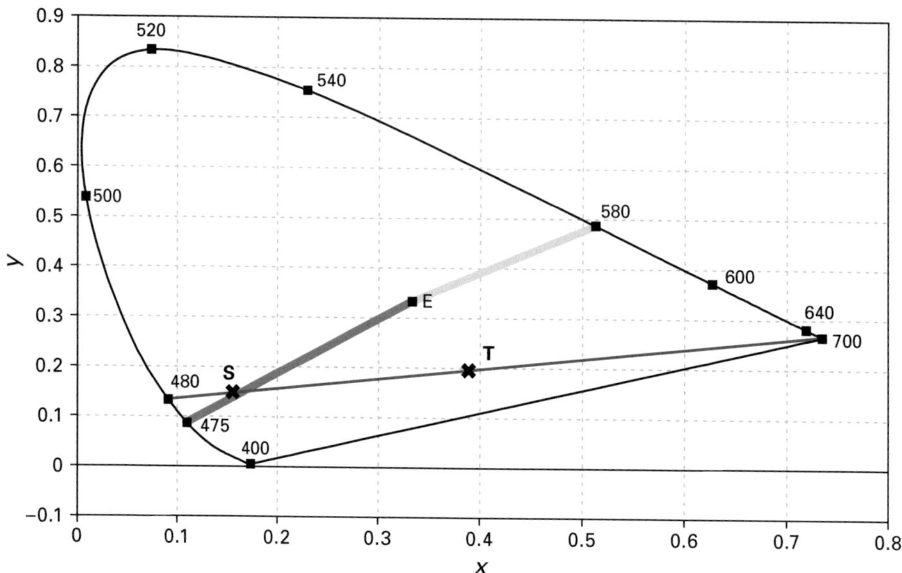

Figure 11.4
S looks to a normal as T looks to a protanope. See plate 17 for a color version of this figure.

11.2.2 Problems for the Reduction View

However, the reduction view is not entirely free from difficulty. This section discusses three problems. The first two are not at all serious; the third is more difficult.

Problem 1: White, Gray, and Black According to the reduction view, various chromatically colored stimuli will be perceived by a protanope as white—spectral light around the neutral point of 495 nm, for instance (see figure 11.2, plate 15). However, light of this wavelength is bluish green (or so normal trichromats say), not white. Because protanopes lack a long-wavelength receptor type, reds look dark to them and are often mistaken for blacks: "On one occasion, a gentleman seeing a lady . . . in church wearing what seemed to him 'a *black* bonnet' asked her 'for whom she was in mourning, and surprised her greatly by the question, for her bonnet was of crimson velvet.'"[23] A protanope will misperceive a variety of chromatically colored stimuli as white, gray, or black. Isn't it unacceptable to suppose that the protanope is in error?

No, it isn't. Misperception itself is not a problem—normal trichromats will also sometimes misperceive chromatic stimuli as achromatic. And it is hardly surprising that dichromats will be more error prone than trichromats: if one wants to detect whether stimuli are white, gray, or black, three receptor types are better than two (and, by the same token, four are better than three). Further, two is good enough. Viewing natural scenes, a dichromat will not usually misperceive chromatic stimuli as achromatic.

Problem 2: The Missing Shade of Violet As can be seen from figure 11.4 (plate 17), stimuli that lie below a confusion line ending at 475 nm (which cluster at the violet end around 400 nm) do *not* lie on a confusion line that intersects the curve representing the protanope's gamut. A protanope will be able to see these stimuli, yet on the reduction view, there is no way for them to appear! How can this be?

This problem can be solved by extending the protanope's gamut past the spectrum locus in the chromaticity diagram. These stimuli will have the same hue but will be more saturated than any color seen by someone with normal color vision. There is thus no corresponding color in the normal gamut, and the procedure given above for finding corresponding colors will fail for some colors in the protanope's gamut. Another way to put the point is to observe that the spectrum locus for protanopes need not occupy the same position in the chromaticity diagram as it does for normal trichromats. And on the reduction view, it won't, reducing normal trichromatic vision to protanopic vision results in a shift in the appearance of the spectrum toward greater saturation.

Problem 3: Hue Misperception According to the reduction view, various purple and turquoise stimuli will be perceived by a protanope as blue (see figure 11.4, plate 17).

Similarly, various orange and olive stimuli will be perceived as yellow. Since purple and turquoise stimuli are not blue, and orange and olive stimuli are not yellow, protanopes will misperceive a variety of stimuli as blue or yellow. Again, isn't it unacceptable to suppose that the protanope is in error?

This is more of a puzzler. A protanope is supposed to have a functioning yellow-blue channel, although he lacks one of the front-end transducers. Consider a satsuma, which (being orange) biases the yellow-blue channel in the yellow direction. A normal trichromat does *not* misperceive the satsuma as yellow. So, one might think, a protanope won't misperceive it as yellow either—he has a functioning yellow-blue channel that is in (approximately) the same state as the trichromat's. Yet, according to the reduction view, the satsuma looks yellow to a protanope.

Relatedly, the reduction view *does* lead to widespread misperception of natural scenes, as can be confirmed by looking at various dichromatic simulations. And since a protanope sees only *unique* yellow or *unique* blue, the precise hue of practically every yellow and blue object will be misperceived. Naturally occurring blue and yellow objects are not usually unique yellow or blue (the clear daylight sky being a notable exception). Bananas, for instance, are greenish when unripe and a little reddish when ripe. Normal trichromats will disagree on whether an object is unique blue—Tye's "puzzle of true blue" (2006a); red-green dichromats pose the puzzle of true blue in spades![24]

Further, the reduction view does not confine widespread misperception to human dichromats. Most mammals are dichromats, and according to the standard account of the evolution of primate trichromacy, around thirty million years ago gene duplication transformed the single longer-wavelength photopigment of our dichromatic ancestors into two, our current L and M photopigments. (For reviews, see Jacobs and Rowe 2004 and Surridge et al. 2003.) This added a new red-green opponent channel to the existing yellow-blue channel. On the reduction view, human red-green dichromacy can accordingly be thought of as something of an atavistic glimpse of our distant evolutionary history. If misperception is widespread among human red-green dichromats, then presumably it also is among the many dichromatic mammals who share versions of the ancient yellow-blue system of color vision.

The threat of widespread error will of course not bother eliminativists like Hardin, who think that nothing is colored and therefore that error is ubiquitous. But one might have thought that the standard theory of color blindness, like other standard theories in visual science, would take no official position on these philosophical debates.[25]

11.3 The Alien View

The reduction view, as we have seen, has much in its favor. But the problem of hue misperception provides some motivation for the rival alien view, on which the dichromat's gamut is not a subset of the normal trichromat's—specifically, a protanope does not see unique yellow and unique blue, but some other hues entirely. The alien view—if not in the version just mentioned—occasionally surfaces in the philosophical literature. Hacker, for instance, claims that red-green dichromats see "Rubies, emeralds, and clouds [as] *gred*," where gredness is a color not seen by the rest of us (Hacker 1987, 152).[26]

It would be hasty to embrace the alien view simply because of the problem of hue misperception. Is there anything more to be placed on the pro side of the ledger? We consider two related arguments. The first argument attempts to derive the alien view from three premises. The first two are particularly important: that the relations of similarity and difference between the hues are essential to them ("Essence," for short), and (something close to) the thesis Johnston has called "Revelation" (1992, 138). The second argument takes off from the failure of the first.

Let us begin by elaborating the first two premises of the first argument.

11.3.1 Essence and Revelation

Essence—the first premise—is, in one version or another, something of a consensus view among philosophers. David Armstrong, in a defense of a physicalist theory of color, claims that the resemblance relations among the hues are internal and so hold in every possible world (1987, 44). Hardin, although entirely unsympathetic with Armstrong's physicalism, agrees that "the hues have certain characteristics necessarily" (1993, 66), and clearly thinks that examples include resemblance relations, for instance that "red [is] more like orange than like blue" (120). Other equally prominent examples are easy to find.[27]

Turn now to the second premise, Revelation: as Johnston explains it, it is the thesis that "[t]he intrinsic nature of canary yellow is fully revealed by a standard visual experience as of a canary yellow thing" (1992, 138; see also Hilbert 1987, 37–38), and likewise for the other colors. Revelation, at least as officially stated—we will come back to this later—implies the slightly weaker thesis of Revelation$^{(-)}$, that anyone who can see canary yellow (for example) is in a position to know the intrinsic nature of canary yellow. Revelation$^{(-)}$ is the second premise. It is weaker than Revelation because it does not imply anything about the source of the knowledge of the nature of canary yellow; in particular, it does not imply that the source is "a standard visual experience as of a canary yellow thing."

Revelation and Revelation$^{(-)}$ are powerful solvents. Whatever "intrinsic nature" precisely comes to, if the colors are physical properties, then this is paradigmatically part of their intrinsic nature.[28] Since seeing canary yellow does not put one in a position to know that canary yellow is a physical property (such-and-such reflectance type, say), Revelation$^{(-)}$ (and so Revelation) imply that color physicalism is false. Perhaps partly for this reason, Revelation and similar theses are, unlike Essence, quite controversial.[29]

11.3.2 The First Argument for the Alien View

Essence unaided does not imply the alien view. By Essence, it is in the nature of unique yellow that it is more similar to orange than to green; by itself this implies nothing about whether there could be someone who can see only unique yellow and unique blue. Even if we add in Revelation$^{(-)}$, the desired conclusion that such a perceiver is impossible doesn't follow. By Revelation$^{(-)}$, someone who can see unique yellow is in a position to know everything about its nature. Hence, someone who can see unique yellow is in a position to know that it is more similar to orange than to green. Fine, but why does such a person need to be able to *see* orange and green? Why can't he be a red-green dichromat, as the reduction view has it, and be able to see only unique yellow and unique blue?

However, it does seem plausible that someone who can see *only* unique yellow and unique blue (and, hence, not orange or green) is *not* thereby in a position to know that unique yellow is more similar to orange than to green.[30] (Red-green dichromats might well know this, but not, presumably, simply by reflecting on their color experience.) So why not add this as a third premise? By Essence and Revelation$^{(-)}$, someone who can see unique yellow is in a position to know everything about its nature. Hence, someone who can see unique yellow is in a position to know that it is more similar to orange than to green. And since, by the third premise, this is not possible if one sees *only* unique yellow and unique blue, it is not possible to see only these hues. Red-green dichromats must see other hues, so the alien view is true.

There two problems with this argument. The first, and simplest, is that both Revelation and Revelation$^{(-)}$ are far too strong to be plausible. In fact, our formulation of Revelation, although it follows the letter of the quotation from Johnston above, does not accurately capture what he has in mind. Johnston does not mean that *simply* seeing canary yellow in isolation is enough to discern its nature. A diverse range of experiences is needed—which may well include seeing canary yellow things next to red things, other shades of yellow, and so on. *This*, more modest, version of Revelation does not imply Revelation$^{(-)}$ and so is of no help to the alien view.

Discussion of the second problem can be postponed, because the second attempt to establish the alien view runs into similar sand.[31]

11.3.3 The Second Argument for the Alien View

The basic idea behind the first argument was that if one sees unique yellow (say), one must have information about the other colors, information one cannot have if one does not actually see those colors. In particular, if one sees unique yellow and unique blue, one has information about orange and green, information one can only possess if one actually sees orange and green (or, at least, other colors). The second argument for the alien view tries out this basic idea in a different way.

We know that the colors exclude each other simply on the basis of visual experience: if something is yellow, it is not blue or green; if something is canary yellow, it is not sky blue; if something is sky blue, it is not navy blue. More to the point, if something is unique yellow—yellow with no hint of red or green—then it is not red.

How is this knowledge to be explained? One obvious and attractive proposal is that there must be something about the way colors are visually encoded that allows us to infer that they exclude one another. Consider, as an analogy, the names "Bill" and "Ben," which refer to certain individuals who happen to be, respectively, the man who has (exactly) seven cats and the man who has (exactly) six cats. Bill and Ben "exclude" each other: if the visitor at the door is Bill, he isn't Ben. However, information about the visitor encoded using "Bill" is not going to allow us to draw this conclusion: we might know that the visitor is Bill and yet still reasonably wonder whether the visitor is Ben. On the other hand, switching from names to descriptions makes a useful inference available: "The visitor is not the man who has six cats" is a logical consequence of "The visitor is the man who has seven cats."

If one sees unique yellow and also sees orange, one is then in a position to know that if something is unique yellow, it is not orange. How could these two colors be represented in a way that would permit this conclusion to be drawn? The way *unique yellow* is defined gives a clue: as Hardin puts it, unique yellow is "a yellow that is neither reddish nor greenish" (1993, 39). Suppose that when one sees an object *x* as unique yellow (that is, when one sees unique yellow), *x* is represented as being yellow but neither reddish nor greenish. And suppose that when one sees an object *y* as orange (that is, one sees orange), *y* is represented as, to use Hardin's phrase, "some degree reddish and also [as] some degree yellowish" (39), with these two components of *chromatic strength* being roughly equal. Then this would explain how seeing these colors puts one in a position to know that they exclude each other: given this kind of chromatic perceptual coding plus a bit of logic, it follows that if something is unique yellow it is not orange.

Having independently motivated this claim about the way colors are represented, we can make a second attempt to reach the alien view. If unique yellow is represented as a yellow that is neither reddish nor greenish, then seeing unique yellow by itself should put one in a position to know that it is not reddish, hence not reddish and yellowish in roughly equal proportion, hence not orange. But someone who sees only unique yellow (and unique blue) could not know solely on the basis of color experience that if something is unique yellow, then it is not orange. So, someone who sees unique yellow and unique blue must also see other colors, and therefore the alien view is true.

The chief problem with this argument lies in the last step, which combines one premise with another that varies inversely in plausibility. Why do we think that seeing unique yellow and unique blue alone is *insufficient* to know that unique yellow excludes orange? Because that information is apparently not present in the experiences of seeing unique yellow and unique blue. Contrariwise, if we have convinced ourselves by some philosophical argument that seeing unique yellow supplies the information that it excludes orange, then we have no reason to maintain that someone who sees only unique yellow and unique blue is not thereby in a position to know that unique yellow excludes orange.

Basically the same problem afflicts the first argument for the alien view. One of its two premises, Revelation$^{(-)}$, was derived from Revelation (in the strong form given above) and has no apparent support without it. Suppose we have convinced ourselves of the truth of Revelation: the nature of unique yellow is completely exposed by an experience of seeing that color, and so seeing unique yellow is sufficient to know that it is more similar to orange than to green. The third premise of the argument was that someone who sees only unique yellow and unique blue is not in a position to know this fact about the essence of unique yellow. However, establishing Revelation, and so Revelation$^{(-)}$, removes any reason to believe the third premise.

11.4 The Reduction View Revisited and Revised

The only obstacle to the reduction view is the problem of hue misperception. In this section, we argue that it can be removed.

11.4.1 Opponent Processes and the Reduction View Again

Let us return to the argument in section 11.1.5 that leads from the opponent-process theory to the reduction view and make it more explicit.

(11.1) A red-green dichromat's chromatic information (or misinformation) about a stimulus is supplied solely by his (essentially normal) yellow-blue channel, since he has no functioning red-green channel.

(11.2) Either the dichromat's yellow-blue channel is biased toward yellow, or it is biased toward blue, or it is in balance.

(11.3) If a perceiver's yellow-blue channel is biased toward yellow, the channel is supplying the information (or misinformation) to the perceiver that the stimulus is yellow; similarly for the other two cases.

By (11.2), there are three cases. Consider the first: a dichromat's yellow-blue channel is biased toward yellow. By (11.3), the dichromat is receiving the information that the stimulus is yellow; by (11.1), the dichromat has no other chromatic information. So, the total information concerning the hue of the stimulus is that it is yellow (in fact, unique yellow). Assuming that this information is present in how the stimulus looks, the stimulus will look yellow. Similarly for the other two cases. Hence:

(11.4) To a red-green dichromat, things look yellow, blue, or achromatic.

The problem with this argument is that premise (11.3) is false. Recall the satsuma of section 11.2.2. When viewed by a normal trichromat, her red-green channel is biased toward red, and her yellow-blue channel is biased toward yellow. The information in these two chromatic channels is supposed to be combined at some stage in perceptual processing to produce the perception of orange. A protanope looking at the satsuma is getting information just from his yellow-blue channel, since his red-green channel—if indeed he has one—is inoperative. According to premise (11.3), the information supplied by his yellow-blue channel is that the satsuma is yellow, and that's why it looks yellow to him. Similarly, if we consider a tritanope looking at the satsuma, the information supplied by his red-green channel is that the satsuma is red. Returning to the trichromat, her perceptual information about the hue of the satsuma is the combination of the information in both channels; that is, the information in the protanope's yellow-blue channel and the tritanope's red-green channel. Now, "combining" the information that *p* with the information that *q* is just to conjoin these two propositions. But then we get the result that the perception of orange is the perception of yellow-and-red, which is wrong—an orange satsuma is neither yellow nor red.

Our assumption that when the yellow-blue channel is biased toward yellow, it supplies the information that the stimulus is yellow might be objected to as oversimplified: we have ignored the fact that the yellow-blue channel supplies information about *degrees* of yellowness and blueness. (Something like this complication is required, because when a stimulus is seen as slightly reddish yellow, say, the red-green channel is slightly biased toward red, and the yellow-blue channel is more heavily biased toward yellow; cf. Hardin on "*x* percent red and *y* percent yellow" [1993, 120].) But adding this complication makes no difference. Suppose that having some "degree of yellowness" or having "nonzero percent yellow"—whatever these expressions mean, exactly—entails being yellow. Then the argument from the opponent-process theory to the reduction view is essentially unchanged, and so is our objection against premise (11.3). Suppose, on the other hand, that having some "degree of yellowness" or having "nonzero percent yellow" does not entail being yellow. Then premise (11.3) is false to begin with.

If the yellow-blue channel is "biased toward yellow" it does not—contrary to what the terminology suggests—signal that the stimulus is yellow. What does it signal instead? In a normal trichromat, the channel is positive just in case the stimulus looks to have a hue from that half of the hue circle that starts in the yellowish reds next to unique red, and runs through orange, yellow, and yellow-green, stopping just short of unique green: in short, just in case the stimulus looks (even a tiny bit) *yellowish*.[32] Hence, if the yellow-blue channel is biased toward yellow, it signals that the stimulus is yellowish, a superdeterminable of the determinable yellow. Similarly, if the channel is negative, it signals that the stimulus is bluish. The yellow-blue channel is misnamed—it should really be called the *yellowishness-bluishness* channel.

11.4.2 The Revised Reduction View

If the flawed argument from the opponent-process theory is repaired to accommodate the fact that the yellow-blue channel supplies information about yellowishness and bluishness, not yellowness and blueness, then it supports the *revised* reduction view. On the standard reduction view, a red-green dichromat sees objects as either unique yellow or unique blue. On the *revised reduction view*, a red-green dichromat sees objects as either yellowish or bluish.

As advertised at the beginning of this chapter, the revised reduction view might as well be called the *revised alien view*. Yellowishness is not an *entirely* alien hue—something is yellowish iff it is either yellowish red, or orange, or yellow, or greenish yellow, or yellowish green. But it is alien*ish*: normal trichromats never see this hue without seeing more determinate hues like orange and yellow. On the revised reduction view, a red-green dichromat sees yellowishness unaccompanied.

Switching to the revised reduction view removes the problem of hue misperception (section 11.2.2). To red-green dichromats, satsumas look yellowish, not yellow; and since they are yellowish, they are not misperceived. The ancient "yellow-blue" system of color vision is not excessively error prone after all, since the hues it detects are the more inclusive yellowishness and bluishness, not the relatively exclusive yellowness and blueness.

11.4.3 Hue Magnitudes

The revised reduction view can be further elaborated and explained with the aid of the hue magnitude account of the visual representation of colors in Byrne and Hilbert (2003).[33] In the terminology of that paper, there are four *hue magnitudes*, *R*, *Y*, *B*, and *G*, which come in degrees like length and temperature. *Yellowishness* was used earlier to stand for the property of having some degree or other of the *Y* magnitude; since context will disambiguate, we can use *yellowishness* also to stand for the *Y* magnitude, and similarly for the other hue-ish terms.

When one sees an object as unique yellow, it is represented as having a degree of yellowishness that is 100 percent of its *total hue*, the sum of its degrees of all the hue magnitudes. When one sees an object as orange, it is represented as reddish to a degree that is roughly 50 percent of its total hue, and yellowish to the remaining degree.

This explains, incidentally, our knowledge of color exclusion (compare the proposal discussed in section 11.3.3): if something is 100 percent yellowish, it can't be roughly 50 percent yellowish, and so it isn't orange.

The hue magnitude account is independent of the opponent-process theory, which is not an account of the content of visual experience, but the two make a nice fit. The yellow-blue channel provides information about degrees of yellowishness and bluishness, and the red-green channel provides information about degrees of reddishness and greenishess. Together, these two channels contribute, to the content of visual experience, the relative proportion of yellowishness (bluishness) and reddishness (greenishness).

Consider, again, a normal trichromat and a protanope looking at our satsuma. Suppose that the trichromat's experience represents the satuma as having reddishness that is 40 percent of its total hue (the sum of the four hue magnitudes), and yellowishness that is 60 percent of its total hue. The protanope has no functioning red-green channel, so the only two magnitudes represented by his visual experience are yellowishness and bluishness. For him, the "total hue" of the satsuma is a quantity that is the sum of its degrees of yellowishness and bluishness: call that the satsuma's total *reduced* hue. Combining the hue magnitude account and the revised standard view, the protanope's visual experience represents, not that the satsuma has yellowishness that is 100 percent of its total (unreduced) hue, but rather that it has yellowishness that is 100 percent of its total *reduced* hue (the sum of its yellowishness and bluishness). The satsuma, then, is not represented as *unique yellow*, but rather as having some degree or other of the yellowishness magnitude—that is, as *yellowish*.[34]

11.4.4 But Is the Red-Green Channel either Inoperative or Absent?

The above argument for the revised standard view makes use of an assumption first introduced in section 11.1.5, namely that in protanopes and deuteranopes, the red-green channel either is present but makes no contribution to color perception, or is absent. Is that assumption plausible?

Suppose, on the contrary, that a functioning red-green channel is present: it develops in every human being whether there are appropriate cone types to supply it or not, and its output contributes to color perception. One possibility is that the channel output is permanently biased—toward greenishness for protanopes and reddishness for deuteranopes. If so, then the standard reduction view is wrong: the two hues seen by protanopes are yellow-green and blue-green, and the two deuteranopic hues are orange and purple. Another (arguably more likely) possibility is that the dichromatic visual system adapts to the unvarying red-green signal by setting the red-green channel output permanently to neutral. This would vindicate the standard reduction view: when a protanope looks at a satsuma, his experience represents that the satsuma has yellowishness that is 100 percent of its *total* hue; that is, that the satsuma is unique yellow.

If either of these possibilities obtain, then dichromacy is a kind of pathology, not just a less discriminating form of color vision than the trichromatic condition. On the first possibility, for instance, a protanope's color-vision system signals greenishness, no matter whether the seen object is greenish or not. This is a matter not merely of error, but also of complete insensitivity: veridical perception of greenishness is an accident, like a stopped clock that reads the correct time. And similarly on the second possibility, veridical perception of an object that is neither reddish nor greenish will be a fluke.

However, the available evidence supports the conclusion that congenital dichromats, at least, have no functioning red-green channel. First, a general rule of visual processing is that unchanging inputs are ignored. For example, if an image is stabilized on the retina, it fades away (Martinez-Conde et al. 2004). This is why we don't perceive in ordinary circumstances the shadows cast on the retina by the veins in the eye. If dichromats have a red-green channel, there is no reason to think its output would contribute to their color experience.

Second, congenital dichromats might well not possess a red-green channel at all. Research on nonhuman color vision indicates that the processing of chromatic information in particular is not set by a preexisting wiring diagram, but instead is driven by the kind of stimuli available. In some South American monkeys, M- and L-cone pigments are coded by two alleles on the X chromosome. The males and homozygous females thus have only two cone pigments and are dichromatic; heterozygous females have three and are trichromatic (Jacobs et al. 1993). (That is, heterozygous females are both retinally and functionally trichromatic.) Mice are normally dichromatic, but if an allele for a third cone pigment is inserted on the X chromosome, the heterozygous females will be trichromatic (Jacobs et al. 2007). Insertion of a third cone pigment gene in some of the cones of the retina of (normally dichromatic) adult male squirrel monkeys results in functional trichromacy as soon the pigment is expressed (Mancuso et al. 2007). This suggests that opponent channels emerge partly in response to the cone inputs, as opposed to being created independently.

11.4.5 Unilateral Dichromacy Again

According to the revised reduction view, red-green dichromats see yellowishness and bluishness, and not trichromatic hues like unique yellow and unique blue. Nevertheless, as we have seen, unilateral dichromats will describe stimuli presented to their normal eye exclusively as yellow and blue, and will accept matches between stimuli presented separately to their normal and dichromatic eyes. Is this a problem for the revised reduction view?

Not really. First, the revised reduction view does not obviously apply to unilateral dichromats. Central color processing and representation in unilateral dichromats, as opposed to bilateral dichromats, has developed in response to a mixture of dichromatic and trichromatic inputs. Because of this fact, it is a plausible conjecture that color is always represented by unilateral dichromats using the two-dimensional hue code of normal trichromats. If so, then some stimuli presented to a unilateral dichromats dichromatic eye really do look unique yellow or unique blue. This would explain why they accept matches between the two eyes, and use "yellow" and "blue" to label stimuli presented to the dichromatic eye. That the central mechanisms of unilateral dichromats may differ from those of bilateral dichromats presents a real problem in interpreting the data from unilateral dichromats, and has been appealed to, for example, to explain why the known cases of unilateral tritanopia deviate from the predictions of the standard reduction view (Alpern et al. 1983).

Second, even if unilateral dichromats are (improbably) cyclopean bilateral dichromats, and therefore see the bilateral's colors with their dichromatic eye, the matching and description data are consistent with the revised reduction view. Unless very careful measures are taken to ensure that subjects accept only complete perceptual matches, the fact that a subject accepts a match between two stimuli does not establish that they are perceptually identical, only that they are similar in a salient respect (see, for example, Arend and Reeves 1986). If we suppose that yellowishness and unique yellow are saliently similar, then matching experiments pose no difficulty. And given that the unilateral dichromat has just normal color vocabulary to describe stimuli presented to his dichromatic eye, it is not surprising that "yellow" and "blue" are the words of choice.

11.5 Summary Conclusion

This chapter has argued for the revised reduction view: red-green dichromats see the world as having two superdeterminable hues, yellowishness and bluishness. More cautiously: *if* dichromatic vision is a reduction of normal trichromatic vision, then the revised reduction view is true.

The colors we normal trichromats see are either determinables like yellow, orange, and blue, or determinates like canary yellow, coral, and navy blue. Red-green dichromats do not see any of these colors. In that sense, the vulgar are vindicated: red-green dichromats are not just merely color deficient—they are color-blind.

Acknowledgments

Thanks to an audience at Florida State University, and to Justin Broackes, Jonathan Cohen, and Mohan Matthen for helpful comments. We dedicate this paper to Larry Hardin for all he's done to promote empirically informed discussions of color by philosophers.

Notes

1. See Sharpe et al. 1999, table 1.5 (this gives figures only for red-green deficiencies; as Sharpe et al. discuss, other kinds of deficiency are exceptionally rare).

2. *Anomalous trichromacy* is a less severe defect that comes in two varieties, corresponding to each of the two varieties of red-green dichromacy. Although we are focusing on red-green dichromacy, some of the data we report also cover anomalous trichromats.

3. If a dichromat's color experiences are a subclass of the normal kind, then there is no obvious barrier *in principle* to knowing what dichromatic experience is like. But if a dichromat's color experiences are quite different from the normal kind, then (according to many philosophers), we can never know what they are like. Relatedly, a dichromat can never know what the full range of normal color experiences are like. Recall Fred, the forgotten hero of "Epiphenomenal Qualia"—subsequently eclipsed by his co-star Mary. "Fred's optical system is able to separate out two groups of wavelengths in the red spectrum as sharply as we are able to sort out yellow from blue . . . We are to Fred as a totally red-green color-blind person is to us" (Jackson 1982, 274).

4. See Chisholm 1957, ch. 4; Jackson 1977, ch. 2; Thau 2002, 226–231.

5. One reason is given by Hurvich: dichromats report seeing colors "of the same general nature as the grayness of 'night vision'" (1981, 244). "Same general nature as" should be construed as "similar to," not as "identical with": the grayness of "night vision" is not the same as the grayness of "day vision" (see note 25 below).

6. Hardin reads Boynton very differently than we do, citing the section referred to in support of the standard reduction view (1993, 146).

7. See, in particular, www.vischeck.com, which uses the algorithm of Brettel et al. 1997. See also Brettel's page: http://www.tsi.enst.fr/~brettel/colorblindness.html.

8. *Geranium zonale*, as Dalton calls it, is the horsehoe cranesbill, now named *Pelargonium zonale* (Hunt et al. 1995, 987n4).

9. However, rods and cones do interact (Stabell and Stabell 1998; Buck et al. 2006; Thomas and Buck 2006); what's more, a recent study reports "distinct color appearances mediated exclusively by rods" (Pokorny et al. 2006; see also note 21).

10. For instance, that cones of the same type have exactly the same spectral sensitivity. This simple assumption is actually too simple: for this and other complications, see MacLeod 1985.

11. Stimuli like colored papers can be assigned tristimulus values provided the illumination and viewing conditions are specified.

12. See, e.g., Hardin 1993, 116–119. A color space is a comprehensive representation of the relations of similarity among colors. The representation is spatial in the sense that degrees of similarity are represented by distances in the space. Actual color spaces constructed by color scientists are motivated by concerns that often lead to a nonuniform relationship between distance in the space and perceived similarity among the colors. The various versions of the color solid are examples of color spaces, each of which makes its particular departures from the ideal because of the purposes for which it was constructed (see again Hardin 1993). In addition, there are variations among normal individuals in the relations of similarity and difference that they will perceive. At a sufficiently detailed level of description, there may not be any two observers who share the same color space.

13. An equal energy light has a flat spectral power distribution.

14. The subjects were not (exclusively) dichromats but had either a "strong deutan" or "strong protan" deficiency. That is, the subjects were either deuteranopes, protanopes, or had extreme forms of *deuteranomaly* or *protanomaly* (the two varieties of anomalous trichomacy). See note 2 above.

15. To say nothing of the notorious "unique green" (Hardin 1993, 79–80).

16. A nice illustration of Chevreul's circle is at http://webexhibits.org/colorart/simultaneous .html.

17. Including (approximately) Dalton: "My yellow comprehends the red, orange, yellow and green of others; and my blue and purple coincide with theirs" (quoted in Sharpe et al. 1999, 29). On the other hand, according to Kaiser and Boynton, "Few dichromats can be convinced that their color vision accords with the theoretical description [the standard reduction view] just given" (1996, 453).

We asked a colleague who is a red-green dichromat (or severe anomalous trichomat) to describe a reproduction of Chevreul's circle. He replied: "All of the colors are out of my comfort zone—I feel like I'm guessing all the way round. Noon: reddish [in fact this segment is green]. Shading off to green (??) at 2. Very unsure about this. Shading off to something darker at 4, but I can't tell if it's the same color or not. Getting reddish around 5. 6 red, and I'm more confident about this than my previous guesses. Blue starts getting mixed in around 7:30. 9–10 definitely has blue in it, but it's not pure blue. Probably one of those purple/mauve colors that I'm lousy at. 11 starts getting reddish again."

18. One might reasonably ask what "biased toward red" (e.g.) is supposed to mean; this is taken up in section 11.4.1.

19. For useful discussions of differences in the processing of signals from the two eyes of unilateral dichromats, see MacLeod and Lennie 1976 and Alpern et al. 1983.

20. Unilateral tritanopia is a different story. There are no well-characterized cases of congenital unilateral tritanopia, and the cases of acquired unilateral tritanopia that have been reported do not fit any simple version of the reduction view (Graham et al. 1967; Alpern et al. 1983). In light of the very small number of cases, it would be a mistake to place any great weight on this difficulty. Taken at face value, the reported cases seem to conflict with the predictions of the standard reduction view. On the other hand, unilateral tritanopes are willing to use standard color terms to characterize the appearance of stimuli presented to their tritanopic eye, and they are willing to accept matches between stimuli presented to the dichromatic eye and the normal eye. These last two points provide support for the truth of some form of the reduction view, although not for the standard version in particular.

21. A fourth problem for the reduction view is posed by a cluster of empirical results. In color-naming experiments, putative dichromats use red and green (or all eleven basic color terms) in systematic and repeatable ways (Scheibner and Boynton 1968; Wachtler et al. 2004). It is not entirely clear how much of the color-naming performance can be explained on the basis of

color-naming strategies adopted by dichromats living in a trichromatic culture. In addition some putative dichromats are trichromatic for larger stimuli (Smith and Pokorny 1977). Although the interpretation of the results of these and similar experiments is not straightforward, it may be the case that inputs from the rods are capable of driving an extra color channel (Buck et al. 2006). For reasons of space we will set these complications aside.

22. See, in addition to Brettel et al. 1997 and Viénot et al. 1995, Capilla et al. 2004.

23. This is from George Wilson, *Researches on Color Blindness: with a supplement on the danger attending the present system of railway and marine colored signals* (1855), quoted by Hurvich (1981, 241).

24. See also Cohen et al. 2006; Tye 2006b; Byrne and Hilbert 2007b; Tye 2007; and note 15 above.

25. But what about scotopic vision, when the only input is from the rods? Don't cats (and everything else) look gray in the dark? And since most things aren't gray, isn't scotopic vision infected with massive error? No, it isn't: viewed scotopically, objects look light and dark, but not gray. Turning the lights down is not like changing a color image to a black-and-white one.

26. However, Hacker's version of the alien view has nothing to be said for it. He has mistakenly taken the fact that dichromats confuse some reds, greens, and grays to indicate that dichromats see all red, green, and gray objects as having the same hue.

27. See Johnston 1992; Clark 1993.

28. And so (we assume) are the similarity relations between colors (Johnston 1992, 152).

29. For more discussion of Revelation, restoring some omitted details, see Byrne and Hilbert 2007a.

30. Cf. Boghossian and Velleman: "the experience of seeing something as red does not by itself reveal that the property now in view has a yellower neighbor (orange) and a bluer neighbor (violet)" (1991, 129).

31. There is also a third problem, which is that the argument is in danger of proving too much. Consider Johnston's version of Essence ("Unity"), which is not restricted to hue: "Thanks to its nature and the nature of the other determinate shades, canary yellow, like the other shades, has its own unique place in the network of similarity, difference, and exclusion relations exhibited by the whole family of shades" (1992, 138). Substituting Unity for Essence, the conclusion of the argument is that dichromats can't even see achromatic colors.

32. Cf. Bradley and Tye 2001, 472; in Byrne and Hilbert 1997a, 278–280, "yellowish" was defined less inclusively.

33. See also Byrne 2003.

34. Similar considerations show that the "black-white," or "achromatic," channel is misnamed: it should be called the "dark-light" channel. If it were contributing information about the degree

of grayness of the stimulus, then (since every stimulus affects the achromatic channel) everything would look gray to some degree.

References

Alpern, M., K. Kitahara, and D. H. Krantz. 1983. Perception of colour in unilateral tritanopia. *Journal of Physiology* 335: 683–697.

Arend, L., and A. Reeves. 1986. Simultaneous color constancy. *Journal of the Optical Society of America A. Optics and Image Science* 3: 1743–1751.

Armstrong, D. M. 1987. Smart and the secondary qualities. *Metaphysics and Morality: Essays in Honour of J. J. C. Smart*, P. Pettit, R. Sylvan and J. Norman, eds., 1–15. New York: Blackwell.

Boghossian, P., and J. D. Velleman. 1991. Physicalist theories of color. *Philosophical Review* 100: 67–106. Page reference to the reprint in *Readings on Color,* vol. 1, *The Philosophy of Color*, A. Byrne and D. R. Hilbert, eds. Cambridge, MA: MIT Press, 1997.

Bonnardel, V. A. L. 2006. Color naming and categorization in inherited color vision deficiencies. *Visual Neuroscience* 23: 637–643.

Boynton, R. M. 1979. *Human Color Vision*. New York: Holt, Rinehart, and Winston.

Bradley, P., and M. Tye. 2001. Of colors, kestrels, caterpillars, and leaves. *Journal of Philosophy* 98: 469–487.

Brettel, H., F. Viénot, and J. D. Mollon. 1997. Computerized simulation of color appearance for dichromats. *Journal of the Optical Society of America A. Optics and Image Science* 14: 2647–2655.

Buck, S. L., L. P. Thomas, N. Hillyer, and E. M. Samuelson. 2006. Do rods influence the hue of foveal stimuli? *Visual Neuroscience* 23: 519–523.

Byrne, A. 2003. Color and similarity. *Philosophy and Phenomenological Research* 66: 641–665.

Byrne, A., and D. R. Hilbert. 1997a. Colors and reflectances. *Readings on Color,* vol. 1, *The Philosophy of Color*, ed. A. Byrne and D. R. Hilbert, 263–288. Cambridge, MA: MIT Press.

Byrne, A., and D. R. Hilbert, eds. 1997b. *Readings on Color.* Vol. 1, *The Philosophy of Color.* Cambridge, MA: MIT Press.

Byrne, A., and D. R. Hilbert. 2003. Color realism and color science. *Behavioral and Brain Sciences* 26: 3–21.

Byrne, A., and D. R. Hilbert. 2007a. Color primitivism. *Erkenntnis* 66: 73–105.

Byrne, A., and D. R. Hilbert. 2007b. Truest blue. *Analysis* 67: 87–92.

Capilla, P., M. A. Díez-Alenjo, M. J. Luque, and J. Malo. 2004. Corresponding-pair procedure: a new approach to simulation of dichromatic color perception. *Journal of the Optical Society of America A. Optics and Image Science* 21: 176–186.

Chalmers, D. J. 2002. *Philosophy of Mind: Classical and Contemporary Readings*. Oxford: Oxford University Press.

Chisholm, R. M. 1957. *Perceiving: A Philosophical Study*. Ithaca, NY: Cornell University Press.

Clark, A. 1993. *Sensory Qualities*. Oxford: Oxford University Press.

Cohen, J., C. L. Hardin, and B. P. McLaughlin. 2006. True colours. *Analysis* 66: 335–340.

Dalton, J. 1977. John Dalton's discovery of his color blindness. *Applied Optics* 16: 520.

Edridge-Green, F. W. 1911. *The Hunterian Lectures on Colour-Vision and Colour-Blindness*. London: Kegan Paul, Trench, Trübner.

Graham, C. H., Y. Hsia, and F. F. Stephan. 1967. Visual discriminations of a subject with acquired unilateral tritanopia. *Vision Research* 6: 469–479.

Hacker, P. M. S. 1987. *Appearance and Reality*. Oxford: Blackwell.

Hardin, C. L. 1993. *Color for Philosophers: Unweaving the Rainbow*. Expanded ed. Indianapolis, IN: Hackett.

Hilbert, D. R. 1987. *Color and Color Perception: A Study in Anthropocentric Realism*. Stanford, CA: CSLI.

Hunt, D. M., K. S. Dulai, and J. K. Bowmaker. 1995. The chemistry of John Dalton's color blindness. *Science* 267: 984–988.

Hurvich, L. M. 1981. *Color Vision*. Sunderland, MA: Sinauer.

Jackson, F. 1977. *Perception: A Representative Theory*. Cambridge: Cambridge University Press.

Jackson, F. 1982. Epiphenomenal qualia. *Philosophical Quarterly* 32: 127–136. Page reference to the reprint in D. J. Chalmers, *Philosophy of Mind: Classical and Contemporary Readings*, Oxford: Oxford University Press, 2002.

Jacobs, G. H., J. Neitz, and M. Neitz. 1993. Genetic basis of polymorphism in the color vision of platyrrhine monkeys. *Vision Research* 33: 269–274.

Jacobs, G. H., and M. P. Rowe. 2004. Evolution of vertebrate colour vision. *Clinical and Experimental Optometry* 87: 206–216.

Jacobs, G. H., G. A. Williams, H. Cahill, and J. Nathans. 2007. Emergence of novel color vision in mice engineered to express a human cone pigment. *Science* 315: 1723–1725.

Jameson, D., and L. M. Hurvich. 1978. Dichromatic color language: "reds" and "greens" don't look alike but their colors do. *Sensory Processes* 2: 146–155.

Johnston, M. 1992. How to speak of the colors. *Philosophical Studies* 68: 221–263. Page reference to the reprint in *Readings on Color,* vol. 1, *The Philosophy of Color,* A. Byrne and D. R. Hilbert, eds. Cambridge, MA: MIT Press, 1997.

Judd, D. B. 1948. Color perceptions of deuteranopic and protanopic observers. *Journal of Research of the National Bureau of Standards* 41: 247–271.

Kaiser, P. K., and R. M. Boynton. 1996. *Human Color Vision*. 2nd ed. Washington, DC: Optical Society of America.

Krastel, H., and J. D. Moreland. 1991. Colour vision deficiencies in ophthalmic diseases. *Inherited and Acquired Colour Vision Deficiencies: Fundamental Aspects and Clinical Studies*, D. H. Foster, ed., 115–172. Boston: CRC Press.

MacLeod, D. I. A. 1985. Receptoral constraints on color appearance. *Central and Peripheral Mechanisms of Color Vision*, D. Ottoson and S. Zeki, eds., 103–116. London: Macmillan.

MacLeod, D. I. A., and P. Lennie. 1976. Red-green blindness confined to one eye. *Vision Research* 16: 691–702.

Mancuso, K., J. Neitz, W. W. Hauswirth, T. B. Connor, and M. Neitz. 2007. Gene therapy treatment of color blindness in adult primates. *Journal of Vision* 7: 15.

Martinez-Conde, S., S. L. Macknik, and D. H. Hubel. 2004. The role of fixational eye movements in visual perception. *Nature Reviews Neuroscience* 5: 229–239.

Pokorny, J., M. Lutze, D. Cao, and A. J. Zele. 2006. The color of night: Surface color perception under dim illuminations. *Visual Neuroscience* 23: 525–530.

Pole, W. 1859. On color-blindness. *Philosophical Transactions of the Royal Society of London* 149: 323–339.

Ruddock, K. H. 1991. Psychophysics of inherited colour vision deficiencies. *Inherited and Acquired Colour Vision Deficiencies: Fundamental Aspects and Clinical Studies*, D. H. Foster, ed., 4–37. Boston: CRC Press.

Scheibner, H. M. O., and R. M. Boynton. 1968. Residual red-green discrimination in dichromats. *Journal of the Optical Society of America* 58: 1151–1158.

Sharpe, L. T., A. Stockman, and H. Jägle. 1999. Opsin genes, cone photopigments, color vision, and color blindness. *Color Vision: From Genes to Perception*, K. R. Gegenfurtner and L. T. Sharpe, eds., 3–51. Cambridge: Cambridge University Press.

Shepard, R. N., and L. A. Cooper. 1992. Representation of colors in the blind, color-blind, and normally sighted. *Psychological Science* 3: 97–104.

Sloan, L. L., and L. Wollach. 1948. A case of unilateral deuteranopia. *Journal of the Optical Society of America* 38: 502–509.

Smith, V. C., and J. Pokorny. 1977. Large-field trichromacy in protanopes and deuteranopes. *Journal of the Optical Society of America* 67: 213–220.

Sperling, H. G. 1991. Vulnerability of the blue-sensitive mechanism. *Inherited and Acquired Colour Vision Deficiencies: Fundamental Aspects and Clinical Studies*, D. H. Foster, ed., 72–87. Boston: CRC Press.

Stabell, B., and U. Stabell. 1998. Chromatic rod-cone interaction during dark adaptation. *Journal of the Optical Society of America A. Optics and Image Science* 15: 2809–2815.

Surridge, A. K., D. Osorio, and N. I. Mundy. 2003. Evolution and selection of trichromatic vision in primates. *Trends in Ecology and Evolution* 18: 198–205.

Thau, M. 2002. *Consciousness and Cognition*. New York: Oxford University Press.

Thomas, L. P., and S. L. Buck. 2006. Foveal and extra-foveal influences on rod hue biases. *Visual Neuroscience* 23: 539–542.

Tye, M. 2006a. The puzzle of true blue. *Analysis* 66: 173–178.

Tye, M. 2006b. The truth about true blue. *Analysis* 66: 340–344.

Tye, M. 2007. True blue redux. *Analysis* 67: 92–93.

Viénot, F., H. Brettel, L. Ott, A. B. M'Barek, and J. D. Mollon. 1995. What do color-blind people see? *Nature* 376: 127–128.

Wachtler, T., U. Dohrmann, and R. Hertel. 2004. Modeling color percepts of dichromats. *Vision Research* 44: 2843–2855.

12 What Do the Color-Blind See?

Justin Broackes

12.1 Color Blindness: A Guide and Test for Theories of Normal Vision

What do the color-blind see? Can we tell? Does it matter? The theory of normal human color vision is one of the triumphs of nineteenth-century science, emerging in Helmholtz and J. C. Maxwell in the 1850s, out of suggestions from Tobias Mayer and Thomas Young and others half a century and more earlier.[1] The trichromatic theory—the view that a normal perceiver can, in light mixing, match any given color with some combination of three suitably chosen primaries, and that this is a sign of our having three main receptor types at work in human color vision—has constant confirmation in the fact that color photography with three emulsions works (cp. Maxwell 1855, 136–137), as does color television with three phosphors. The psychophysics and theory of color measurement have been marvelously systematized;[2] the "opponent theory" first proposed by Ewald Hering as a radical alternative to Helmholtz's theory has been modified and in some manner joined up with it—notably by Hurvich and Jameson, who in the mid-1950s took the outline of G. E. Müller's model of color vision, devised an experimental "hue cancellation" procedure as a quantitative measure of certain features in the structure of our experienced color space, and found a physiological counterpart for that structure in the findings of the new electrophysiology of Granit and Hartline (and Kuffler and Svaetichin), and later in the reports of De Valois and De Valois on the macaque visual system. There are other stories to be told about the physiology of the receptor pigments of the eye and the genes that code for them, about the various kinds of cells in the retina and elsewhere in the visual system, and now increasingly about color processing in the brain. It is a topic on which psychologists, physicists, biologists, and neurophysiologists—not to mention paint manufacturers, dyers, and makers of photographic equipment—have reason to be proud and glad of the convergence of interests and views.

Color blindness might at first seem just a peripheral abnormality. But it has often been both a guide to the nature of normal color vision and a test application for theories of it. It promises to provide cases where the various components of a complex process that are either hard or impossible to separate artificially are found already separated in nature.[3] The great physicist J. C. Maxwell, having just graduated in mathematics from Cambridge in 1854, did his first research comparing the color matching of ten normal perceivers with that of two color-blind people (whom we would now call protanopes), to identify the sensation that was "wanting" in the color-blind but present in the normal perceivers (see Maxwell 1855). And the color-blind are often a test of a particular theory—in both its more empirical and more theoretical or philosophical aspects. If normal vision involves (as Helmholtz and Maxwell thought) three basic sensations, say, of red, green, and violet,[4] produced by three receptor, or nerve, types, then we might expect a person missing one of these types to have sensations corresponding merely to the other two. Someone missing the "red" receptor, or nerve, type ("red-blind," in Helmholtz's terms) would have sensations only of green and violet (along with the blue that results from combining them, which would—surprisingly, indeed—be experienced on looking at white things)—all at high saturation, since there would be no third color sensation ever to desaturate the other two. On the other hand, if (as Hering believed) color vision involves some kind of distinct red-green, yellow-blue, and light-dark processes, then we might expect color blindness to involve the loss of one of those three dimensions. And we might hope to find some kind of empirical confirmation—though it would no doubt be indirect—of one or other of these views of the color-blind, and hence in turn of some or all of Helmholtz's or Hering's general view.

For 150 years and more, the main camps, whatever their many differences, have agreed on one thing: that the main groups of color-blind people—those who confuse reds and greens and browns, and are today classed as dichromats[5]—have in some sense no perception of red and green at all, but only of yellow and blue:

> All colours appear to the color-blind as if composed of blue and yellow. (Maxwell 1855, 140)[6]

> [A red-green blind person] sees as colourless what to others appears in one of the two fundamental colours, red and green; while, in any mixed colour containing red or green, he sees only the yellow or the blue. (Hering 1878, 107; cp. Hering 1880, esp. 102–103)

The two main camps managed, in their different ways, to agree on this. For Hering, the color-blind had sensations simply of yellow or blue. For Helmholtz and Maxwell, they had (to take the case of the "red-blind") sensations of green and violet, but *called* them yellow and blue—and were relevantly sensitive only to yellow and blue. And the dominant view has been the same over the course of the last century (see figure 12.1). To take just two representative speakers:

> The color perceptions of both protanopic and deuteranopic observers are confined to two hues, yellow and blue, closely like those perceived under usual conditions in the spectrum at 575 and 470 mμ. (Judd 1948, 247)

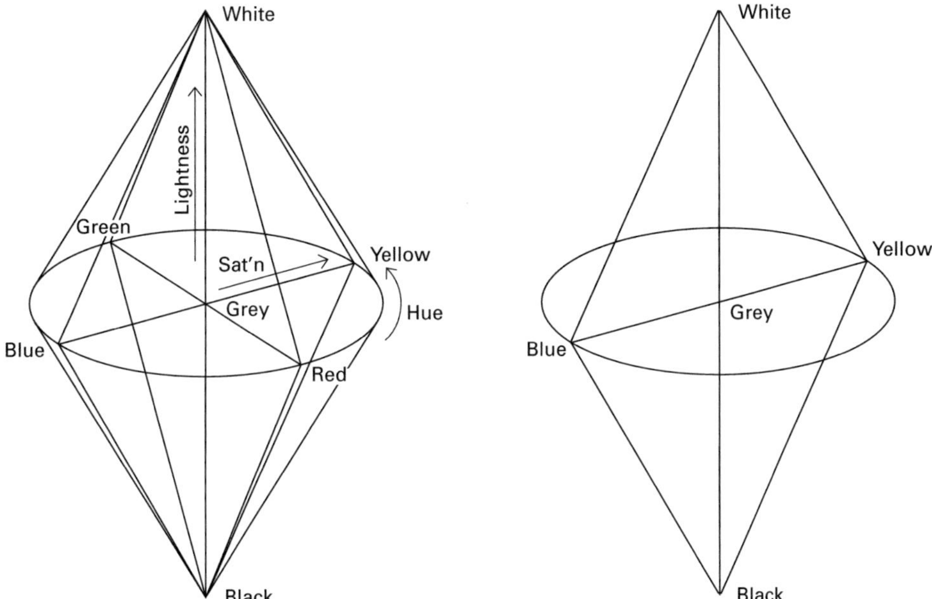

Figure 12.1

Collapse of color space in dichromats according to the standard model. (Left) Illustrative representation of the experienced color space of normal trichromats—as a double-cone (see, e.g., Ostwald 1931, vol. 1, ch. 7; De Valois and De Valois 1975, 148; Hurvich 1981, 11; Rossotti 1983, 146). Variation is possible in three dimensions: variation in hue is indicated by circular motion about the central vertical axis, saturation by distance from the central axis, lightness for surface colors (or brightness for lights) by height along the central axis. (For the notion of *saturation*, see *Note on terminology* in note 13.) Labels mark the position of unique (i.e., phenomenally pure, apparently unmixed) hues, which mark out two intersecting cardinal axes of phenomenal color-space: red-green and yellow-blue. (The diagram is intended only as illustrative; for other kinds of color-order system, see, e.g., Zelanski and Fisher 1989 ch. 6; Hunt 1987, ch. 4 [1998 ch. 7]; Kuehni, this volume.) (Right) Illustrative representation of the experienced color space of protanopes and deuteranopes, according to the standard model—three dimensions collapse to two. Variation is possible only in lightness and saturation of a single yellow or blue hue. (The circle at middle height is no part of the dichromat color space—it is included to show the perspective and to indicate something of what has been lost by comparison with the space of normal trichromats.)

> Two types [of dichromat] fail to experience red and green hues, both of which are presumably seen as grays. As a result, the world appears to them in various shades of blues, yellows, and grays, as though the red/green dimension of the color solid had been eliminated. (Palmer 1999, 104)[7]

The deficiency, we are told, is not a local one confined, for example, just to certain reds and certain greens, or to those colors under certain circumstances: it amounts to the complete loss of a whole dimension. Orange and purple must lose their red component, turquoise and chartreuse lose their green, and the whole world of color collapse from three dimensions to two.

This view might seem to have almost ideal credentials. It fits what some of the color-blind themselves have reported (e.g., Scott 1778, 612–613: "I do not know any green in the world . . . ; but yellows . . . and all degrees of blue, except those very pale . . . , I know perfectly well."). In Helmholtz's version, it seemed confirmed by some elegant mathematical modeling of the confusion patterns of the color-blind (which I will take up in section 12.3). Hering's view was perhaps phenomenologically more plausible in allowing the color-blind to perceive white as white, rather than as blue (in the case of the "red-blind") or purple (for the "green-blind"). And Hering's view could claim empirical vindication in the 1880s and 1890s with some studies of color-blindness in just one eye, which reported (as we shall see in section 12.4)—using the good eye to calibrate the experiences of the bad—that a "red-green color-blind" eye indeed yielded experience only of yellow and blue. What is now the standard synthesis (promoted in different ways by Hurvich and Jameson, and others)—combining three cone types in the retina (as in Helmholtz) with some kind of opponency (as in Hering) in the later processing—might seem to inherit all the advantages of each party. And its proponents have eagerly reaffirmed the claim: "The red/green deficient sees only yellows and blues in addition to white" (Hurvich 1981, 244).

12.1.1 First Reasons to Be Skeptical about the Standard Theory

This view of the color-blind is, however, I think, very unlikely in the end to be true. I shall mention right away two difficulties. First, many dichromats clearly *talk* of seeing more than just varieties of yellow and blue. One might suspect they were merely aping other people's words without understanding them; but (as we shall see) at least some of them talk as if they both recognize and experience a much larger range of colors. Dalton, who was, it seems, a deuteranope (Hunt et al. 1995), believed that in the spectrum—for example, looking at light from a prism—he saw only yellow, blue, and perhaps purple (Dalton 1798, 90), but he reported seeing plenty of other colors under other conditions, especially by the light of candles or an oil lamp. Crimson by day, he says, has a blue tinge; but by candlelight it "becomes yellowish *red*"; "Pink by candle-light seems to be three parts yellow and one *red* or a *reddish yellow*" (Dalton 1798, 92, my emphasis). Dalton confuses reds and browns and greens, but he does not talk as if he had just a single kind of color experience produced more or less indifferently by all of them (and, in the standard model, by gray or pale yellow). Rather, it seems, reds and browns sometimes look green, greens sometimes look red or brown—and the experiences in the two cases are different: "A decoction of Bohea tea, a solution of liver of sulphur, ale, &c. &c., which others call brown, *appear to me green.*" By contrast, "Green woollen cloth, such as is used to cover tables, *appears to me* a dull, dark, *brownish red* colour" (1798, 92, my emphasis). Dalton does not talk as though he had just a single pair of hues. Things may, it seems, look red or look green to him, and not just yellow or blue—neither of the first two seeming merely a variety of the latter two (or of gray). He talks as if he had a system of colors much like other people, but with a tendency to get experiences of one color where a normal person would get experiences of another. Reports from other people point to a similar condition. The reports are of course not to be taken automatically at face value; but we shall have to investigate whether they might in fact be true.

Secondly, there is experimental evidence from many directions that the color-blind do not regularly make the kinds of confusion that they are supposed, on the standard model, to make. Dorothea Jameson (Jameson and Hurvich 1978) devised a fascinating experiment with the Farnsworth D-15 test, in which two out of three protanopes, having just produced a characteristically confused ordering of the little colored caps in the Farnsworth test, then go on to give the color names of the caps no less correctly than a normal trichromat. It seems that many dichromats can often, notwithstanding their failings, in some way recognize red caps as red and green caps as green. (The Farnsworth test involves fifteen little caps, each containing a circular patch of color, about 12.5 mm in diameter, set in a black plastic surround, to be placed in order of color similarity, starting with a reference cap, which is blue.) There has in fact been a stream of other reports of red-green discrimination in certified dichromats (see section 12.5): a selection would include Nagel (1905, 1908, 1910); Scheibner and Boynton (1968); Smith and Pokorny (1977); Nagy and Boynton (1979); Nagy (1980); Montag and Boynton (1987); Montag (1994); Crognale et al. (1999); and Neitz et al. (1999). The evidence seems to be that a majority of dichromats who confuse red, green, and yellow spectral lights when they are presented in small fields of 1° or so,[8] actually manage to distinguish those colors when they are presented in larger fields, especially of 10° or more. They may do better with surfaces rather than spectral lights, and (when presented with surfaces) with fairly bright rather than dim illumination, and with higher saturation rather than lower. Additional viewing time may also help, along with other factors, which that I explore in section 12.6—for example, seeing things in various kinds of illumination, and letting the eyes rove over the object, seeing it both with direct and with peripheral vision.

Discriminating red and green is one thing; actually seeing—or having sensations of—red and green is another. Jameson and Hurvich, having shown that their two out of three protanopes did well at labeling the red and green caps, insist nonetheless that the success is merely a matter of what we might call "judgment" or "inference," rather than sensation: the dichromats are using some "*rule* . . . for correlating red vs green hue *names* with lightness," while it remains true that "only two hues [namely, yellow and blue] and an achromatic locus are *available*" to them (Jameson and Hurvich 1978, 154, 151, my emphasis; the protanopes are supposedly using the lightness rule "if dark, then red" [1978, 154].) But such rules would not be very effective, and the proposed one doesn't fit the subjects' actual performance. (There are greens that are darker than many reds and reds that are lighter than many greens, and Jameson and Hurvich themselves admit that the one cap in that category in the Farnsworth test was still correctly named by their two protanopes.) And other investigators have reached very different conclusions: Nagel concludes that the majority of dichromats have a sensation of red, and some "remnant" perhaps of green (1908b, 23; 1910). Boynton and Scheibner say, "We must now agree when [the protanope] says that the sensation of red is not unknown to him" (1967, 220). Both sides in this debate make claims that need to be investigated and tested, and we need to ask (as I do in sections 12.7 and 12.9) what kinds of empirical evidence might succeed in deciding between the rival views. But among the many options, we will need to consider the possibility that, even if the information from a dichromatic eye has in some sense only two color dimensions, still the color-blind may, by using cues within that information stream, succeed not only in gathering information about the "missing" dimension of color, but also in synthesizing three-dimensional color experience.

12.1.2 Methods and Accessibility: General Importance of the Issues

The issue is philosophically important, I believe, rather than being merely a question of straight scientific fact, so to speak. First, it raises questions about the ways in which one group of human beings can learn about the experiences of another group with a significantly different sensory system. It is of course quite standard to examine and try to learn from other people's discriminatory and classificatory behaviour, but I shall be particularly interested in how *what people say and do* can tell us about the *structure* and *interrelations* among the colors they see—the patterns of similarity and difference, which colors seem to "contain" others (as we say orange contains red and yellow), which colors are "opposite" to or contrast strongly with others, and so on. This will be important in investigating whether the color space of dichromats really is merely two-dimensional (cp. sections 12.7 and 12.9 below). Secondly, the debate provides some interesting cases in which to examine the relation of *information* to *perceptual experience*, and the relation of the complexity of *receptor systems* to the complexity of the *information* that is gathered with them. We will need to consider, for example, whether the color-blind who correctly call red things red are really *seeing* them as red, or merely "working out" or "judging" that they are (see section 12.7). And if, as I shall argue, there is a good case for saying that at least some dichromats see red and green in addition to yellow and blue, then we will need to reflect on how it can be that in some sense a two-dimensional input can yield something like three-dimensional information and experience. (I offer some mathematical modeling for this in section 12.6). We may have reason to doubt some of the simpler conceptions of the relation of qualia, or "sensations," to their causes.

A third reason for philosophical interest is that we need to keep in mind the question of what in the world colors, primarily, *are*. To be very quick on an issue that I develop a little further (in section 12.8) but must mainly leave for another occasion, we may need both what might be called a more "dynamic" conception of *colors* and a more dynamic conception of our *perception* of them. The case for the latter idea may be an easier one to make: just as feeling the weight of something (to the extent that we do so at all) often involves going through a process of interacting with it—*weighing* it in the hand or lifting it up and down; seeing what it can do by testing it out under a variety of slightly different manipulations—and just as there are things one can identify by touch in the dark but only by *moving* or *rolling them around* in one's hand, so also seeing the shininess of a small piece of paper typically involves not just catching a glimpse of it under a single condition of uniform illumination, but rather letting one's eye rove over it, or moving one's head a little to see how it catches the light at various angles. And the perception of colors may often operate in a similar way. The larger debate on these issues is not for the present place.[9] But if that is more our picture of perception in general, with us harvesting information over time and space in order to synthesize or update our picture of those portions of the world we are navigating our way through (not to mention the closer contact of trade and exchange at ports of call), then it will, I think, seem less strange to allow, as I think we will find we must, that a color-blind person—while lacking at any one moment much of the information available to the normal sighted—may yet succeed over time in compensating in large part for that deficiency. And such a conception of the *perception of color* may in turn both influence and be influenced by our philosophical conception of what *colors* themselves *are*—whether they are, for example, sensations, or dispositions to produce experiences in us, or dispositions to change incident light, or whatever else.

I might mention a fourth and last question, in the epistemology as well as sociology of scientific thought. If, as I think we shall see, there has been evidence over more than two hundred years, often from respected and substantial figures—like Dalton in 1798, George Wilson in 1855, Wilibald Nagel in 1905–1910, Smith and Pokorny in 1977, Robert Boynton and his collaborators repeatedly between 1967 and 1990—that the standard theory of dichromats cannot be correct, at least on what is indiscriminable from what, and perhaps also on what the experience of dichromats is, then we must ask how it is that that standard theory has been repeated again and again, as if it were simply straightforward fact—as if it just *had to be* fundamentally right, even as the best evidence pointed to its being incapable of being strictly true. I find myself wondering whether the situation may not have some of the structure of the early stages of a Kuhnian scientific revolution, where people stand by an established theory, while there is countervailing evidence that is in a sense well known and *noted* but that is held at a distance as being fundamentally unimportant, because there seems no other "paradigm" available within which it could be give a more significant place.[10] Whether the present situation is of that kind will depend, of course, on whether any such new picture or theory develops and meets with success.

I appeal on a number of points to the history of work on color perception over the last two hundred years, much of it published in German. I do so with some delight—partly because I think we benefit from reexamining the empirical bases of the views that have been dominant since Maxwell and Helmholtz in the 1850s. But I do it also because this is a field in which there is a lot that we would not learn at all (including matters of, so to speak, straight scientific fact) if we did not learn it from the history. The dominant views since the 1850s have left little room for even envisaging that with just two color receptor types, perception of more than two principal hues might be possible at all; the dominant trends in psychology have had little time for debates about the experience of the color-blind without a firm experimental grounding, and have had limited interest in devoting the ingenuity that would be required to devise experiments that might actually tell us about it. For this reason it is, I think, particularly interesting—though there are risks, of course, too—to go back to descriptions of color blindness from a time before that model had fully taken hold, for example, to Dalton and George Wilson, and to rare independents like Wilibald Nagel, who know the standard theory as well as anyone but are unashamed to talk of their own evidence and experience as conflicting with it. And, as we shall see from our study (in section 12.4) of Judd's 1948 survey of some eighty years of unilateral cases of color blindness, even what look like authoritative and balanced later reviews of a subject may give practically no sign of the variety of conflicting views and evidence that actually exists in the original reports.

12.2 Some Terms and Tools

12.2.1 Types of Color Blindness: Diagnosis

I present a table from Wyszecki and Stiles (1982), as a summary of the main forms of color blindness (table 12.1). Normal color vision in humans is standardly taken to be trichromatic, evidence for which comes from the fact that, with a set of three suitably chosen lights, we seem to be able, by some mixture of different quantities of those three, to match any colored light presented to us. The main forms of color-blindness that I shall be concerned with are forms of *dichromacy*, where there is such poor

Table 12.1
Salient properties of color defectives

Characteristic	Protanomalous	Deuteranomalous	Protanope	Deuteranope	Tritanope	Rod-Monochromat
Color discrimination through the spectrum	Materially reduced from red to yellowish-green but to a varying degree in different cases		Absent from the red to about 520 nm	Absent from the red to about 530 nm	Absent in the greenish-blue (445 to 480 nm)	No color discrimination
Neutral point (i.e., wavelength of monochromatic stimulus that matches a fixed "white" stimulus)	None	None	490–495 nm	495–505 nm	568 and 570 nm	All wavelengths
Shortening of the red (i.e., reduced luminous efficiency of long wavelengths)	Yes	No	Yes	No	No	Yes
Wavelength of the maximum of luminous efficiency curve	540 nm	560 nm	540 nm	560 nm	555 nm	507 nm
CIE 1931 chromaticity of the confusion point (dichromats only)	—	—	$x_{pc} = 0.747$ $y_{pc} = 0.253$	$x_{dc} = 1.080$ $y_{dc} = -0.080$	$x_{tc} = 0.171$ $y_{tc} = 0$	—
Percentage frequency of occurrence						
among males	1.0	4.9	1.0	1.1	0.002	0.003
among females	0.02	0.38	0.02	0.01	0.001	0.002
Cone types thought to be present	M, M′, S	L, L′, S	M, S	L, S	L, M	None
Standard view of range of hues experienced	Y, B weak R, G	Y, B weak R, G	Y, B	Y, B	R, G	None

Note: From Wyszecki and Stiles (1982, 464, footnotes omitted). Final two rows added by present author, following (1) e.g., Nathans (1999); Neitz and Neitz (2000); (2) e.g., Judd (1948) (for protans and deutans) and Hurvich (1981, chs. 16–17).

discrimination (at least with small fields) that it takes only a pair of suitably chosen lights (e.g., red_{650} and $blue_{460}$), rather than the normal three, to get by some mixture an acceptable match for any presented light. The natural explanation is that dichromats lack one of the standard cone types, as is now well-confirmed: *protanopes* (the "red-blind," in the language of Helmholtz and Maxwell) lack the L cones, the long-wave receptor type, while *deuteranopes* (the "green-blind") lack the M cones, the medium-wave receptor type. Definitive diagnosis of these conditions depends, as we shall see, on tests with a Nagel anomaloscope or similar apparatus. But the general symptoms are these: protanopes and deuteranopes typically confuse reds and greens; in a spectrum (i.e., a real spectrum produced, e.g., by a prism) they mostly see only *yellow* (from the red, orange, yellow, and yellow-green parts of the spectrum) and *blue* (from the blue-green, blue, and violet parts), with the central green zone looking to them more or less colorless, near what is known as the *neutral point* (at ~490–495 nm for protanopes, ~495–500 or 505 nm for deuteranopes). Protanopes have what is sometimes called a *shortened spectrum,* in that, lacking L cones, they see much less of the extreme red end of the spectrum than normal trichromats do. There is a third form of dichromacy, *tritanopia*, where the S cones are missing: the tritanope too is said to see only two hues, standardly taken to be red and green. But the condition is extremely rare, and I shall say almost nothing more about it here.

There are less extreme forms of color-blindness known as *anomalous trichromacy*. The two main varieties, known as *protanomaly* and *deuteranomaly*, involve a person's making similar but milder confusions to those found in the corresponding forms of dichromacy, *protanopia* and *deuteranopia*. Reds and greens are confused in ways resembling the dichromat confusions, but testing reveals that the condition is one of *trichromacy*, in that a set of three lights is typically needed in order to match any given light, though it is *anomalous*, in that the matching is somewhat different from normal. Until recently, there were a number of competing hypotheses about the nature of these conditions (see Wyszecki and Stiles 1982, 462–463; Mollon 1997). The accepted view today (see, e.g., Nathans 1999; Neitz and Neitz 2000; Deeb 2005) is that a *protanomalous* person is lacking L cones (like a protanope), but has two varieties of an M-cone pigment type with slightly different wavelengths of peak sensitivity (we might call these M and M′ cones), so he can compensate to some extent for the absence of the L cones. A *deuteranomalous* person has no M cones but has two varieties of the L-cone type (L and L′), which, again, allows some compensation for the deficiency. There is much variation in the degree of severity in cases of anomalous trichromacy; the defect is often said (though there are certainly exceptions) to be less severe the larger the separation between the peak absorptions of the two varieties of the remaining L- or M-cone type. There are extremely interesting questions about how we should characterize what anomalous trichromats see—but I shall confine myself here to the more severely affected groups, the dichromats.

The main tools of diagnosis, from the 1880s onwards, can be divided into two kinds: those for the practicing doctor and those for the experimental scientist. For the doctor, there are pseudoisochromatic plates like those of Stilling (Leipzig 1877–1879, rev. 1922, 1952) and Ishihara (1917), which are still in use today. These are like a pointillist painting made up of colored dots in which certain patterns and shapes are visible or salient for one kind of observer but not for another. (In one Ishihara plate, the normal read the figures 29, and the red-green blind read 70.) The easily administered confusion tests with wools or colored papers were by the late nineteenth century known to be very imperfect—for the telling reason that too many color-blind people, even dichromats, proved capable of "passing" them.[11] Improved tests used small colored caps to be placed in order of color similarity, as with the Farnsworth 100-hue and D-15 tests (Farnsworth 1943).[12]

For the experimental scientist, more precise investigation—both of color blindness and of normal color vision—involves metameric color matching: that is, seeing how one color can be matched with a mixture of other colors that is different in physical composition. In his first researches, Maxwell used spinning discs or tops to produce an optical mixture of variable-sized segments of colored papers that could be attached to the top (Maxwell 1855). But much more revealing was the "spectral apparatus" that Maxwell soon had made, which displayed two adjacent fields of light: one field of spectral light,[13] to be compared with a second field, containing a mixture of two or more kinds of spectral light in adjustable combinations (see Maxwell 1860, and figure 8; cp. Nagel 1914, 63–80). A later version of the same kind of apparatus was the colorimeter of W. D. Wright, which was used in gathering much of the data on which the CIE (Commission Internationale de l'Éclairage) colorimetric system was based (Wright 1927, 1946, chs. 3–4). In the earlier days, the fields were, for technical reasons, bound to be small or desaturated or impure—and a virtue was made, in much of the quantitative work, of the near necessity of small fields of 1° or 2° or less. Nagel's anomaloscope (as we shall see) uses small fields; but in other work, Nagel is a rare exception, expressing his delight in 1910 at the new apparatus he had built in Berlin, which allowed large fields of 20°—and he sometimes even used fields of 1/2 to 3/4 m^2 viewed from a distance of 50–75 cm, which might extend over an angle of 70° or more (Nagel 1910, 6–7).

The easily administered tests, like the Ishihara plates and the Farnsworth D-15 cap test, have the important weakness that they may make clear, for example, that a person has some kind of protan deficiency without determining whether he is a protanope or merely protanomalous. A differential diagnosis depends on the Rayleigh matches that he makes—that is, the proportions of red and green light in a mixture that he accepts as matching a given yellow. (Think of the way yellow is produced on a TV screen, by a normal combination of the red and green phosphors. Then add the fact—first pointed out by the physicist Lord Rayleigh in 1881—that some color-blind people [the anomalous trichromats] need abnormal proportions if they are not to find the mixture either "too red" or "too green"; while—pretty much as Helmholtz and Maxwell had earlier said—others [the true dichromats] are so indiscriminating that they accept almost any red-green mixture [abnormal or normal] as "yellow," just as long as it is bright enough.) These variations can be tested in many ways, but for relative simplicity and convenience, the physiologist Wilibald Nagel perfected, in about 1907, an anomaloscope that it is still a standard tool today, along with more recent designs that operate on similar principles (see figure 12.2). Nagel designed the apparatus to present one field of yellow_{589}, which could be adjusted in brightness, and a second field containing a mixture of red_{671} and green_{535} in adjustable proportions. Anomalous trichromats are identified as those who need either more red or more green than normal, in order to match the yellow; while dichromats are those who accept almost *any* mixture of the red and green, as long as the yellow is suitably adjusted to match it in brightness. (A rough explanation for the dichromat confusions is this: the S cones have practically no sensitivity in the green-yellow-red part of the spectrum, above ~530 nm; so dichromats, presented with the stimuli from that part—namely, 535, 589, and 671 nm—will have only one cone type to perform the task of discriminating them: hence a suitable adjustment in brightness should be able to make any two of these stimuli seem equivalent.) When authors describe someone as a *dichromat*, the standard criterion—the more or less definitive test—is this acceptance of red_{671}, green_{535}, and yellow_{589} (or similar lights) as more or less interchangeable and indiscernible in 1° to 2° fields in an anomaloscope, as long as the fields are suitably adjusted in brightness.

12.2.2 Chromaticity Diagrams

The standard model of color blindness can probably be understood only with a grasp of the principles behind chromaticity diagrams and colorimetry in general; but I shall say just a few things here that, with some examples, will, I hope, be adequate for now.[14] A newcomer who wants to move on to the later discussion of color blindness may find that it is enough, for a first impression, just to examine figures 12.3 and 12.4, with their captions, and come back to the main text later.

I later make much use of the *x*, *y* chromaticity diagrams of the CIE (Commission Internationale de l'Éclairage), which provide a very helpful framework for presenting a lot of color information. The particular form of these diagrams was standardized in 1931, on the basis of color-matching data that had been gathered in the 1920s by W. D. Wright and others. They are part of a whole color measurement system that has mathematical characteristics that it would be inappropriate to explore here. But the basic idea may be understood as an application of the same principles as in a Maxwell triangle (figure 12.3). Taking three reference lights (for example, blue, green, and red of 460, 530, and 650 nm), we may use a triangle to represent the various proportions

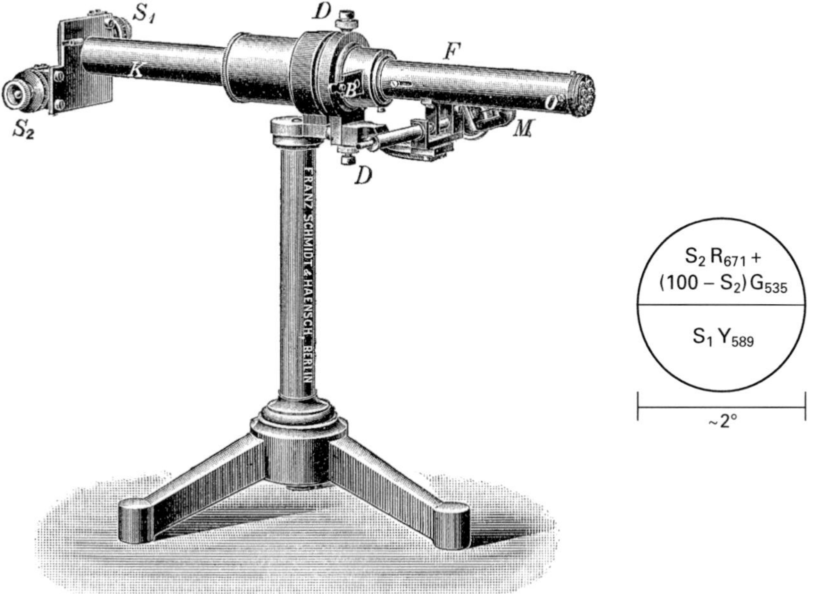

Figure 12.2

Nagel anomaloscope model II (Nagel 1914, 74; cp. 1907c; 1908b, 35–37). When pointed at a light source farther to the left (not illustrated), the instrument presents a circular field, the lower half of which is illuminated with spectral yellow light (e.g., 589 nm), and the upper half with spectral red or green (e.g., ~671 and ~535 nm), or a mixture of the two in any desired proportion. (See diagram on right.) The instrument is built on the principle of a spectroscope that uses prisms to create a spectrum from a light source, and slits to select desired portions from that spectrum. The light source is placed at the extreme left on the axis of collimator tube K (Kollimator; a gas or petrol lamp was typically used, or now an electric bulb, set about 18 cm from the end of the tube); the prisms are in the center of the apparatus; the resulting image is viewed through lens O (Okular) of the telescope F (Fernrohr). At the far end of tube, the left screw (S_2) controls the mixture in the upper half-field (from pure green_{535} at a setting of 0 to pure red_{671} at 100); the right screw (S_1) controls the lightness of the yellow_{589} (from 0 for darkest to 100 for lightest). For the visual angle, see Pokorny et al. (1981, 25), referring to Nagel anomaloscopes available at the time; Farnsworth had earlier had a model II with an adjustable diaphragm (controlled by B [Blende] in the illustration), giving visual angles of 1° 15′, 2° 10′, or 3° 15′ (Willis and Farnsworth 1952, 29).

Some illustrative examples of settings (figures in italics correspond to settings on the screws S_1 and S_2: they should not be read as exact quantitative measures (cp. Nagel 1908b, 36–37)):

Normals: e.g.,
$\mathit{55}\ R_{671} + \mathit{45}\ G_{535} = \mathit{14}\ Y_{589}$

Protanomals (need more red, and the red light is of relatively low luminance to them): e.g., $\mathit{70}\ R_{671} + \mathit{30}\ G_{535} = \mathit{4}\ Y_{589}$

Deuteranomals (need more green): e.g., $\mathit{42}\ R_{671} + \mathit{58}\ G_{535} \approx \mathit{14}\ Y_{589}$

Protanopes and deuteranopes are so undiscriminating as to accept practically *any* ratio of R_{671} and G_{535} as matching the Y_{589}, given suitable adjustments in brightness.

in any combination of those three lights. We can "measure" the color of any given light by specifying the mixture of our three reference lights that would (for a standard human perceiver) be a match for it; and we can "measure" the color of a surface by doing the same for the light that it reflects under some particular kind of illumination. The theory and practice of this are a subject in themselves; the relevance of this to color blindness is that the same diagrams as are used to specify the colors of things to normal trichromats are also often used to specify the patterns of confusion found in dichromats—and they are a promising, though sometimes misleading, guide to what dichromats do and do not see.

Where do the colors of the spectrum—that is, the ideally narrow-wave bands of light coming from a prism or diffraction grating—belong in such a diagram? Answer: spectral lights are higher in saturation than anything that can be produced by a mixture of other lights, so they cannot be mapped within or on the boundaries of a triangle like the one in figure 12.3. But spectral lights stand in the same kinds of relations to the lights we already have in our triangle as those lights stand in to each other, and they can be given a place on the same plane, but outside the triangle. This will be clearer from an example. To match a spectral yellow_{580}, we can find a mixture of the red_{650} and green_{530} primaries that is a good match in hue; but any such mixture will be less saturated than the pure yellow_{580}.[15] If, however, we take a little of the blue primary to desaturate the yellow_{580}, then we can get an apparently perfect match. We discover experimentally an equivalence such as this:

$$10.0\ Y_{580} + 0.11\ B_{460} = 2.88\ R_{650} + 3.26\ G_{530}$$ [16]

We may then rephrase the color-matching information in the following form, and treat the right-hand side as specifying an equivalent for our 10.0 units of Y_{580}:

$$10.0\ Y_{580} = 2.88\ R_{650} + 3.26\ G_{530} - 0.11\ B_{460}.$$

And that allows us to fix a position in the diagram for Y_{580}, just outside the main triangle, slightly higher and to the right of the $R_{650}G_{530}$ line.

An experimenter could go on to do similar tests for the other kinds of light in the spectrum—seeing what quantities of our R, G, and B primaries are needed to match light of a range of wavelengths, e.g., 400 nm, 410 nm, and so on, through the visible spectrum (or, more exactly, seeing how each light, if desaturated with one of the primaries, can be matched with some combination of the other two). This is the kind of work that was done by Maxwell (1860), and redone with greater accuracy in the 1920s by J. Guild and W. D. Wright. The kinds of results found are illustrated in figure 12.3—and with suitable averaging, smoothing, and other processing, they became the basis of the standard system of colorimetry adopted and published by the CIE in 1931.

It is a high-level fact of the human visual system, known as Grassmann's third law, that (within limits) if lights *a* and *b* match each other, then they will behave

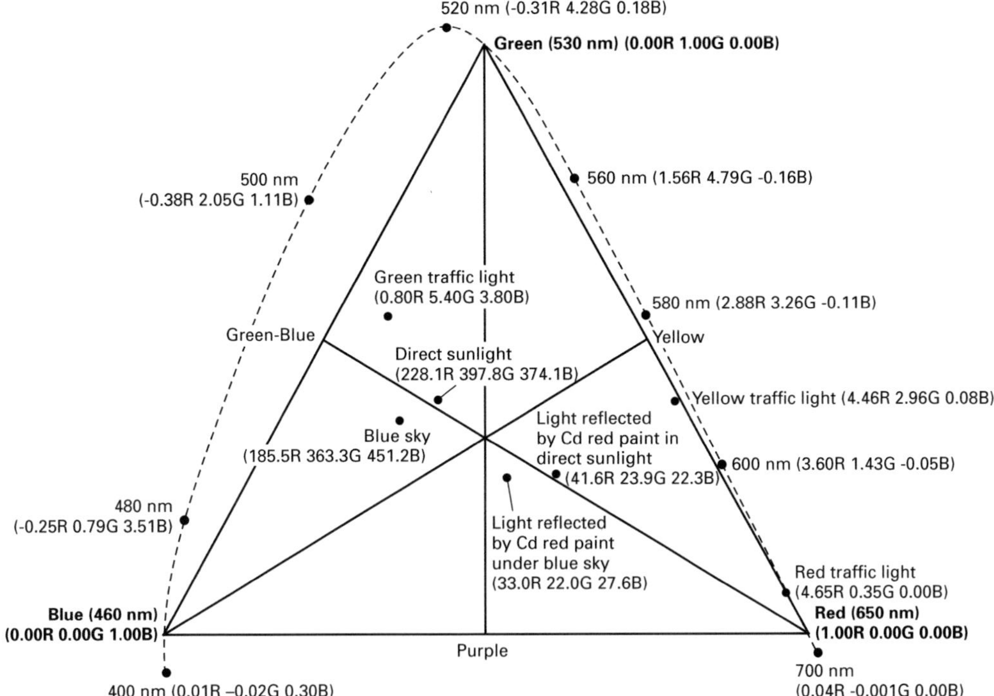

Figure 12.3

Maxwell triangle: mixtures of three "primaries" in different proportions and the mixtures required to match various standard light sources. Three reference lights, or primaries (red_{650}, $green_{530}$, and $blue_{460}$), are represented at the three vertices. Mixtures of those lights in different proportions are represented by points on the plane whose "closeness" to the three vertices represents the relative amount of the corresponding three kinds of light in the mixture. (A mixture of n units of light A and m units of light B will be represented at a point M on the straight line between them, such that $AM/MB = m/n$. In general, for a mixture of several kinds of light, if we associate with each component a weight proportional to the number of units of that light and place that weight at the point representing the component, then the resulting mixture will be represented at the center of gravity of the resulting system of weights.) Points mark mixtures found to match various common kinds of light (traffic lights, blue sky, etc.) and (on the dotted-line curve) light of different kinds through the visible spectrum from 400 nm to 700 nm. (In parentheses are the representative amounts of the three primaries needed to match the various lights. Absolute intensities are only illustrative.) The three primaries and the relations between units for them are set as in the W. D. Wright system (e.g., 1946, 125–135). White would normally be in the center of a triangle like this, but it is here displaced to the left and slightly up (to a position roughly between "direct sunlight" and "blue sky"), because of the relatively large size of Wright's units of red_{650}. The positions of points on the curve (and the tristimulus values in parentheses) are my own recalculation by means of a linear transformation of CIE *XYZ* functions. Positions of common objects within the triangle (and tristimulus values) are illustrative only: for railroad signals, see Judd (1952, 174); for light reflected from paints, see section 12.6. For the general principles of such representations, see Newton (1704/1730), 154–158; Mayer (1775); Maxwell (1855, 1860); Hunt (1987, 58–60).

indistinguishably when used as constituents in further mixtures: if *a* matches *b*, then *a* mixed with any further light *c* will match *b* mixed with *c*, regardless of what differences there may be in the wavelengths of light composing *a* and *b*.[17] Hence, if we have set up the diagram treating one set of lights (e.g., R_{650}, G_{530}, and B_{460}) as primary, with further lights (e.g., 400, 405, 410 nm) being defined in relation to them, we may go on to recognize that each of the first set of "primaries" can itself be matched by a combination of other lights (e.g., R_{640}, G_{510}, and B_{440}) in appropriate quantities—and each could be replaced, for any mixing-and-matching purposes, with the equivalent mixture; hence, if we know the amounts of the first primaries needed to match some particular light *L*, we can also calculate what quantities of the second set (e.g., R_{640}, G_{510}, and B_{440}) would be needed to match it. In effect, we can translate between one set of "primaries" and another. And we may treat a diagram like this as representing a space of *colors* of lights (or strictly, their *hues and saturations*, brightness being a third dimension not here represented): two lights will have the same position on the diagram if and only if they are indistinguishable in hue and saturation; the color of any mixture of lights will be determined on similar center-of-gravity principles; and we can jettison any original assumption of any particular set of lights as "primaries."

A Maxwell triangle is standardly an equilateral triangle. But there is nothing compulsory about the exact positioning on the page of the vertices that represent whatever primaries or other reference points we may adopt. In fact, the same set of color-matching data can be presented in different diagrams, according as we place the vertices in a different layout—stretching and rotating (or indeed reflecting) the triangle in various ways—and changing, if we wish, the size of the units for each of the reference lights. A diagram based directly on our real R, G, and B primaries needs, as we have seen, negative quantities to specify certain highly saturated lights; but we can lay out the same data differently, with axes placed to avoid any negative quantities. This was done in the classic CIE 1931 system, which transformed information originally gathered in terms of red, green, and blue primaries in terms, instead, of new *X*, *Y*, and *Z* primaries, which might be thought of as imaginary supersaturated red, green, and blue lights (see figure 12.4).[18]

Our first Maxwell triangle separated off the *relative* amounts of our three primaries from the absolute amount: our CIE *x*, *y* chromaticity diagram—which can be thought of as a projective transformation of the Maxwell triangle—is similar. It separates off *x* and *y*, which together capture something like the hue and saturation of a light, and which together are said to constitute its *chromaticity*,[19] leaving *luminance*, the psychophysical correlate of brightness, as a third dimension to be specified separately and represented by a third variable, *Y*.[20]

It should be remembered that the data behind the classic CIE diagrams are simply "the average colour matching properties of the eyes of 17 British observers" (Hunt

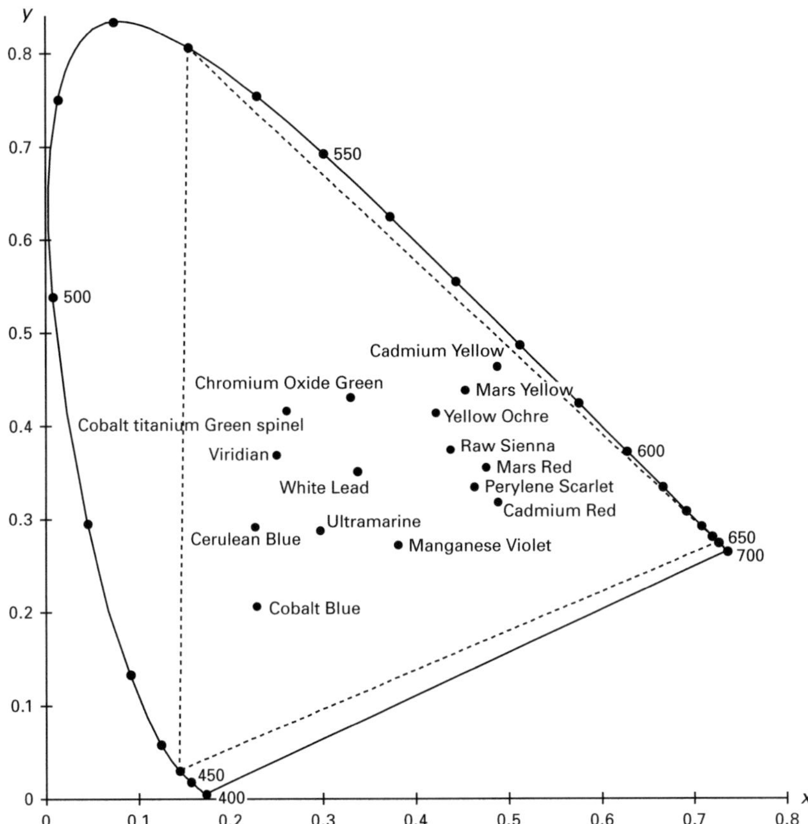

Figure 12.4

CIE 1931 *x*, *y* chromaticity diagram, showing locus of spectral hues (on the plectrum-shaped curve) and chromaticity coordinates for fifteen standard pigments when illuminated by direct sunlight (of correlated color temperature 5500 K: for explanation, see *Note on terminology* in note 13). The diagram can be thought of as an alternative mode of presentation of the same kind of color-matching information as in figure 12.3, where B_{460}, G_{530}, and R_{650} were our "primaries." The transform of our original Maxwell triangle appears here in dotted lines; and the geometry of the plectrumlike curve indicates what proportions of such primaries would be needed to match light of the various wavelengths that make up the visible spectrum. Similarly, the position of each of the fifteen pigments (or, strictly, of light reflected from them when illuminated with standard illuminant D_{55}, approximating direct sunlight [5500 K]) represents the proportions of such primaries in a mixture that would match light from that pigment. One may also think of the new representation as specifying the character of any given light, by giving the proportions of some new *X*, *Y*, and *Z* primaries that would be needed to match it, where *X*, *Y*, and *Z* can be thought of as imaginary supersaturated red, green, and blue primaries, which would be represented respectively at (0, 1), (1, 0), and (0, 0) in this figure. See text for further details. (CIE diagram: e.g., Wright 1969, ch. 4; Hunt 1987, chs. 2–3. Chromaticity of 15 pigments: author's calculation, from pigment reflectance functions in Mayer 1991, 65–134, and relative spectral power distribution function for CIE illuminant D_{55} (e.g., Wyszecki and Stiles 1982, 8–9), sampled at 20 nm intervals from 400 to 700 nm.)

1987, 44). Even when the color-matching functions are now published at up to twelve decimal places, for spectral lights taken at 1 nm intervals (as in Wyszecki and Stiles 1982, 725–735), they derive from a small number of people who do not agree perfectly among themselves. There are some known faults with the color-matching data (see, e.g., Wyszecki and Stiles 1982, 330–332; Stockman and Sharpe 1999) and there are more sophisticated color-matching functions that have been developed since, by the CIE and others. But the basic CIE 1931 framework is mathematically developed and still serves as a fundamental reference system in color science. The inaccuracies have been described as "negligible in most practical situations" (Hunt 1987, 44); and I shall develop my own discussion here in terms of the 1931 system. It would obviously be possible to use more exact functions, and for certain purposes it would be better. But it would not be of much significance in the present context: first, I shall be discussing classic conceptions of color blindness that were developed in accord with just the kind of color-matching functions that are used in the CIE 1931 system. And secondly, if we used any of the modified systems that I am aware of, the points of principle that I shall be discussing would arise in much the same way.

So far we have been mostly representing lights, rather than surfaces; but the system can be applied in a similar way to represent the character of *light reflected* from a surface under a particular kind of illumination: we assign it a point in the diagram (as with the pigments in figure 12.4) that represents the character of a mixture of reference lights that would be (for a normal human perceiver) indiscernible from the light reflected from the relevant surface in that context of illumination.

It is worth asking precisely what is represented by position in a chromaticity diagram. People are sometimes tempted to say, "How the object looks." But strictly, the diagram represents not the sensations of subjects, but stimuli in the world—though grouped according to a relation of phenomenal indiscernibility (to standard human perceivers). A light stimulus is given a particular position in the diagram according to the proportions of red, green, and blue primaries (or any equivalent to them) that would be needed for an apparent match or substitute for it (for standard human perceivers).

One might be tempted to say, then, "The chromaticity diagrams tell us *which things look the same*, though not how those things look." But that is only true in a limited way. *In any one context*, two light stimuli with the same position in the diagram (and the same luminance) will look the same, and can be substituted for one another without the color appearance changing; but stimuli with the same chromaticity and luminance may look quite different, if they are in different contexts—because of adaptation and "reading" of the general scene, and other factors yet.

An example should make this clear. A blue card in yellow light may be reflecting light to the eye that matches a certain mixture of our primaries—a mixture that is (we may suppose) in fact white, more or less equivalent to CIE standard illuminant C. If so, it will be mapped at position $x = 0.310$, $y = 0.316$. Similarly, a yellow card in blue light may also be reflecting light that matches that same combination of primaries (and, similarly, illuminant C); it too will be mapped at $x = 0.310$, $y = 0.316$. And yet, it may be (if a viewer adapts to the two different illumination conditions) that the first card *looks blue* (as indeed it is); and the second card *looks yellow* (as it too is). They are mapped at the same position in a CIE diagram because the light reaching the eye from the two of them would be matched with the same combination of reference lights. But that fact does not fix the things' respective appearance: one card looks blue, and the other looks yellow. And two such lighting environments may even be both present at one time in different parts of one person's visual field: think of sitting inside by a window, reading by reflected light that is mostly sky blue, while things on the window-ledge outside can be seen bathed in a pool of yellow sunlight.

This will be of some importance. It is sometimes tempting to think that if the *chromaticity space* of an observer is restricted in some special way, then her *experience* is automatically restricted in a corresponding way. (So, if we found that the chromaticity space of dichromats reduced to a line, we might think their experienced color space, aside from brightness, would similarly reduce to a line.) But there is no need in general for this to be true—given the remarkable adjustments that the visual system makes to adapt or compensate for abnormal conditions and "restriction" or "bias" in its input—and in some cases it is definitely false. Color constancy is one well-known form of such adjustment. But it may be worth mentioning some more extreme cases too—and in particular the impressive demonstrations of Edwin Land (see, e.g., Land 1959a, 1959b), where—to take just one example—if we have two suitable color-separation photographic transparencies of the same scene (one taken through a red filter, the other through a green filter, both on black-and-white film) and project them superimposed, one with red light and the other with white, then viewers report seeing the original scene, not just (as we might expect) in varieties of red, white, and black, but in virtually all color categories, including red, yellow, green and (though less successfully) blue. (Land's scenes included a bowl of fruit on a patterned tablecloth, and a still life with packs of Kleenex, Shredded Wheat, Jell-o and Campbell's Tomato Soup.) The basic phenomenon—indeed Land's original observation in 1955 (Bollo 1959)—was that, starting with a standard projection of red, green, and blue "separations" onto a screen, one can turn the blue projector off and still (with some adjustment of the brightness levels) get an impressive range of color impressions; and he found that the two projector lights can even be varied considerably in color and brightness from the original red and green, without the effects being lost.[21]

Land explored further varieties of such phenomena, using a *dual monochromator* that allowed the two color separations to be illuminated (or transluminated) with narrow wavebands of almost monochromatic light. With, for example, orange_{615} and green-yellow$_{560}$ as illuminants, one might expect people to see nothing but orange and greenish-yellow and the yellowish hues in between, all at virtually maximum saturation since there would be nothing to desaturate the other two. But people actually report seeing things in virtually every color category, including blue and white. The phenomenon is even more remarkable if illustrated in a chromaticity diagram (see figure 12.5). One short *straight line* in chromaticity space, on either side of yellow—from yellow-red$_{615}$ to green-yellow$_{560}$—ends up producing a range of experience (including white) that is almost wholly outside what is usually associated with that line in the diagram.

I mention this partly because, if it turns out that some dichromats experience more than just varieties of yellow and blue, then their condition will be structurally parallel to that of normal trichromats presented with Land's demonstrations: whatever differences there may be in the exact causal factors at work, it would seem that in both cases, 2-D color input is producing more than the standardly expected 2-D range of experience. Land's subjects are in effect turned into temporary dichromats by the loss of a third color signal; but they evidently do not end up with experience of merely two main hues—and it would be nice if this could guide us in understanding the experience of dichromacy. Unfortunately, I do not think we have much understanding of the Land phenomena themselves: they are certainly not explained by Land's own retinex theory (Land 1977) nor, I think, by the critics (Judd 1960; Walls 1960), who have invoked low-level mechanisms of adaptation and contrast, and the Helson-Judd theory of "color conversion" (Helson 1938; Judd 1940). (Those accounts have implausibilities in themselves, and they make no sense of the great increase in vividness and range of color seen when the two images are exactly in register.[22]) I suspect we will not have a good understanding of the colors seen by dichromats until we have a greater understanding of the colors seen with Land phenomena too—and the latter, I think, remain extremely puzzling. Some of the factors I model in section 12.6, such as those involving the macula, are of no relevance to the Land phenomena; but others may have a role in both kinds of case—like the fact of individual surfaces being simultaneously presented under the modulating effects of a variety of kinds of illumination and spatial orientation—though we are only at the beginning of an understanding, I suspect, of how such factors actually operate.

There are limitations to chromaticity diagrams like these. They can be accurate only to the extent that normal human color vision is genuinely trichromatic. If instead, for example, rods—which are usually thought to be involved mainly in vision at night—play a significant role also at higher light levels, then two samples that stimulate the cones in similar ways, and are (on the basis of colorimetric data) assigned the same position in x, y, Y space, may after all (even in the very same context) look different from each other. I shall not consider that fundamental issue here, though it is certainly not to be dismissed.[23] Equally, if (as is the case) in the population there

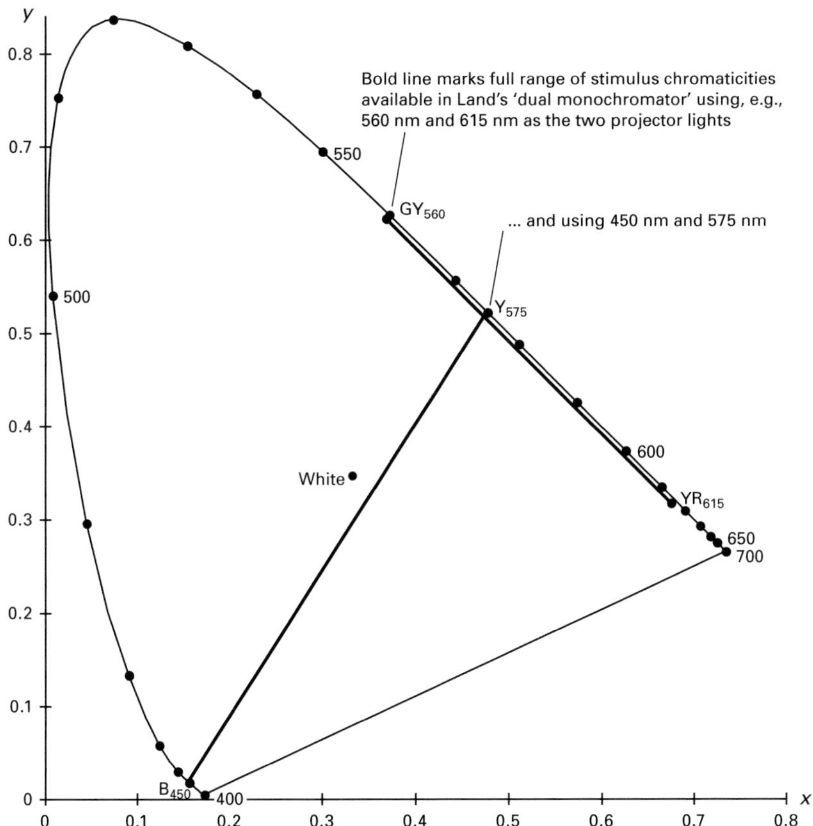

Figure 12.5

Chromaticities of stimuli available in Land's two-projector experiments: another collapse to a line. When viewers are presented with two black-and-white color-separation transparencies, projected in Land's dual monochromator with yellow-red$_{615}$ light and green-yellow$_{560}$, respectively, the chromaticities available fall merely on a straight line between those two projector colors; but the viewers report seeing not just varieties of yellow, but red, green, gray, white, and even some blue—colors wholly outside the range of experiences normally associated with that line in chromaticity space. With blue$_{450}$ and yellow$_{575}$, the stimuli range "objectively" from that blue to that yellow, via a slightly pinkish white; but viewers report seeing not merely white, yellow, gray, and blue, but also red, orange, brown, and green. (One might note the similarity between Land's yellow$_{575}$-blue$_{450}$ line and the yellow$_{575}$-blue$_{470}$ line that Judd (1948) thinks captures the full range of color supposedly seen by dichromats.) This diagram illustrates experiments reported in Land 1959b, 641. The representative white is D_{55}.

are variants of the three main cone pigment types, then the color-matching functions (and hence the layout of spectral hues and other stimuli in a chromaticity diagram) will not be exactly the same for the various groups with the various types of receptor. And if there are individuals who not only have L, M, and S cones, but also more than one variant of some of the cone pigment types, then their color matching may be impossible to represent solely in a three-dimensional space. I shall try, however, to abstract from such issues here. I shall argue that the evidence suggests that the people standardly classified as dichromats do not, as the standard model implies, see merely two principal hues: they seem to see more colors than they "should." And I shall investigate ways in which this might be achieved, even if those subjects are indeed strictly—as far as the number of retinal receptor types is concerned—dichromats.

12.2.3 Brief Statement of the Standard Theory

There are two main claims to what I shall be calling the standard theory of dichromat perception: (1) that dichromats make confusions among stimuli that lie along confusion lines like those marked in the diagrams that follow. For the relevant form of dichromat, (a) all the lights represented on one confusion line are indiscernible (except perhaps in brightness), and (b) they produce only a single kind of color experience (with a similar proviso). One might say a whole confusion line collapses to a *point*, representing a single color perceived by the dichromat (with variation only in brightness) in place of the whole variety of colors that normal trichromats would perceive given the variety of stimuli represented on the line.[24] What is more, it is thought that the single color points for each of the many confusion lines all line up neatly on a straight line (rather than, for example, forming an arc or curve): thus, (2) dichromats see colors lying on a single straight color axis, standardly taken to run from yellow_{575} to blue_{470}, via neutral.[25] The full range of colored things, both lights and surfaces, looks either yellow or blue, in varying degrees of (perhaps vanishing) saturation and brightness.

The two parts of the theory can be held separately.[26] But the main claims go naturally together and might easily seem absolutely secure achievements of color science. The main support for (1) lies in some theoretical considerations in Helmholtz and some empirical work in Pitt (1935). Support for (2) comes partly from general theoretical considerations about opponency in the visual system, but above all from reports of some rare cases of color blindness in one eye, where the other eye is near-enough normal for it to be used as a reference point in identifying the perceptions of the weaker eye. The combination of these views—which is shared by Judd, Le Grand, Hurvich and Jameson, and many others—is illustrated in figure 12.6. My own view is that we have strong reason to doubt the foundations of both parts of the theory. But I shall go relatively briefly over those reasons (in sections 12.3 and 12.4), since the core of the present discussion (in section 12.6) is, rather, to propose some hypotheses about *how*, if dichromats do escape the consequences of these arguments, they may be achieving their success.

12.3 The Confusion Diagrams and Their Experimental Basis

Where do confusion diagrams like this come from? There is an extremely interesting theoretical argument in Helmholtz that I can only touch on here;[27] for practical empirical support, Pitt (1935) is usually cited as giving something like definitive evidence. Unfortunately, that evidence is not quite what it appears to be.

The confusion diagrams like those in figure 12.6, are extremely attractive and powerful. They systematize a huge amount of material (whether that material is strictly accurate or not) and allow us to make good sense of, for example, the neutral point and patterns of common confusion. The diagrams seem to show particularly clearly the relation of dichromatic systems to trichromatic systems (as *reduction* systems, rather than as, with anomalous trichromacy, *transformation* or *modification* systems [for these terms, see Nagel 1908a]). And if a person asks why the loss of a long-wave receptor should affect perception of green as well as red, the relevant diagram offers a helpful answer: things lying on any protanope confusion line are discriminated by normal trichromats only thanks to their possession of the L cones (which effectively tells them *how close* [on the diagram] a stimulus is *to the protan confusion point,* i.e., to the point representing the L cone): hence, if a green and a red lie at opposite ends of a single confusion line, someone lacking an L receptor cannot be expected to be able to discriminate them. In general, we should expect loss of "red" cones to affect the recognition of greens no less than of reds—which is one reason why *L cone* is a better term than *red cone,* and why von Kries's term *protanopia* is better than the older term *red-blindness.*

But are the diagrams in fact correct? How much are they based on relatively direct evidence, and how much on elaborate theory? The main source is Pitt's 1935 study, which presents diagrams that (when converted into the CIE system) are equivalent to those given in figure 12.6 (except for the labeling of the color regions and the hue line that I have inserted, following Judd 1948). Pitt's diagrams (see figure 12.7) even include the chromaticities for dozens of paint colors in some standard illuminant. And one might easily by tempted to read off from them such claims as that a deuteranope would find indistinguishable (except perhaps for a difference in lightness) the purple and light green paints in Pitt's confusion class 19, or the yellow, orange and red in class 27.

The fact is that Pitt's experimental data on dichromat confusions involved only lights, not surface colors, and among lights, only two rather restricted sets. What Pitt tested was simply which blue_{460}-and-red_{650} mixtures could be found to match (for a relevant dichromat) each of a series of spectral lights, about thirty in number, selected

a)

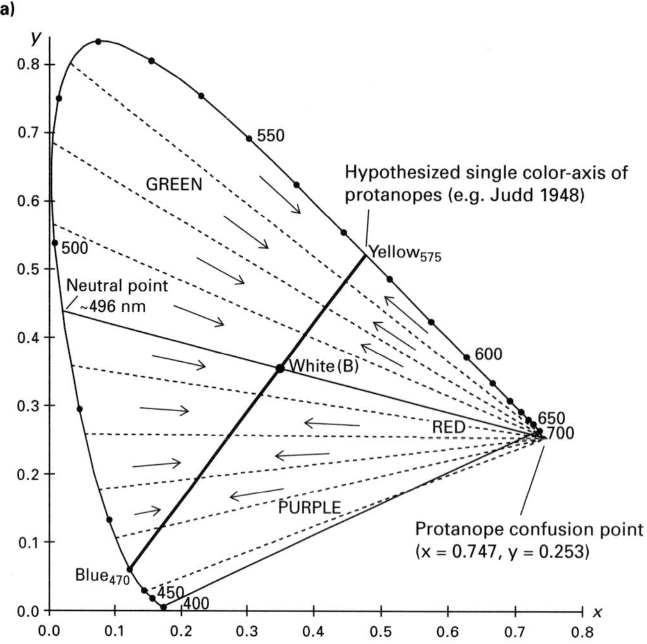
y
x
550
600
650
700
500
450
400
GREEN
RED
PURPLE
Hypothesized single color-axis of
protanopes (e.g. Judd 1948)
Yellow$_{575}$
Neutral point
~496 nm
White (B)
Blue$_{470}$
Protanope confusion point
(x = 0.747, y = 0.253)

b)

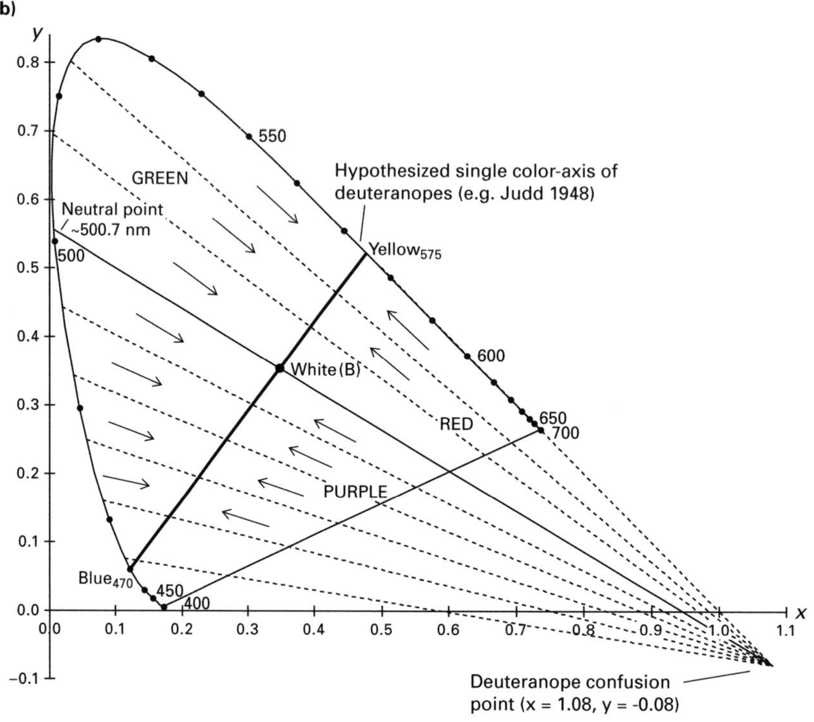
y
x
550
600
650
700
500
450
400
GREEN
RED
PURPLE
Hypothesized single color-axis of
deuteranopes (e.g. Judd 1948)
Neutral point
~500.7 nm
Yellow$_{575}$
White (B)
Blue$_{470}$
Deuteranope confusion
point (x = 1.08, y = -0.08)

Figure 12.6
Standard theory of dichromacy: confusion lines (dotted), and how stimuli in CIE 1931 chromaticity space are supposed to look to protanopes (a) and deuteranopes (b). These diagrams combine two ideas: (1) the use of confusion lines to indicate sets of stimuli supposedly confused and (2) the claim that each confusion class is experienced by a dichromat as merely a more or less desaturated variety of a unique yellow and blue. Dotted lines indicate sets of stimuli confused by the relevant kind of dichromat. The *confusion point* (or *convergence point*) can be interpreted (for followers of Maxwell and Helmholtz) as corresponding to the *primary sensation* (red for protanopes [a], green for deuteranopes [b]) that is present in normal trichromats and absent in the relevant form of dichromat. (More abstractly, it represents an imaginary supersaturated red or green stimulus that, if presented to a normal trichromat, would stimulate exclusively the cone type that the relevant type of dichromat lacks—a red or green that the dichromat is wholly "blind" to.) The *neutral point* is the point in the spectrum that looks neutral to the relevant type of dichromat, or that can be matched with *white* (here taken as CIE illuminant B). The dichromat's color experiences are supposedly merely those that a normal trichromat would typically get from colors lying on the line joining blue_{470} and yellow_{575} (marked as a bold line). The normal chromaticity plane is divided into two zones by the solid line running between the confusion point and the neutral point. Stimuli on the line look neutral; those above (and to the right) look yellow; those below (and to the left) look blue. We may think of all the colors represented in the diagram as "collapsing" onto the yellow-blue axis along the confusion lines, as indicated by the arrows: the relevant dichromat is supposed to see merely the color represented at the projection onto that single axis. (For confusion diagrams like these, see, e.g., Le Grand 1957, 332; Hsia and Graham 1965, figure 4; Hurvich 1981, 194; Wyszecki and Stiles 1982, 464 and 466. They all derive from Pitt 1935, 47 and 49, transposed into CIE *x*, *y* diagram form by Judd 1943, 305, and 1945, 202: cf. also Sharpe et al. 1999, 28; and see section 12.3 below. For an earlier version of the idea, see Maxwell 1855, figure 2, and 1860, esp. figures 10 and 11; for a fine diagram with one "fan" of confusion lines for protanopes and a second "fan" for deuteranopes, see von Kries 1882, 141. For the specification of the yellow-blue color axis, see Judd 1948 and section 12.4 below; again, there is an ancestor of the idea in Maxwell 1855, esp. 139; 1860, 440. The arrow representation is my own.)

at intervals through the spectrum (1935, 14–17, esp. figure 4)—as viewed in the Wright colorimeter, in which two adjacent fields of light are compared, each being a rectangle of about 1° by 2° (Pitt 1935, 8; Wright 1946, chs. 3 and 4). The data all concern small fields, not large, and *lights*, not surface colors, and (among lights) only two particular ranges. And at relatively low levels of brightness.[28] To draw any conclusions about confusions of paints—or even larger or brighter fields of light—would be to draw consequences from Pitt's data only in conjunction with large theoretical assumptions (e.g., that real dichromats fit the Helmholtz-Maxwell model): it is not an experimental finding. And (as we shall see in section 12.5) there is plenty of evidence that those theoretical assumptions are actually untrue of many, indeed most, of the dichromats of the real world. Whatever the explanation may be, it seems that the confusions

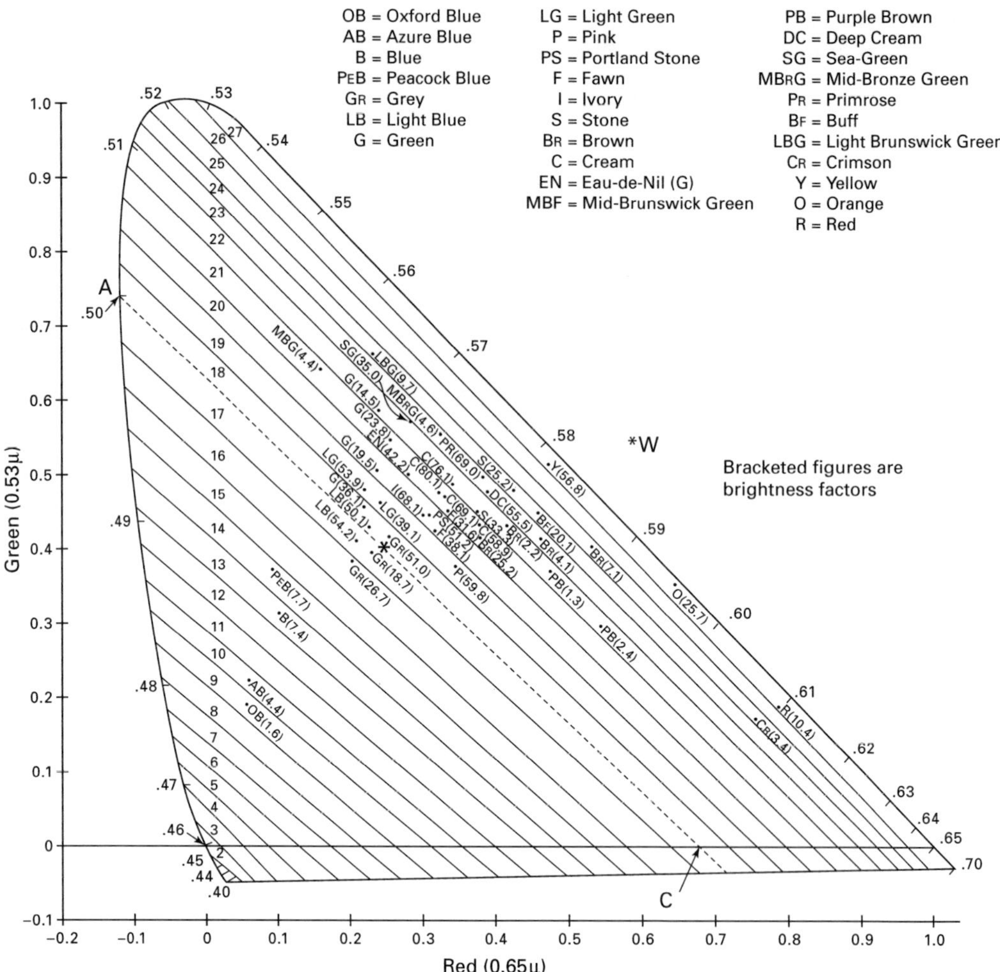

Figure 12.7

Trichromat chromaticities and deuteranope confusions: Pitt 1935, 47. The diagram can be thought of as specifying (in W. D. Wright's system) the proportions of three lights, red_{650}, green_{530}, and blue_{460} (represented at (1, 0), (0, 1), and (0, 0) respectively) needed in a mixture to match any given light stimulus (for a normal trichromat). The plectrumlike curve shows the mixtures required to match pure monochromatic light of various wavelengths through the spectrum (and, along the straight line at the bottom, running from 400 to 700 nm, the purples produced by mixing violet and red from the extremes of the spectrum; cp. Wright 1946, chs. 3 and 4; cp. figure 12.3 above.) Each of the many confusion lines is supposed to join stimuli that (except perhaps for brightness) would look the same to a deuteranope: Pitt has spaced them at appropriate distances to yield twenty-seven confusion classes. The experimental data behind the diagram are the deuteranope matching of spectral hues (on the plectrum curve) with particular mixtures represented in these diagrams are just not usually made if the fields are larger than about 4°, or under other more favorable conditions. Research like Pitt's has given us genuine guidance on the patterns of insensitivity found in dichromats under rather extreme and special conditions—which in turn are a guide to the sensitivities of the various types of cone. But it would be a mistake to take that as telling us anything very firm, either about the discriminatory capacities of dichromats under more relaxed and ordinary conditions, or about the character and limits of the full range of their color experience.

There is a second problem—small in itself, but indicative of larger difficulties of principle. There is a decent level of agreement among experimenters on the position of the protanope confusion point—usually placed at about $x = 0.747$, $y = 0.253$, just beyond the red end of the spectrum.[29] But when it comes to a deuteranope confusion point, researchers seem only to come up with data that imply quite different positions—disagreeing with each other and with themselves so much, indeed, that one might wonder if there is any such point to be found at all. Table 12.2 shows a small selection of values from different sources.[30]

If we want any provisional or illustrative figures—and we will need some for the modeling in section 12.6—then the figures from Wyszecki and Stiles in the final row of table 12.2 are probably as good as any. (The figures are derived from Pitt's data, via Judd 1945, 202;[31] my own analysis of Pitt's data, in row three, agrees well with these figures.) But it would be a mistake to attach any great reliability to them.

Given that good experimenters seem only to come up with such conflicting views, we might wonder if there is really a well-defined question to which they are giving such different answers. The five deuteranopes in Scheibner and Boynton (1968), for example—to mention yet another available guide—agree well among themselves on their neutral point, which implies (by my calculation) a deutan confusion point at about $x = 1.62$, $y = -0.62$; but they agree dreadfully with Pitt's deuteranopes. One

of blue_{460} and red_{650} (on the horizontal axis, $\text{green}_{530} = 0$). For example, 480 nm is matched (by Pitt's deuteranopes) with a mixture in the proportions 0.205 R_{650} 0.795 B_{460}; 494 nm is matched with 0.50 R_{650} 0.50 B_{460}; and 500 nm is matched with 0.68 R_{650} 0.32 B_{460}—and the dotted line joining those last two points also goes through the point for white (CIE standard illuminant B, marked with an asterisk * near the center of the diagram): 500 nm is a neutral point for Pitt's deuteranopes, a part of the spectrum that looks white. Letters (decoded in the list at the top right) mark the chromaticities of some forty-four paints (as identified for normal trichromats: figures supplied to Pitt by the UK National Physical Laboratory). The confusion lines converge at about $r = 3.15$, $g = -2.28$ in this representation. Transformed into the CIE 1931 system, the diagram takes the form of the x, y diagram for deuteranopes (figure 12.6b), and the lines converge at about $x = 1.08$, $y = -0.08$. (Pitt 1935, 49, also gives a similar diagram for protanopes, which corresponds, in the CIE system, to figure 12.6a.)

Table 12.2
Some rival views on deuteranope confusion points

Researcher	Deuteranope confusion point
What Pitt (1944) takes his 1935 data to indicate*	$x = 1.28$, $y = -0.25$ (i.e., $r = 9.5$, $g = -8.5$)
What Pitt (1944) would like his 1935 data to indicate	$x = 1.413$, $y = -0.413$ (i.e., $r = \infty$, $g = \infty$)
What Pitt's data actually indicate (JB's calculation from six pairs of data points)	$x = {\sim}1.06$, $y = {\sim}{-0.075}$ (i.e., $r = {\sim}3.15$, $g = {\sim}{-2.28}$)
Farnsworth (1954), using color chips (as reported by Hsia and Graham 1965, 209)†	$x = 0.90$, $y = 0.00$
Farnsworth, for a range of subjects (as reported by von Schelling 1960)‡	*varying on a line between* $x = 1.35$ $y = -0.35$ and $x = 1.75$ $y = -0.75$
Nuberg and Yustova (1958, 2:480)**	$x = 1.70$ $y = -0.70$
Vos (1978); Walraven (1974); Smith and Pokorny (1972)††; cf. Wyszecki and Stiles (1982, 614–615)	$x = 1.4000$, $y = -0.4000$
Estévez 1979 (see Hunt 1987, 62–63; 1998, 61–62)	$x = 2.306$, $y = -1.305$ ($u' = -0.534$, $v' = 0.680$)
Wyszecki and Stiles 1982, 464	$x = 1.080$, $y = -0.080$

* The first three lines in this table derive from the fact that in his original 1935 study, Pitt gives no specification for the confusion points at all; in his 1944 paper, he gives coordinates ($r = 9.5$, $g = -8.5$), but they agree neither with his diagram nor with the data it was based on; what is more, having admitted that the lines actually converge (at a point he incorrectly identifies), Pitt goes on to propose that we should *treat* the confusion lines nonetheless as being parallel in the coordinate system of his diagram. Pitt is motivated here by his conviction that in deuteranopia, "two of the fundamental sensations are identical," or have *fused* (1944, 106)—a view that had supporters at the time but has practically none today. Pitt seems unaware that there is nothing particularly special about the confusion lines being parallel in the particular coordinate system he is using: lines parallel in one coordinate system will be nonparallel in another system, and it makes no sense to suppose that a convergence of lines (in one fairly accidentally chosen system) indicates the "loss" of a primary sensation, whereas parallelism (in that system) indicates "fusion" of two sensations instead. Quite aside from the empirical disagreements between Pitt and other researchers, Pitt's readiness to abandon his own evidence, and the theoretical confusion with which he does so, are hardly confidence inspiring.

† These figures are utterly unbelievable: the deuteranope confusion point would be a purple (rather than some kind of green, which would surely, if anything, be needed, if it is to correspond to the receptivity of the missing M-cone type).

‡ Farnsworth puts the variation down to variations in macular pigmentation. There is good logic in this, but it opens up a multitude of further problems. If we suppose (following Pitt 1935) that the neutral point (matched with illuminant B) is ~500 nm for deuteranopes with average macular pigmentation, then at a rough estimate, we might expect a deuteranope with only *half* the

Table 12.2
(footnote continued)

normal density of macular pigmentation to have a neutral point (matching that same illuminant B) at ~496 nm, and one with *one-and-a-half times* the normal density to have a neutral point at ~504.5 nm (my own calculation from the square root of the Wyszecki and Stiles function for macular transmittance [1982, 721], and the profile of illuminant B [1982, 759].) And on one standard method of deriving a confusion point—by joining, in a 1931 CIE chromaticity diagram, the neutral point and illuminant B, and seeing where that line crosses the line at –45° that joins the spectral greens, yellows, and reds that dichromats confuse in an anomaloscope—we would get deuteranope confusion points ranging from about $x = 0.743$, $y = 0.257$ to $x = 2.283$, $y = -1.283$! It should be obvious that variation in the macula can wreak havoc with any attempt to get a single standard deutan confusion point. However, this is only the beginning of our problems: on similar principles, we would expect a similar variation in macular pigment density (ranging from 1/2 to 1 1/2 times the average) to shift the protanope neutral point (matching illuminant B) between about 491.4 nm and 500 nm—which would imply a confusion point shifting between $x = 0.626$, $y = 0.374$ and $x = 1.014$, $y = -0.014$. Both those extremes are absurd. (The first is close to the position of orange_{600}, which is certainly not invisible to protanopes with low macular pigmentation, and the second is below the line $y = Y = 0$, so would represent a red of negative luminance, i.e., a *blue-green,* which cannot be the "missing primary" of a protanope with high macular pigmentation.) What this shows is that the shape and proportions of the spectral locus in a 1931 CIE diagram cannot be taken to fit the color matching of people with even moderate deviations from average macular pigmentation, and that the line at –45° ($x + y = 1$) cannot be taken as a dependable confusion line for them either. We have a lot more work to do.

** Nuberg and Yustova's figures—derived using just one confusion line and a method of Maxwell's—are described by Scheibner and Boynton (1968, 1157) as "probably one of the most accurate determinations."

†† These are not experimental studies: Vos (1978), Walraven (1974), and Smith and Pokorny (1972) all build on Vos and Walraven (1971), which is a derivation of König receptor primaries, based among other things on the authors' choice of $x = 1.40$, $y = -0.40$ as the deutan confusion point, itself based largely on Nimeroff (1970). Nimeroff's article is actually a survey: it arrives at a weighted mean of the various deuteranope confusion points reported by investigators (principally Pitt for eight subjects and Nuberg and Yustova for twelve). Nimeroff's weighted mean was, incidentally, $x = 1.53$, $y = -0.53$, not $x = 1.40$, $y = -0.40$: Vos and Walraven evidently take it that a certain degree of indeterminacy attaches to Nimeroff's figures, so they are at liberty to *choose* a figure within the range $x = 1.53 \pm 0.17$, $y = -(0.53 \pm 0.17)$, and they give some complex reasons for their particular choice (802, 811). Nimeroff (1970) himself, incidentally, presents $x = 1.40$, $y = -0.40$ as his own (1969) reevaluation of the confusion point implied by Pitt's data—on the basis of treating the confusion lines in Pitt's deuteranope diagram as being *parallel*—which is simply incorrect, as anyone knows who gets a photocopier, reduces Pitt's diagram to 25 percent size and extends the lines to see where they meet. It should be obvious that none of this constitutes a basis for anything but the most approximate view.

possibility is that in such disagreements the conditions or the task are in some way different in the various studies. Maybe the subjects' macular pigmentation is different; maybe the adaptation level is different; the size or brightness of the fields; the amount of time the fields are exposed; the manner in which the two fields were presented (side-by-side or successively; with a surround or not). And there are different mathematical methods: for example, most theorists, like Pitt, ignore brightness and concentrate only on the chromaticities of lights confused and therefore need two confusion lines to get a confusion point; but Yustova employs a method of Maxwell's that makes inferences from a single confusion line, taking into account the exact brightness levels in a match, and calculating the position on that single line of a stimulus that would have zero luminance (to the dichromat). But if the variations in experimental and mathematical method can move the deuteranope "confusion point" from $x = 1.08$, $y = -0.08$ to $x = 1.70$ $y = -0.70$ and again to $x = 2.306$ $y = -1.305$, then we must ask whether there is even a well-defined topic that is the subject of their disagreement. The implied differences over which light stimuli a deuteranope will in practice confuse are often relatively small, but the placing of the M-fundamental is importantly different. We must ask whether single and definite confusion points are there to be found at all: in view of the genetic variation within the class of "normal" observers, we should be aware that the "normal" chromaticity diagram may need to be different for one subgroup and for another; among dichromats, we may be able to find a definite confusion point only for a subgroup of people who all have the same receptor types; and, to the extent that additional factors are at work (e.g., the macula, and perhaps—though, as we shall see, the candidates are not all equally likely—rods or additional cone types, or other retinal inhomogeneities), we must expect people to be unhappy with a match that they "ought" to find acceptable, when those additional factors allow a pair of supposedly matching fields actually to be differentiated. Above all, we must remember that if the evidence I have mentioned before is correct, the kinds of confusion illustrated in these diagrams are indeed made with small light fields of 1° or so, but they simply do not usually occur with larger fields—or even, reliably, with smallish fields of surface color. So these diagrams are going to be of little use in telling us the *full range* of "what a dichromat sees." This is not to object to the attempt scientifically to separate out relevant factors for separate study; it is to say that the component factors need much more careful separation than they have received so far, and that we cannot expect that, when we put all the factors together, the confusions of dichromats will be very thoroughly captured by a traditional confusion diagram. I leave for an endnote a similar cry of protest from a distinguished researcher of nearly fifty years ago.[32]

12.4 The Unilateral Cases—and Nagel's Remarkable Reports

Supposing we read the confusion diagrams as implying the collapse of the range of hues to just two, how would we know which hues they were? Studies of color blindness in one eye are supposed to tell us. In those rare cases where a person has one color-blind eye and another that is normal or near-normal, the hope is that the more normal eye can be used to calibrate or specify the perceptions had with the other eye: the person might, for example, *point* to color samples (e.g., on a color chart) that look to the normal eye the same in color as certain controlled stimuli presented to the color-blind eye; or he might simply *tell* us what color particular things look when he sees them with his color-blind eye—and the fact that he has one good eye could serve as evidence that he means by the color terms pretty much the same as normal perceivers do. As it happens, the classic survey of such cases by Deane Judd (1948) reaches the firm conclusion that the main groups of dichromats see just a unique yellow and a unique blue, corresponding to 575 and 470 nm.[33] As a report selecting, out of forty original articles, spanning eighty-nine years and several languages, a core group of ten for detailed evaluation, Judd's article might seem a paradigm of diligent judgment. On deeper investigation, it turns out, I think, to be nothing of the kind.

It is important to acknowledge immediately the remarkable fact that there are dichromats who themselves talk of seeing only yellow and blue (and neutral) in the spectrum, or indeed under any circumstances at all. (Scott 1778 and Pole 1859 are, respectively, perhaps the first and the most remarkable among them.[34]) But we have already seen evidence from dichromats who believe they see more than those two hues (like Dalton 1798 referring to what he saw in candlelight: section 12.1.1), and we will have to consider whether they might in fact be right. And we must remember that we cannot assume that whatever is true of *some* protanopes or deuteranopes is also true of all: there may be individual differences, or important subgroups that have not yet been properly distinguished—varying in genetic endowment, or in the ways they may have developed means to compensate in part for their receptor loss.

Table 12.3 is a version of Judd's table of the ten most reliable reports. Judd gives the impression that—with only three small deviations that he can deal with (marked in my table with asterisks)—the original articles all point in the direction of what I shall call the "yellow-and-blue view" of dichromat experience. I shall save a fuller evaluation for another occasion. But if we go back to the original reports, we get a very different impression. One author in the group, Samuel Hayes, reaches—from a review of the literature and the investigation of his own subjects—a conclusion diametrically opposite to Judd's: "The statement that sensations of blue and of yellow alone are possible to the partially color-blind *cannot be reconciled* with our findings" (Hayes 1911, 402, my emphasis).

In fact, the main point of Hayes's study is to claim that his own unilateral color-blind subject (a unilateral protanope, he believes) sees green in addition to yellow and blue, as also do many other protanopes: "green (as a specific color quality different from yellow and gray) is included in the color system of her protanopic (right) eye" (380).[35] As for the reports in the earlier literature, Hayes says: "Our review of the monocular cases . . . showed that, with the exception of von Hippel's, they reveal nothing but

Table 12.3

"Cases of unilateral defect of vision giving information regarding protanopic and deuteranopic color perceptions" (Judd 1948, 254), intended to show that deuteranopes and protanopes see only a yellow of ~575 nm and a blue of ~470 nm

	Classification of the defect		Indicated perceptions	
Author, date	Author's own	Present [i.e., Judd's]		
a. Unilateral dichromacy				
Hippel (1880)	Deuteranopia	Protanopia	Yellow	Blue
Holmgren (1881)	Protanopia	Protanopia	Greenish yellow	Violet blue
*Hippel (1881)	Deuteranopia	Protanopia	*Like 589 nm	Blue
Hayes (1911)	Protanopia	Protanomaly	Yellow	Blue
*Sloan (1947)	Deuteranopia	Deuteranopia	Like 5Y 5/5 [colored paper with dominant wavelength 575 nm], also [spectral light of] *584 nm	Like 3PB 5/5 [colored paper with dominant wavelength 478 nm], also [spectral light of] *452 nm
b. Dichromatic fovea, with near-normal periphery				
Nagel (1905)	Deuteranopia	Deuteranopia	Yellow	Blue
c. Unilateral anomalous trichromacy				
von Kries (1919)	Deuteranomaly	Deuteranomaly	Like 573 nm	Like 464 to 480 nm
d. Unilateral acquired deficits				
Hering (1890)	Approach to deuteranopia	Approach to deuteranopia	Yellow	Blue
Hess (1890)	Approach to deuteranopia	Approach to deuteranopia	Like 575 nm	Like 471 nm
*Goldschmidt (1919)	Protanomaly	Protanomaly	*Orange, yellow	Blue

Note: From Judd (1948, 254), with modifications. I have replaced Judd's "do" (i.e., "ditto") with whatever it refers to, and I have listed author and date together in col. 1; I have replaced mμ (and the inaccurate μ) with nm. The four categories (a, b, c, and d) and the ordering of the various rows under them are my own: Judd simply lists the ten reports in chronological order. I have marked with an asterisk (*) the authors and relevant claims in three reports that Judd describes as 'exceptions' to his hypothesis.

Judd's reports under the column for "Author's own" classification of the defects are in four cases inaccurate, in that Hering and his followers Hippel and Hess wholly rejected the concepts of protanopia and deuteranopia (or their ancestors red- and green-blindness) in favor of the general notion of red-green blindness (with shortening of the spectrum at the red end, if admitted at all, put down to the media of the eye). Hering's own description of his subject's diseased eye is as "almost red-green-blind" (1890, 15); Hess describes his own subject as having a "red-green sense" that has "almost completely disappeared" (1890, 35). What is more, Hering diagnoses his patient as suffering from an atrophy of the optic nerve, and the Hess case is similar—so neither case is likely to have been fundamentally very similar to deuteranopia.

meagre experimentation, glaring contradiction, and theoretical bias" (1911, 404). My own view is that Hayes is much too generous to treat von Hippel's case as an exception to this general judgment. We shall come back to this in a moment.

The cases surveyed in Judd's review are actually of four very different types. The last group, (d) the acquired deficits, is, I think, of absolutely no value to us. Even if we knew in those cases what color sensations were or were not had with the affected eye, it would be entirely unclear what implications that might have for ordinary cases of congenital dichromacy. In two of the three cases that Judd studies, the deficits derive apparently from some kind of optic nerve atrophy or neuropathy (and in the third case, from a gunshot wound): those are certainly not fundamentally similar to a lack of L or M receptors in the retina, and we would need independent reasons to be justified in taking the former kind of pathology to tell us anything much about the precise limits of experience in the latter kind of deficiency. What is more, even if there were a physiological similarity, an *acquired* deficit might or might not tell us about experience in a *congenital* case.[36]

What might really help us are (a) the supposed cases of unilateral dichromacy. The best evidence Judd has for his yellow-and-blue view is probably the case of a seventeen-year-old man who was the subject of three reports, by Hippel (1880, 1881) and Holmgren (1881). His right eye suffered from a marked "red-green blindness" (and a bad squint uncorrected by several operations[37]),while the other eye was apparently normal. But there are disagreements between Hippel and Holmgren (and between the two of them and Judd)—and both original authors show the strain of making the case conform to their theoretical predilections. They both say that the man sees only two hues with his weak eye, but do not agree on which hues they are. Hippel, as a supporter of Hering, wants the hues to be yellow and blue; Holmgren, as a supporter of broadly Helmholtzian ideas, expects them to be a variety of green and of violet. Hippel (1881, 50) reports that the colors experienced correspond to 589 nm and 450–460 nm (which are actually a slightly *orange* yellow and a violet blue). Holmgren (1881, 306) claims instead that they are a *greenish* yellow and an "indigo-violet." Judd, some sixty-five years later, seems to think he can, so to speak, split the difference on the yellows and bleach any violet tinge out of the blues—and declares the colors to be unique yellow and unique blue. I think it is doubtful, on closer investigation, that the boy saw only two hues with his dichromatic eye. The best evidence that Hippel presents is simply that yellow and blue are the only colors that the subject reports seeing, when presented, under controlled conditions, with particular stimuli of various colors. Spectral lights, when presented in isolation in a spectroscope in 1 nm widths, are described only as yellow or blue (1880, 180). Certain annular portions of rotating discs that look green and red to the normal eye are described as merely looking yellow and blue to the dichromatic eye (182). There are similar results with Dor's plates and some pieces of red and green glass.

But there is counter-evidence in the article that suggests an opposite conclusion. Under other conditions, the man reports plenty of other colors with his supposedly dichromatic eye. Looking at a full spectrum, he reports "*red*," "yellow or *green*," and "blue" (Hippel 1880, 180, my emphasis). When asked to name the colors in the index to Radde's scale (a color-order system of the late nineteenth century), he performs the task "fairly correctly" when presented with the whole index page of forty-two different samples (which he certainly cannot have done using just the terms "yellow" and "blue"). When the thirty colors in the higher saturation series in Radde's index are presented to his color-blind eye, separately and out of sequence, he uses the terms "red," "green" or "greenish," "yellow," and "blue" pretty much for red, green, yellow, and blue and calls none of the greens red or reds green.[38] He makes some serious mistakes: he calls the orange-red and orange samples "yellow" and calls some of the greens "yellow." His color vision is obviously seriously deficient. But he still seems to have a far from negligible capacity for red-green discrimination, and his use of the terms is surely prima facie evidence that things sometimes looked red or greenish to him with his color-blind eye. (Remember the premise of the whole project: that the subject knew from his good eye what the standard meaning of these terms was.) There are many further problems that I shall not go into here (especially with the spectroscope tests that Hippel reports), but it should be obvious that the claims of Hippel and Holmgren are nowhere near being established by the evidence they give. They are conclusions they could feel entitled to draw only given a commitment to their own general *theories*—though the theories were incompatible with each other, and neither was in any position to claim evidence-trumping authority.

Perhaps the most interesting detail here, however, is in the man's performance with the low-saturation samples (nos. 31–42) in Radde's index: when they are presented individually and separately, he fails to recognize any red or green tinge in any of them. But there is one exception: he describes 31, neutral gray (along with the pale orange-red 32), as "light green." If this is true, it may well be a case of a kind of color *induction* and it may be a sign of processes at work more generally: it may be, not that the relevant aspect of color is *bleached out* of the dichromat's experience, but that it is, so to speak, *painted in*, and sometimes painted in incorrectly. It would be a mistake to draw any firm conclusion from the fragmentary information in the present case; but the hypothesis would fit quite well with what we know of some of the other operations of our sensory systems. The partially deaf sometimes become hypersensitive to sound, the wholly deaf may hallucinate sounds; there are, it seems, something like automatic gain controls in some of our sensory systems that turn up the amplification (as well as working to harness ancillary sources of information) where there is low amplitude or reduced variation in sensory signal input. In cases of anomalous trichromacy, there is a phenomenon of *exaggerated* contrast noted by Nagel (1908b, 24–25); and recently—with normal trichromats—Richard Brown and Don MacLeod have talked of "gamut expansion" in cases where pale color patches are seen against a uniform gray background. (Color patches that look pale when seen against a brightly colored background, for example, look richer in color when seen against a neutral gray; Brown and MacLeod 1997, 845; for some caveats, see also Faul et al. 2007. This may be a by-product of constancy mechanisms for dealing with fog and similar conditions that dilute colors generally, or of central adaptations to a general weakening of peripheral sensory input.) The message of the anomalous trichromacy cases may be not how much is *lost* (the implication therefore being, as Judd believes, a total loss of red-green perception in cases of full dichromacy) but, rather, how much is *saved*—as the visual system works to compensate for information loss, sometimes (as when pale gray is reported as looking green) inaccurately.

I shall pass quickly over the Hayes (1911) case—which may or may not be a genuine case of unilateral dichromacy. The subject is clearly reported to see green with the affected eye, so if the eye is (as Hayes tells us) protanopic, then it seems a counter-example to Judd. On the other hand, there is certainly a possibility that the eye was merely protanomalous. (Hayes seems to have a slightly impressionistic grasp of the distinction between protanomaly and protanopia. And he was certainly in no position to make a firm differential diagnosis between the two conditions with the equipment he had available—spinning tops, but no anomaloscope or spectral mixing apparatus.) If the eye is only protanomalous, then it belongs with category (c)—the cases of anomalous trichromacy—and, as we shall see, under that heading too, it lends no support to the yellow-and-blue view. One can conclude very little from the case, but either way, it gives no support to Judd.

The third and last case of unilateral dichromacy that Judd mentions might seem more helpful and more rigorous. Sloan and Wollach (1947 and 1948)[39] announce that their experiments show that "the color perceptions of the [subject's] dichromatic right eye include only blues, yellow, and the white-black series" (1948, 508)—which sounds like support for Judd. Unfortunately, the experiments show no such thing. To compare the perceptions of the subject's two eyes, the investigators run two main series of experiments, one with Munsell papers (i.e., pieces of colored paper meeting specifications in the Munsell system) and the other with spectral lights in a photometer. The tests using colored Munsell papers that the investigators report involve only four narrow ranges of color samples: four blues, three yellows (or rather, browns), three bluish greens, and three purples, selected with the aim of discovering which of the yellows and blues looked the same to the two eyes, and which of the blue-greens and purples look more or less neutral to the dichromat eye. With spectral light stimuli—selected for the same purposes—the main tests involve only four blues, three yellows, and three bluish greens. How on earth is that supposed to tell us the *full range* of the subject's color experience with the eye in question? The authors are not *testing* whether the experience with the bad eye included merely two hues; they are *assuming* that it included only two hues and are simply trying to identify *which* blue and *which* yellow the two supposed hues might be.

The authors do tell us that "[1] In the spectrum and [2] in the Munsell 100-hue series [i.e., the colors of the Farnsworth-Munsell 100-hue test], *the subject reported that* with the right eye *he saw only blues and yellows*" (1948, 509, my emphasis). But we are not told of any effort to make an experimental test of whether that very general report from the subject was actually true. Even if it was true, it would not establish the conclusion the authors want. We know (1) that some people (like Dalton) report seeing only yellow and blue (and maybe purple) *in the spectrum*, but report other hues *in other circumstances*. What is more, (2) the Munsell papers that Sloan and Wollach used in their experiments were all are of medium-low saturation and lightness: they are (approximately) the colors that Farnsworth used in his 100-hue test. But those colors were chosen by Farnsworth not for bringing out the maximum range of color sensation in a dichromat, but as samples that would present color-blind people with maximum difficulty, to facilitate quick diagnosis of their weaknesses. So even if it were true that the subject saw only yellows and blues (1) from the spectrum and (2) from the Munsell colors presented to him, that would not tell us that he *never* saw any other hues under more favorable conditions. The Munsell papers presented were, as we have seen, of medium-low lightness and saturation (as is evident from the fact that with the good eye, the subject described some of them as "pinkish-blue" and "brown").[40] Where were the bright reds and vivid deep greens?! We know from other sources (see section 12.5) that if there is any residual perception of red and green among dichromats, then these medium-low saturation colors are among the kinds of sample *least* likely to reveal it. The Munsell paper samples were two-inch squares, which at 40–60 cm would subtend about 5°–7° of visual angle; whereas we know that any residual red-green discrimination is likely to be better when the samples are larger. Finally, we are told that the illuminant for the papers was "artificial daylight (Macbeth daylight lamp)" (Sloan and Wollach 1948, 507). One should not be misled by the label. The light of a Macbeth daylight lamp is not much like ordinary daylight: it is a small desk lamp with a metal dish reflector behind the tungsten filament bulb, and a blue filter in front, which cuts out nearly all of the red, orange, and yellow light, and much of the green, so as to transmit a spectral profile like that of daylight of ~6000 K. The result is a hard and dim light compared with that of a 100-watt bulb or real daylight; and it shines directly upon the samples to be viewed, with minimal reflection from walls, table, and so forth. It has, for diagnostic purposes, the significant advantage, in comparison with redder and brighter lights, of increasing the number of errors made by color-blind people on standard tests (Schmidt 1952). Precisely for that reason, it is not a good light in which to investigate any residual but weak capacities for color perception.[41] If Sloan and Wollach's subject had had any perception of red and green, the method of investigation could hardly have been better designed to *avoid* finding it.

I shall say little about (c), the cases of unilateral protanomaly and deuteranomaly, except that it is completely obscure what consequences one might draw from them about the experience of protanopes and deuteranopes. Judd assumes that if in anomalous trichomacy the experience of certain hues is *impaired*, then in full dichromacy the experience of those same hues will be entirely *absent*: but this link proposition is a theoretical claim that may or may not be true. We know indeed that certain models—derived from Helmholtz and Hering—imply this; what we need to know is whether those theoretical models are in fact dependable and empirically secure. As it happens, in the only clear report of a unilateral anomalous trichromat that Judd takes up—von Kries's 1919 report on a man with one deuteranomalous eye—it was, it seems, *not* the sensations of red and green that were maximally impaired, but rather those of green and yellow, followed by red and blue, in that order. What is more, the reduction in the intensity of sensation of both red and green was (so von Kries says) "surprisingly slight" (150). On Judd's own principles, the conclusion to draw from the case would be that dichromats, if they experienced merely two hues, would experience red and blue! But von Kries's report might also be taken as evidence that there is no neat segmentation of hues into pairs, with one pair totally disappearing in cases of dichromacy.

As for Hayes's case, if the affected eye is after all merely protanomalous (rather than protanopic, as Hayes had thought), then it is unsurprising that some, perhaps weakened, sensation of green is maintained; but we could only conclude that this sensation of green would be wholly *absent* in cases of dichromacy given the link proposition I mentioned earlier—and that is a claim of high theory that is put in doubt by much of the evidence. In the absence of that link idea, Hayes constitutes not the slightest support for Judd's main thesis; and either with or without that idea, von Kries constitutes counter-evidence for that thesis.

This leaves (b), the remarkable case in Nagel (1905) of the train driver Sch., who is described as having a dichromatic fovea and trichromatic periphery. (This is a case, therefore, not of one dichromatic eye and one normal eye, but of two eyes, each apparently with one dichromatic region—centrally—and another nearly normal region, in the periphery.) It is something of an outrage that Judd presents this as being a report of a person who has sensations of *only blue and yellow* with the dichromatic part of his eye. There are three things to say about the report and its evidential value.

1. The article in no sense reports yellow and blue as the sole "indicated perceptions" for the dichromatic part of the eye. There is indeed a comment on the train driver's perceptions of yellow, which we shall come to in a moment, but the article contains not the slightest mention at any point of the presence or absence of perceptions of blue—indeed it contains no occurrence of the word "blue" (or *blau* or any suitable equivalent) at all.[42]

2. On the subject of yellow, the one relevant comment is Nagel's remark that it is "theoretically important . . . that Sch. calls spectral red and yellow-green, in foveal vision, yellow" (1905, 100). But this must be taken with Nagel's report that Sch., on other occasions, also called those same spectral stimuli either *red* or *green* (1905, 94).[43] It is perfectly possible (Nagel's report is not entirely explicit) that the man had a variety of sensations from stimuli in the long-wave part of the spectrum—of yellow, red, and green, depending on context, size, relative lightness, and other factors.
3. Most importantly, Nagel's article actually describes not one case but two—the train driver Sch. and Nagel himself, presented as an ordinary case of deuteranopia—though one that is far from fitting the standard theoretical conceptions about that condition. Having finally been forced to conclude that Sch.'s periphery must be normal or near normal (on the basis of his excellent practical performance), Nagel comments, "It must be said that in establishing such a diagnosis [of a normal or near-normal periphery] the most extreme caution is called for, since one might be tempted to contemplate something similar in case of a very large number of dichromats, especially deuteranopes" (Nagel 1905, 97). What Nagel means is that many dichromats perform excellently when tested on more than just a small foveal region of vision, without that being a reason to say they have a trichromatic periphery after all. And here Nagel presents the most remarkable finding of the whole article: that dichromats like himself can in large fields recognize red, in addition to yellow and blue.

The claim is of great importance, and it leads to some yet more important further statements that get not the slightest attention from Judd. Nagel says that, for a green light, he can always produce in the laboratory—though it is not easy—another color that a dichromat will confuse with it: namely, some kind of yellow (or brown) or white (or gray). (Figure 12.6 above may make the expected equivalences clearer.) On the other hand: "Not so with Red. I find it absolutely impossible to get a satisfactory match between vivid Red and Green or Brown, as long as the . . . field size [is] of at least 10°" (Nagel 1905, 97). A few years later, he has additional evidence:

> Among 30 dichromats of both types (protanopes and deuteranopes) that I have examined since [a few years ago], I have found *not one* who displayed a dichromatic system also in *large-field vision*. Without exception they were able with large areas to *recognize the colour red with complete certainty* in all nuances, even at quite low saturation. With areas of a size of, say, 1/2 to 3/4 square metres, observed from a distance of 1/2 to 3/4 m, the so-called dichromat recognizes such low degrees of saturation of red (even when mixed with unsaturated yellow or blue), that he seems at first to do only slightly worse than the normal. (Nagel 1910, 6, my emphasis)

If red is recognized at various levels of brightness and saturation—and is not matchable with any mixtures of yellow and blue—the obvious implication is that it is not producing merely some variety of sensation of yellow or blue. And Nagel says so explicitly in the same article: he knows very well what it is like to see the world only in yellow and blue. But that is not at all his normal experience. He can bring about that extraordinary reduced condition by wearing red filter capsules for half an hour or so: having flooded the eye with so much red, he finds, on taking off the filters, that his sensation of red is "exhausted" for about a minute:

It is a most remarkable experience, to see the environment thus in the way in which the dichromat is supposedly bound to see it, so others say, but in which I have *never seen it before*: totally without the colour red. Red objects appear, even when they are quite large, exactly as they otherwise look only in foveal view—brown [i.e., dark yellow], yellow, gray or blue, according to the exact shade of the red. The effect lasts for about a minute after wearing the filter capsule for half an hour, and two minutes after exhaustion for one hour. The red sensation then returns slowly afterwards and is normal again after about five minutes. (Nagel 1910, 10, quoting from Nagel 1907d, my emphasis)

It is puzzling that this second article ("Farbenumstimmung beim Dichromaten," *Zeitschrift für Sinnesphysiologie* 1910) is mentioned in the bibliography to Judd's 1948 paper. One might be tempted to say that Judd shows not the slightest sign of having understood it (whether or not he might ultimately accept it). Judd talks in all innocence of the unlikelihood that his yellow-and-blue view might even ever be *challenged*.[44] In fact, the existence of a challenge is not an unlikely possibility but an actual and evident fact—staring him in the face from the pages of one of the articles he makes a pretense of disinterestedly surveying. He says not one word about these challenges: it is not that he mentions them and answers them casually—he simply shows no sign of even seeing them.

Judd is a serious author who made very substantial contributions to color science. A Cornell Ph.D., he spent most of his career at the U.S. National Bureau of Standards, during a period when much of the most exciting work on color vision arose out of technical demands and practical needs; he was one of main developers of the colorimetric systems of the CIE and may have been the first to plot an (x, y) diagram (Judd 1952, 108n).[45] But there is something bizarre about his treatment of this topic. It is serious in style, the work of a person of authority, and clearly the result of a lot of hard work. But it contains strange misunderstandings.[46] It has the air of presenting a range of rival views—acknowledging, for example, that Holmgren and Hippel disagree on which two hues their seventeen-year old subject may have seen—but it simply never takes seriously the no less present disagreement on the prior question of whether dichromats can be expected to see merely two hues at all. On that, Nagel and Hayes were already explicit opponents, and von Kries had presented evidence (in the 1919 article) that could be used to raise doubt. Judd talks of them as if they either already were or could easily be turned into supporters of his own cause.[47] They were not, and to pretend that they were is to treat them with something like indifference or disdain. The condescension is puzzling given that Nagel and von Kries, at least, were figures of intelligence, distinction, and scientific independence of mind, with medical training as well as many of the instincts of a physicist, who had spent years of their lives working with actual cases of color blindness. And in neither technical nor theoretical domains had there been any particularly radical advance between their day and Judd's. The anomaloscope that Nagel had invented was still at least as good as any of its rivals forty years later (Willis and Farnsworth 1952). And the "zone" theory, a later version of which Judd supported, had itself been proposed by von Kries in 1882, with not merely the general idea of a neural processing stage subsequent to the (largely chemical) activity of the retina, but even the precise and prescient conception of quotients or log differences to capture the structure of subjective color—with something like (L / M) or β (log L – log M), for example, to correspond to the opposition of red and green (von Kries 1882, 62, 163–164; 1905, 269).

Let us look again at Judd's table and ask what would be put in the last two columns if we recorded not "*indicated* perceptions" but "*reported* perceptions." ("Indicated" seems in Judd to mean "either reported, or implied by something (perhaps minimal) in the report taken together with my own theory.") Any sign of support for Judd's yellow-and-blue view, I think, disappears. To take (a), the three cases that have a claim to be unilateral dichromats: The Hippel/Holmgren/Hippel patient certainly reported things as red, greenish, and light green; Hayes's subject G. W. certainly reported things as looking green (though she may or may not have been truly protanopic); and Sloan and Wollach in their investigations employed methods that, if their subject had had any sensation of red and green with his dichromat eye, might almost have been designed to fail to reveal it. Meanwhile, (b) Nagel's report is completely silent about the general range of sensation of his train driver; and when he does make general claims about dichromats (himself included), he actually says they recognize and have sensations of red (Nagel 1905, 1907, 1910). We might add that (c) von Kries's anomalous trichromat retained the sensation of red, it seems, more strongly than that of yellow, and in any case lost sensation of both red and green less than one might expect. And we might remember that (d) the acquired deficits are of no primary evidential value for us on congenital dichromacy. Altogether, these reports seem to me to constitute a case not of moderate or *weak* evidence for the yellow-and-blue view, but of *no* evidence for it at all. What we have instead is a number of authors firmly wedded to a yellow-and-blue theory—Hering, and his followers Hess, and Hippel, and later, Sloan and Wollach and Judd himself—sometimes to the point of stamping out the evidence that they themselves had let slip, that it could not quite be correct.[48] There may indeed be some truth in that view; but I think it is entirely unclear from the evidence we have seen what truth exactly it might be.

I shall mention only the three most impressive cases of unilateral color blindness that I know from later years, though they are worth examining in more detail than I can give here. The Graham and Hsia case (e.g., Graham et al. 1961) certainly seems to involve the perception of merely yellow and blue; but it is not a case (in the affected eye) of any of the recognized forms of congenital dichromacy.[49] It is in fact entirely obscure quite what the underlying nature or cause of the subject's color blindness might have been—and how it might be a guide to the experience of people with any of the more usual kinds of dichromacy.

The unilateral case reported by MacLeod and Lennie (1976) is extremely interesting: it seems to be a case of a deuteranopic eye that provides sensation of two principal colors, but the colors reported are a *greenish* blue and a *reddish* yellow (corresponding to 473–474 nm and 610 nm), which fits the theories of neither Helmholtz nor Hering (nor Judd, who wanted blue_{470} and yellow_{575}). And, more remarkably, the authors find that the nearest thing to a "neutral point" for the subject's dichromatic eye is probably greenish, rather than pure white: so (with some other evidence too) they conclude that the subject's sensations using that eye are reduced to the colors represented on a line in normal chromaticity space running between 473–474 nm and 610 nm, but it is not a straight line but an arc "bowed toward the green corner" (698). (The line would therefore include green-blue, a variety of greens—some very desaturated—yellow, orange, and orange-red.) This is good evidence against standard views. But when the authors hold that the full range of sensations with the dichromat eye is confined merely to a line in chromaticity space, they go beyond, I believe, what the investigations they report can establish. There is the familiar shortcoming (and similar limitations affect the other two recent reports): that the main investigation involved spectral (or narrow-band) lights in small fields—and it is not at all clear whether, with nonspectral (i.e., mixed) lights, a wider range of test intensities, surface colors, larger fields, and (a very different but imporant issue) color presentations involving more complex and varied contexts and "natural" environments, a wider range of color experiences might not have been had.

The third case (Alpern et al. 1983b) is an acquired tritanopia in one eye. As an acquired tritanopia (following a serous chorioretinopathy two years earlier), the case is unlikely to tell us much about cases of congenital protanopia or deuteranopia (at least, in advance of our having a much better theoretical understanding of the full variety of cases). But it does have two particularly interesting features, which may alert us to possibilities to test for in other cases. One is that the sensations had with the tritanopic eye (when identified by means of stimuli presented to the normal eye) follow, it seems, not a straight line in chromaticity space between red and green (as the traditional theories claim), but a curve that bends back on itself—there are not just two principal hues experienced, but a whole range, including red, orange, yellow, and blue, though, it seems, no green at all.[50] Secondly, the case is described as one where change of *brightness* in a stimulus by itself produces change in *hue* sensation. Violet_{422} and green_{523}, we are told, lie on a single confusion line—that is, have the same chromaticity—for the tritanope eye. But, at the intensities used in the actual testing, the latter looked a *greener* blue than the former—thanks to, the authors tell us, its being brighter.[51]

Though Alpern et al. imply that *level of brightness* can by itself affect hue sensation, they unfortunately do not study brightness as a separate variable in itself. (They notice it only as an incidental by-product of the different brightnesses of the stimuli produced with their apparatus.) Other investigators have given evidence that suggests such effects may be more widely found. Pitt says, of his dichromats, "various decreasing intensities of a green (.53 μ) were called yellow, yellow, reddish-yellow, red, red" (1935, 33). Pitt does not take the reports at face value: he thinks that, when the dichromat *says* he is seeing yellow, reddish yellow, and red (as the intensity of a $green_{530}$ stimulus is progressively reduced), all that can be happening is that he is experiencing the same pure yellow hue at progressively lower levels of brightness—this one hue being successively "called by different colour names" (33). But in the absence of any special theoretical precommitment, it would seem equally open to think that in Pitt's experimental situation, the $green_{530}$ stimulus actually produced (when bright) sensations of *yellow*, and (when less bright) sensations of *orange*, and finally (at lowest luminance) sensations of *red*. This is just the kind of thing that Alpern et al. provide firm evidence of with their own trianopic subject, and it might well occur more widely. It is something that we should test for further.

It may be worth adding that the studies of unilateral dichromats suffer from the same kinds of weaknesses as have vitiated much experimentation on dichromats in general. I shall discuss these more fully later (in section 12.9) in the light of some more adventurous empirical findings (surveyed in section 12.5), but some of the more general points are these: we need more evidence on performance with large fields as well as small, surfaces as well as lights, at varying levels of saturation, and with such "dynamic" factors as those of seeing a single surface in a variety of kinds of illumination. With such factors ignored, it can only be hazardous to make any claim about the limits within which a unilateral dichromat's experience is (as Judd puts it: 1948, 247) "confined."

I have already mentioned that we need to know more about the *cause* and nature of these unilateral conditions before we can take a view on how similar the underlying physiology might be to ordinary cases of congenital dichromacy. But even if we had an ideally good account of the perceptual experience of a unilateral dichromat and had reason to think the retinal physiology was reasonably close to that of an ordinary bilateral dichromat, there would remain important questions about how much the former could tell us about the experience of the latter. The reason is that the presence of a second, normal or near-normal eye may itself have an effect on the kind of experience yielded by a dichromatic eye: there may be influences on neural development and processing in a unilateral case (e.g., to minimize disparity in the experiences associated with the two eyes) that are simply not present in an ordinary bilateral case.

We have found *not one case* of a representative unilateral dichromat who we can be confident had sensations only of yellow and blue. (Two cases might seem to come close, but the Sloan and Wollach investigation involved only smallish fields and medium-low saturation surfaces, chosen from a tiny range of hues, seen under dim (indeed mesopic) conditions—and it can be no guide to the subject's full range of sensation. And the Graham and Hsia case resembles no known variety of congenital dichromacy at all.) But even if we had found such a case, or came to find one in the future, it would be quite unclear—in the absence of a much better general understanding than we actually have of the nature of both unilateral and bilateral deficits—precisely what conclusions we might draw from it about the experience of ordinary bilateral dichromats.

12.5 The Many Reports of Residual Red-Green Discrimination among Dichromats

I have already mentioned the evidence in Jameson and Hurvich (1978) that dichromats make distinctions that, on the standard theory of dichromacy, they should not be able to do. The evidence is not at all isolated; there is evidence from the 1850s and from the early years of the twentieth century that was rather shockingly neglected thereafter. And there is recent evidence too, from the last forty years. In fact, odd as it may sound, it is about as close as anything can be to being an established fact in this subject that dichromats—that is, people classified as such on the basis of, above all, the Nagel anomaloscope—are often quite capable of making distinctions between colors that the standard theory implies should be indistinguishable for them (and that, in the Nagel anomaloscope, genuinely are indistinguishable for them). I shall here just mention a few of the findings. What is striking is, first, how much disagreement there is among experimenters on what might account for these abilities, and, secondly, the extent to which authoritative writing on the subject of color blindness still perpetuates the standard theory of it.

I shall here survey just a handful of the most important studies, made by experimenters in the distinguished laboratories of Robert Boynton at UC San Diego (and earlier at the University of Rochester) and of Joel Pokorny and Vivianne Smith at Chicago. The experiments study performance at three main tasks: the color naming of spectral lights (Scheibner and Boynton 1968; Nagy and Boynton 1979), the color naming of surface color samples (Montag and Boynton 1987; Montag 1994), and the matching of spectral lights with mixtures of other lights, as, for example, in Rayleigh matches (Smith and Pokorny 1977; Nagy 1980). The color-naming studies are all unanimous on the phenomena, though they disagree on the explanation of them: as long as the stimuli are larger fields (e.g., 8° rather than 1°), presented for longer periods of time (e.g., 5 seconds rather than 30 ms), then nearly all dichromats prove able to name the colors with some success in the red-green dimension as well as in the yellow-blue dimension—not with the same accuracy as normal trichromats, but with a degree of success that should be totally lacking if the standard theory were right. The color-matching experiments tell a similar story: while dichromats (almost by definition) accept a Rayleigh match of Y_{589} with R_{670} or with G_{545} as long as the fields are small (e.g., 1° in size), it seems (as, e.g., in Smith and Pokorny 1977) that they actually reject any such match when the fields are larger, e.g., 8° (which is still only the equivalent of about an 8 cm diameter circle viewed at 60 cm). That is, the "dichromats" are behaving under these conditions in the same way as some form of trichromat.

How do they manage it? There is no agreement among the authors of these studies, and indeed the evidence does not suggest a single explanation. Rod intrusion? The suggestion is rejected in Scheibner and Boynton (1968), supported after all in Nagy (1980; at least for certain conditions) and in Montag and Boynton (1987), but rejected again in Montag (1994)—for the fairly impressive reason that (in surface color-naming tasks), "Using high light levels so that the rods are saturated does not impair performance" (Montag 1994, 2137).

The other explanation that the experimenters usually turn to instead is that the dichromat may after all have some cones of the type usually supposed to be missing, or perhaps a variant like those found in anomalous trichromats. Scheibner and Boynton (1968) attribute their subjects' success in naming of 3° spectral fields to cones "of the type usually considered missing" (1968, 1158). Nagy and Boynton talk of a "weak residual third cone mechanism" (1979, 1259; cp. Nagy 1980), having ruled out a rod explanation on the ground that their subjects do only slightly worse at giving the color names of 12° spectral fields five minutes or so after a light bleach than when they have not had a light bleach at all—whereas rods would be expected still to be out of service at that point. (When you "bleach" the pigments of the eye by exposing it to an extremely bright stimulus [in this case, 4 minutes of white light], the cones recover faster than rods: the authors expect the cones to be back in action 5 to 8 minutes later, while the rods would still be saturated.) Thus, Nagy and Boynton (1979) conclude, many people standardly classified as dichromats are, in a sense, not dichromats at all. Montag and Boynton (1987), however, examining dichromats' color naming of surface samples, take precisely the opposite view, having found that their one protanope subject did rather badly after a bleach after all. But then Montag (1994) explicitly retracts his 1987 conclusion and goes back once again to the hypothesis of some cones of the supposedly "missing" type—on the ground that increased light levels (not involving a bleach, but just ordinary bright lighting) do not significantly impair dichromats' performance.

I shall make just a couple of comments. First, it is a little odd that rods and cones are mostly treated as rival and mutually exclusive hypotheses. There seems no a priori reason why we should expect just one factor to be *the* explanation of all the phenomena in question. Rods and cones might both play a role, perhaps under different conditions, or perhaps sometimes operating together. More importantly, it is odd to suppose that rods and cones exhaust the options for possible explanations here: and among the

thing, to look at it in a variety of different kinds of light, or from a variety of points of view—seeing how one and the same surface catches the different kinds of light in the environment in various different ways. They test the surface out—putting it through its paces, so to speak, to see what it can do. So we might inquire: Might seeing a surface color *under two different illuminants* provide a dichromat, who at first seemed to lack information on the red-green dimension, with information about that third dimension, rather in the way that seeing a scene *from two different points of view* provides a person who is looking with only one eye with information on the third spatial dimension? And might some such process even cause some of the things around to *look* red, for example—rather than merely being *judged* to be red—just as the right kind of looking from different directions, even with just one eye, seems actually to cause some things in the environment to *look* nearer, rather than merely being *judged* to be nearer? And might it even be the case that a single stimulus or input that, in a sense, contained only two dimensions of color variation might be taken or "read" by people—given its own complexity, and the structure and regularities of our world—as a presentation of things varying in three color dimensions? There is reason for thinking that the answer might be yes to all these questions, though whether the cues that I investigate here are cues that are actually *used*, as well being *usable*, is a further question I shall have to leave for later experimental investigation.

12.6.1 Fifteen Pigments, in Two Different Illuminants: Normal Trichromats and Protanopes

Let us suppose the standard account of the mechanics of vision in the dichromat: suppose that protanopes, for example, lack L cones and receive information only from M and S cones. We will expect them to confuse colored lights according to the standard confusion lines (cp. figures 12.6, 12.7; look ahead to figure 12.11). They should confuse light corresponding to cadmium red and viridian, or to Mars red and cobalt titanate green spinel. In practice, however, we find they are surprisingly successful at identifying the colors of things that they are supposed to confuse. So how do they do it? What might be the methods or mechanisms by which they achieve what limited success they actually achieve?

The clue that I shall follow up is the fact that a red and a green thing that look indistinguishable in one kind of illumination may behave differently as the illumination changes. As we shall see, light sources differ substantially in physical character: figure 12.8 shows how different, for example, the direct light of the sun and the light of the remainder of the sky typically are—the former is yellowish, the latter bluish. For illustrative purposes, in some mathematical modeling I shall concentrate on two particular kinds of light, labeled D_{55} and D_{75} by the CIE, which approximate to direct sunlight (of correlated color temperature 5500 K) and to the sky without sun (7500 K). (For correlated color temperature, see *Note on terminology* in note 13.) They are far

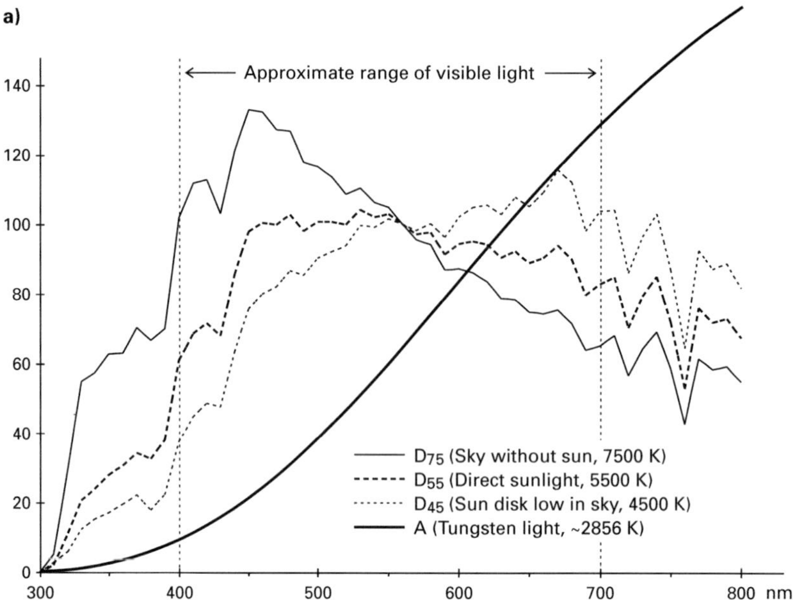

Figure 12.8

Representative natural illuminants. (a) Spectral power distribution of some representative illuminants approximating daylight of correlated color temperature 4500 K, 5500 K, and 7500 K (CIE standard illuminants D_{45}, D_{55}, and D_{75}), and the light of a tungsten lamp of about 2856 K (standard illuminant A). Note how much red is in the sun's light when it is low in the sky, and how much blue is in the light of the sky, and how the differences between the curves are all the more striking if we concentrate on the visible part of the spectrum, from 400 to 700 nm. (My graph from standard CIE basis functions; see, e.g., Hunt 1987, 186–195.)

enough apart for interesting effects to show up, but much greater natural differences do occur in our environment, especially if we include the light of a fire or an incandescent light bulb.

When there is a change in illumination, it may be that some objects respond to that change differently from others, and this is a sign of differences in color among those objects. For example, suppose the light changes from the yellow of direct sunlight (D_{55}) to the blue of the remainder of the sky (D_{75})—perhaps a cloud covers the sun, or a person picks up an object and turns it so it catches the light from a different part of the sky. Then it may happen that a red and a green thing that were indistinguishable earlier will come to look different: the red thing will have darkened in relation to other things, while the green will not. (Green is more or less equally "close" to a blue illuminant and to a yellow; red is always "farther" from a blue than from a

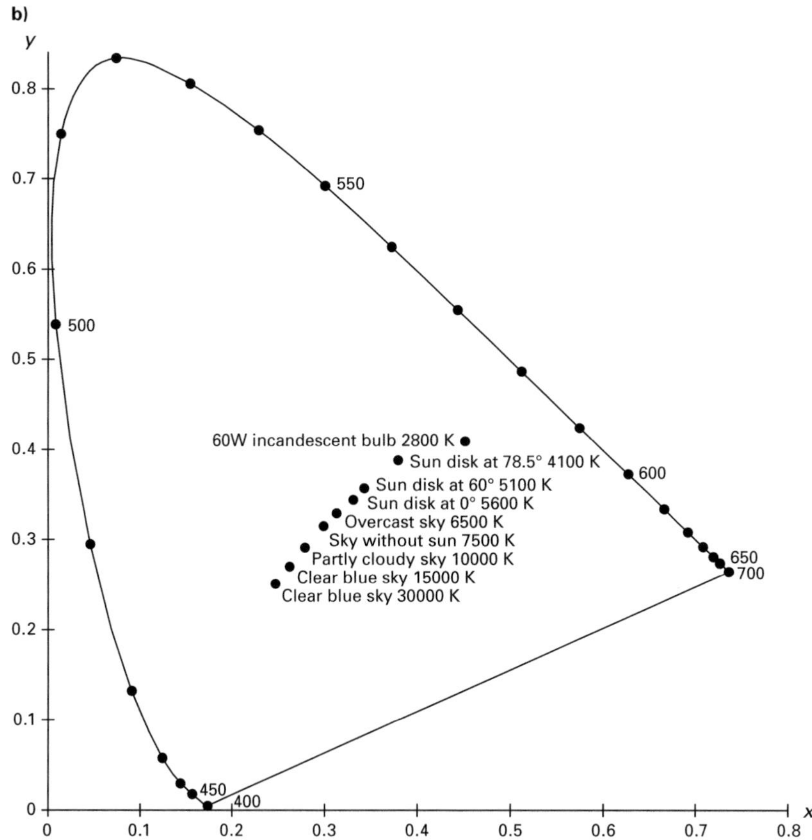

Figure 12.8
(continued)
(b) Chromaticity of representative light sources (mapped from data in Wyszecki and Stiles 1982, 28 and 8–9). Note that natural light (except for low temperature sources) varies mostly along something very close to a straight line running between approximately unique blue and unique yellow (~470 nm and ~575 nm). The direction of that line, incidentally, is very similar to that of the variation produced by a yellow filter such as the macula (see section 12.6.5).

yellow.) If a pair of red and green things initially looked equal in lightness, then the red will now look darker. This is, as we shall see, a general phenomenon for normal trichromats. And this might be a dichromat's clue to recognizing or "recovering" the red-green dimension that he is standardly supposed to be missing. This is the hypothesis that I shall model here. Of course, we do not usually see the illumination in the way that we see objects that are reflecting that light (though we should not forget that we can look at the sky and, even if not for long, at the sun and other sources). But provided there is a good range of colored things in the environment, we will have good guides to the color of the illuminant; and it is a fact of everyday life that, over a wide range of changes in illumination, we manage quite decently to adjust to, and keep track of, the changes—not merely "discounting" or "eliminating" the color of the illuminant (on the model of either Helmholtz or Hering), but *keeping track of* and recognizing, in some general way, *both* the colors of things *and* the colors of the light.[55]

To test the hypothesis—that difference in behavior of red and green things under changes in illumination contains information that could be a clue to their color for protanopes and deuteranopes—I shall briefly examine such changes of illumination for normal trichromats, and then focus in detail on the case of protanopes; deuteranopes will appear in section 12.6.6.

One remarkable fact that may simplify the task of recovering the "missing" dimension of color for dichromats is that the main changes in illumination in our environment lie pretty much on a single axis, running between yellow and blue: only at quite low temperatures do we find such departures as the relatively orange light of a 60 watt light bulb, for example. Purple and green ambient light are simply not much found in our environment; if there were any need for a constant reminder of the position of unique blue and unique yellow to recalibrate our visual systems as we and the environment change, indeed one could take the main axis of change in natural lighting as a pretty good guide.[56] (I think of the significant fact that people place unique hues pretty constantly throughout their lives, unaffected by changes in the color of the lens of the eye.) As we will see later, once we have taken account of changes in illuminant along the YB-axis, there are residual changes in retinal stimulation that we can attribute either to the change in the illuminant or to the nature of the objects seen.

For color samples, I have taken fifteen pigments, using their spectral reflectance curves as in Mayer (1940/1991; see, e.g., figure 12.9). For the spectral composition of the two kinds of illuminant, I use CIE functions for standard illuminants D_{55} and D_{75}, corresponding to direct sunlight of 5500 K and sky without sun of about 7500 K (cp. Hunt 1987, 186–188; Wyszecki and Stiles 1982, 8–9). Multiplying these together (I sample each function at 20 nm intervals from 400 to 700 nm), we can calculate the

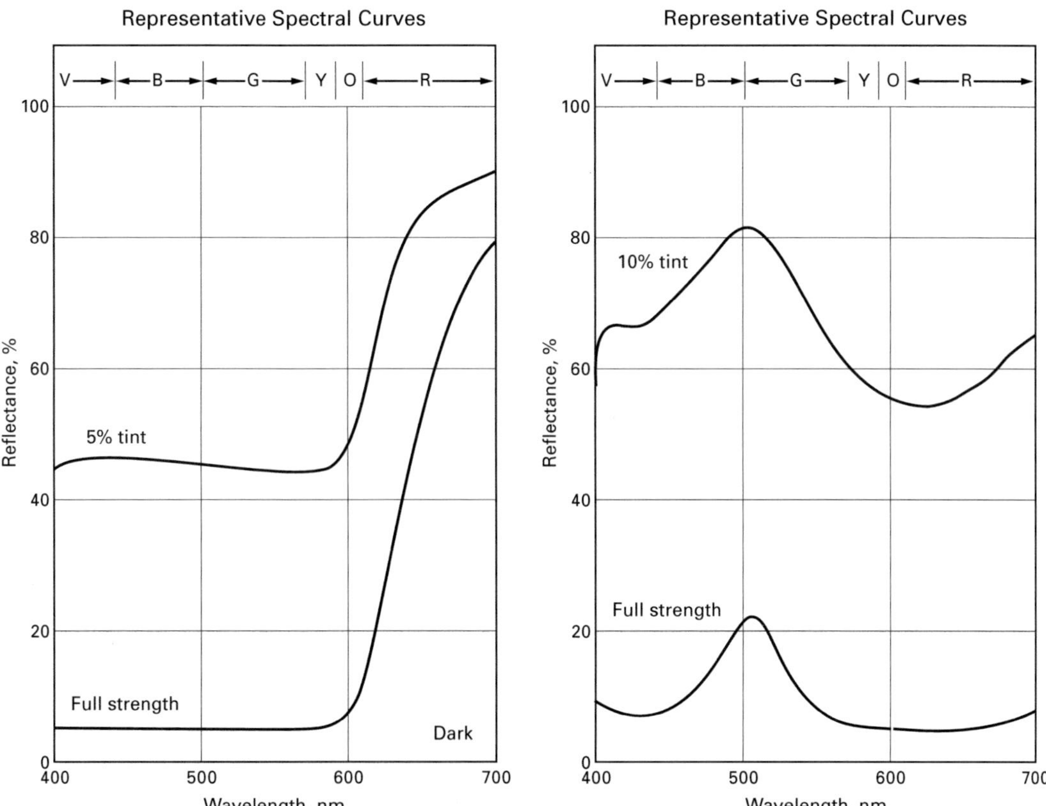

Figure 12.9
Spectral reflectance curves for two sample pigments. (Left) Pure cadmium red (PR108). (Right) Viridian (PG18; Mayer 1940/1991, 80 and 98). In each graph, the lower line gives the reflectance function for the pigment at full strength, the upper one for the pigment in a *tint*, or mixture with white. The present study considers the pigments only at full strength.

spectral character of light reaching an observer's eye. Using the standard CIE color-matching functions, we can calculate *X*, *Y*, *Z* values for the light stimulus component at each sampled wavelength at the eye, and sum these to get *X*, *Y*, *Z* values for the complex total stimulus, and thereby get *x*, *y* values for it (see, e.g., Hunt 1987, ch. 2, for the method). And we can do these calculations for each pigment, first as seen in direct sunlight, and then as seen in sky without sun. The shift (as the light changes from D_{55} to D_{75}) may in general be characterized as a blue-ening: thus in figure 12.10, a line on the (*x*, *y*) plane marks the shift in the *x*, *y* values for each pigment type, and it points roughly toward the bottom left, signaling an increase in the blueness of the stimulus at the eye as the illuminant changes in that way.

What is more remarkable than the direction of these general chromaticity shifts is the pattern of changes in *luminance* that take place at the same time. To separate out the effects of the color of the illuminant from those of its total luminance, I shall assume we are dealing with a case where the D_{55} and the D_{75} are matched in luminance:

Figure 12.10
Changes in chromaticity and luminance of fifteen pigments, as illuminant changes from D_{55} to D_{75}: normal trichromats. An approximately perspectival view of a 3-D representation of tristimulus values: the (*x*, *y*) plane can be read as horizontal, with points on it representing the chromaticity of the pigments; from each point there is a line rising approximately vertically out of the plane, the height of which represents the luminance of the pigment above a certain "floor" represented by the (*x*, *y*) plane (and set here at about 4% of the luminance of white). Each "skyscraper" represents a pigment first in D_{55} (with the righthand vertical wall), and then in D_{75} (with the lefthand vertical wall). Note how the roofs of the skyscrapers slope in different directions in different parts of the diagram: in the bottom left and upper middle (for blue and green pigments), the lefthand wall of each skyscraper is taller than the righthand wall: the pigments have higher luminance in D_{75} than in D_{55}. But the skyscrapers in the right and lower right of the diagram (for yellow and red pigments) have the opposite slope: the pigments have higher luminance in D_{55} than in D_{75}. As the light becomes bluer, there is a surge in luminance for pigments close in color to the new illuminant. Values for the proportionate luminance increase are given in parentheses: green pigments have a positive increase in luminance (3.8%, 3.2%, 0.9%); the red pigments have a negative increase (–5.9%, –4.5%, –4.6%).

Note: The perspective is as if looking down from a height, directly above the point $x = 0.315$, $y = 0.13$, and the lines of the skyscrapers all converge on a point (in line with this point, and behind the paper, so to speak), which can be taken to represent $X = Y = Z = 0$: the diagram can therefore be thought of as a 2-D drawing of a 3-D representation of the CIE *XYZ* tristimulus space. For any line coming out of the floor, the distance (in the 2-D drawing) from the origin ("behind" $x = 0.315$, $y = 0.13$) to the point where the line emerges from the floor represents a "floor" luminance of 15 (where, for comparison, lead white has a luminance of ~369, and ultramarine ~16): and "heights" above the floor along that line are represented proportionately. The metric for heights is therefore strictly not the same as in a perspective representation.

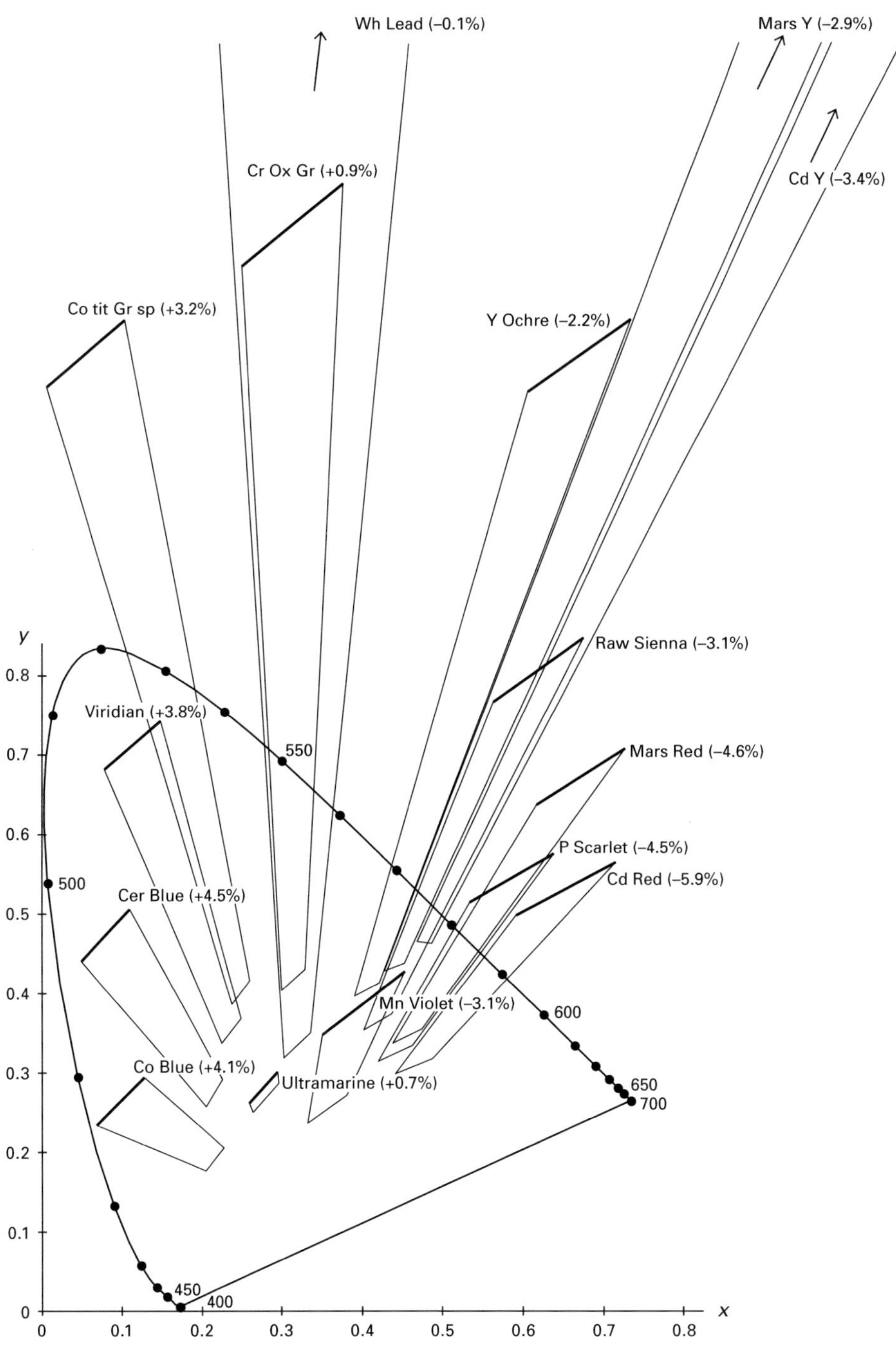
Wh Lead (–0.1%)
Mars Y (–2.9%)
Cd Y (–3.4%)
Cr Ox Gr (+0.9%)
Co tit Gr sp (+3.2%)
Y Ochre (–2.2%)
Raw Sienna (–3.1%)
Viridian (+3.8%)
Mars Red (–4.6%)
P Scarlet (–4.5%)
Cd Red (–5.9%)
Cer Blue (+4.5%)
Mn Violet (–3.1%)
Co Blue (+4.1%)
Ultramarine (+0.7%)
550
500
600
650
700
450
400
y
x
0
0.1
0.2
0.3
0.4
0.5
0.6
0.7
0.8

some things get lighter in appearance, while others get darker, but the average is constant. The change from D_{55} to D_{75} is a blue-ening; unsurprisingly, therefore, blue things benefit from the change more than others: the blue pigments increase in luminance under the change, while the yellow ones decrease (cobalt blue gains 4.1% in luminance; cadmium yellow loses 3.4%). But more important for us is the fact that *green* things also tend to get lighter, being relatively close to the blue illuminant, while *red* things get darker, being relatively distant from it. The effects are illustrated by the heights of the "skyscrapers" in figure 12.10 for normal trichromats. The proportionate increase in luminance as the light changes from D_{55} to D_{75} is, for the three green pigments, positive: +3.8%, +3.2%, and +0.9% (for viridian, cobalt titanate green spinel, and chromium oxide green). The proportionate increase in luminance for the red pigments is negative: –5.9%, –4.6%, –4.5% (for cadmium red, Mars red and perylene scarlet).[57] It is indeed the case that red and green things interact differently with the light as the sun goes in (behind a cloud) and we have sky without sun, or when we tip the things so they catch the light of a different part of the sky.

So far I have been talking of these changes as they affect the normal trichromat; but the luminance functions for dichromats are different (particularly in the case of protanopes: lacking the L cones, they see much less brightness in the red end of the spectrum). So let us move now to the case of the protanope and calculate the change in luminance under the same change of illumination.[58] The results appear in figure 12.11: there is a clear difference in the behavior of green and red things under the change. I have again modeled a case where the overall average luminance is the same before and after the change. It turns out, just as in the trichromat case, that blue pigments gain most in luminance, followed by green, while the yellow and red lose luminance. If we suppose a protanope with samples of cadmium red and of viridian that initially look indistinguishable to him in sunlight such as D_{55}; then, if he can bring them into bluer (but similarly bright) skylight such as D_{75}, the viridian will lighten by 2.9%, whereas the cadmium red will *darken* by a similar percentage, 2.3%. The difference is not huge, but it would be a good clue for a protanope to use: among reds and greens that are initially indistinguishable, those things are *greenish* that become relatively light as the incident light becomes bluer, while those things are *reddish* that become relatively dark.

I have made the proposal so far under a supposition that is really unnecessary. On this supposition, the information needed to tell red and green things apart is not available to the dichromat at one instant, but it is available *over time*, when there are suitable changes in illumination. But, of course, there is no necessity that changes like this be successive in the world rather than just successive in our experience of the world: the viewer may (while the illuminants in the environment remain unchanged) simply move the object or move her head so that the light reflected by it to the eye

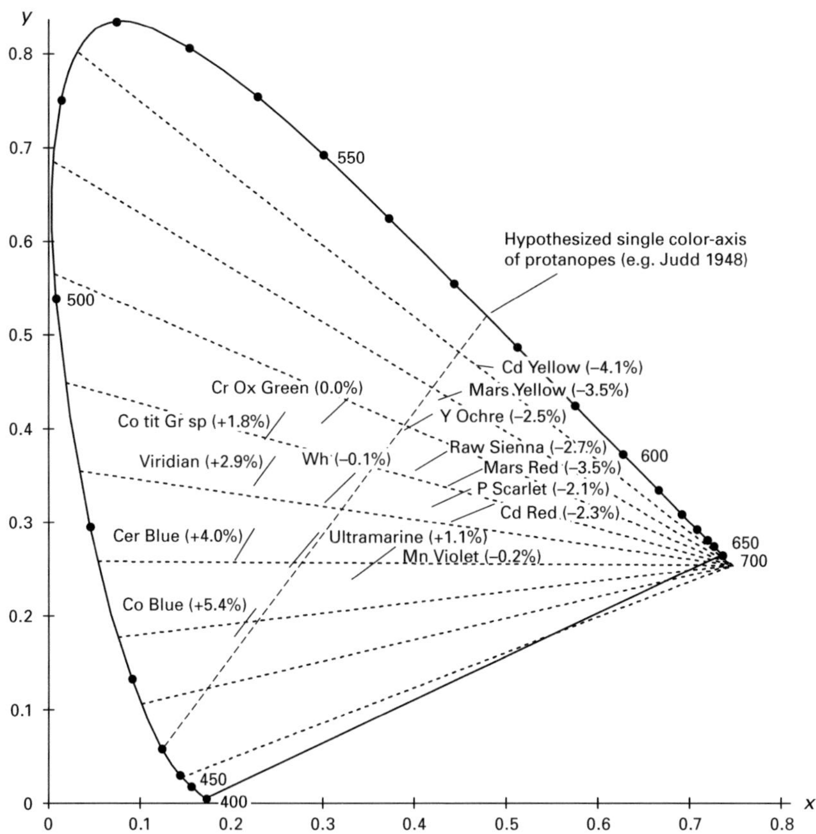

Figure 12.11

CIE 1931 chromaticity coordinates for fifteen pigments, with and without direct sunlight: protanopes. For each pigment, a line indicates the shift in chromaticity between the pigment in D_{55} (at the right or upper-right end of the line, labeled with the color name) and the pigment in D_{75} (at the left or lower-left end). Proportionate increase in luminance under the change is given in parentheses. As the light changes from D_{55} to D_{75}, the green pigments show an increase in luminance of 2.9%, 1.8%, 0.0%; the red pigments show an increase of –2.3%, –2.1%, –3.5%. Of red and green things that seem confusable, green things lighten most, while red things darken.

comes from, say, the blue sky rather than the sun directly—or from the window rather than a light bulb. And it is not necessary even that the changes be successive at all: if, for example, there is a uniform surface, different parts of which (as seen from the present point of view) catch light from different sources in different ways, then the kinds of variation I have been talking of may manifest themselves even at one single moment. A single large expanse of green, seen by the light of a window, may in one region have a bright patch of direct sunlight on it, while another expanse of red, in the same scene, has a corresponding patch of sunlight falling on it—the difference between the variegation in the one expanse and the variegation in the other providing at one instant the same information as we initially thought of as being gathered over time. And of course curved surfaces often provide more such variation than flat ones, as they face light coming from a wider range of directions. Our ability to use this kind of variation as a guide to color depends, of course, on our being able to recognize uniformly colored expanses as precisely that, and to recognize or take account of their shapes: we need, with the red and green expanses, to be able to take the appearance of highlight as the effect of sunlight falling on a uniform surface in a nonuniform manner (i.e., being reflected from it at a variety of angles to the present viewpoint), rather than being instead, say, the effect of a nonuniform surface coloration being seen under exceptionally uniform illumination. But we do manage to recognize and keep track of such things quite well in our environment; and—though the topic is one for another occasion—we live in a world to which we are generally well-enough attuned that what we take to be the case in such circumstances is mostly not too far from what is actually the case.

12.6.2 Recovering 3-D Color from 2-D Receptor Information: A First Projection

Let us work with the idea that at any one time from any one region, the protanope receives only 2-D color information, and assume that the information can be mapped in terms of luminance and the one-dimensional line that runs from 470 nm to 575 nm on the standard CIE diagram. (I don't believe for a moment that the protanope actually has experience only of the two hues of yellow and blue that this line spans; but it will do no harm to accept the proposition that one dimension of the protanope's input data can be mapped on such a line.)

We may picture the information available as having collapsed from a solid to a plane; abstracting from luminance, the hue stimulus space has collapsed from a plane to a line. The result of the collapse can be pictured as in figure 12.12.

Thus, aside from lightness, cadmium red would for the protanope be a near-metamer for viridian, and cobalt titanate green spinel would apparently lie in hue (or, rather, just saturation) between Mars red and perylene scarlet. Perhaps most remarkably, the saturated pigments viridian and cadmium red would be supposedly almost

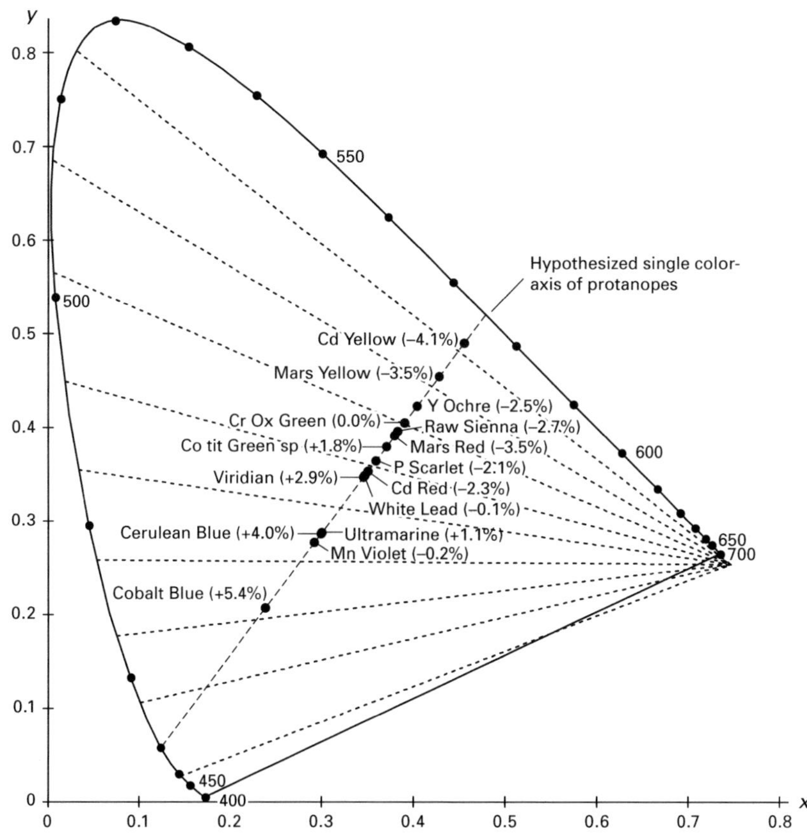

Figure 12.12

CIE 1931 chromaticity coordinates for fifteen pigments, in D_{55} direct sunlight: standard model collapse for protanopes. In parentheses are given the percentage increases in protanope luminance for each pigment under a change of illuminant from D_{55} to D_{75} (of equal total luminance): greens generally get lighter; reds get darker.

indistinguishable from some achromatic gray. And, as the light changed from D_{55} (yellow) to D_{75} (blue), apart from varying in relative lightness, the samples could change only by moving to other projections along that same axis, exhibiting more varieties of the unique blue and yellow that supposedly fix the limits of the dichromat's color world.

From this kind of input, how could one recover information about the "missing" red-green color dimension? Here is a first attempt. Noticing the way the reds become darker and the greens lighter under our change of illuminant, we might try taking the magnitude of the luminance change under the change of illuminant as a direct sign of the missing color dimension. This can be done in a variety of ways, but it takes only one example (see figure 12.13) to show the problems in principle with all of them.

The difficulty is that the yellows are shifted too far toward the red, and the blues are shifted too far away from it: the projection had taken no account of the fact that as the light shifts from yellow to blue (e.g., from D_{75} to D_{55}), this affects not just reds and greens, but also yellows and blues. Yellows will darken, and blues lighten—and these variations have nothing to do with the missing RG-axis.[59] We should not be taking the luminance increase of cobalt blue as a sign of greenness in it, or the lack of such increase in cadmium yellow as a sign of redness. Rather, we should discount the overall variation in lightness that goes with position on the YB-axis, and then see how much further variation in lightness remains. This is what I attempt in a second projection.

12.6.3 A Better Projection

We need to distinguish the change in luminance due to the yellow-blueness of a sample from the change in luminance due to its red-greenness. This is not particularly difficult conceptually, but I have adopted a relatively simple mathematical method that fits the phenomena reasonably well, rather than exploring more sophisticated methods that might fit it better.

I first calculate the distance in the YB dimension of each sample from *N* (imaginary pure white). (*N* corresponds also to the color of the illuminant. If there is a reasonable range of colored objects in the environment, then it should not be particularly hard to identify.) We may suppose in general that a pigment's position alone on this single YB-axis is itself partly responsible for a change in luminance under the kind of change of illuminant that we have been considering, but to what extent? One reasonable suggestion is this: almost completely, in the case of colors near the extremes of this YB scale, for a reason I shall explain. In this particular case, take the two samples with the most extreme values on the one input color dimension, namely (in my data) cobalt blue and cadmium yellow. Assuming that the samples have been chosen from an environment reasonably rich in samples that differ fairly broadly within color space

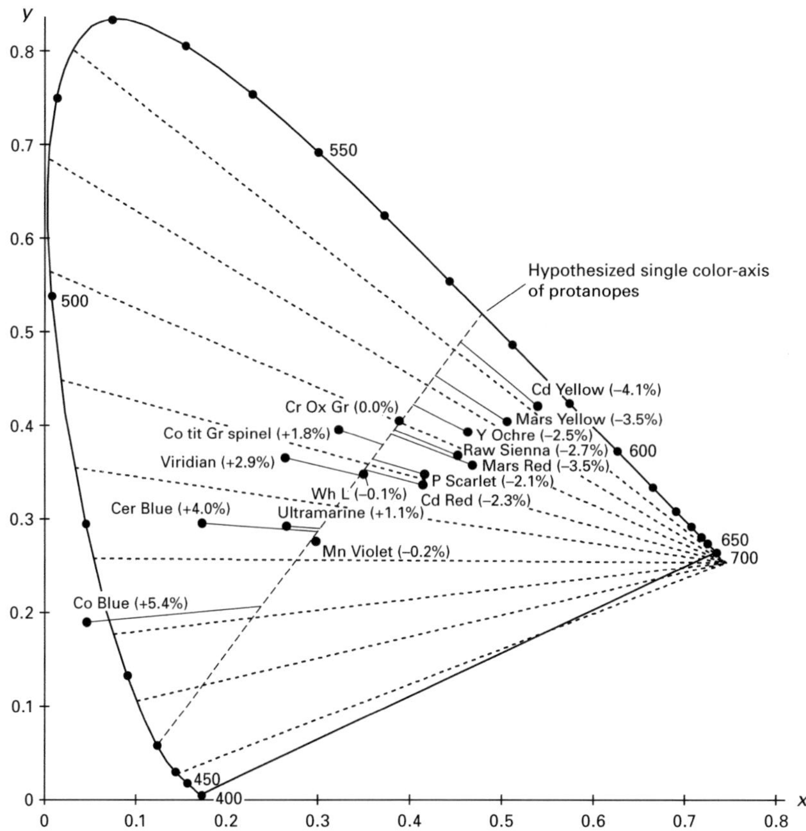

Figure 12.13

One crude projection into CIE 1931 space for fifteen pigments, in direct sunlight (D_{55}): protanopes. Hue-axis values have been projected away from the confusion point in proportion to the increase of luminance (under the change from D_{55} to D_{75}). The problem is that the projection makes the yellows come out too red and the blues too green—and both kinds come out as too saturated.

and that fall (in CIE *x*, *y* chromaticity space) within a closed curve such as a circle, we may suppose that if in *x*, *y* space the two samples are at extremes on the YB-axis, then they also lie fairly close to the YB-axis—that is, they have relatively small values on the "missing" hue axis orthogonal to this one.[60] This enables us to calculate a factor for the percentage increase in luminance attributable to the yellow-blueness of samples (which I will call S′) and to work toward the new projection in figure 12.14.

If we set up a scale on the YB dimension, taking white as the origin of the scale and measuring distance in the units of the CIE *x*, *y* chromaticity diagram (look back to figure 12.12), then in the single YB dimension, cadmium yellow (illuminated by D_{55}) lies at +0.1843 from the position of pure white; and, as the illuminant changes (to D_{75}), it increases in luminance by –4.1%. Cobalt blue lies on the axis at –0.1722 and increases in luminance by +5.4%. If we attributed all of the increase in cadmium yellow's luminance (under the illuminant change) to its position on the YB-axis, we could calculate the *increase in luminance* (under the change of illuminant from D_{55} to D_{75}) *attributable per unit of YB-ness* as –4.10% / 0.1843, that is, –22.48% (treating a –4.10% increase in luminance as due solely to a value of 0.1843 on the YB-axis). And for cobalt blue, we could calculate 5.4% / –0.1722, that is, –31.43% (treating a 5.4% increase in luminance as due solely to a value of –0.1722 on the YB-axis). I propose (though I would not claim any particular theoretical reason) that we attribute not quite all of the increase in luminance of cadmium yellow to its YB-ness—and therefore try out –22% as the factor we are seeking. Taking the factor as –22%, we attribute a –4.05% increase in luminance for cadmium yellow as due simply to its position near the Y end of the axis (0.1843 × –22% = –4.05%); similarly we attribute a 3.79% increase in luminance for cobalt blue to its position near the blue end of the axis (–0.1722 × –22% = 3.79%). This of course leaves some residual lightness variation to be attributed to other factors—that is, in my rather simple model, to position in the red-green dimension, which the dichromat is "missing": this is partly a good thing, given that the cadmium yellow and cobalt blue samples are (in CIE *x*, *y* space for normal trichromats) actually not *on* the YB-axis, and it will be good to allow some of the loss of luminance in the cadmium yellow to be due to some orange-redness in its hue; and some of the gain in luminance in cobalt blue to be due to some greenness in it. No doubt other and better functions could be devised for discounting the change in lightness attributable solely to YB-ness in any sample.

Alternatively, we might propose saying that, within any one confusion class, the average increase in luminance (under the kind of change of illuminant we have been examining) can be put down to position on the YB-axis; and the deviation from that average can be put down to position on the "missing" RG-axis.

If we then use this factor (–22% multiplied by the value on this YB-axis) as the degree of change attributable simply to YB-ness, then—in general, for any sample—when we have discounted this element, there remains some further change in light-

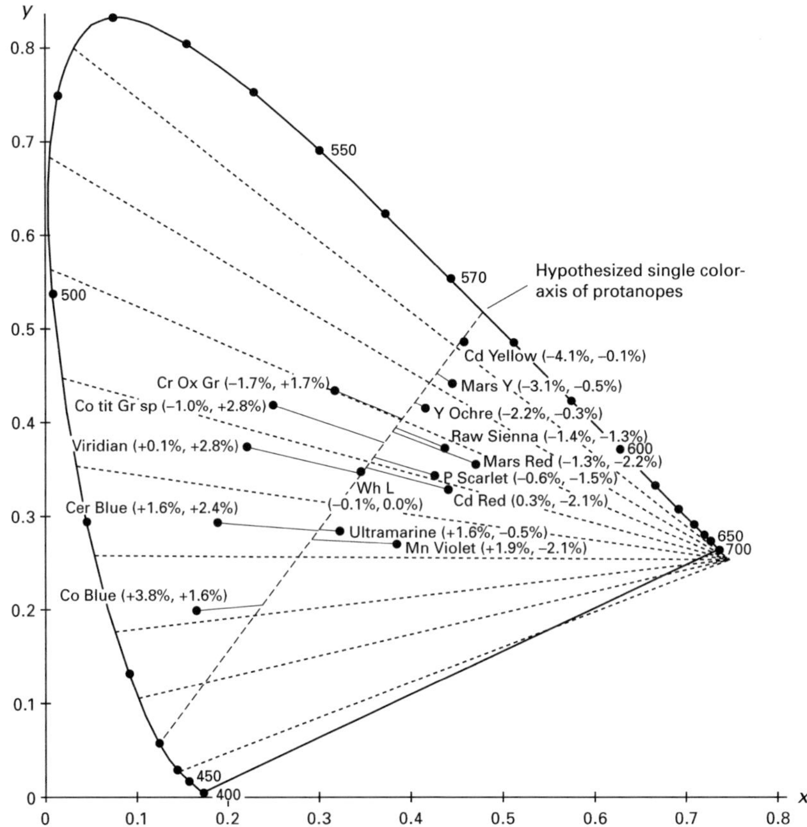

Figure 12.14

Recovering the "missing" dimension of red-greenness for protanopes: improved projection into CIE *x*, *y* space for fifteen pigments. The pattern of recovered chromaticities compares well with the actual chromaticities (cf. figures 12.11 and 12.4). For each pigment, we start out knowing (*a*) its total percentage change in luminance under the illuminant change from D_{55} to D_{75} and (*b*) its distance (and direction) from pure white on the YB-axis—as illustrated in figure 12.12. (For viridian, *a* is 2.9%, *b* is –0.004; for cadmium red, *a* is –2.3%, *b* is 0.012.) We have estimated factor S′ (*percentage change in luminance attributable per unit of YB-ness*) as being –22% (on the basis of the values of *a* and *b* for our most extreme yellow and blue, cadmium yellow and cobalt blue). For each pigment, we multiply *b* by S′ to get the pigment's *percentage change in luminance attributed to YB-ness* (0.1% for viridian, –0.3% for cadmium red). We subtract that figure from *a*, which leaves the pigment's *residual percentage change in luminance*, i.e., its *percentage change in luminance attributed to RG-ness* (2.8% for viridian, –2.1% for cadmium red). To get projection distances for the graph we need a second factor S″ (*percentage change in luminance attributable per unit of RG-ness*), also taken here as –22%. Points that have "collapsed" onto the YB-axis (as shown in figure 12.12) are here projected toward (or away from) the protan confusion point by a distance given as *percentage change in luminance attributed to RG-ness* / S″. (For viridian, this distance is –0.13, i.e., 2.8% / –22%, in the units of CIE *x*, *y* space, for cadmium red it is 0.09, i.e., –2.1% / –22%.) In parentheses are given *percentage change in luminance attributed to YB-ness* and *percentage change attributed to RG-ness* for each pigment: they can be taken as a measure of coordinates for the pigment's YB-ness and RG-ness in the model.

ness attributable to its RG-ness. Thus, when the first factor has been applied (–22% × value on YB-axis), the three reds have residual increases in lightness of –2.1%, –1.5%, and –2.2%; the three greens have residual increases of +2.8%, +2.8%, and +1.7%. Such residual differences can be attributed to the redness vs. greenness of the samples, and they can be used as a guide for recovering this "missing" dimension.

We may then project our results into the "missing" dimension in the CIE x, y space. I take the standard protanope confusion point $x = 0.747$, $y = 0.253$ as the origin (cp. Wyszecki and Stiles 1982, 464; cf. section 12.3 above); points on the YB-axis are projected by different amounts according to the degree of residual increase in lightness attributable to the "missing" dimension; those with positive values are projected beyond the YB-axis (i.e., out into the upper left of our diagram); those with negative values (for residual lightness increase) are projected backwards (or negatively) from the YB-axis (i.e., into the lower right of the diagram). But how are we to calculate the amounts of the projection? One attractive idea is to suppose there is a constant factor that is *the increase in luminance* for any pigment *per unit of RG-ness*; given such a factor, we may calculate from the *residual increase in luminance* attributed to any pigment its *degree of RG-ness*. But what is a suitable factor?

One reasonable hypothesis is to suppose that the *luminance increase per unit in the other dimension (RG-ness)* (we may call this S″) is equal to the *luminance increase per unit of YB-ness* (which I have called S′, and have taken to be –22%). And this turns out to be a pretty good hypothesis. Given the residual luminance increase attributable to RG-ness, we can calculate a coordinate for something like RG-ness and use that as the distance by which a point on the YB-axis should be projected along a line toward the confusion point. Thus, for example, cadmium red has a residual luminance increase of –2.1%; this implies a value in the "missing" RG dimension of ~0.09 (which is –2.1% / –22%), and if we then project the point representing cadmium red on the YB-axis by a distance of 0.09 (in the units of CIE x, y space) in the direction of the protanope confusion point, then it moves from about (0.35, 0.35) to (0.44, 0.33). Conversely, viridian, with a residual luminance increase of 2.8%, is given a value of –0.13 in the "missing" RG-dimension (since –0.13 = 2.8% / –22%), and it is projected a slightly greater distance in the opposite direction. The results can be seen in figure 12.14.

This is really pretty good as recovery of color information: it is not far from being the converse of the original "collapse" of the chromaticity plane to a single line. And it shows that the protanope has the information necessary to recover pretty much the "missing" red-green dimension of color space, even if it is not accessible to him in the normal way. The mean discrepancy between the recovered coordinates and the original coordinates for the fifteen different pigments in D_{55} is about 0.026 (in the units of the CIE x, y space), which is less than four-fifths of the distance between the two closest pigments in our selection, cadmium red and perylene scarlet in D_{55}.

There could obviously be other hypotheses for the recovery of the "missing" dimension, and we could try small adjustments to the various constants employed in the model. It turns out that our value of –22% was not a bad one to choose for the *luminance increase per unit of YB-ness* and *per unit of RG-ness*. (Small changes in the values for S′ and S″ produce relatively small improvements in the recovery.[61]) There are of course more radical ways in which one might probe the proposal and make improvements. That 0.1 (in CIE *x*, *y* units) of displacement from the YB-axis along any protanope confusion line should count as representing the same degree of redness or greenness—regardless of which part of the *x*, *y* diagram one is in—is surely not correct. That we should be talking of red-greenness, rather than, say, red-turquoiseness, is debatable. (The complementary to the red of the protanope confusion point is not green but a blue-green of about 493–495 nm.) The most evident weakness of the algorithm is with the blues: cerulean blue and cobalt blue are 0.04 and 0.06 too far to the left—one might say that the projection had gone too far. (But ultramarine has crossed to the wrong side of the YB line, so there's another problem too.) Meanwhile, at the other end of the YB scale, cadmium yellow is 0.03 too far to the left (as similarly is cadmium red): one might say the projection had not gone far enough. One way to reduce these problems is to take a warmer color instead of *N* as our zero point (when setting up a scale along the YB line): if we use a white which (in D_{55}) projects onto $x = 0.350$, $y = 0.353$ (and take S′ and S″ as –23%), we find the mean discrepancy reduces to 0.0186. And there must be other well-motivated improvements one could make.

But such issues are, I suspect, of secondary importance. There are going to be limits in principle to the perfectibility of any recovery method like the one employed here. Information on the "missing" dimension of, for example, RG-ness is no doubt only imperfectly captured in the kinds of cue I am studying.[62] The claim is not that perfect recovery of the "missing" color dimension is available with the relatively slight information from a single illuminant change. It is that we can see that an approximate recovery is possible, and that it is not implausible that the dichromat visual system might in fact operate on some parallel or similar principles—if, that is, the more central parts of the dichromat system are in some way set up or primed for some kind of trichromatic experience. It might even be a reasonable speculation that if this happens at all, then it is thanks to the operation in the dichromat of visual information-processing systems that operate similarly in the *trichromat* too—the suggestion would be that there are many ways in which color constancy mechanisms operate in normal trichromats (some of which show up, I think, most remarkably in Land's red-and-white projections), and dichromat recovery of color might be a by-product of such systems. (For a survey of the achievements and shortcomings of our understanding of color constancy, see Maloney 1999.) Except that that no doubt puts the evolution of this the wrong way around: trichromacy in the retina emerged out of genetic variations in dichromatic systems (see, e.g., Regan et al. 2001); but, for receptor variations (in what had previously been a dichromatic system) to be of any value, there needed to be already some readiness in the remainder of the system to benefit from the new variety of inputs (rather than simply regarding it as "noise" or inaccuracy to have a deviant variety of a standard receptor type). So perhaps the dichromat, recovering 3-D color, corresponds to a rather interesting moment: that of the evolution of trichromacy where the remainder of the visual system has already in some way developed to the point of tracking, though imperfectly, more than simply two dimensions of color, though it lacks the third receptor type that would provide a huge improvement for this process.

The recovery in the present model is not perfect; and whether its weaknesses and strengths match the weaknesses and strengths of those protanopes who succeed best in recognizing the colors of things is a matter for further investigation. But we see already that cadmium red and viridian, which at a single glance might seem indistinguishable to the protanope, can be distinguished. The information is available in the protanope's perceptions to allow a pretty good recovery of color in something like an RG dimension as well as the YB dimension. Whether this method—seeing how things behave under variations in illuminant—is a method that is actually used by those protanopes who prove successful at identifying colors is a matter for further empirical work. Of course, we have also seen that there are some protanopes who seem to remain bad at red-green discrimination even when given large-field samples and good amounts of time: The individual differences will surely be worth investigating.

12.6.4 Other Kinds of Lighting Change

I shall not examine other kinds of lighting change in detail, but it may be worth mentioning just one change in the opposite direction: from direct sunlight (D_{55}) to a yellower (rather than bluer) illuminant, like CIE standard illuminant A, which represents tungsten light of ~2856 K—approximately, the light of a 100 W incandescent tungsten bulb. It turns out that the behavior of the colors under the change is similar to that which we observed with the change from D_{55} to D_{75}—only, as one would expect, in the opposite direction: as the light becomes yellower, the greens in general have negative increases in protanope luminance (–6.1%, –2.2%, +4.0%), while the reds have positive increases (8.5%, 9.7%, 17.9%). And a similar approximate recovery of the "missing" RG dimension could be achieved by similar methods to those we have already used.

12.6.5 Seeing with and without the Macula

The last case that we should look at may be the most important: the effect of the macula. The fovea, the ~2° central area in the retina with the greatest density of cones, is covered with the *macula lutea* (or "yellow spot"), a yellow filter that extends over a little more than ~4° (see, e.g., Le Grand 1957, 351–354; and figure 12.15). When we

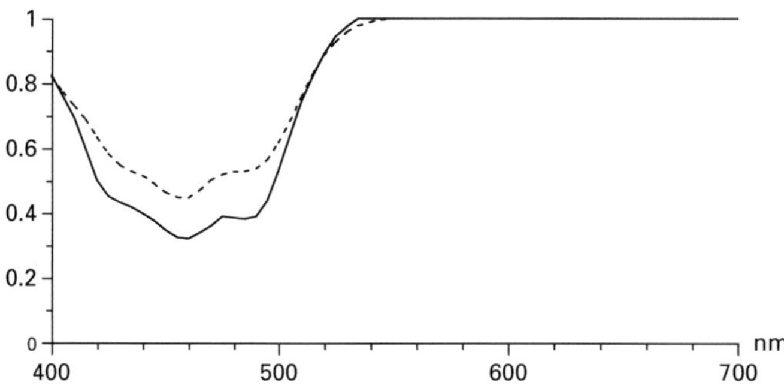

Figure 12.15
Transmittance of the macular pigment. Full line gives the transmittance function according to Wyszecki and Stiles (1982, 719–721). Dotted line gives the transmittance function according to Bone, Landrum, and Cains (1992; my calculation from the log density function available at, e.g., http://cvrl.ucl.ac.uk/database/data). For the modeling later in this section, I use the Wyszecki and Stiles function.

look at an object and then at something else a little to the side of it, we see the object first through the macular filter, and then with para- (or peri-)foveal vision unfiltered by the macula. We are usually unaware of the effects of the filtering;[63] but seeing both with and without this filter gives us a constant stream of additional information on the colors of things. We might almost say that, as their eyes rove over objects, normal trichromats are getting five- or six-dimensional color information all the time. (Filtered and unfiltered S cones have very different response patterns, with peak sensitivity wavelengths [λ_{max}] separated by about 10 nm; filtered and unfiltered M cones differ less, but still significantly; and the difference is very small in the case of L cones. See figure 12.16.) Of course, we must remember that there are no S cones at the very center of vision, but they become most common at about 1° of eccentricity (Stockman and Sharpe 1999, 63)—which is still well within the macular region. So deuteranopes (who have L and S cones) might be thought of as having at least one extra dimension of information, and protanopes (who have M and S cones), two extra.

The difference between seeing the same thing with and without the yellow macular pigmentation is remarkably similar to the difference between seeing it in yellowish and then bluish light. It is unsurprising, therefore, to find that the effects are similar to those we discovered with the change in illuminant from D_{55} to D_{75}. (Look ahead to figure 12.17 and compare figure 12.11.) One might even be struck by how similar is the direction of change, at an average angle of 234° in CIE *x*, *y* space (to compare with 224° for the change from D_{55} to D_{75}).

The effect of removing the macular filter (e.g., looking at an object with parafoveal vision) is substantial. But it is in some ways hard to model and evaluate: we are obviously in no position to remove and replace the macula at will; and there are large interpersonal variations in the degree of pigmentation of the macula. But we have functions at least for the general profile of filtration it provides, and we can model the effect of seeing something without the macula by treating the macula as if it were a filter in front of the eye, and calculating the effects of removing that filter. (The ordinary CIE 1931 2° color-matching functions can be taken to deal principally with foveal vision, which is mediated by the macula; to model the effect of a stimulus seen without that filter, mathematically, we treat the stimulus as if it were multiplied by the reciprocal of the transmittance of that filter.)

Obviously, with the removal of any filter, the total luminance of the light stimulus will increase, and (since the macula is a yellow filter) in our case, the luminance of the blues will increase more than that of the yellows, and the luminance of the greens more than of the reds. To clarify the differential variations among the different hues, I shall pretend that the total average luminance is the same with the macula and without. (If you like, we can read this as modeling the phenomenological fact that constancy mechanisms actually result in more or less equal apparent luminance across much of the visual field: there is no question of a peripheral view of an object being phenomenally brighter than a foveal view of it.)

Modeling in that way (figure 12.17) the effects of removing the macula, we see that for normal trichromats, our green pigments generally have a positive increase in luminance (+9.2%, +3.9%, –2.8%), while the reds have a negative increase (–5.1%, –4.4%, –6.2%). That is, as we saw before with the change from D_{55} to D_{75}, the greens gain in relative luminance (except for chromium oxide green, which has a large yellow component), while the reds lose. For protanopes the pattern of increase in luminance is similar: the greens increase (+8.5%, +1.9%, and –5.6%), and the reds decrease (–2.8%, –3.0%, –6.6%).

There may seem one puzzle or exception to the general rule I have proposed: while in general the greens increase in luminance in the absence of macular pigment, chromium oxide green actually *decreases* (the increase is –2.8% for normal trichromats, –5.6% for protanopes, and –3.3% for deuteranopes). But this is easily understood: chromium oxide green is a relatively yellowish green, and (unlike our other green pigments, but like, e.g., grass) it actually reflects a large proportion of red light; hence, as the macular filter is removed (and the light loses proportionately in its red and yellow components), it decreases in luminance. But still, if we consider reds and greens *within a single standard dichromat confusion class* (i.e., a red and a green of roughly equal YB-component), then it is clear that our greens always have a larger luminance increase than our reds under a blue-ening change in illuminant, or the kind of removal of macula that we have been considering. We may still say that a relatively large rise in luminance when an object is seen without macular filtering is a good sign of *greenness* in it.

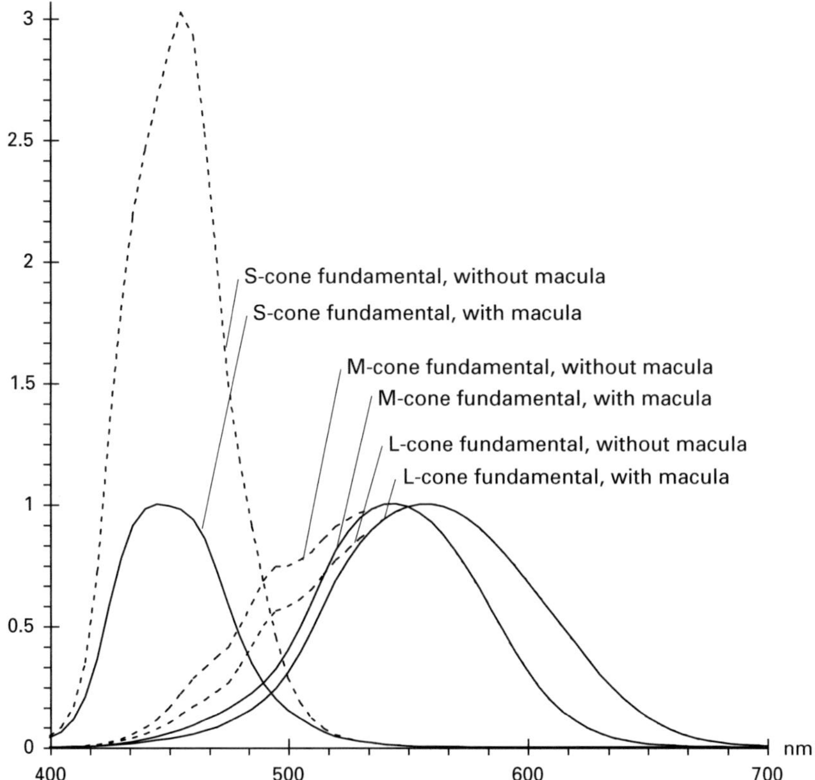

Figure 12.16

Hypothesized response functions for L, M, and S cones with and without the macula: König fundamentals derived from the 1931 color-matching functions (Wyszecki and Stiles 1982, 607); and calculated responses of such L, M, and S cones in the absence of macular pigment (broken lines: my calculation from macular transmittance function in Wyszecki and Stiles 1982, 721). The calculated responses of the S cones without the macula must of course be interpreted with discretion; but short-wave light is a much stronger stimulus for unfiltered S cones than for the central S cones behind the macula, and their maximum response is at ~455 nm rather than ~445 nm.

Our model for the presence and absence of macular filtering delivers a very similar pattern of luminance changes to what we found with the lighting change from D_{55} to D_{75}. The same methods should therefore work, whereby a dichromat might in large part recover 3-D color information from 2-D stimuli—first discounting the luminance increase that is due to its position on the YB-axis, and then treating the residual luminance increase as due to its position on a RG-axis. The task is slightly harder, however, and we cannot expect the recovery to be as accurate as with the previous case of D_{55} and D_{75}. The reason is that, though in gross terms a higher increase in luminance under the modeled change is a decent sign of relative greenness (or lack of redness), there are plenty of exceptions in detail. From figure 12.17 we see that raw sienna is less red (or more desaturated) than Mars red (while being more or less equal to it in yellow-blueness), but its increase in protan luminance under the envisaged change is not higher but lower (–7.0% for raw sienna, –6.6% for Mars red). Perylene scarlet, while less red than cadmium red, does not have a higher increase in protan luminance under the change, but a lower one (–3.0% rather than –2.8%). Recovery of the "missing" color-dimension from the kind of information about macular filtration that we have considered here is therefore almost bound to have inaccuracies of detail.

It is worth remembering, however, that we are not aiming to find cues or information sources that would make the color-blind *infallible* in "recovering" or identifying the colors of things: they obviously have no such ability. And it could yet be that the strengths and weaknesses of the kind of recovery system that I have been modeling might turn out to coincide very well with the strengths and weaknesses that dichromats in practice display. If so, it would be strong support for the hypothesis that change in luminance as the eye roves over objects is indeed an operative sign of red-greenness for the color-blind.

Mathematical modeling, of course, cannot decide whether the kinds of cues that I have modeled are the ones actually used by dichromats to recover the "missing" dimension where they do so: it can show that the relevant information cues are available, but not that they are in practice used. To decide on whether they are used, and in what ways, is a matter for further empirical research.

12.6.6 Deuteranopes

There are no new points of principle raised for the case of deuteranopes; the same pattern of change in response to the illuminant change emerges. For deuteranopes, as for protanopes, within any one confusion class, the greens have the highest percentage increase in luminance under the shift from D_{55} to D_{75}, while the reds have the lowest—and this relative luminance increase or lack of it can be used as a sign of the greenness or redness of things seen. (Increases in deuteranope luminance under the change from D_{55} to D_{75} are 3.7%, 3.1%, 0.7% for the greens; –6.2%, –4.8%, –4.9% for

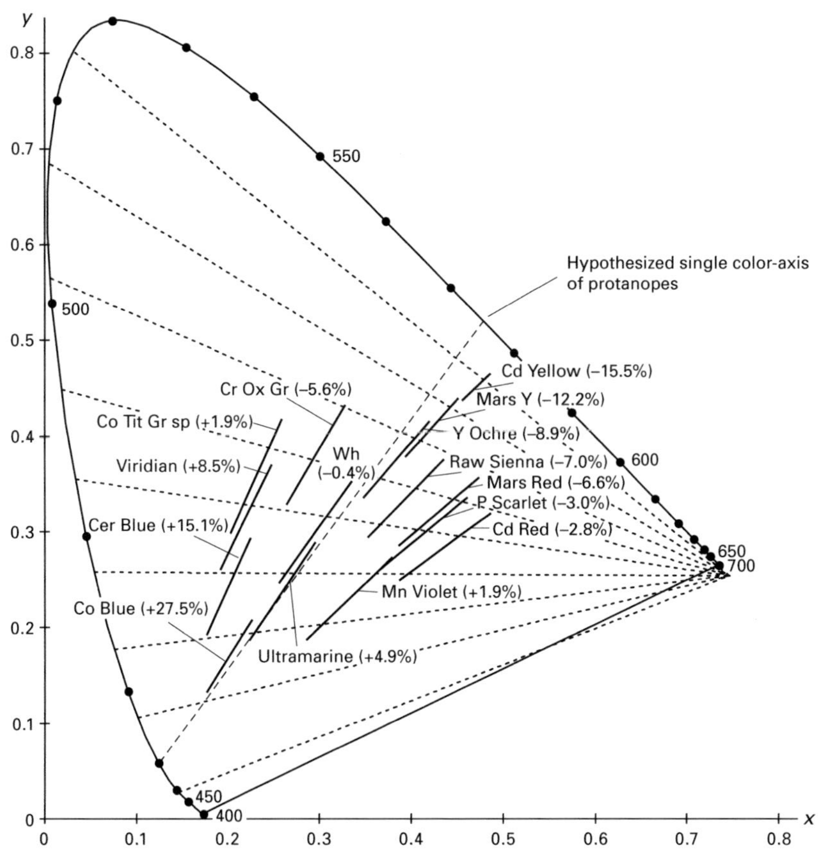

Figure 12.17

CIE 1931 chromaticity coordinates for fifteen pigments, with and without macular filter: protanopes. Figures in parentheses give percentage increase in protanope luminance. The effect of absence of the macula on protanope luminance is substantial: with removal of the macular filter, the greens increase in luminance (by 8.5%, 1.9%, and –5.6%), more than the corresponding reds (which increase by –2.8%, –3.0%, –6.6%).

the reds.) And something similar occurs again with the presence and absence of the macula (the greens increase by 9.0%, 3.6%, and –3.3%; the reds by –5.7%, –5.0%, and –6.8%).

12.7 Is This Just "Inference" to "Judgments," Not Real Perceptual Experience? Color and "Aspect Shift"

Some would say that the kinds of factors I have been studying might be in play, but they could only result in *judgments*, not in real *sensation* or *perception* of the trouble-some colors. To use the factors I've mentioned in identifying colors might seem no better than the old idea in Maxwell (and earlier, Seebeck 1837) of giving the color-blind a pair of glasses, with one green lens and one red lens, to enable them to distinguish the things they otherwise confuse. Maxwell claimed that the method yielded perfect discrimination but no *sensations* of red and green:

> By furnishing Mr X. with a red and a green glass, which he could distinguish only by their shape, I enabled him to *make judgments* in previously doubtful cases of a colour *with perfect certainty*. I have since had a pair of spectacles constructed with one eye-glass red and the other green. These Mr X. intends to use for a length of time, and he hopes to acquire the habit of discriminating red from green tints by their different effects on the two eyes. *Though he can never acquire our sensation of red*, he may then discern for himself what things are red, and the mental process may become so familiar to him as to act *unconsciously* like *a new sense*. (Maxwell 1855, 141, my emphasis)

The effects I have been investigating are very much the equivalent of seeing something with and without a filtered glass—only my filter is either permanently at work already inside the eye (in the case of the macula), or continually at work outside it (if changes in lighting can be thought of as changes in the filtration of our main illuminant, the sun). Whatever the consequences may be of merely putting on and taking off glasses, however, it seems to me that we should be slower to dismiss the idea that the factors earlier mentioned might ever have the effect of yielding real experience of red and green. They have of course been in action from birth, and they feed into a brain that is, we may suppose, normally predisposed toward three-dimensional color experience. But what we really need is some good evidence on the issue from subjects' behavior and reports.

How exactly could we get it? If (as I would agree) even perfect accuracy in the classification of the colors of things at sight is not always enough to show that a person has sensations of red and green, what additional indications might establish this? There are, I think, many kinds of evidence that can help (some of which I'll touch on in section 12.9), but I shall here consider one particular phenomenon that is, I suspect, extremely significant, though I have never seen it reported in the psychological literature.[64]

I hope I can be forgiven for describing my own case. Of course there are reasons sometimes to be skeptical of such reports; but there are reasons also for thinking that a reflective color-blind person may know things about his own case that a normal-sighted experimenter will not know unless he is told. In any case, there is a fine tradition represented in Dalton (1798), Nagel (1905, etc.), and Koffka (1909) writing on their own color-blindness, not to mention William Pole (1856, 1859), Friedrich Schumann (1904), and others.[65]

I confuse certain reds and greens and am protanomalous.[66] Unsurprisingly (being an anomalous trichromat), I have on occasion quite definite sensations of red (e.g., from fire engines) and of green (e.g., from grass). More surprisingly, however, there are things that look definitely red to me at one time and definitely green at another. And there is a particularly remarkable way in which this can occur. I may see an object and take it to be red (this has happened to me with the dark painted walls of a dining room, with a Lederjacke seen in the window of a shop in Salzburg, and with a multitude of things large and small). A moment later, I may realize I cannot be sure after all—with reds (or greens) like this, I know I sometimes make mistakes. I may then lift up the object and move it around, or (if I can't move it) move my head to see it from different angles in different lighting; I may take it over to the window (or look at another portion of that same wall, closer to the window); and then suddenly, perhaps, I realize that it is green. At that point, if not just before, the object comes to *look* green. Earlier, it had looked red; now it looks green; and these are two quite different ways a thing may look. And once I recognize that the object is, so to speak, ambiguous for me, I find (and this is the really remarkable phenomenon) that I have some limited ability to make the appearance *shift*—not instantly at will, but repeatedly, with some effort. Taking the object back to the place where it had first looked to me red, I still *see* it as green, just as I now take it to be. But then I may remind myself, "A moment ago, it looked red in just these circumstances; so surely you should be able to see it as red again!"; and in that case, perhaps after a moment, the thing may after all shift back to looking red. Conversely, if I tell myself once again that the object really is green, as I later discovered over by the window, then I can get it to shift back again to looking green. It is like a case of aspect shift, where by *wanting* to see the rabbit in the duck-rabbit drawing—or wanting to see the Necker cube one way rather than another—we find that the appearance changes (see figure 12.18). A difference, however, is that in the ordinary aspect-shift cases, at least one standard view is that the thinker has "the same sensations" whether she sees the duck or the rabbit, one cube layout or another; whereas, in the color case that I have just described, even the "sensation" changes.[67] Indeed the apparent *lightness* may also change: when I recognize a word on a page as printed in red, it comes to seem *darker* and more clearly delimited than when I see it as green, as well as coming forward from the plane of the paper. One might say that this was a case of an *ambiguous stimulus*, rather than an ambiguous appearance.

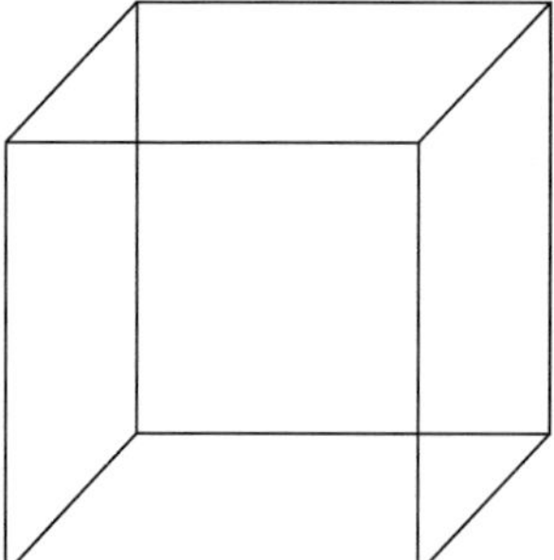

Figure 12.18
Necker cube. A two-dimensional stimulus that can be taken or "read" in two different ways as a representation of a three-dimensional scene: one might talk of the two-dimensional image being "projected" into three-dimensional space in two different ways. See Necker (1832) for the original observation of the "sudden and involuntary change in the apparent position of a crystal or solid represented in an engraved figure": Necker's example was actually of a rhomboid figure (336), which is in some ways more dramatic than the now standard example of a cube.

I mention this because, though I am an anomalous trichromat, when the kind of experience I am describing here occurs, I am in some respects in the same position as a *dichromat.* Whatever information may be coming from the retina is apparently compatible with the object's being red, and also with its being green: whatever 3-D color information may come from the eye at other times, at this moment the information available is, in color terms, only two dimensional, and it is compatible equally with two very different 3-D-color layouts in the world—with there being a red thing there, and with there being a green thing there.[68] Perhaps the simplest comparison is with the Necker cube: it is literally a case of a two-dimensional stimulus that can be treated or *taken* in more than one way three-dimensionally.

An ambiguity in color stimuli may seem a rare thing, but there are several comparable phenomena. The Gestalt psychologist Wilhelm Fuchs (1923, e.g., 278) describes cases where a circle that is actually gray can come to appear either yellow or blue, depending on whether it is grouped or taken with one group of circles, that are yellow, or another, that are blue.

And yet we should not think that these kinds of color shifts are found, so to speak, only pathologically—either in cases where color blindness reduces the number of dimensions of color information, or (as in Fuchs's case), when we are actually misperceiving, for example, a gray thing as blue or yellow. We might compare some other kinds of sudden change of appearance. A classic example (Gelb 1929) is the black spinning disc of 10 cm radius, illuminated with an arc lamp whose conical beam is adjusted to fit precisely the outline of the disc and no more. The result, to an unsuspecting person who enters the otherwise weakly lit room, is that the black disc looks white or pale gray. "*This impression of white is*—under the given conditions—*absolutely compelling*" (674). But the impression changes, or flips, as soon as someone puts even a small but genuinely white piece of paper into the path of the arc lamp and holds it a few centimeters in front of the disc: "At that moment we see the disk *black*, the little piece of paper *white*, and both indeed as *strongly illuminated*" (Gelb 1929, 674). We might design a similar case with color: a dark blue disc, set up to fit precisely the beam of light coming from a yellow spotlight, might initially look a pale gray to someone seeing it for the first time; but when a white card is introduced into the path of the spotlight, suddenly the disc should (one would expect) be recognized as dark blue, and the card as white, and both of them be seen as in a special yellow beam of light. This latest phenomenon is different from the cases usually described as aspect shift, in that the viewer may have relatively little control over how to see the scene: once the white card has been brought into the path of the light, it may be very hard to "take" or see the scene in the original way again. But with a special effort of will and "imagination," once the card has been withdrawn, one might be able to "think away" the existence of the beam of light. If one imagined oneself back into the state of mind one had been in on first entering the room (or perhaps imagined an authority saying, "It actually *is* a pale gray disc after all: while you looked away, we replaced the dark blue disc with a pale gray one and turned off the spotlight"), then perhaps one could regain the original perception of the card as pale gray. And again, the possibility of the *shift* comes from what we might call the fact that we are "taking" the world, and representing it, to have features in more dimensions than we are reliably getting information about: with just three dimensions of color input, the normal trichromat sees, in a sense, *six* dimensions of colors in the scene—three dimensions of object color and three dimensions of illuminant color (cp. Katz 1935, 190, quoted in note 55). And of course, that is bound to result in ambiguities and occasional shocks to our expectations.

When things look red or green to me in this alternating way, this is, I am sure, not just a matter of *judgment*, but of *appearance* or sensation. What makes me so sure? What can I say to make it plausible to others that I really am at different times *seeing* red and green—rather than merely "judging" that there are things of those colors in front of me? One revealing piece of evidence is the fact that in many cases (e.g., when dealing with colored words printed on a page), when I see something as red, it seems to *come forward*, and when I see it as green, it seems to *recede*: this is surely not just a matter of judgment.[69] (The word will also look *paler*, i.e., lighter and less saturated, when I see it as green—and that too seems a matter of appearance rather than judgment.) But the best proof may be in the experience of *contrast* effects. I may be looking at an autumn tree. At first it looks to me unremarkable—with some variation in color, but nothing particularly striking and no notable color contrasts. And then I come to wonder if the foliage might be an autumnal mix of red and green. Looking and trying to tell, suddenly I realize, the tips of the leaves are red, indeed a rich throbbing red; and the most central part of the leaves nearer the stalk is green, a good rich green; and as the appearances become definite, the two colors are suddenly in strong contrast with each other. One and the same scene, viewed from the same position, now looks utterly different. And this is clearly not just a matter of judgment: the red and the green are shouting at each other—there is a visual contrast that was not apparent before. Judgment may play a role—I may notice a red precisely because someone has told me that it is there—but the role then is in *stimulating* what ends up as being a clear difference in the visual *appearance*.

Such cases suggest, I think, the need to recognize the considerable *time* it may take for determinate impressions to form in color-blind people,[70] and—perhaps more radically—the need to recognize (as present in a period before the full impression of color has developed) *indeterminate* color experiences. "It's some kind of yellow, but not a pure yellow—I'm not sure if it's a reddish yellow or a greenish yellow. At this moment it doesn't look reddish yellow, nor does it look greenish yellow. I can't tell quite what it is; let me try to get a better view of it." *Looking reddish yellow* is different from *looking greenish yellow*, and both are also different from *looking a dark but desaturated unique yellow*—and the present indeterminate experience is identical with none of them.

12.8 Philosophical Conceptions of Color: Seeing the Same Things with Different Sensory Equipment

There are of course among philosophers and cognitive scientists many different conceptions of colors. Some talk as if *red* or *redness*, for example, might be taken as literally a certain sensation or kind of sensation; others—more promisingly—take it to be a disposition in objects to produced a certain sensation or kind of sensation. One might initially think that whenever the illumination on an object changes, a perceiver's sensations will change, in parallel with the changes in the stimulus at the eye. It is a definite advance to pay some attention also to the phenomena of color constancy, that is, to the fact that there is some sense in which, even when there are large changes of illumination, though of course only within limits, a red book will continue to look red. But we must be careful about the characterization of the phenomenon: it is not that the visual system wholly and literally "discounts the illuminant" and yields an awareness of, merely, the constant color. The wall in front of me looks in a sense a single shade of pale yellow or magnolia, but almost every square centimeter in another sense presents a different appearance, as the light from the several windows in the room falls differently upon it, modified also by shadows from the curtain rails, window frames, and other objects, and by reflections from other things around. It would be a mistake to say that the visual system in such circumstances yields awareness merely of the constant colors of things; rather, it yields complex experiences in which we are aware, or able to become aware, *both* of the constant colors *and* of the varying illumination.[71] The content of the perception might be said to be of a wall of a more or less *uniform* shade of magnolia, with the *light falling differently* upon it in different places.

It seems to me a further step forward to see that that particular magnolia surface has not merely a disposition to produce a particular sensation S_1 under some particular lighting condition C_1, but also a disposition to produce a slightly different sensation S_2 under slightly different lighting condition C_2, and so on. And then one might say that the characteristic feature of that kind of magnolia color was not straightforwardly *the disposition to produce sensation* S_1 (or even S_1 *under condition* C_1), but rather *the disposition to produce a range of sensations* S_1–S_n, *under a range of circumstances* C_1–C_n.

There are arguments for going one stage further again, however.[72] We may characterize the relevant character of the surface in terms of its disposition, not to produce a range of *kinds of sensation* under a range of kinds of illuminations, but to reflect a range of *kinds of light* under a range of kinds of illumination. That is, the color of the surface would be characterized by the way it *changed* the light falling on it in the process of reflecting it—for example (to take pure cadmium red pigment: see figure 12.9), reflecting only about 6% of incident light in the blue-green-yellow-orange parts of the spectrum (from ~400 to ~600 nm), but sharply increasing proportions through the orange and red (from ~20% at 620 nm to 80% at 700 nm),. The relevant character of the surface would be given by its spectral reflectance function, and the color of an object would be a little like the elasticity of a spring: when a weight of 2 is put on a spring with elasticity *e*, it stretches to 2*e* (assuming some suitable units), when a weight of 5 is put on it, it stretches to 5*e*, and when a weight of 0.3 is put on, it stretches to 0.3*e*. (Similarly, the character of the light reaching the eye from a cadmium red pigment will continually change as the incident light upon it changes; but there will be a constant function taking us from the character of the incident light to the character of the reflected light—and the types of constant function will be characteristic of the various types of color.) The elasticity is a single feature of the thing, manifested in a pattern of activity fully exhibited only under a range of different conditions. So also, I believe, the color of a surface is a single complex feature of it that is fully manifested only as the object is seen in a variety of lighting conditions: to get a full appreciation of the color of a surface, one has, so to speak, to put it through its paces and see what it can do to a variety of kinds of light. Not that one cannot, of course, in some sense get a pretty good impression of the color of something just at a glance—just as someone with some experience of springs may form a good impression of the elasticity of a spring seeing it just once, for example, with a standard 1 kilo weight on it. But what he is seeing at that time, if he sees the elasticity, is a power that will only fully manifest itself through a range of different actions (e.g., as he pulls the weight down a bit and sees how the spring bounces, or as he replaces the first weight with another and sees the results). And so also with color: I may take it at a glance that what I have in front of me is a piece of white reflective card, but I will be able to confirm that attribution to the thing only if it goes on to react suitably, as I pass my hand over it and see the pattern of shadows cast, or as I bring it close to the relatively orange light of a table lamp. If, on the other hand, it actually doesn't change its appearance, but goes on obstinately looking unvaryingly as it did at first, then I know that I have a strange glowing object before me—not a white reflective surface at all. And actually, even if it is glowing, it should still reflect some light from things around and show some, even if relatively slight, variation in appearance from the different kinds of incident light—so if it doesn't do so, it is, even more weirdly, a black body, emitting light (at the high temperature associated with white) and absorbing any light that falls upon it.

Where does this leave our present debates about color blindness? Let me start with a first approximation. Suppose, as in the standard view, whether in Helmholtz's version or Hering's or some other, that dichromats lack a certain kind of color sensation because they lack a certain kind of color receptor. In that case, the protanope—lacking the red receptor, or perhaps even the "red-green" color system—will simply be unable to see red, if red is either a (certain kind of) sensation or a disposition to produce such a sensation. But if red is, for example, a *disposition to produce a range of sensations*, or if the redness of surfaces is—as on the suggestion of the last paragraph—a *disposition to change the incident light* in certain ways, then one might actually be able to allow the protanope in some sense still to see red. Even if we suppose (as we are currently doing) the absence of "the sensation of red," it might be that the protanope could still recognize or represent red. Suppose one of the "signatures" of a red thing is to darken as the light changes from (yellowish) sunlight to (bluish) skylight: that signature could be picked up on by the protanope. Lacking the "sensation" of red, he might nonetheless be able to recognize red things. And the lack of "the sensation" might in some sense not matter much. People wearing sunglasses can *see* the whiteness of a sheet of paper without having "a sensation of white" from it: the pattern of appearance, even if in some sense entirely brownish, is still the pattern of appearance of a white thing—and it is white that the thing looks to be. If we view a surface color as a power to produce a range of appearances under a range of conditions, then, one might say, the person with sunglasses may see only *one subset* of those appearances, and still be said to see white. (But one would have to admit that he doesn't see the same manifestations of the whiteness as a person looking at the object without sunglasses.) If on the other hand we view a surface color as a power to change the light in various ways, then we might even say that the person with sunglasses saw *the very same* manifestations of whiteness as the person without. The two of them might even be watching the same pattern of changes in the world, whether mediated or not through the sunglasses, as one of them, for example, moves her hands over the objects at a distance of a couple of centimeters, watching the shadows cast, and moves the objects around to catch the light from different sources in different ways. It could be no less true that they saw the same colors and same patterns of variations in light reflectance as that they saw the same shapes and sizes of the objects, even though one viewer was obviously closer, and the other a little farther away. I have often worked with this kind of conception in this discussion: allowing that the dichromat may lack a certain kind of color sensation, I have been interested in how he might nonetheless *see* colors—making use of color information available through variation in illumination over time and space, or with and without the macula.

But a further step can be taken. One might conjecture that, if the information for color identification is available, perhaps in some sense the "sensation" of red itself might also be available. Must we assume that the processes in the brain that underlie the sensation of red only ever occur in response to the standard textbook pattern of activity involving both L and M cones (e.g., positive firing in some L-M opponent-process system)? Might the "sensation" of red not occur, not merely in response to some such particular momentary stimulus, but also in response to a suitable *pattern* of stimuli—for example, the pattern of *darkening-under-a-change-from-sunlight-to-sky-light*? And might not the dichromat, in that case, be lucky enough to be able to enjoy some sensation of red despite the absence of what in normal trichromats is its most obvious standard cause?

A third option—besides saying that the dichromat might *see* red and green while having rather different "sensations," or alternatively saying that he might after all be able to have the normal "sensations"—might be to say that the notion of a sensation, or *quale,* of red might need to be rethought altogether. Not that there is no such thing as experience of white, or of red; but the idea that there is an identifiable quale of, say, white, or of red_{27}, may not make much sense, any more than the notion of a quale of, say, shininess. Of course, one can experience the shininess of something, but that experience comes from having a range or *pattern* of experiences, not from having just one. And maybe the idea of a quale of white or even red_{27} makes no more sense than that of a quale of shininess. But then, what of the experiences that make up this range or pattern? Would they not be still describable as qualia? Maybe. But there might still be questions about the coherence of identity conditions for such qualia in ways that do not arise with other things or other kinds of experience. This is not the occasion for debate on that; but at the very least, the investigation of philosophical conceptions of color has already given us two promising lines of thought that seem to leave room for the idea of dichromats having genuine sensory experience of colors, even if they do not see them in quite the ordinary way that the normal sighted do.

12.9 Some Experiments for the Future

I would like to propose a principle of some importance, though it may sound naïve at first:

We need to spend more time investigating what the color-blind *can* do, rather than what they can't.

From the late nineteenth century, color-vision testing was mostly driven by the need for tests to prevent people from having jobs on the railways and at sea, where they might be a danger to others, or in the textile or paint industries, where they might be slow or unreliable; while on the other hand, there were young men hoping that a bit of coaching (or vitamin A) might save them from rejection. While much effort was put into contriving conditions in which failures of the color-blind become clear (with small and desaturated samples, and lights like the Macbeth lamp), the relative success of those people under other conditions was often treated as little more than an annoyance—either insignificant, random, or fraudulent. We know, however, that the successes are not random and they certainly need not be fraudulent—they are persistent and repeatable enough, even under controlled conditions, to be a headache to people designing diagnostic tests. And we owe ourselves a better understanding both of what they amount to and of how they occur. I would like, therefore, to propose we make investigations of the color-blind particularly under those conditions where they tend to do better rather than worse. We should:

1. Have more experiments using *surfaces*, rather than *lights*.[73]
2. Investigate with *large* samples, not just *small*.[74]
3. Use *saturated* as well as *desaturated* samples.[75]
4. Test how subjects improve (or deteriorate) when given *more time* to look at samples.[76]
5. Investigate performance under *varying illumination*: not just brighter and darker,[77] but also deliberately of different spectral composition (e.g., yellowish and bluish), especially where one object can be seen under a *succession* of different kinds of light, and the "dynamics" of a surface color can be exhibited and tested. We should examine cases where objects are seen *simultaneously* under a *variety* of kinds of illumination coming from different directions.

Two further proposals relate to issues that came up when considering contrast effects and indeterminate color impressions (in section 12.7). We should, with the color-blind:

6. Investigate perceived *structure* and color *relations* (e.g., being *darker than, redder than, whiter than*, and *more saturated than*), rather than just *color-matching* and *color-recognition*.[78] We should investigate, in various color arrangements and juxtapositions,

what things are seen as *contrasting with* other things and (though this is a further issue) as *coming forward from* or *receding from* them.
7. When subjects are being asked to describe the colors of things, check not only the *degree of confidence* in the judgments they make, but also the degree of *indeterminacy* in their color impressions. If subjects are forced (for example) to use a basic color term in response to a presentation, let them be allowed not only to say if they feel unconfident in their judgment, but also to report if a change in circumstances (e.g., more time, or looking from closer up) results in a more determinate color impression.

One last principle is, I suspect, extremely important, though I cannot do more than mention it here. We should:

8. Pay attention to the *individual differences* within the present classificatory groups, which may tell us about (a) subgroups with different genetic endowments and different receptor systems in the eye, and (b) the different extents to which people can learn to compensate for deficiencies in their sensory systems.[79]

Using these principles and others, there are many questions that we should be able to begin to give a good answer to. Some of them arise particularly out of the issues of sections 12.6–12.7: what is the *structure* of the color space of dichromats—and is it two-dimensional or three-dimensional? (Of course, the answer may not be the same for all dichromats, in view of proposal 8 above.) One helpful way to investigate that, I think, would be to investigate a subsidiary issue: how widely is what I called "aspect shift" in color found among dichromats? And where it does occur (for example, with alternations between what are reported to be appearances of red and green), is the idea of *projection* into a third color dimension (perhaps as in section 12.6) helpful in modeling how it occurs?

There are many experiments that I have begun to sketch that should contribute to just such questions, but I shall mention just three here. The first is best illustrated with a diagram (see figure 12.19).

Proposed Experiment Series (12.1) Let us take a sequence of colors at fairly high saturation and medium and equal lightness (for the relevant type of dichromat being tested), ranging from yellow to red (sequence A in figure 12.19). According to the standard model, dichromats should see this sequence of increasingly red samples as actually a sequence of increasingly desaturated forms of yellow tending toward gray (as in sequence C).

1. So let us ask, when participants are presented with sequence A, "Do you see this sequence of samples as *getting redder*?" (Alternatively, or in addition, do the samples look to be *getting less saturated*, that is, *losing intensity of color, coming to look more like a neutral gray*?) Try again with a sequence of samples going from *yellow* to *green*

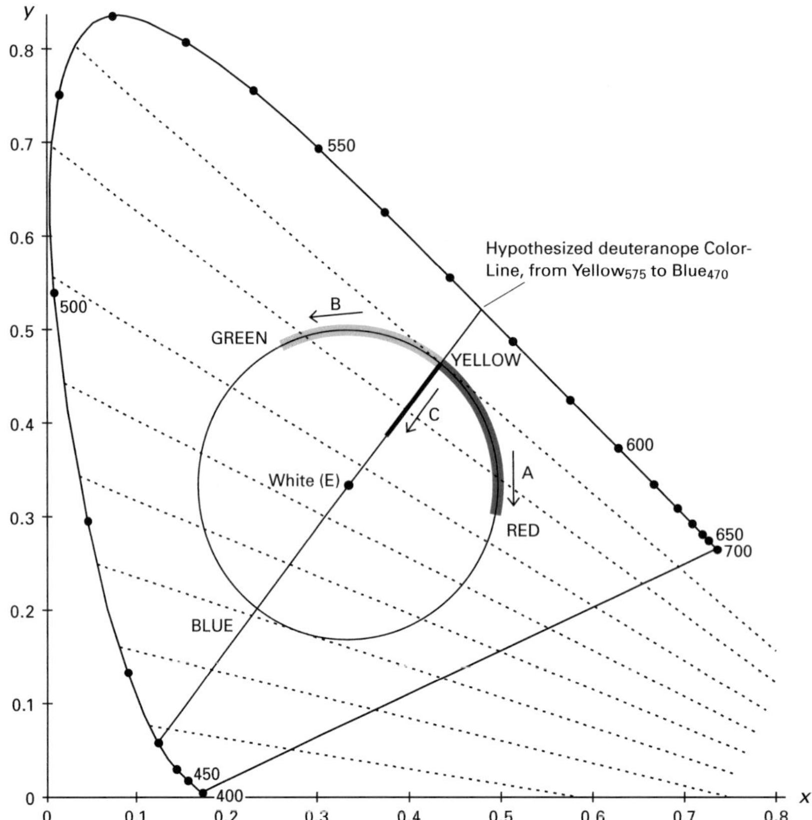

Figure 12.19

Appearance of colors from a color atlas to a deuteranope, according to the standard model: some experimental tests. The circle is a rough illustration of the chromaticities of a range of pigment samples in a color atlas at fairly high saturation. (Lines of equal saturation are, strictly, not circles in CIE *x*, *y* space: cf., e.g., Wyszecki and Stiles 1982, 857. But a circle should be good enough for illustrative purposes.) The standard confusion lines (dotted) indicate how these samples would (for a theoretical deuteranope) project onto the standard yellow-blue hue line. I have marked three sequences of color samples: sequence A runs on the periphery of the circle, approximately from unique yellow to unique red; B runs, again on the periphery, approximately from unique yellow to unique green; sequence C is a series of increasingly desaturated yellows, running along the diameter of the circle, in the direction of a neutral gray. In the standard theory, sequences A and B should look more or less equivalent to C, approaching a neutral gray. We should test whether certified dichromats do in fact find these sequences equivalent. If not, then how do they describe the differences between then? If some of the colors in sequence B are described as looking increasingly green, then we should see if there is any tendency to treat that color green as a "unique" color, that is, as a color of which some instances can be phenomenally pure or unmixed. (And if so, we should ask the subject to point out an instance of this pure green.) With tests like these, we can learn much about the structure of the experienced color space of the dichromat.

(sequence B). Try again with a sequence from *yellow* to a desaturated *yellow-gray* (sequence C). The three sequences should, in the standard model for deuteranopes, all yield the same answers. Much of the evidence (reviewed in section 12.5) already suggests we should expect a rather different outcome—that, as long as the fields are reasonably large, these various sequences will not in fact be seen the same way. But the details of an empirical investigation would I think be of interest.

2. If the traditional view is correct, then a strong red should look a desaturated yellow or (for protanopes, even a gray). So let us ask (showing a deep red and a supposedly equivalent desaturated yellow), "Which of these samples is most highly saturated [or 'has a deeper color']?" And (showing a series of reds), "Which of these is the deepest or purest red (i.e., the least yellowish)? Is it [namely, the one chosen as most pure red] a *strong* (or *saturated*) color, or a relatively weak (or desaturated) color?" (We surely should be able to rule out that something might at one and the same time look a saturated red and a desaturated yellow!) Then (presenting the participants with what they have chosen as the purest or deepest of reds), "Is there any sense in which this sample looks yellow? Is it in fact just a yellowish red?" We could present participants with an extension of sequence A beyond unique red into the purple region: "You've described the earlier sequence as a series of samples of increasing redness [supposing this is true]. Do you find the redness continues [in the further region A′, the extension of the sequence] to increase? Or is there a pure 'unique' red among all these samples? And if so, please indicate which it is." Depending on the outcome, we might have very good confirmation that some dichromats are quite capable of distinguishing hue variation from mere saturation variation—that is, that they have (along with brightness) three dimensions of color vision after all. It is worth noting that, in the phenomenal color circle, unique red *stands out* or is, for normal perceivers, salient; but there is nothing particularly special, in the sequence of progressively more desaturated yellows, about any (supposedly dichromat-equivalent) desaturated dark yellow or gray—so it really is particularly implausible to think that correct labeling (if it occurs in our dichromats) of samples as "unique red" can be dismissed as just a matter of "judgment" or "inference" (labeling a particular shade of grayish-yellow "red').

3. Show sequence A (which we might loosely call "the reds") and sequence C ("the yellows"), which (according to the standard theory) should be indistinguishable to the relevant group. Ask subjects, "Do these two sequences of colors look the same?" Have one additional sample, which actually belongs in the red sequence between two other samples. Ask the subject to place the additional red sample next to the samples that most nearly resemble it (either among the reds or among the yellows). Of course we really know which outcome to expect if the reports surveyed in section 12.5 are correct: if the fields are reasonably large, the subject should place it among the reds; but again, the empirical details as they emerge could be very interesting.

All of this could be repeated with larger and smaller samples, and (in slightly different form) with suitable randomizing of the color presentations—trying to find out the relevant dimensions of color space for the relevant groups of dichromat. It might well be that dichromats do better seeing certain colors correctly when they appear in relevant sequences rather than when out of context.

Proposed Experiment Series (12.2) Jameson and Hurvich say that, when the subjects in their 1978 experiment (see section 12.1.1 above) were asked to give the color names of the Farnsworth caps and did so so remarkably correctly, there was *no change in their experience* as a result of the questioning procedure. The authors give no particular basis for their claim, and it would be worth checking, particularly in view of the fact—unmentioned by Jameson and Hurvich—that the color-blind often report that it takes time for determinate color experiences to "develop." My own experience while doing the Farnsworth D-15 test under a Macbeth lamp[80] was that the caps were extremely hard to identify or put in any color order, and I mixed up reds and greens in a typically protan way. At slightly higher levels of brightness (with "daylight" fluorescent lighting), the colors continued at first to be hard to identify; but when the conductor of the test asked, "what color is that?" I suddenly found (quite to my surprise) that I could answer almost immediately and (as the experimenter later confirmed) almost entirely correctly. I then went on to put the caps in virtually the correct order; and I found (as I would put it) that the caps were coming to *look* the colors I was describing them as being, which they had not done before. The greens looked green, the reds and pinks looked red and pink, whereas earlier they (and indeed the blues too) had looked an indeterminate medium-dark color—I'd had the feeling only of seeing something that I could not "place" or identify in color. (And, note, the change did not occur simply with the brighter illumination; the really decisive change occurred only when the experimenter began to ask the questions.) And whereas, at the earlier time, I had placed cap 2 between caps 14 and 13, I later came to place it between caps 1 and 3: the *patterns of apparent similarity* among the caps had changed.[81]

It may well be that hearing and trying to answer the question, "What color is it?" itself provokes either a different way of looking or a different mode of brain processing, and that this occurs not just in me but in dichromats. Dichromats often report that it takes time for some of their more determinate color experiences to develop. We should see whether we can replicate this stage of the Jameson and Hurvich experiment, to see whether other subjects report a richer color experience developing as a result of the extra time and perhaps better illumination and the request to give the color names of the caps. By examining patterns of similarity that dichromats find among the caps, we might perhaps find that, even if when first presented with the test caps under the Macbeth lamp, a subject had what could indeed be described as a two-dimensional color space; yet perhaps under better conditions (when asked the color names, with better lighting, larger samples, etc.), he might prove later to have a three-dimensional color space.

Proposed Experiment Series (12.3) On the subject of "aspect shift": Our first task would be to find out how widely the phenomenon can be found among dichromats as well as anomalous trichromats—how much there are "ambiguous" stimuli that are reported to "shift" in color appearance for them. A second task would be to try to identify exactly which are the reds and greens particularly liable to be "ambiguous" (perhaps for each relevant subgroup), and what kinds of shift in appearance they undergo. I shall not pause here over those tasks; but if we had some success with them, there is one more unconventional line of inquiry we might explore. Suppose we have identified a particular dark green that is (for the relevant group) liable to be seen *either* as a dark green *or* as red. Let us give a dichromat a ball *A* of the relevant dark green, in lighting condition *C*. Suppose it initially appears red to him; he then moves it around, takes it to the window, comes after all to see it as green; he brings it back to the earlier condition *C* and still sees it as green. We might take it that *tracking* the object is important to its having a relatively settled appearance: the appearance in *C* is influenced by the history of the object's having being seen in other circumstances, which have (by cues perhaps like those I modeled in section 12.6) revealed its true color and given it a relatively settled appearance. But then, with the subject looking at ball *A* still in circumstances *C*, suppose we produce another (in fact practically indistinguishable) ball *B* of the same dark green. Now, does it ever happen that ball *A* is seen as green by the subject, while ball *B* is seen as red? Or are they always seen as exactly the same color?

Suppose they are sometimes seen differently. What if we now take the two balls from the subject? (Presumably ball *A* continues to look green, and *B* to look red.) We put them behind our back, shuffle them, and then present them to the subject once again. Suppose he does not know which of the two is *A* and which is *B*; does he then see them both as red, both as green, or (perplexedly and perplexingly) see one as red and the other as green? (After all, he's currently under the impression that that's what his environment really contains: one red ball and one green.) And in that case, if an aspect shift can be brought to occur, does it ever happen that the subject shifts from seeing one as red and the other as green to seeing the other as red and the one as green?!

The options can seem almost too absurd to contemplate. But the question is only a little more far fetched than some of the cases in Fuchs and Wertheimer, and we really know far too little about aspect shift to know what the answer will be.

12.10 Conclusion: The Metatheory of the Debate

At every stage of the evolution of this subject, at least since Dalton first gave it a really thorough description, there has been evidence of great authority—whether it was ultimately to be accepted or not—against the yellow-and-blue theory. While John Herschel (in 1827 and 1833) was arguing that Dalton saw only yellow and blue, Dalton's own 1798 report had already given a strong indication that he sometimes saw green things *as red* and red things *as green* (where *looking red* and *looking green* certainly did not seem the same as each other, nor the same as some variety of *looking yellow* or *gray*), and Dalton himself thought that he often saw the colors correctly in candlelight. As Maxwell was coming to publish his first, brilliant, paper on the subject (1855), he also came across George Wilson's articles in the *Edinburgh Monthly Journal of Medical Science*, based on Wilson's studies of some seven hundred men in the Edinburgh garrison and twenty or so people who had written in response to a request for personal reports of color blindness that he had made in the pages of the *Athenaeum*. And Wilson believed his evidence pointed to a conclusion quite opposite to the two hue theory that Maxwell supported.[82] While Helmholtz advanced a version of the two hue view, the editors of the third edition of his book had different views on the matter: von Kries expressed reservations,[83] and Nagel, as we have seen, produced dramatically opposing evidence—both from his own case and from many dozens of other dichromats. One might have thought that, from the early twentieth century, the view that dichromats see more than just two hues would be at least a respectable minority view, to be mentioned if also rebutted by those advancing the more standard view; but in the pages of Pitt, Wright, and Judd, there is scarcely a mention of it. Pitt (1935, 34) mentions Mary Collins, who in turn quotes from R. A. Houston's (1922) paper; but Collins and Houston are, I am afraid, not the equal of Nagel or even George Wilson,[84] and in any case, Pitt gives them no serious discussion. Judd (1948) talks of Hayes (1911) and Nagel (1905), but as if they actually *supported* his view—which is the very opposite of the truth.[85] Curiously, even though Boynton was a collaborator on many important studies of residual red-green discrimination in dichromats, still, in his own 1979 book, when he feels called upon to deliver a general statement on the experience of dichromats, he reaches again for the standard yellow-and-blue theory.[86] Jameson and Hurvich (1978) have evidence of red-green discrimination in their dichromats, but still state—offering no particular evidence, as if it were simply *established fact*—that their dichromat subjects can have perceptual experiences only along the yellow-blue axis and must therefore be using a rule like "if it's dark then call it red"—a rule that they themselves admit could not account for the behavior of their subjects.[87]

It is almost as though there were a kind of blindness in the supporters of the standard view even to the *possibility* of an alternative. My suspicion is that these authors know that opposing evidence exists; but they do not take it seriously—for example reporting it honestly and giving reasons to reject it—because, in their view, it has literally no chance of being true. To suppose that, with just two dimensions of input, dichromats might have three dimensions of color experience would, for these authors, be to believe in a kind of *magic*.[88] In a way, this is the view even of many of the dissenters of recent years like Nagy, Montag, and Boynton: when they find residual red-green discrimination in dichromats, they also almost invariably go on to speculate on the *extra receptor type* that, they suppose, must be at work—whether rods or residual cones of the "missing" or "forbidden" (the terminology is remarkable) type. My own view is that it is not particularly plausible that either rods or a few remaining cones of the supposedly "missing" type are responsible for the red-green discrimination that has been found. Further hypotheses have been suggested—like variations in pigment density (whereby one pigment could produce different sensitivity profiles in variants of a single cone type)—and these hypotheses are both interesting in themselves and promising as possible factors in the phenomena we are studying. But a controlling assumption in almost all of the work has been that for a three-dimensional space of color, we have either to find or to postulate three kinds of receptor—that two-receptor types can provide at most a two-dimensional space of color experience. A central project of this discussion has been to question that assumption. Merely questioning it is of little value, however, without some conception of *how* it could actually be false, a model of how three dimensions can come from two. This is what I have attempted in section 12.6, suggesting that what matters is not so much which receptors we have as what we do with them—an idea I have attempted to link with a more dynamic conception both of color vision and of colors themselves (section 12.8). Suggestions of actual underlying mechanisms will have to come later.

So, what do the color-blind see? I do not know, and I suspect it will take some time for us even to follow through the kind of experimental tests I sketched in section 12.9 that I think would prepare us for an answer. But to hazard a guess, on the basis of the evidence we already have: quite possibly pretty much what normal perceivers see. Only not, of course, always at the same time or under the same circumstances as normal trichromats—and particularly not with small samples and in poor lighting. And probably not with anything like such fine discrimination as normal perceivers. But as for the broad structures of the color space of the normal trichromat—hues (forming a circle), saturation, and brightness—I would be very surprised if the majority of dichromats did not have all of that.

But—and this is another theme of the whole discussion—I have to say that I do not think any of us has much knowledge at the moment on the matter. There is virtual certainty that there is red-green discrimination among many dichromats—though it is time that this were regarded not as a peripheral detail but a significant truth (and it is only one of several reasons for having reservations about confusion diagrams like Pitt's; see section 12.3). But whether many or most of the discriminating dichromats also have *sensations* of red and green may be genuinely uncertain. The opponents (like Judd) mostly oppose the claims for extremely bad reasons, as we have seen (section 12.4). But at least some of the supporters could surely be wrong too: Pole (1859) declared that he had been quite convinced for thirty years that he could see red and green but later decided he had been deluding himself. I do not think Pole's reasons for saying this are good ones; but he might conceivably have been right, and there are likely to be important individual differences. The balance of probability seems to me mainly with the reports of Nagel and Dalton indeed, but we certainly need more empirical work to find out how widely their condition is shared—whether many dichromats have red and green experience or only a few, and if so, how exactly they manage it.

People sometimes talk as if color vision were something on which the main work was done in the mid-nineteenth century and redone, with greater accuracy, at the start of the twentieth, and psychologists know just about all they need to know—with the exception of the genetics and neurophysiology that will show how the processes long known about are realized and come into existence. Meanwhile the psychophysics of color can be thought of as working with early twentieth-century equipment, while the fruits of newer technology, whether in computing or genetics or functional MRI, have transformed experimentation to a much greater extent in other areas of psychology and cognitive science. These contrasts seem to me rather overdrawn. But in fact my suspicion is that large issues remain quite unexplained—the Land phenomena, the character of dichromat discrimination and experience, the modeling of the experience of anomalous trichromats, much of color constancy in normal trichromats—and to work out explanations of those phenomena, we are going to have to rethink much of the work that has been taken for granted for so long. I am no fan of the idea that scientific revolutions overturn what came before—on the contrary, I would say that they often preserve the core of what had earlier seemed established, but transpose it into a new key, or put it in a new frame. What Newton, holding a particle theory of light, had said about "Fits of easy Reflexion and easy Transmission" (1704/1730, 348) was rephrased a century later in terms of wave activity and has been rephrased once again by people doing quantum electrodynamics; but much of Newton's quantitative and explanatory work can be kept, though rephrased in new terms. I have a suspicion that we will not have an explanation of the phenomena in Land's red-and-white presentations and of what (if Nagel is right) seems to be 3-D color experience on the part of dichromats, until we put at the core of our models of vision the dynamic and the varying *series* of experiences, rather than the static pairs of stimuli that were the main object of so much experimentation with spectral apparatus—and that are actually the main experimental basis we have for our standard conceptions of color blindness. If we have success in that direction—some of which, I suspect, will rather significantly depend on adventurous use of the technologies of computing, genetics, and MRI—then I rather expect that everything will look different, and yet the old things in Helmholtz and Maxwell will also in a sense look much the same, rather as the theory of complex numbers left the theory of real numbers looking both the same and different. An emphasis on the dynamic, on contexts and change, is, in a sense, a commonplace of the theory of vision in general, in spatial vision and object recognition, for example. It has not been a commonplace in the central theory of color vision—though it has had its place in what seemed subfields of constancy and contrast phenomena—because color has so often seemed the *simple* out of which the other complexes are formed. But maybe that too is an appearance that could change.

Acknowledgments

I dedicate this chapter to Larry Hardin, with thanks and admiration for both his personal guidance and his professional leadership in this field. He has set new standards for philosophers working on color: the lucidity, rigor, and excitement of his work are an example. For comments and discussion, I am grateful to Ted Adelson, Patrick Cavanagh, Mike D'Zmura, Rainer Hertel, Ken Knoblaugh, Donald MacLeod, John Mollon, Ken Nakayama, Kevin O'Regan (particularly for first getting me thinking about the macula many years ago), Steven Pinker, Michael Tarr, Bill Warren, and Billy Wooten; also to audiences at Brown, UI Chicago, MIT, and UC San Diego, and in Oxford, Cambridge, Paris, and Bielefeld. I am particularly grateful to have had the invitation to reply to Dorothea Jameson on this subject at a conference at the University of Pennsylvania in 1994, where I first began to develop these ideas. Philosophers who have helped with comments and discussion include Sean Aas, Alex Byrne, Jonathan Cohen, David Hilbert, Christopher Hill, Ben Jarvis, Nenad Miščević, John Morrison, Joshua Schechter, and Jonathan Westphal. I owe a special debt to Mohan Matthen for advice and encouragement, without which this chapter would not have been.

Notes

1. A basic reading list of classics would include Helmholtz (1867, 1924, sections 19, 20); Maxwell (1855) and (1860); and (for those with mathematical interests) Grassmann (1853) and the very remarkable Mayer (1775). For more recent theory, excellent introductions, and masterly surveys, of very different types, see Evans (1948); Le Grand (1948, vol. 2 of 1945–1956; tr. 1957/1968); Boynton (1979); Wandell (1995); and Mollon (2003).

2. For the work of W. D. Wright and others that fed into the standard colorimetric systems of the CIE (Commission Internationale de l'Éclairage), which I use in later sections, see Wright (1944/1969); a helpful introduction is Hunt (1987, with later editions 1991, 1998). On technical issues, the standard authority is Wyszecki and Stiles (1982).

3. "As there is little hope of detecting it [the exact nature of the "anatomical contrivance" at work in the eye] by dissection, we may be content at present with any subsidiary evidence which we may possess . . . furnished by those individuals who have . . . a variety of . . . Colour-Blindness" (Maxwell 1855, 137). On the basis of the confusion points of the various kinds of dichromat (see section 12.3), together with the CIE color-matching functions for the normal observer, Wyszecki and Stiles, in the 1967 edition of their book, calculated the sensitivity profile of each of three normal cone types: see Wyszecki and Stiles (1982, 604–608); cp. König and Dieterici (1892); Judd (1952, 104–107); Vos and Walraven (1971).

4. Blue rather than violet was sometimes taken as primary, but I mostly do not mention that option separately here.

5. I explain the classifications further in section 12.2. I occasionally talk of *dichromats*, where what I mean is strictly the 99.9% of dichromats who are protanopes or deuteranopes, setting aside the very rare cases of tritanopia.

6. Cf. John Herschel (1827, section 507, p. 434); see also his letter to John Dalton, 22 May 1833 (in Wilson 1855, 60). "All hues for the color-blind eye can be obtained by mixing yellow and blue" (Helmholtz 1867, 295; 1896, 362; 1924, 2:148).

7. For similar views, see Le Grand (1957, 329; Fr. edn. 2:333; 1968, 344); Boynton (1979, 380); Hardin (1988, 145–146); Neitz and Neitz (2000, 695); and, hypothetically and with some philosophical differences, Byrne and Hilbert, this volume.

8. The size of a visual stimulus is often given in terms of the *visual angle* it subtends at the eye: a circular cap of diameter 1 cm at a distance of 60 cm, for example, subtends an angle of about 1°. For present purposes, I describe 1° stimuli as small; by large fields, I mean those of 8° to 10° or more. For *spectral light*, see note 13.

9. There are many directions in which people have taken these issues further; for one line of development, see O'Regan and Noë (2001) and Noë (2004, esp. ch. 3, which takes up some proposals from Broackes [1992]).

10. Kuhn (1962/1970, esp. chs. 6–8) argues that a scientific theory is typically not rejected simply because (as we might say) the theory "conflicts with the evidence." On the contrary, we live all the time with anomalies and recalcitrant evidence that "normal science" may ignore, or treat as unimportant, or try to explain (whether with epicycles, or ideally, with success). Rejection typically occurs only when some more attractive theory comes along that gives theoretical significance to recalcitrant phenomena that had previously seemed unimportant or incidental.

11. Pitt, on Abney's wool test: "while most of the dichromats were readily detected while using this test, others managed to get through it without making a fault" (Pitt 1935, 31–32).

12. The Report of the Committee on Color Vision 1892 (see Royal Society 1890) gives in the appendices (368ff) a survey of the means of practical diagnosis available at the time; for more recent methods, see Pokorny et al. (1981), or Fletcher and Voke (1985 chs. 5–8).

13. *Note on terminology*: *spectral light* is light from a portion of the spectrum (e.g., a narrow band specified, approximately, as 652 nm ± 1 nm), produced, for example, by passing sunlight through a prism (or by reflection or transmission from a diffraction grating) and selecting a suitable portion by means of a narrow slit in a metal assembly. I talk at times of "red light," "yellow light," and so forth, meaning light from the red part of the spectrum (or the yellow part of the spectrum, etc.). A mixture of red and green light (with no light from the yellow part of the spectrum) may indeed also look yellow: but when that is what I have in mind, I give further specifications. I at times specify the wavelength (in nanometers) of a light by using a subscript number, for example, red_{671} (for red light of 671 nm) or green_{520} (for green of 520 nm); and sometimes I abbreviate the main color terms by using just the initial letter, e.g., Y_{575} (for yellow light of 575 nm), or B_{440} (for blue of 440 nm). All color terms used in such labels are intended as rough indications, not exact specifications.

It may be helpful to give explanations of two other terms. "*Saturation* may best be defined as the percentage of hue in a color. In common speech the saturation of a given color is described by such words as pale or deep, weak or strong, in connection with the name of some hue. The concept is roughly parallel to that of the purity of a chemical compound or the concentration of a solution" (Evans 1948, 118).

Correlated color temperature: At successively higher temperatures, a black body glows reddish orange (e.g., at about 1300–3500 K), yellow (~4000–5500 K), white (~6000 K), and finally blue (7000 K and above). (A black body is defined as one that absorbs all the radiation incident upon it, unlike, e.g., a gray body, which reflects some of that radiation.) Natural light sources (even if not strictly "black") often emit a similar energy distribution to that of a black body, and their character may be specified by giving their *correlated color temperature*, i.e., the temperature of a black body that is the closest fit to their own energy distribution, as, e.g.: candles (~1800 K), 60 W incandescent bulb (2800 K), sun disc at 60° (5100 K), sky without sun (7500 K), clear blue sky (15000–30000 K) (Evans 1948, 26–27; Wyszecki and Stiles 1982, 28 and 8–9).

14. For a classic presentation of chromaticity diagrams, see Wright (1944/1969, chs. 3 and 4); Hunt (1987, chs. 2 and 3) is a good textbook account; a popular introduction is Rossotti (1983, ch. 15).

15. This slightly undercuts the simple account I gave earlier of Rayleigh matches and the principles of the Nagel anomaloscope: if in these color-matching experiments, a mixture of R_{650} and G_{530} is not a perfect match for yellow_{573} unless the latter is desaturated with some blue, then we should not expect any mixture of R_{671} and G_{535} to be a perfect match for Y_{589} in the anomaloscope without some blue to desaturate the Y_{589}. And if the L : S (or M : S) cone response ratios are actually different for R_{671} and G_{535}—as indeed is the case according to, for example, the Stockman and Sharpe (1999) functions, and the König fundamentals derived from the CIE *XYZ* functions—then we might think we there had a basis for dichromats (lacking M or L cones) still to distinguish R_{671} from G_{535} even in the anomaloscope. This is all true, but the effects are small and on this basis, the difference for a deuteranope between R_{671} and G_{535} would still only be the same as the difference between R_{671} and R_{671} very slightly desaturated with some white. The issue is worth further study, however; and it is worth remembering that the newer anomaloscopes that use LEDs to simulate the Nagel primaries are not doing quite the same thing as the original model.

16. This should be read: 10.0 units of Y_{580} *mixed with* 0.11 units of B_{460} *matches* 2.88 units of R_{650} *mixed with* 3.26 units of G_{530}. The + sign signifies additive light mixture; the = sign signifies matching. The units used here for the R_{650}, G_{530}, and B_{460} primaries are set on similar principles to W. D. Wright's system (so that a match for $Y_{582.5}$ requires equal numbers of units of R_{650} and G_{530}; and a match for G_{494} requires equal numbers of units of G_{530} and B_{460}: cp. Wright 1946, 125–135), but the values are my own recalculation from CIE *XYZ* functions. The units for Y_{580} are arbitrary.

17. Grassmann (1853, 78; 1854, 260); wrongly listed as Grassmann's second law in Hunt (1987, 206); and completely travestied in the English translation of Helmholtz. Helmholtz himself states the principle right: "Colours that appear the same produce [i.e., with some third colour] mixtures that appear the same" ("gleich aussehende Farben gemischt gleich aussehende Mischfarben geben," Helmholtz 1867, 283, my translation). Unfortunately, the published translation has *"Colours that look alike produce a mixture that looks like them"* (Helmholtz 1924, 2:133)—which is not what Grassmann or Helmholtz either means or says, and which is a much weaker principle that would not serve the purposes either author has for it. The translation (for which, shockingly, the Harvard psychologist L. T. Troland seems, with E. J. Wall, to have had responsibility [Helmholtz 1924, 1:vi]) makes the same mistake a second time (1924, 2:148), thereby making nonsense of an important argument in Helmholtz (discussed below: see note 27).

18. Given R, G, and B values within a system like W. D. Wright's (or our earlier Maxwell triangle), we can define *X*, *Y*, and *Z* values by a linear transformation like this:

$$\begin{pmatrix} X \\ Y \\ Z \end{pmatrix} = k \begin{pmatrix} 2.6495 & 0.3120 & 0.3754 \\ 1.0000 & 1.6248 & 0.0796 \\ 0.0000 & 0.0796 & 2.1546 \end{pmatrix} \begin{pmatrix} R_{650} \\ G_{530} \\ B_{460} \end{pmatrix}$$

The system is set up so the *Y*-value of a stimulus is equal to a measure of its *luminance* (the psychophysical correlate for brightness), a function for which (for spectral stimuli) had been previously defined by the CIE in 1924.

19. The variables *x*, *y*, *z* are defined as the *proportions* of *X*, *Y*, and *Z* in a mixture: $x = X / (X + Y + Z)$; $y = Y / (X + Y + Z)$; and $z = Z / (X + Y + Z)$. Since $x + y + z = 1$, we need only represent *x* and *y* directly in a diagram: that fixes *z* as well.

20. In addition, the spectral locus above about 630 nm (which is a straight line, corresponding to the fact that the S cones seem inactive in that range) is set in the *x*, *y* diagram to coincide with the line at –45° from (0, 1) to (1, 0).

21. Similar demonstrations had been presented by Ralph Evans at the Optical Society of America in March 1943 (Walls 1960, 32), and similar processes seem to be involved in the limited, but at their best remarkable, successes of two-color movie film systems of the Technicolor Corporation (1917, 1922, and 1927) and, later, Cinecolor, which continued into the 1950s.

22. It seems to me very likely that the success of Land's images comes precisely from their involving ordinary objects—with expanses of more or less uniform surface color presented as illuminated differently in different parts of the image, and with highlights and shadows. Indeed it may turn out to be somewhat ironic that after presenting his remarkable red-and-white experiments,

Land looked, in later work, to Mondrians to illustrate the fundamental principles of color constancy at work in his earlier phenomena. It may be precisely the opposite: that the colors of Mondrians, if photographed and displayed in a red-and-white presentation, would actually produce a far narrower range of color impressions than Land's original red-and-white demonstrations did. It may be that it is precisely because the original demonstrations were of more "natural" three-dimensional scenes that the colors in them were (to the extent that they were) recognizable. But that is a hypothesis that needs to be tested.

23. Rods have been given a significant role in perception particularly of the red part of the spectrum, at over 580 nm, and especially around 645 nm—which occasioned revisions of the color-matching functions by Stiles and Burch (1959). See Wyszecki and Stiles (1982, 333–343, 354–371).

24. I should perhaps point out immediately the invalidity of the move from (a) to (b) here: even if lights *a* and *b* on a single confusion line are indistinguishable when (for example) *a* is increased in intensity so as to appear as bright as *b* does, that does not imply that, in advance of any such adjustment of intensities, *a* will look of the *same hue* (or saturation) as *b*: *a* might, for example (for all that is implied by the chromaticity diagrams) look *red* and *b* look *yellow*—though if *a* was increased in intensity, it would come to look yellow like *b*. For a report of something like this actually happening (in a case of acquired unilateral tritanopia), see Alpern et al. (1983b, 688–689). We are told that violet$_{422}$ and green$_{523}$, for example, though they lie on the same tritan confusion line, looked to their subject's tritan eye of *different* hues: violet$_{422}$ looked a fairly saturated blue$_{\sim 483}$, whereas green$_{523}$ looked a more desaturated variant of blue$_{489}$, which is greener in hue. (I read these figures, rather approximately, from figure 2 at 688.) The authors attribute the hue difference merely to the fact that (in their experimental conditions) the violet$_{422}$ was less bright than the green$_{523}$. Wachtler et al. (2004) effectively maintain (1)(a) without (b): they propose a model, on which the experienced color-space of dichromats collapses to a plane, but it is not a vertical slice, nor a flat surface slicing through at an angle, but a curved surface that bends through the normal trichromat's color-space—thus yielding different hue-sensations from physically similar stimuli at different intensities. MacLeod and Lennie (1976) propose a version of (1)(a) and (b) without (2): they believe the sensations from their subject's dichromatic eye are restricted to a line in chromaticity space, but it is an arc that curves through several hue-regions (see end of section 12.4 below). MacLeod, this volume, explores this option further: section 6.5.

25. Helmholtz and Maxwell of course hold (2) in a different form: the single axis of sensation runs, in the case of the "red-blind," between green and violet or blue (and, in the case of the "green-blind," between red and violet or blue), so it will pass through several hue regions, all at virtually maximum high saturation. (By contrast, Hering's line between yellow and blue via white might be said to involve no change of hue at all, but only positive and negative saturations of a single hue.) It might be more precise to talk of a single *hue-and-saturation* axis, but I shall talk of a single color axis (having in mind the aspects of color other than brightness), hoping that the flexibility in ordinary talk of color makes this accurate enough as well as less cumbersome.

26. (For reasons given in note 24) Alpern et al. (1983b) effectively reject (1b) and (2) for their tritanopic subject, but still insist on (1a).

27. Helmholtz (1867, 295; 1896, 362; 1924, 2:148). The second German edition is the one to read: it contains some improvements over the first in details of the argument, but they are omitted in the third edition; and the English translation (which is done from the third edition) also contains a complete misunderstanding of Grassmann's third law (see note 17 above). For those who read French but not German, the French translation (from the first German edition) is much better than the English. The core of the argument is that, if red and green lights R and G are indistinguishable for color-blind person *p*, then (by Grassmann's third law) R and G, and any intermediate shade produced by a mixture of them, will all have the same mixing potential for *p*: they can all be substituted one for another in any mixture with further kinds of light, and the change will be indiscernible to *p*. But one of the intermediate shades produced by a mixture of R and G will in fact be equivalent to some shade that could be produced solely from yellow and blue (it will effectively be neutral in red-greenness): let us call it S. In that case, all colors that are normally produced by R, G, and blue, can instead (for *p*) be produced *from yellow and blue alone*. (We simply substitute S—which is producible from yellow and blue alone—for whatever R and G constituents would normally be involved.) And the same is true (by parallel reasoning) of colors normally produced by red, green, and yellow. In short, all colors whatever can for *p* be produced (or matched) by mixtures of *merely yellow and blue*. (Strictly, note, this doesn't immediately tell us that they all *look* merely yellow and blue to *p*; that is a conclusion requiring further evidence, drawn, typically, from the unilateral color blindness cases.) To be very brief on what goes wrong here: we know from anomalous trichromacy that there are cases where a red and a green R and G are confused in some particular circumstances but not in all, and that even the regular confusing of them is not sufficient for us to conclude that R and G will have identical mixing potential (especially if negative quantities, or proportions >1, of either might be involved). In general, we know from anomalous trichromats that confusions can occur without the whole chromaticity space collapsing to a line. And something like this may be true for dichromats as well as anomalous trichromats. For example, R and G may be confused in small fields but not large, or at low brightness but not high brightness, or (to speak approximately) when relatively desaturated but not otherwise—and if we set the standard of indistinguishability as, not indistinguishability in *some* circumstances, but indistinguishability in *all* (even with large fields, high saturation, etc.), then the empirical evidence today suggests that this standard is one that, with the dichromats of the real world, real reds and greens mostly do not actually meet.

28. Wright himself said that the brightness of his colorimeter fields "was more limited than could have been wished" (Wright 1946, 50).

29. Wyszecki and Stiles (1982, 464); though I should mention that the neutral points of the protanopes in Scheibner and Boynton (1968) imply a protan confusion point at $x = 1.03$, $y = -0.03$, or even (if we include one "discrepant" subject) $x = 1.28$, $y = -0.28$. This is theoretically absurd, in that it would place the red receptor in a part of the CIE diagram that corresponds not to red, but to a red of negative luminance, i.e., a *green-blue*! (One of the details of the CIE system is that line $y = Y = 0$ represents zero luminance, and things with Y-values less than zero would have negative luminance.)

30. Where authors have given values in other systems (e.g., the W. D. Wright system or the CIE 1976 u', v' system), I have converted the figures to the CIE 1931 x, y system and given the orginal

figures in parentheses. In some cases I have not seen the original sources but have followed an authoritative intermediate source: obviously this puts me in no position to judge the accuracy of the figures or the methods employed, but it is still, I hope, of some help in showing the variety of conflicting voices around. For two other surveys of many very different reports of confusion points, see von Schelling (1960) and Nimeroff (1970).

31. More perplexingly, Judd had previously suggested $x = 1.000$, $y = 0.000$ (1943, 305)—and repeated the suggestion at (1952, 105)—which makes little sense in that it would imply (in placing the confusion point on the line $y = Y = 0$) that the M cones made no contribution to luminance.

32. Dean Farnsworth takes up the issue of confusion points in the opening article of the opening issue of *Vision Research* (1961), the journal which he founded but never lived to see published. Farnsworth points out the huge spread among Pitt's six deuteranopes in the spectral light they match with white (illuminant B), which in turn implies huge differences in the confusion points each subject had, if he had one at all; except that their confusions don't in fact imply a single point. Farnsworth points out how "the six real existing individuals with apparently impossible responses were averaged into one theoretical individual with reasonable responses" (1961, 3): thus was born the canonical deuteranope of Pitt's famous confusion diagrams. The spread is not mere experimental error; it may best be explained, Farnsworth believes, by individual variations in ocular pigmentation, e.g., in the macula; such individual differences mean that, even if the receptors of (for example) all human deuteranopes were the same, we should not expect their neutral points to be the same. (Any white will have its hue modified according to the pigmentation in the eye even if a spectral hue will not.) Even more troublingly, Farnsworth adds that, given ocular pigmentation, we should not even expect confusion lines to be *straight* (given the varying results to be expected when a filter interacts with light of one hue but at a series of graded levels of saturation). Joining points corresponding to pairs of lights confused by a non-averaged dichromat, *we will not get a set of lines that intersect at a point.* "We can make sense of this individual's data only if we assume that his confusion line with white is curved through the two points" (1961, 5). Farnsworth's conclusion is that if we want to use such diagrams to discover the "missing sensation" of the dichromat, then what we really need is "a color-mixture diagram at the *retinal* level. The immediately apparent way to approach this is to devise means of measuring the quality and variety and *range* of pigmentation in normals and in color defectives" (5, my emphasis). In the absence of that, we can only expect Pitt's "enticingly simple diagrams" to continue to be the object of "uncritical acceptance and a consequent remarkable number of misleading beliefs" (1). I enthusiastically agree.

33. "A review of the rather considerable literature . . . shows that the color perceptions of both protanopic and deuteranopic observers are confined to two hues, yellow and blue, closely like those perceived under usual conditions in the spectrum at 575 and 470 mμ" (Judd 1948, 247, quoted in section 12.1).

34. Interestingly, Helmholtz (1867, 298; 1924, 2:151–152) thinks a "red-blind" person will *talk* of seeing only yellow and blue in the spectrum, but his sensations will actually be of *green* and violet or blue. Maxwell takes the same view.

35. As far as I can see, Judd believes he can stop Hayes's report from being evidence against his own view by downgrading the subject's deficiency: Hayes calls it protanopia (1911, 397), but Judd reclassifies it as protanomaly (1948, 253)—because of the subject's seeing certain mixtures on a spinning top as "much too green." But we should be wary of assuming that anyone who "sees green" (or "sees red") cannot be a dichromat, when the very matter at issue is whether dichromats are or are not capable of seeing red or green: it would be a case of reasoning in a circle. Of course, a protanope will surely not reject a brightness-adjusted Rayleigh match in a Nagel anomaloscope as "too green" (compared with the Y_{589})—indeed this is the standard criterion for distinguishing a protanope from a protanomalous person. But the case with spinning tops—with surface colors and larger fields—may well be different. And when the lightness of the two fields in an anomaloscope is not the same, then even certified dichromats sometimes describe the darker field as "greener" than the other. The criterion for dichromacy is not that no greenness is ever reported, but that it may be canceled by adjustments of lightness. In the end, I too have doubts about whether Hayes's subject G. S. was or was not a protanope; but if she was, then (because we are told she had sensations of green) she is at least a putative counter-example to Judd's only-yellow-and-blue claim, and certainly no support; and if she was not, then she falls into my category (c) of anomalous trichromats and again, though for different reasons, is no help to him.

36. Once we recognize the possibility of systems developing to compensate in part for color blindness (whether or not we will decide that the possibility is actually realized), then a priori we might expect acquired color blindness (especially in just one eye) to be less debilitating than congenital color blindness (in that an acquirer at least has had some period of normal trichromatic color input, which might help in guiding or calibrating any compensation system); alternatively, we might expect an acquired color blindness to be more debilitating (if some forms of compensation depend on neural changes that can occur early in neural development, but not later). Without direct empirical knowledge of *both* kinds of case, there is no obvious way to say how far the one may or may not be a guide to the other.

37. This fact makes me wonder if neurological disease outside the eye may not have been the cause of both conditions—in which case the color blindness really belongs in category (d) and is of even less evidential value to us. Certainly nothing in Hippel and Holmgren gives any information on the exact cause. It should be said that even with the most recent report of supposedly congenital unilateral color blindness (MacLeod and Lennie 1976), we do not have independent information on the genetics or (e.g., from reflectance densitometry) the actual cone equipment of the retina.

38. Colors 1–30 in Radde's index form a hue circle with the following main reference points: 1. cinnabar (i.e., slightly orange red); 4. orange; 7. yellow; 10. yellow-green; 13. grass green; 16. blue-green; 19. blue; 22. violet; 25. purple; 28. carmine (i.e., deep red, inclining perhaps to purple). (I am grateful to Rainer Hertel for letting me see a copy from the Staats- und Universitäts-Bibliothek, Hamburg.) They were described by Hippel's subject (when presented individually and out of sequence) as 1–15: yellow; 15–16: greenish or yellowish; 17–22: blue; 23–24: light blue; 25–30: red (Hippel 1880, 181; sample 15 is indeed double-counted by Hippel).

39. Judd mentions only "Sloan 1947," apparently forgetting the co-author Lorraine Wollach. The 1947 publication is an abstract: the main report (1948) appeared only later, and Judd does not mention it; but he no doubt knew plenty about the experimentation, since he is thanked for having advised the authors on it (Sloan and Wollach 1948, 509).

40. The Munsell papers were of value and chroma 5/5. The maximum for value in the Munsell system is 10/ (for white). There is no general maximum for chroma, and different ranges of chroma will be practically realizable depending on the hue (and value) in question; but in many cases, Munsell papers are available at chromas up to /10, and in some cases /12 or /14.

41. One might compare the well-known weaknesses of the CIE standard source C, which combines a tungsten filament lamp (2856 K) with a deep blue filter to yield light of correlated color temperature 6774 K, which supposedly matches daylight: "The illuminance levels that normally can be achieved with such sources [as standard sources B and C] are relatively low and confined to small areas of illumination, making them *unsuitable as sources for visual inspection*. Their relative spectral radiant power distributions, closely correspondent to [the mathematically specified functions that define] CIE standard illuminants B and C, *do not represent adequately* spectral distributions of *daylight*" (Wyszecki and Stiles 1982, 147, my emphasis). If these kinds of illuminant are held unsatisfactory for industrial color evaluation and matching by normal trichromats, then it is puzzling that they should be used in an investigation that pretends to tell us the *full range* of color sensations available to a dichromat eye.

42. Except as part of the compound word "blue-green" (Blaugrün) in two places (97), one of which says something that contradicts both what Judd reports from the article and what Judd himself wants to conclude. Nagel says that he, a dichromat, finds "a match between blue-green and purple *always unsatisfactory*, as soon as the size of the field exceeds 10°" (1905, 97, my emphasis). That is, with large fields of blue-green and purple, Nagel does *not* (as Judd would like) see merely the blue component—and he has found "numerous other deuteranopes" with very similar behavior (97).

43. When presented with a circular 3° to 4° field, one half of which showed R_{680} light and the other half YG_{550}, Sch. saw them both as yellow (as indeed Judd would like) as long as the two half-fields were of equal lightness. However, "If the one half was darker than the other, then Sch. mostly *said the halves were differently coloured*, sometimes the darker one green, the lighter one yellow or red, sometimes the darker one red, the lighter one yellow" (1905, 94, my emphasis).

44. "The probability of such contradiction arising may be estimated from the fact that the eight [unilateral or similar] cases recorded so far . . . fail to contradict it" (Judd 1948, 255). Judd talks also of how "a reasonable search of the literature has failed to uncover any reliable evidence against this indication" (255; the "indication" here being, strictly, a slightly weaker claim, that red-green confusers see yellow and blue, without it being specified exactly which yellow and blue they are).

45. He was president of the Optical Society of America in 1953, and recipient of many awards from technical and engineering societies. For some biography, see Judd (1979, iii–vii).

46. It is puzzling that Judd cites Goldschmidt (1919) at all: the case is not of unilateral color blindness but of a unilateral scotoma with peripheral loss of acuity, following a gunshot wound to the left forehead and temple. The color blindness (an anomalous trichromacy "with red-weakness") showed up with "similar results" in both eyes (199). (Judd may be hoping that the exacerbation of the protanomaly when combined with the effects of the gunshot wound will tell us what a more severe color blindness—protanopia—would involve. But the causal factors in the two cases are so fundamentally different that it is anyone's guess what inferences one might be able to make from the one kind of case to the other.) Yet more oddly, the one relevant passage that Judd quotes from that article seems, if anything, to imply that the subject saw *green*—whereas Judd reads it as saying almost the opposite of that. Goldschmidt reports: "H. had the habit of fairly regularly naming a number of colour hues incorrectly, especially yellow and violet ones, in particular, *as if the green in the spectrum seemed spread out into the yellow*, and the blue into the violet" (1919, 205, my emphasis). What that surely says is that the man talked as if certain yellow parts of the spectrum *looked green to him*, and certain violet parts looked blue. Judd quotes part of the passage, but reads it as implying (by a line of inference that I can only guess at) that the dominant impression from the long-wave part of the spectrum was actually "orange-yellow" (Judd 1948, 252), the latter being a disappointment to Judd, since he would have preferred the dominant impression to be of unique yellow.

47. "Hayes' conclusions are sound if translated into accepted terminology" (Judd 1948, 253). This is an extraordinary statement. Hayes's "main contention" is "that people properly classed as color-blinds have some sensations of red and green" (1911, 399)—and there is no sense in which that claim can be "translated" into agreement with Judd's own view other than by ignoring what it actually says. What Judd no doubt has in mind is that he can undercut Hayes's opposition by downgrading the diagnosis of the bad eye in Hayes's subject from protanopia to protanomaly (in which case it would not be surprising that it provided sensations of green); but while that might at a pinch be called a change in terminology, it too is actually, and would be understood by all parties as, a substantive change of diagnosis, not a purely verbal change.

48. In his first article, Hippel reports his patient as naming the colors of Radde's index, when presented separately, with a variety of color terms, including "greenish," "red," and "light green," as well as "yellow" and "blue" (Hippel 1880, 181). A year later, Hippel takes it all back:

> I must add a modification to the reports made by the patient in my first communication, in so far as they relate to the naming of the individual colours in the Index to Radde's Plates. Since this person had previously never attempted to compare the colour sensation of his right eye with that of the normal left eye, he was therefore, at the time of the first experiments with the right eye alone, precisely *in the same situation as a bilateral color-blind person.* As the latter employs the expressions Red, Yellow, and Green, which he is familiar with, *in complete confusion*, so also did my patient at first. Hence [sic] the names that he applied to the colours in the first trials give us not the slightest information on the sensations he had of them. This changed very soon with him, *once he became aware of his defective sense of colour* in the right eye and began then to compare the sensations of the two eyes with each other. *The terms that he now chooses* for the individual colours of Radde's Scale *correspond completely to the sensation he has*; they are as follows: 1–14 yellow, 15 grey, 16 and 17 dark grey, 18 slightly blue, 19–21 blue, 22 and 23 dark grey, 24 blackish, 25–27 dark grey, 28–30 yellow-grey, 31 grey, 32 darker grey, 33–36 yellow, 37 grey-yellow, 38 dark grey, 39 light blue, 40–42 blue-grey. (Hippel 1881, 51–52, my emphasis)

What Hippel curiously seems to forget is that his patient had *not* used the expressions "red," "yellow," and "green" "in complete confusion": on the contrary, he had used them with considerable (though imperfect) sensitivity to the red-green variations among things. And it is bizarre that Hippel thinks his patient knew how to apply the color terms perfectly correctly in relation to experiences caused by his left eye, but somehow not in relation to experiences caused by his right eye (until, that is, he had been instructed in the special process of the "comparison" of sensations). One can indeed misreport impressions. But the reasons Hippel gives for thinking his patient was doing so are bad ones, and I shudder at the thought of this seventeen-year old being retrained by his distinguished doctor (we are not told how) until he properly confined himself to the approved vocabulary of yellow, gray, and blue.

49. Graham and Hsia mention some significant differences (1958, 47), and MacLeod and Lennie (1976, 691–692) point out more: to match 500 nm light, so they report, Pitt's deuteranopes needed a 1:15 mixture of B_{450} and R_{650}, but the Graham and Hsia subject seems to have needed a 5:1 mixture—i.e., some seventy-five times more blue than an ordinary deuteranope! And the reported neutral point (502 nm) is completely incompatible with what Graham et al. publish as the subject's color-mixing functions.

50. Scanning the spectrum from 700 to 400 nm, the subject's experience pretty much divides into three: a region varying in hue, with red, orange, and some yellow all represented (and all at fairly high saturation); a zone of yellowish hue and fast diminishing saturation, reaching a neutral point (around 567 nm), and no sensation of green at all; and finally a blue region (with no violet) where there is only a small amount of variation in hue sensation (between sensations corresponding to 470 and 490 nm), but the saturation first increases sharply to a maximum (with stimuli in the ~488–482 nm region), and then decreases to a second near-neutral point as the stimuli approach 400 nm.

51. See note 24. It cries out to be studied further, how *lightness* variation in a stimulus may also change the *hue* experienced by dichromats (such changes being already quite interesting enough, though no doubt largely different in nature, in the case of normal trichromats: see the curved lines of constant hue in, e.g., Wyszecki and Stiles 1982, 420–424, 670–672).

52. Compare the comments on the many different factors at work in depth perception at the start of section 12.6.

53. I am grateful to John Mollon for the observation. Unsurprisingly, such sufferers tend to be more distinguished as bowlers than batsmen, and sometimes particularly as captains. Others cricketers who had lost the sight of one eye are William Clarke (1798–1856), first-class underarm spin bowler, who played for Nottingham, the MCC, and his own All-England Eleven, and Eiulf Peter Nupen (1902–1977), who has been described as a deadly bowler on the matting pitches on which South African cricket was played in his time. Baseball seems to have been slightly less forgiving of such losses, but Thomas Jacob Sunkel was a major league pitcher—for the Cardinals in 1939 and later the New York Giants and Brooklyn Dodgers—despite a cataract that had left him effectively without sight in one eye.

54. R. Descartes, *La Dioptrique* (1637), Discours 6. (AT 6:130–147, translated in CSM 1:167–175); G. Berkeley, *An Essay towards a New Theory of Vision* (1709), sections 1–51; Wheatstone (1838).

For a historical treatment of these and many other contributions on the issue, see Wade (1998, ch. 6). For a person using only one eye to recognize the shapes of solid things, "The motion of the head is the principal means he employs" (Wheatstone 1838, 380)—though Wheatstone (I think unfortunately) takes it that while this supplies "accurate information" on depth, it does not supply "that vivid effect arising from the binocular vision of near objects."

55. As Katz said, "The objects, with their colour qualities, are apprehended as belonging in a *chromatically illuminated* field" (1935 [1930], 190)—a point beautifully illustrated by the demonstrations of Lotto and Purves (e.g., 2002). See also the discussion of the blue card disc seen in a yellow spotlight in section 12.7 below.

56. On the other hand, normal trichromats do more or less equally well with illuminant color variation in the direction of green as they do with variation along the YB-axis—though they do slightly less well with variation in the direction of red (see Delahunt and Brainard 2004, and refs.); whether this is true also of dichromats is something that I do not know has been investigated.

57. I give the pigments (both green and red) in order of increasing yellowness (i.e., from bottom left to top right in the chromaticity diagram), so that the respective members of each group of three correspond: corresponding members fall more or less on a traditional confusion line.

58. For this model, I use the hypothesized König fundamentals, dichromat color-matching functions, and luminous efficiency functions, V_p and V_d, devised by Wyszecki and Stiles (1982, 463–471, 604–608). I have doubts about the confusion points derived from Pitt (1935), and these functions will be subject to similar reservations; but they are quite good enough, I think, to illustrate the general principles in the circumstances.

59. The same point is valuably employed in Golz and MacLeod (2002) on color constancy.

60. If a series of points is scattered within a 1-unit radius circle centered on (0, 0), then if a point has an *x*-value of 0.2 or 0.4, then it may have any of a very wide range of *y*-values; but if it has an *x*-value of 0.95 or 0.97, then it must have a *y*-value much closer to zero.

61. If we keep S″ equal to S′, then we get an improved recovery if we set them equal to –21.4% rather than 22%, but that lowers the mean discrepancy between recovered and original coordinates (for our fifteen pigments) only from 0.0226 to 0.0224. There is, I think, no special reason why the *luminance increase per unit of RG-ness* should be the same as *luminance increase per unit of YB-ness* in CIE *x*, *y* chromaticity space (that is, why S″ should be the same as S′). Lengths that are equal in one color system may not be equal when transformed into another system, and most theorists believe that, for example, the CIE 1976 *u*′, *v*′ chromaticity space has a metric that fits better with the perceived distances and differences among colors than does the 1931 *x*, *y* space. But (given the *x*, *y* space and the general structure of the algorithm) it turns out to be not much worse than more complex alternatives. If we allow S″ to differ from S′, then setting S′ to –20.3% and S″ to –23.5%, we can reduce the mean discrepancy to 0.0197. But that too is a relatively small improvement.

62. We know perfectly well, for example, that two metameric greens that under illuminant D_{55} belong in the same position in *x*, *y* chromaticity space will behave slightly differently as the light

changes—and may look different from each other in D_{75}. But in that case any algorithm of the kind I am envisaging here would be almost bound to end up "recovering" slightly different chromaticities for the two things in D_{55}—which would, *ex hypothesi*, be false. But still, the differences in behavior between those metameric greens would pale in comparison with the differences in behavior between either of them and red things—and that is the kind of difference I am working on here, and somewhat more subtle varieties of it.

63. There are rare circumstances where the macula has a noticeable effect. Maxwell describes one, suggested to him by the physicist George Stokes: "It consists in looking at a white surface, such as that of a white cloud, through a solution of chloride of chromium made so weak that it appears of a bluish-green colour. If the observer directs his attention to what he sees before him *before his eyes have got accustomed* to the new tone of colour, he sees a pinkish spot like a wafer [i.e., a light thin crisp cake, baked between wafer-irons, typically pink or orange-red] on a bluish-green ground; and this spot is always at the place he is looking at. The solution transmits the red end of the spectrum, and also a portion of bluish-green light near the line F [i.e., about 486 nm]. The latter portion is partially absorbed by the spot, so that the red light has the preponderance" (Maxwell 1870, 230, my emphasis; von Kries makes the same point in Helmholtz 1924, 405). For the effects of this ~4° "Maxwell spot" in viewing large and even fields, see Wyszecki and Stiles (1982, 133).

64. When first writing this, I had never heard anyone else talk of it—except for one philosopher, who had heard me describe it and had replied enthusiastically that the description fitted his own experience too. Since then, one other person has reported the same to me (when I talked at a symposium at the Laboratory of Integrated Neuroscience at UIC in 2007). The New Zealand artist Will Furneaux can be found (in a post at www.conceptart.org dated November 20th, 2008), saying: "Sometimes . . . (I think because it's your brain that interprets signals from your eyes) I actually "see" red when in actual fact it's green, and then *when I realise it turns green*" (my emphasis). It seems that the phenomenon may be quite common, while being unrecognized in the academic literature.

65. Dalton and Nagel were deuteranopes (on Dalton see Hunt et al. 1995); Koffka was protanomalous; Pole was a protanope. Friedrich Schumann counts as deuteranomalous by the anomaloscope, but describes himself as lacking any "green process" in the cortical center for light sensations (1904, 12–13). Pole (1859) is an extraordinary case, a convert, one might say, and proselytizer for the yellow-and-blue view that was just at the time becoming established. A civil engineer by training, talking to the distinguished Fellows of the Royal Society (among whom is Sir John Herschel, a great promoter of the yellow-and-blue view, and to the company of whom Pole would two years later himself be elected Fellow), he declares that for thirty years he had *believed* he saw red and green along with other colors, but that he has now come to recognize that it was only various tones of yellow and blue that he had ever been perceiving (together with a sensation of green, which he supposes to be the mixture of yellow and blue sensations, and which would be experienced on looking at white and other neutral things). With Pole's personal confession being apparently almost ideal support for Herschel, Herschel in a further paper of his own congratulates Pole—while also correcting some of his confusions—as having given "the only clear and consecutive [i.e., consistent] account of that affection which has

yet been given" by a color-blind person who knows the theoretical accounts of it and is "*in a position to discuss his own case scientifically*" (Herschel 1859–60, 73, my emphasis). I must consider Pole's arguments on another occasion. (His evidence to "prove" that his sensations are confined to yellow and blue and their mixture [1859, 328] comes only from the confusions he makes—which are certainly not conclusive on the matter.) But the extraordinary interplay of intellectual styles and social and academic precedence (and interdependence) here—as also, over the subsequent century, the influences of technological changes and public safety imperatives, especially in shipping and the railways—make clear how a serious assessment of the textual and intellectual evidence on this subject cannot be separated from the larger social and cultural history.

66. I display strong protan deficiencies on the Farnsworth D-15 test; on the Oculus anomaloscope, I behave as protanomalous; genetic tests indicate the absence of a gene for L receptors. I am grateful to John Mollon for the Farnsworth and anomaloscope tests and to Dr. Ulrike Dohrmann (Institut für Humangenetik, Albert-Ludwigs-Universität, Freiburg) for the molecular-genetic analysis.

67. It may be of interest to readers of Pylyshyn (1999) that this case—if I am not misdescribing it—seems a clear counter-example to his thesis of the "cognitive impenetrability of visual perception": it seems to be *thought,* or *cognition,* that influences the second color shift (when I have returned to the earlier lighting environment, and then, with some effort of will, get the thing to "flip" back from green to red). Pylyshyn argues that in the Necker cube case, the shift in appearance is caused not by cognition directly, but only by cognition influencing which parts (and especially corners) of the cube image the viewer focuses attention on (Pylyshyn 1999, section 6.4, following Kawabata 1986; Peterson and Gibson 1991; it may be worth mentioning that Pylyshyn's claim here was originally made by Necker himself [1832], and opposition to it was voiced by Wheatstone 1838, 382). It would be interesting if Pylyshyn or others were tempted to run a similar line for the color case, arguing, perhaps, that to get the surface to look red, I would need to focus on one particular pattern of shading and highlights, and to get it to look green, I would need to focus on some other pattern. I would be fascinated if there were support to be had in that manner for the kind of suggestions I have made in section 12.6; but for the time being, the case seems a prima facie challenge to Pylyshyn's thesis.

68. It may be significant that, in at least some such cases, there is a flipping between the thing's looking deep red and its looking deep green, with apparently no option of seeing the thing as (for example) merely a moderately saturated red or, simply, a grayish dark yellow. The situation seems to be of a *bistable,* rather than *multistable* stimulus (where "multi" means >2)—perhaps like the surface depicted in a picture that can be seen as concave or as convex, but not—or not easily—simply as *flat.* I wonder what might be the cues here that force, so to speak, the projection to alternately quite distant points in color space and not those in the middle.

69. Nor is just an accidental or personal matter. Light traveling from air or a vacuum into a dense medium is refracted differently according to its wavelength, as we see with a prism. The refractive index of the lens of the eye is greater for blue light than for red, and the focal length smaller. So when a green object is effectively in focus on the retina, a similarly clear image of a

red thing at the same distance will be behind the retina—and to bring it into focus, one will have to focus as if the red thing were closer than it really is.

70. As Mr T., one of George Wilson's informants, tells him, "the individual berries . . . appear to me for the first few seconds rather black than red, and *only gradually assume their red hue*" (Wilson 1855, 30, my emphasis).

71. Cp. Katz (1935 [1930], 190), quoted above in note 55.

72. I have developed the arguments in Broackes (1992, esp. secs. 3.1 and 3.3).

73. Dalton thought that he saw only yellow and blue (and perhaps in some degree purple) in the spectrum but could see red and green surfaces under candlelight. Yet plenty of writers (Graham et al. 1961 are an example) move without any argument whatever from "*x* sees only two hues in the spectrum" to "*x* sees only two hues." (See section 12.5 on Montag and Boynton 1987 and Montag 1994.)

Where we do continue experimenting with light stimuli, rather than surfaces, I would particularly recommend a study of the effects (e.g., on dichromats' hue sensation) of variation in lightness of stimuli. Is it the case that red lights if brighter look "more yellow" to the dichromat? (Always?) What are the conditions (of brightness, contrasting surround, or whatever else) under which dichromats take yellow light to be green? With a unilateral case, the way hue sensation may vary with stimulus brightness should be particularly carefully examined. Alpern et al. (1983b) show the importance of the issue (see end of section 12.4 above and note 24), but do not study it per se.

74. There have been recent "large-field" experiments (see section 12.5), but the spectral light fields are of up to 8° (Smith and Pokorny 1977) and 12° (Nagy and Boynton 1979; Nagy 1980), which pales beside Nagel's use of 20° and even 60°–85° (Nagel 1910, 6–7). The "large-field" surface samples of Montag and Boynton (1987) and Montag (1994) are in fact 3.8 cm squares—subtending an angle of only about 4° at 60 cm. By contrast, Nagel used (among other things) papers of 50 cm × 70 cm (1910, 8)—some 240 times larger.

75. By contrast, for his dichotomous B-20 test, Farnsworth proposes, "The saturation should not be high enough to permit the factor of *hue recognition* in color experienced applicants to interfere with apparent serial arrangement according to hue" (1943, 575, my emphasis). Farnsworth seems to suggest, incidentally, that *recognizing* the hues can get in the way of classifying them according to their *appearance*. I would be tempted to investigate the alternative hypothesis that recognizing the hue of a sample may actually change the appearance of that sample to a subject—so the sample may look different after it has been recognized from how it looked before (and the post-hue-recognition classification may be perfectly true to the post-hue-recognition appearance).

76. "Finally, it is also not a matter of indifference, how long the period is during which the colour is acting upon the eye. In order for the red sensation and the remnant of the green sensation that may be present to develop in the colour-blind person, a considerable amount of time must elapse if he is looking at a large coloured surface, and all the more, the smaller the surface

is and the paler and the more impure the colour is. With a normal person it makes almost no difference at all. Even if a red and a green light flash up for just a fraction of a second, he recognizes the colour immediately; the recognition time is almost unmeasurably short" (Nagel 1908b, 23). Compare also Wilson (1855, 30; Mr T., quoted in note 70); Guttmann (1907–08); and Montag (1994).

77. Nagy and Boynton (1979) considered two different illumination levels in their surface-color-naming experiments, and Montag (1994), three and more levels (investigating rod contributions).

78. Boynton made a good case for going beyond color-matching experiments—"Class A" experiments, to use G. S. Brindley's term—and investigating color naming as well (see, e.g., Boynton and Scheibner 1967, 206, 211, 220). My proposal is to take a further step and attend not just to basic color classifications (e.g., red, green, yellow, blue), but also to the higher-level (but no less perceptible) interrelations perceived among colors: and to do this for the color-blind as well as the normal trichromat.

79. Among the deuteranopes of Smith and Pokorny (1977), for example, ten refused an 8° large-field match of Y_{589} with G_{545}, but one (CM) accepted it. Was the difference in appearance between the relevant half fields less for CM than for the others? Or was he simply more tolerant in his acceptance of approximate matches? (And were there differences at the retinal level between him and the others—or only at the stages of later processing?) These questions can be investigated, even if the evidence is hard to assess. (One can ask, for example, on the first issue: "On a scale of 1 to 10, how well do these fields match?"; not just "Do these fields match?") Of the two protanopes in Montag and Boynton (1987), one (P2) behaved quite similarly to a normal trichromat when presented with larger fields (3.8 cm square surface colors, viewed at 60 cm) and given up to five seconds for his responses; but the other (P1) used the term "green" quite indiscriminately even under those conditions—as much for reddish things as for greenish.

80. Under the direction of John Mollon.

81. My first ordering of the caps (under a Macbeth lamp, ~200 lux) was: P, 15, 1, 14, 2, 13, 3, 11, 4, 12, 10, 6, 7, 8, 9. My later ordering (under Sylvania Luxline Plus F58W/860, ~500 lux) was P, 1, 2, 3, 4, 5, 6, 7, 8, 9, 11, 10, 12, 13, 14, 15. When the illumination was changed back to the first condition (Macbeth lamp) and the caps were reshuffled, my ordering was P, 1, 2, 3, 4, 5, 6, 8, 7, 9, 11, 10, 12, 14, 13, 15. I am most grateful to the experimenter for these tests; responsibility for the nonstandard suggestions I am developing from them is of course entirely my own. There are phenomena here that I think are worth examining rigorously, but the present comments are meant as only a first glance at some of the material that would have to be examined.

82. "My own strong conviction, which is at variance with the opinion of some distinguished writers on optics, is, that red and green are both visible in favourable circumstances to the majority of the subjects of chromato-pseudopsis. . . . All the colour-blind persons whose vision I have formally tested, could in favourable circumstances occasionally distinguish red from green" (Wilson 1855, 46–47). Maxwell responded in a letter, which Wilson included as an appendix to the book version of his publication.

83. "All in all, one should really say, therefore, that the factual basis for the much mentioned view, according to which dichromats are red-green blind and yellow-blue seeing, is thoroughly scanty" (von Kries 1905, 166).

84. Among other problems, Collins makes no serious use of the distinction between dichromats and anomalous trichromats where she needs it, though she knows it as one of Nagel's special interests.

85. Even more strangely, Judd (1948, 255) says "a reasonable search of the literature has failed to uncover any reliable evidence against" his view that dichromats see only yellow and blue—whereas Judd himself has cited George Wilson's book and three articles of Nagel's, and Nagel's work at least (even if one might think of it as something that in the end was outweighed by opposing evidence) could not by any good judge be dismissed as simply unreliable.

86. "On the basis of the opponent-color model that has been presented, one can make reasonable predictions about what dichromats should see, assuming that the replacement hypothesis is correct and that the visual pathways are normal except for the r-g channels. For spectral colors, there is a division at the neutral wavelength that should cause shorter wavelengths to appear blue and longer ones yellow" (Boynton 1979, 380).

87. See figure 2 in Jameson and Hurvich 1978, which gives the "Locus of D-15 test colors in perceived color space" for their deuteranopes and protanopes: the colors are all presented as lying on the blue-yellow axis, with absolutely no element of apparent redness or greenness. The diagram is not the product in any way of the experiments reported in the article: it is an expression of a piece of *theory* that, of course, has some evidence to support it, as well as evidence (as we have seen) against. The authors present it as simply straightforward established fact—which it most certainly is not.

88. Some of them also imagine that the issue is almost impossible to investigate scientifically, so their own views—whatever they may be—are safe at least from falsification. The strangest view of all is perhaps that of those who proclaim the issue meaningless ("sensations are not logically comparable between different individuals" [Boynton 1979, 381]), while yet making affirmations—often undauntedly combative or dismissive—on the issue itself. A variant of the first view is Judd's:

> The consequences of this choice [of 575 nm and 470 nm as the sole hues sensed by the protanope or deuteranope] are *bound to appear correct* to binocular red-green confusers. These consequences [presented in eight double-column pages of tables, in which Judd specifies, in the Munsell system, the supposed appearance of some 360 ordinary colors to protanopes and deuteranopes—them all, of course, turning out to be merely varieties of 5Y or 5PB or N] *can be contradicted only* by observers having *unilateral* defects; and the probability of such contradiction arising may be estimated from the fact that the eight cases recorded so far . . . fail to contradict it" (1948, 255, my emphasis).

This is false at every stage: Nagel is a binocular dichromat and can give very good reasons for rejecting the only-yellow-and-blue view—for example, his report (1910, 9–10; see section 12.4 above) of the difference between his ordinary color experience and the extraordinary experience of *really* seeing the world in merely yellow, blue, and neutral, after having "exhausted" the red sensation. It is evident that there is no necessity at all for Judd's two-hue view to "appear correct" to Nagel, the bilateral dichromat. On the contrary, Judd's proposal of a two-dimensional structure of dichromat color space can perfectly well appear incorrect to a dichromat, and the discrepancy can perfectly well be communicated. Even among 2-D spaces, there are ways, I think, to distinguish experience of, for example, yellow and blue from experience of yellow and red. (For example, the former are complementary and contrasting colors, their mixture is a neutral, and the shortest path between them goes through neutral; the latter do not have those features. There is no such thing as yellowish blue, but there is such a thing as yellowish red.) And there are ways to distinguish experience of yellow and blue from experience of, for example, orange and turquoise. (The former, or at least, *certain* yellows and certain blues, are unique hues—appearing to contain no hint of any other hues—and the latter are not.) And it may even be (and I think it is) that experience of yellow and blue can be distinguished from experience of red and green (see Broackes 2007). As for the second part of Judd's claim: the eight unilateral cases that he studied actually give us plenty of reason to suppose that Judd's view *cannot* be correct—but I have said enough on that already in section 12.4.

References

Alpern, M., K. Kitahara, and D. H. Krantz. 1983a. Classical tritanopia. *Journal of Physiology* 335: 655–681.

Alpern, M., K. Kitahara, and D. H. Krantz. 1983b. Perception of colour in unilateral tritanopia. *Journal of Physiology* 335: 683–697. Reprinted in *Readings on Color,* vol. 2, *The Science of Color,* A. Byrne and D. Hilbert, eds., 232–248. Cambridge, MA: MIT Press, 1997.

Berkeley, G. 1709. *An Essay towards a New Theory of Vision*. Dublin. Repr. in Berkeley, *Philosophical Works*, M. R. Ayers, ed., 1–59. London: Dent, 1975.

Bollo, F. 1959. An astonishing new theory of color. *Fortune* (May 1959), 144–148, 195–206.

Bone, R. A., J. T. Landrum, and A. Cains. 1992. Optical density spectra of the macular pigment *in vivo* and *in vitro*. *Vision Research* 32: 105–110.

Boynton, R. 1979. *Human Color Vision*. New York: Holt, Rinehart, and Winston.

Broackes, J. 1992. The autonomy of colour. In *Reduction, Explanation, and Realism*. D. Charles and K. Lennon, eds., 421–465. Oxford: Oxford University Press. Reprinted in *Readings on Color,* vol. 2, *The Science of Color,* A. Byrne and D. Hilbert, eds., 191–225. Cambridge, MA: MIT Press, 1997.

Broackes, J. 2007. Black and white and the inverted spectrum. *Philosophical Quarterly* 57: 161–175.

Brown, R. O., and D. MacLeod. 1997. Color appearance depends on surround variance. *Current Biology* 7: 844–849.

Byrne, A., and D. Hilbert, eds. 1997. *Readings on Color*. Vol. 2: *The Science of Color*. Cambridge, MA: MIT Press.

Collins, M. 1925. *Colour Blindness.* London: K. Paul, Trench, Trubner.

Crognale, M. A., D. Y. Teller, T. Yamaguchi, A. G. Motulsky, and S. S. Deeb. 1999. Analysis of red/green color discrimination in subjects with a single X-linked photopigment gene. *Vision Research* 39: 707–719.

Dalton, J. 1798. Extraordinary facts relating to the vision of colours. *Memoirs of the Literary and Philosophical Society of Manchester* 5, 28–45. Page references are to the reprint in *Edinburgh Journal of Science* N.S. 5 (1831): 88–99.

Deeb, S. S. 2005. The molecular basis of variation in human color vision. *Clinical Genetics* 67 (5): 369–377.

Delahunt, P. B., and D. H. Brainard. 2004. Does human color constancy incorporate the statistical regularity of natural daylight? *Journal of Vision* 4 (2): 57–81.

Descartes, R. (AT) *Œuvres de Descartes,* Ch. Adam and P. Tannery, eds. Rev. ed. Paris: J. Vrin/ C.N.R.S, 1964–1976.

Descartes, R. (CSM) *The Philosophical Writings of Descartes.* Transl. J. Cottingham, R. Stoothoff, D. Murdoch. 2 vols. Cambridge: Cambridge University Press, 1984–1985.

Estévez, O. 1979. *On the Fundamental Data-base of Normal and Dichromatic Color Vision.* Ph. D. Thesis, University of Amsterdam. Amsterdam: Krips Repro Mepel.

Evans, R. M. 1948. *An Introduction to Color.* New York: Wiley.

Farnsworth, D. 1943. The Farnsworth-Munsell 100-Hue and dichotomous tests for color vision. *Journal of the Optical Society of America* 33: 568–578.

Farnsworth, D. 1954. An introduction to the principles of color deficiency. *Medical Research Laboratory Reports* no. 254, U. S. Navy, New London, 13(15).

Farnsworth, D. 1961. Let's look at those isochromatic lines again. *Vision Research* 1: 1–5.

Faul, F., V. Ekroll, and G. Wendt. 2007. Color appearance: The limited role of chromatic surround variance in the "gamut expansion" effect. *Journal of Vision* 8 (3): 30, 1–20.

Favre, A. 1879. Le traitement du daltonisme congénital par l'exercice chez l'enfant et chez l'adulte. *Gazette hebdomadaire de médecine et de chirurgie,* 2. série, 16: 92–95 and 104–108.

Fletcher, R., and J. Voke. 1985. *Defective Colour Vision: Fundamentals, Diagnosis and Management.* Bristol: Adam Hilger.

Fuchs, W. 1923. Experimentelle Untersuchungen über die Änderung von Farben unter dem Einfluß von Gestalten ("Angleichungserscheinungen"). *Zeitschrift für Psychologie* 92: 249–325.

Gegenfurtner, K. R., and L. T. Sharpe, eds. 1999. *Color Vision: From Genes To Perception.* With foreword by B. B. Boycott. New York: Cambridge University Press.

Gelb, A. 1929. Die "Farbenkonstanz" der Sehdinge. In *Handbuch der normalen und pathologischen Physiologie,* A. Bethe et al., Hsg., 12.1: 594–676. Berlin: Springer.

Goldschmidt, R. H. 1919. Übungstherapeutische Versuche zur Steigerung der Farbentüchtigkeit eines anomalen Trichromaten. *Zeitschrift für Sinnesphysiologie* 50: 192–216.

Golz, J., and D. MacLeod. 2002. Influence of scene statistics on colour constancy. *Nature* 415: 637–640.

Graham, C. H., and Y. Hsia. 1958. The spectral luminosity curves for a dichromatic eye and a normal eye in the same person. *Proceedings of the National Academy of Sciences* 44: 46–49.

Graham, C. H., H. G. Sperling, Y. Hsia, and A. H. Coulson. 1961. The determination of some visual functions of a unilaterally color-blind subject: Methods and results. *Journal of Psychology* 51: 3–32.

Grassmann, H. G. 1853. Zur Theorie der Farbenmischung. J. C. POGGENDORF'S *Annalen der Physik und Chemie* (Dritte Reihe, Band 29) 89: 69–84.

Grassmann, H. G. 1854. Theory of compound colours. *Philosophical Magazine* Series 4, 7 (45): 254–264. (Transl. from Grassmann 1853.) Repr. (with editor's amendments) in *Sources of Color Science*, D. L. MacAdam, ed. Cambridge, MA: MIT Press, 1970, 53–60.

Guttmann, A. 1907–08. Untersuchungen über Farbenschwäche. *Zeitschrift für Sinnesphysiologie* 42: 24–64, 250–270; 43: 146–162, 199–223, 255–298.

Hardin, C. L. 1988. *Color for Philosophers: Unweaving the Rainbow*. Indianapolis, IN: Hackett. (Expanded ed., 1993.)

Harvey, G. 1824. On an anomalous case of vision with regard to colours. *Transactions of the Royal Society of Edinburgh* 10: 253–262.

Hayes, S. P. 1911. The color sensations of the partially color-blind, a criticism of current teaching. *American Journal of Psychology* 22: 369–407.

Helmholtz, H. von. 1867. *Handbuch der physiologischen Optik* (issued as vol. 9 of Gustav Karsten, ed., *Allgemeine Encyklopädie der Physik*. Previously issued in six parts, or Lieferungen, 1856, 1860, 1860, 1866, 1866, 1867.) Leipzig: Leopold Voss.

Helmholtz, H. von. 1896. *Handbuch der physiologischen Optik*. 2nd ed. (posth.; with additional material from Arthur König.) 1 vol. Hamburg and Leipzig: Leopold Voss.

Helmholtz, H. von. 1909–11. *Handbuch der physiologischen Optik*. 3rd ed. rev. and enlarged by A. Gullstrand, J. von Kries, and W. Nagel. Hamburg and Leipzig: L. Voss. Vol. 1, 1909; vol. 2, 1911; vol. 3, 1910.

Helmholtz, H. von. 1924–1925. *Helmholtz's Treatise on Physiological Optics*. Translated from the third German edition. James P. Southall, ed. Vol. 1, 1924; vol. 2, 1924; vol. 3, 1925. Rochester, NY: Optical Society of America.

Helson, H. 1938. Fundamental problems in color vision. I. The principle governing changes in hue, saturation, and lightness of non-selective samples in chromatic illumination. *Journal of Experimental Psychology* 23: 439–476.

Hering, E. 1878. *Zur Lehre vom Lichtsinne: Sechs Mitteilungen an die Kaiserl. Akademie der Wissenschaften in Wien.* 2. unveränd. Abdr. Wien: Carl Gerold's Sohn.

Hering, E. 1880. Zur Erklärung der Farbenblindheit aus der Theorie der Gegenfarben. *Lotos (Jahrbuch für Naturwissenschaften)* 1: 76–107.

Hering, E. 1890. Untersuchung eines Falles von halbseitiger Farbensinnstörung am linken Auge. v. GRAEFE's *Archiv für Ophthalmologie* 36 (3): 1–23.

Hering, E. 1920. *Grundzüge der Lehre vom Lichtsinne.* Sonderabdruck aus dem Handbuch der Augenheilkunde. Leipzig: W. Engelmann. (Four parts, prev. publ. 1905, 1907, 1911, 1920.)

Hering, E. 1920/1964. *Outlines of a Theory of the Light Sense.* Translated by L. M. Hurvich and D. Jameson, of *Grundzüge der Lehre vom Lichtsinne.* Cambridge, MA: Harvard University Press.

Herschel, J. F. W. 1827. Light. *Encyclopaedia Metropolitana.* 2nd div., vol. 4, 341–586. (Part reissued separately, with original pagination.) London: J. J. Griffin.

Herschel, J. 1833. Letter to Dalton, 22 May 1833. At 25–26 in William Charles Henry, *Memoirs of the life and scientific researches of John Dalton.* London: Printed for the Cavendish Society, 1854. Quoted in part, in Wilson 1855, 60.

Herschel, J. 1859–60. Remarks on color blindness. *Proceedings of the Royal Society* London. 10: 72–84. (Also in *Philosophical Magazine,* 4th series, 19: 148–158.)

Hess, C. 1890. Untersuchung eines Falles von halbseitiger Farbensinnstörung am linken Auge. Albrecht v. GRAEFE's *Archiv für Ophthalmologie* 36 (3): 24–36.

Hippel, A. von. 1880. Ein Fall von einseitiger, congenitaler Roth-Grün-blindheit bei normalem Farbensinn des anderen Auges. v. GRAEFE's *Archiv für Ophthalmologie* 26 (2): 176–186.

Hippel, A. von. 1881. Ueber einseitiger Farbenblindheit. v. GRAEFE's *Archiv für Ophthalmologie* 27 (3): 47–55.

Holmgren, F. 1877. *Om färgblindheten i dess förhållande till jernvägstrafiken och sjöväsendet.* In: *Upsala Läkareförenings Förhandlingar,* 1876–1877, 12: 171–251, 267–358. Also issued in book form, Upsala: Berling.

Holmgren, F. 1878. Color-blindness in its relation to accidents by rail and sea. In *Annual Report of the Smithsonian Institution for the year 1877.* Washington, DC, 1878: 131–195.

Holmgren, F. 1881. How do the colour-blind see the different colours? *Proceedings of the Royal Society of London* 31: 302–306.

Hsia, Y., and C. H. Graham. 1965. Color blindness. In *Vision and Visual Perception,* C. H. Graham, ed., 395–413. New York: John Wiley. Reprinted in *Readings on Color.* Vol. 2: *The Science of Color,* A. Byrne and D. Hilbert, eds., 201–229. Cambridge, MA: MIT Press, 1997.

Hunt, D. M., K. S. Dulai, J. K. Bowmaker, and J. D. Mollon. 1995. The chemistry of John Dalton's color blindness. *Science* 267: 984–988.

Hunt, R. W. G. 1987. *Measuring Colour*. Chichester and New York: Ellis Horwood. (2nd ed., 1991. 3rd ed., 1998.)

Hurvich, L. M. 1981. *Color Vision*. Sunderland, MA: Sinauer.

Ishihara, S. 1917. *The Series of Plates Designed as Tests for Color-Blindness*. (16 plates) Tokyo. (7th ed. [complete ed.] with 32 plates, Tokyo: Kanehara, 1936; *Tests for color-blindness*. 10th completely rev. ed., with 38 plates. [Tokyo]: Nippon Isho Shuppan, 1951.)

Jameson, D., and L. M. Hurvich. 1978. Dichromatic color language: "Reds" and "Greens" don't look alike but their colors do. *Sensory Processes* 2: 146–155.

Jerchel, W. 1913. Inwieweit wird das Medizinstudium durch Rot-Grünblindheit beeinflußt? *Zeitschrift für Sinnesphysiologie* 47: 1–33.

Judd, D. B. 1940. Hue saturation and lightness of surface colors with chromatic illumination. *Journal of the Optical Society of America* 30: 2–32.

Judd, D. B. 1943. Facts of color-blindness. *Journal of the Optical Society of America* 33: 294–307. Reprinted in D. B. Judd, *Contributions to Color Science,* David L. MacAdam, ed. Washington DC: U.S. Dept. of Commerce, National Bureau of Standards, 1979.

Judd, D. B. 1945. Standard response functions for protanopic and deuteranopic vision. *Journal of the Optical Society of America* 35: 199–221. Reprinted in D. B. Judd, *Contributions to Color Science,* David L. MacAdam, ed. Washington DC: U.S. Dept. of Commerce, National Bureau of Standards, 1979.

Judd, D. B. 1948. Color perceptions of deuteranopic and protanopic observers. U.S. Dept. of Commerce: *Journal of Research of the National Bureau of Standards* 41: 247–271.

Judd, D. B. 1949. The color perceptions of deuteranopic and protanopic observers. *Journal of the Optical Society of America* 39: 252–256.

Judd, D. B. 1952. *Color in Business, Science and Industry*. New York: Wiley. (2nd ed., with Günter Wyszecki, 1963; 3rd ed., 1975.)

Judd, D. B. 1960. Appraisal of Land's work on two-primary color projections. *Journal of the Optical Society of America* 50: 254–268. Reprinted in D. B. Judd, *Contributions to Color Science,* David L. MacAdam, ed. Washington DC: U.S. Dept. of Commerce, National Bureau of Standards, 1979.

Judd, D. B. 1979. *Contributions to Color Science*. David L. MacAdam, ed. Washington DC: U.S. Dept. of Commerce, National Bureau of Standards.

Katz, D. 1911. *Die Erscheinungweisen der Farben und ihre Beeinflussung durch die individuelle Erfahrung*. (*Zeitschrift für Psychologie u. Physiologie der Sinnesorgane*. Ergänzungsband 7.) Leipzig: Barth.

Katz, D. 1930. *Der Aufbau der Farbwelt*. Zweite völlig umgearbeitete Auflage von [1911]. Leipzig: Johann Ambrosius Barth.

Katz, D. 1935. *The World of Colour*. (Abridged by the author from (1930).) Translated by R. B. MacLeod and C. W. Fox. London: Kegan Paul, Trench, Trubner.

Kawabata, N. 1986. Attention and depth perception. *Perception* 15: 563–572.

Koffka, K. 1909. Untersuchungen an einem protanomalen System. *Zeitschrift für Sinnesphysiologie* 43: 123–145.

König, A., and C. Dieterici. 1892. Die Grundempfindungen in normalen und anomalen Farbensystemen und ihre Intensitätsvertheilung im Spectrum. *Zeitschrift für Psychologie* 4: 241–347. Also publ. separately, Hamburg: L. Voss.

Kries, J. von. 1882. *Die Gesichtsempfindungen und ihre Analyse.* Leipzig: Veit.

Kries, J. von. 1905. Die Gesichtsempfindungen. In *Handbuch der Physiologie des Menschen,* W. Nagel, Hrsg. 3: 109–282.

Kries, J. von. 1919. Über einen Fall von einseitiger angeborener Deuteranomalie (Grünschwäche). *Zeitschrift für Sinnesphysiologie* 50: 137–152.

Kuhn, T. S. 1962/1970. *The Structure of Scientific Revolutions.* 2nd ed., 1970. University of Chicago Press.

Land, E. H. 1959a. Color vision and the natural image: Part I. *Proceedings of the National Academy of Sciences* 45: 115–129.

Land, E. H. 1959b. Color vision and the natural image: Part II. *Proceedings of the National Academy of Sciences* 45: 636–644.

Land, E. H. 1959c. Experiments in color vision. *Scientific American* 200, May, 84–99.

Land, E. H. 1962. Colour in the natural image. *Proceedings of the Royal Institution* 39 (176): 1–15.

Land, E. H. 1977. The retinex theory of color vision. *Scientific American* 237, December, 108–128.

Le Grand, Y. [1945–1956]. *Optique physiologique.* (t. 1. La Dioptrique de l'oeil et sa correction (1945, 1952, 1964), t. 2. Lumière et couleurs (1948), t. 3. L'Espace visuel, (1956)). Paris: Masson.

Le Grand, Y. 1957. *Light, colour and vision.* Translated by R. W. G. Hunt, J. W. T. Walsh, and F. R. W. Hunt. New York: Wiley. (Tr. of *Lumière et couleurs,* vol. 2 of *Optique physiologique*; 2nd ed., 1968.)

Lotto, R. Beau, and D. Purves. 2002. The empirical basis of color perception. *Consciousness and Cognition* 11: 609–629.

MacAdam, D. L., ed. 1970. *Sources of color science.* Cambridge, MA: MIT Press.

MacLeod, D. I. A., and P. Lennie. 1976. Red-green blindness confined to one eye. *Vision Research* 16: 691–702.

Maloney, L. T. 1999. Physics-based approaches to modeling surface color perception. In Gegenfurtner and Sharpe 1999, 387–416.

Maxwell, J. C. 1855. Experiments on colour, as perceived by the eye, with remarks on colour-blindness. *Transactions of the Royal Society of Edinburgh.* vol. 21, pt. 2, 275–298. Reprinted in J. C. Maxwell, *The Scientific Papers of James Clerk Maxwell,* W. D. Niven, ed., 1: 126–154. Cambridge: Cambridge University Press, 1890.

Maxwell, J. C. 1860. On the theory of compound colours and the relations of the colours of the spectrum. *Philosophical Transactions* 150: 57–84. Reprinted in J. C. Maxwell, *The Scientific Papers of James Clerk Maxwell,* W. D. Niven, ed., 1: 410–444. Cambridge: Cambridge University Press, 1890.

Maxwell, J. C. 1870. On Colour-vision at different points of the Retina. In *Report of the British Association.* Reprinted in Maxwell, *The Scientific Papers of James Clerk Maxwell,* W. D. Niven, ed., 2: 230–232. Cambridge: Cambridge University Press, 1890.

Maxwell, J. C. 1890. *The Scientific Papers of James Clerk Maxwell.* W. D. Niven, ed. 2 vols. Cambridge: Cambridge University Press, 1890.

Mayer, R. 1940/1991. *The Artist's Handbook of Materials and Techniques.* 5th ed., 1940. New York: Viking Penguin; London: Faber.

Mayer, T. 1775. De affinitate colorum. In his *Opera inedita,* G. C. Lichtenberg, ed. Göttingen: J. C. Dieterich.

Mayer, T. *Opera inedita: The First Translation of the Lichtenberg Edition of 1775.* Translated by E. G. Forbes. New York: American Elsevier, 1971.

Mollon, J. D. 1997. "... aus dreyerley Arten von Membranen oder Molekülen": George Palmer's legacy. In C. R. Cavonius, ed., *Colour Vision Deficiencies* 13. Dordrecht: Kluwer.

Mollon, J. D. 2003. The origins of modern color science. In *The Science of Color,* Steven K. Shevell, ed., 1–39. Amsterdam and Boston: Elsevier.

Montag, E. D. 1994. Surface color naming in dichromats. *Vision Research* 34: 2137–2151.

Montag, E. D., and R. M. Boynton. 1987. Rod influence in dichromatic surface color perception. *Vision Research* 27: 2153–2162.

Müller, G. E. 1924. *Darstellung und Erklärung der verschiedenen Typen der Farbenblindheit nebst Erörterung der Funktion des Stäbchenapparates sowie des Farbensinns der Bienen und der Fische.* Göttingen: Vandenhoeck & Ruprecht.

Nagel, W. 1905. Dichromatische Fovea, trichromatische Peripherie. *Zeitschrift für Psychologie* 39: 93–101.

Nagel, W., ed. 1905–10. *Handbuch der Physiologie des Menschen.* 4 vols. and suppl. vol. Braunschweig: Vieweg. Band 3: *Physiologie der Sinne,* bearb. von J. von Kries ... (1905).

Nagel, W.1907a. Fortgesetzte Untersuchungen zur Symptomatologie und Diagnostik der angeborenen Störungen des Farbensinns. *Zeitschrift für Sinnesphysiologie* 41: 239–282.

Nagel, W. 1907b. Fortgesetzte Untersuchungen zur Symptomatologie und Diagnostik der angeborenen Störungen des Farbensinns. (Schluss.) *Zeitschrift für Sinnesphysiologie* 41: 319–337.

Nagel, W. 1907c. Zwei Apparate für die augenärztliche Funktionsprüfung. Adaptometer und kleines Spektralphotometer (Anomaloskop). *Zeitschrift für Augenheilkunde* 17 (3): 201–222.

Nagel, W. 1907d. Über experimentelle Überführung trichromatischen Farbensinnes in dichromatischen. *Verhandlungen der physiologischen Gesellschaft zu Berlin,* 15. März 1907; in *Archiv für Physiologie,* 2: 543–545.

Nagel, W. 1907e. Spektralapparat zur Diagnose der Farbenblindheit. *Berliner klinische Wochenschrift* 43: 371.

Nagel, W. 1908a. Zur Nomenclatur der Farbensinnsstörungen. *Zeitschrift für Sinnesphysiologie* 42: 65.

Nagel, W. 1908b. *Einführung in die Kenntnis der Farbensinnsstörungen und ihre Diagnose.* Wiesbaden: J. F. Bergmann.

Nagel, W. 1910. Farbenumstimmung beim Dichromaten. *Zeitschrift für Sinnesphysiologie* 44: 5–17.

Nagel, W. 1914. Methoden zur Erforschung des Licht- und Farbensinnes. *Handbuch der physiologischen Methodik,* hsg. v. R. Tigerstedt, 3. Band 1. Hälfte (*Die sensorischen Funktionen* ...) Leipzig: S. Hirzel.

Nagy, A. L. 1980. Large-field substitution Rayleigh matches of dichromats. *Journal of the Optical Society of America* 70: 778–784.

Nagy, A. L. 1994. Red/green color discrimination and stimulus size. *Color Research and Application* 19: 99–104.

Nagy, A. L., and R. M. Boynton. 1979. Large-field color naming of dichromats with rods bleached. *Journal of the Optical Society of America* 69 (9): 1259–1265.

Nathans, J. 1999. The evolution and physiology of human color vision: Insights from molecular genetic studies of visual pigments. *Neuron* 24: 299–312.

Nathans, J., S. L. Merbs, C.-H. Sung, C. J. Weitz, and Y. Wang. 1992. Molecular genetics of human visual pigments. *Annual Review of Genetics* 26: 401–424.

Necker, L. A. 1832. Observations on some remarkable optical phænomena seen in Switzerland; and an optical phænomenon which occurs on viewing a figure of a crystal or geometrical solid. *Philosophical Magazine* 3rd series, 1: 329–337.

Neitz, M., and J. Neitz. 2000. Molecular genetics of color vision and color vision defects. *Archives of Ophthalmology* 118: 691–700.

Neitz, J., M. Neitz, J. C. He, and S. K. Shevell. 1999. Trichromatic color vision with only two spectrally distinct photopigments. *Nature neuroscience* 2: 884–888.

Newton, I. 1704/1730. *Opticks*. Fourth edition, London: William Innys. Reprinted with Preface by I. B. Cohen and Intro. by E. Whittaker, New York: Dover, 1952, 1979.

Nimeroff, I. 1970. Deuteranopic convergence point. *Journal of the Optical Society of America* 60: 966–969.

Noë, A. 2004. *Action in Perception*. Cambridge, MA: MIT Press.

Nuberg, N. D., and E. N. Yustova. 1958. Researches on dichromatic vision and the spectral sensitivity of the receptors of trichromats. In *Visual Problems of Colour: A Symposium ... 1957*, 477–486. London: Her Majesty's Stationery Office.

O'Regan, J. K., and A. Noë. 2001. A sensorimotor account of vision and consciousness. *Behavioral and Brain Sciences* 24: 939–973.

Ostwald, W. 1918–. *Die Farbenlehre*. 5 vols. Leipzig: Verlag Unesma.

Ostwald, W. 1931[?]. *Colour Science*, trans. J. Scott Taylor. 2 vols. London: Winsor and Newton, s.d.

Palmer, S. E. 1999. *Vision Science: Photons to Phenomenology*. Cambridge, MA: MIT Press.

Peterson, M. A., and B. S. Gibson. 1991. Directing spatial attention within an object: Altering the functional equivalence of shape descriptions. *Journal of Experimental Psychology* 17: 170–182.

Pitt, F. H. G. 1935. *Characteristics of Dichromatic Vision*. Medical Research Council: Reports of the Committee upon the Physiology of Vision, Report 14. London: HMSO. (*Special Report Series*, no. 200.)

Pitt, F. H. G. 1944. The nature of normal trichromatic and dichromatic vision. *Proceedings of the Royal Society* series B, 132 (866): 101–117.

Pokorny, J., B. Collins, G. Howett, R. Lakowski, and M. Lewis. 1981. *Procedures for Testing Color Vision: Report of Working Group 41*: Committee on Vision, Assembly of Behavioral and Social Sciences, National Research Council, Washington, DC: National Academy Press.

Pole, W. 1856. On colour-blindness [abstract]. *Proceedings of the Royal Society of London* 8: 172–177.

Pole, W. 1859. On colour-blindness. *Philosophical Transactions of the Royal Society* 149: 323–339.

Polyak, S. L. 1941. *The Retina*. Chicago: University of Chicago Press.

Pylyshyn, Z. 1999. Is vision continuous with cognition? The case for cognitive impenetrability of visual perception. *Behavioral and Brain Sciences* 22: 341–365.

Radde, O. 1877. *Internationale Farbenskala. 42 Gammen mit circa 900 Tönen*. Hamburg: Stenochromatische Anstalt von Otto Radde.

Rayleigh, Lord: see J. Strutt.

Rossotti, H. 1983. *Colour: Why the World Isn't Grey*. Harmondsworth: Penguin; Princeton, NJ: Princeton University Press.

Royal Society. 1890. *Report of the Committee on Colour Vision*. (This became, almost identically, a Report presented to Parliament, London: H.M.S.O., 1892.)

Scheibner, H. M. O., and R. M. Boynton. 1968. Residual red-green discrimination in dichromats. *Journal of the Optical Society of America* 58: 1151–1158.

Schelling, H. von. 1960. In memoriam: Dean Farnsworth, CDR, MSC, USNR. New derivation for the deuteranopic copoint. *Journal of the Optical Society of America* 50: 645–647.

Schmidt, I. 1952. Effect of illumination in testing color vision with pseudo-isochromatic plates. *Journal of the Optical Society of America* 42: 951–955.

Schumann, F. 1904. Ein ungewöhnlicher Fall von Farbenblindheit. In *Bericht über den I. Kongreß für experimentelle Psychologie in Gießen vom 18. bis 21. April 1904*, F. Schumann, Hrsg., 10–13. Leipzig, im Auftrage [der Deutschen Gesellschaft für Psychologie]: J. A. Barth.

Scott, J. 1778. An account of a remarkable imperfection of sight. In a letter from J. Scott to the Rev. Mr. Whisson ... *Philosophical Transactions of the Royal Society* 68: 611–614.

Seebeck, A. 1837. Über den bei manchen Personen vorkommenden Mangel an Farbensinn. In: J. C. POGGENDORFF's *Annalen der Physik und Chemie* (Leipzig) 42: 177–233.

Shapiro, A. G., J. Pokorny, and V. C. Smith. 1994. Rod contribution to large field color-matching. *Color Research and Application* 19: 236–245.

Sharpe, L. T., A. Stockman, H. Jägle, H. Knau, G. Klausen, A. Reitner, and J. Nathans. 1998. Red, green, and red-green hybrid pigments in the human retina: Correlations between deduced protein sequences and psychophysically measured spectral sensitivities. *Journal of Neuroscience* 18: 10053–10069.

Sharpe, L. T., A. Stockman, H. Jägle, and J. Nathans. 1999. Opsin genes, cone photopigments, color vision, and color blindness. In *Color Vision: From Genes To Perception*, K. R. Gegenfurtner and L. T. Sharpe, eds., 3–51. New York: Cambridge University Press.

Sloan, L. L., and L. Wollach. 1947. A case of unilateral deuteranopia. *Journal of the Optical Society of America* 37: 527.

Sloan, L. L., and L. Wollach. 1948. A case of unilateral deuteranopia. *Journal of the Optical Society of America* 38: 502–509.

Smith, V. C., and J. Pokorny. 1972. Spectral sensitivity of colorblind observers and the cone pigments. *Vision Research* 12: 2059–2071.

Smith, V. C., and J. Pokorny. 1977. Large-field trichromacy in protanopes and deuteranopes. *Journal of the Optical Society of America* 67: 213–220.

Stiles, W. S., and J. M. Burch. 1959. N. P. L. colour-matching investigation: Final report (1958). *Optica Acta* 6: 1–26.

Stilling, J. 1877. *Die Prüfung des Farbensinnes beim Eisenbahn- u. Marine-Personal.* Mit 3 Tafeln. Cassel. Supplemented with:

———*Tafeln zur Bestimmung der Roth-Grünblindheit* (Neue Folge: Lfg. 1, 6 Tafeln) Cassel, 1878.

———*Tafeln zur Bestimmung der Blau-Gelbblindheit* (3 Tafeln), 1878.

———*Tafeln zur Bestimmung der herabgesetzten Farbenempfindlichkeit für Roth-Grün* ... (4 Tafeln; Lfg. 2), 1879.

———*Pseudo-isochromatische Tafeln für die Prüfung des Farbensinnes.* (8 pp., 8 plates.) Kassel u. Berlin, 1883.

Stillings pseudo-isochromatische Tafeln zur Prüfung des Farbensinnes, ed. Geheim-Rat Prof. Dr. E. Hertel. Leipzig: Thieme, 1922, 1926, 1929, 1936.

———*Tafeln zur Prüfung des Farbensinnes.* Herausgegeben von Prof. Dr. Karl Velhagen. Leipzig: Thieme, 1952, 1977, 1989 (ed. Karl Velhagen and Dieter Broschmann). Stuttgart: Georg Thieme.

Stockman, A., and L. T. Sharpe. 1999. Cone spectral sensitivities and color matching. In *Color Vision: From Genes to Perception,* K. R. Gegenfurtner and L. T. Sharpe, eds., 53–87. New York: Cambridge University Press.

Strutt, J. (Lord Rayleigh) 1881. Experiments on colour. *Nature* 25: 64–66.

Vos, J. J. 1978. Colorimetric and photometric properties of a 2° fundamental observer. *Color Research and Application* 3: 125–128.

Vos, J. J., and P. L. Walraven. 1971. On the derivation of the foveal receptor primaries. *Vision Research* 11: 799–818.

Wachtler, T., U. Dohrmann, and R. Hertel. 2004. Modeling color percepts of dichromats. *Vision Research* 44: 2843–2855.

Wade, N. J. 1998. *A Natural History of Vision.* Cambridge, MA: MIT Press.

Walls, G. L. 1960. Land! Land! *Psychological Bulletin,* 57: 29–48.

Wandell, B. A. 1995. *Foundations of Vision.* Sunderland, MA: Sinauer.

Wheatstone, C. 1838. Contributions to the Physiology of Vision.—Part the First. On some remarkable, and hitherto unobserved, Phenomena of Binocular Vision. *Philosophical Transactions of the Royal Society of London* 128: 371–394.

Willis, M. P., and D. Farnsworth. 1952. Comparative evaluation of anomaloscopes. *Medical Research Laboratory Reports* no. 190, 11 (7). U. S. Navy, New London.

Wilson, G. 1855. *Researches on Colour Blindness. With a Supplement on the danger attending the present system of railway and marine coloured signals.* Edinburgh: Sutherland and Knox.

Wright, W. D. 1927. A trichromatic colorimeter with spectral primaries. *Transactions of the Optical Society of London* 29: 225.

Wright, W. D. 1944/1969. *The Measurement of Colour.* 4th ed. London and Bristol: Adam Hilger.

Wright, W. D. 1946. *Researches on Normal and Defective Colour Vision.* With a Foreword by L. C. Martin. London: Henry Kimpton.

Wyszecki, G., and W. S. Stiles. 1982. *Color Science—Concepts and Methods, Quantitative Data and Formulas.* 2nd ed. New York: Wiley. (1st ed., 1967).

Zelanski, P., and M. P. Fisher. 1989. *Color for Designers and Artists.* London: Herbert Press.

Contributors

Justin Broackes Department of Philosophy, Brown University, Providence, Rhode Island

Alex Byrne Department of Linguistics and Philosophy, Massachusetts Institute of Technology, Cambridge, Massachusetts

Paul M. Churchland Department of Philosophy, University of California, San Diego, La Jolla, California

Austen Clark Department of Philosophy, University of Connecticut, Storrs, Connecticut

Jonathan Cohen Department of Philosophy, University of California, San Diego, La Jolla, California

David R. Hilbert Department of Philosophy, University of Illinois, Chicago, Illinois

Kimberly A. Jameson Institute for Mathematical Behavioral Sciences, University of California, Irvine, Irvine, California

Rolf G. Kuehni North Carolina State University, Raleigh, North Carolina

Don I. A. MacLeod Department of Psychology, University of California, San Diego, La Jolla, California

Mohan Matthen Department of Philosophy, University of Toronto, Toronto, Ontario, Canada

Rainer Mausfeld Department of Psychology, Christian-Albrechts-Universität zu Kiel, Kiel, Germany

Reinhard Niederée Department of Psychology, Christian-Albrechts-Universität zu Kiel, Kiel, Germany

Jonathan Westphal Department of English and Philosophy, Idaho State University, Pocatello, Idaho

Index